Trends in Mathematics

Trends in Mathematics is a book series devoted to focused collections of articles arising from conferences, workshops or series of lectures.

Topics in a volume may concentrate on a particular area of mathematics, or may encompass a broad range of related subject matter. The purpose of this series is both progressive and archival, a context in which to make current developments available rapidly to the community as well as to embed them in a recognizable and accessible way.

Volumes of TIMS must be of high scientific quality. Articles without proofs, or which do not contain significantly new results, are not appropriate. High quality survey papers, however, are welcome. Contributions must be submitted to peer review in a process that emulates the best journal procedures, and must be edited for correct use of language. As a rule, the language will be English, but selective exceptions may be made. Articles should conform to the highest standards of bibliographic reference and attribution.

The organizers or editors of each volume are expected to deliver manuscripts in a form that is essentially "ready for reproduction." It is preferable that papers be submitted in one of the various forms of TEX in order to achieve a uniform and readable appearance. Ideally, volumes should not exceed 350-400 pages in length.

Proposals to the Publisher are welcomed at either:
Birkhäuser Boston, 675 Massachusetts Avenue, Cambridge, MA 02139, U.S.A.
math@birkhauser.com
or
Birkhäuser Verlag AG, PO Box 133, CH-4010 Basel, Switzerland
math@birkhauser.ch

Stochastic Processes and Related Topics

In Memory of Stamatis Cambanis 1943–1995

Ioannis Karatzas
Balram S. Rajput
Murad S. Taqqu
Editors

Springer Science+Business Media, LLC

Ioannis Karatzas
Departments of Mathematics &
Statistics
Columbia University
New York, NY 10027-0010

Balram S. Rajput
Department of Mathematics
University of Tennessee
Knoxville, TN 37996-1300

Murad S. Taqqu
Deparment of Mathematics
Boston University
Boston, MA 02215-2411

Library of Congress Cataloging-in-Publication Data
Stocastic processes and related topics : in memory of Stamatis Cambanis, 1943-1995 / Ioannis Karatzas, Balram Rajput, Murad S. Taqqu, editors.
p. cm. -- (Trends in mathematics)
Includes bibliographical references.
ISBN 978-1-4612-7389-9 ISBN 978-1-4612-2030-5 (eBook)
DOI 10.1007/978-1-4612-2030-5
1. Stochastic processes--Congresses. I. Cambanis, S. (Stamatis) 1943- . II. Karatzas, Ioannis. III. Rajput, Balram, 1935- . IV. Taqqu, Murad S. V. Series.
QA274.A1S7692 1998 98-16658
519.2'3--dc21 CIP

Printed on acid-free paper

ISBN 978-1-4612-7389-9

Typeset and reformatted from authors' disks by TEXniques, Inc., Boston, MA

9 8 7 6 5 4 3 2 1

Stamatis Cambanis
1943-1995

Contents

Preface

In the last twenty years extensive research has been devoted to a better understanding of the stable and other closely related infinitely divisible models. Stamatis Cambanis, a distinguished educator and researcher, played a special leadership role in the development of these research efforts, particularly related to stable processes from the early seventies until his untimely death in April '95. This commemorative volume consists of a collection of research articles devoted to reviewing the state of the art of this and other rapidly developing research and to explore new directions of research in these fields. The volume is a tribute to the Life and Work of Stamatis by his students, friends, and colleagues whose personal and professional lives he has deeply touched through his generous insights and dedication to his profession.

Before the idea of this volume was conceived, two conferences were held in the memory of Stamatis. The first was organized by the University of Athens and the Athens University of Economics and was held in Athens during December 18–19, 1995. The second was a significant part of a Special IMS meeting held at the campus of the University of North Carolina at Chapel Hill during October 17–19, 1996. It is the selfless effort of several people that brought about these conferences. We believe that this is an appropriate place to acknowledge their effort; and on behalf of all the participants, we extend sincere thanks to all these persons.

Many of the contributors in this book were speakers at these two conferences; there are others as well. Because of space limitations, it was not possible for us to invite all the students, colleagues and co-authors of Stamatis to contribute, which we very much regret. The topics covered include several aspects of heavy tailed (e.g., stable, semistable and related self similar) processes, the central limit problem and strong law of large numbers, comparison and deviation problems, probability and distribution inequalities, probability density and regression estimation, Markovian property and extreme values, interacting particle approximation, communication networks, the Italian problem, and global dependency measure and prediction.

It took a concerted and collective effort of many people to produce this tribute and we take this opportunity to express our gratitude to them. First our sincere thanks goes to all the contributors for their contributions and for their cooperation, and, in particular, to Gopinath Kallianpur and Ross Leadbetter for their thoughtful biographical article : "Stamatis Cambanis — A Glimpse of His Life and Work." Our grateful thanks to Donald Dawson, Stergios Fotopoulos, Christian Houdré, Norman Johnson, Bob

Kertz, Georg Lindgren, Makoto Maejima, Atma Mandrekar, Brad Mann, Mark Meerschaert, Loren Pitt, Jan Rosiński, David Ruppert, Gena Samorodnitsky, Donatas Surgailis, Alexander Wentzell, Alek Weron, Wojbor Woyczyński, and Jie Xiong for their advice and help in the reviewing process. Our special thanks to the Birkhäuser staff and, in particular, to Ann Kostant and Elizabeth Loew for their excellent cooperation in bringing out this volume in a timely fashion.

Finally we take this opportunity to recognize the special role of Stamatis' life partner, Miranda, a playright and a poet of distinction, whose support and influence played a most significant part in Stamatis' wide ranging impact. We express to her our special thanks for her presence at the two conferences, graciously welcoming the participant with her all too familiar subtle despite the overwhelming grief in her heart.

December 19, 1997

Ioannis Karatzas
Balram S. Rajput
Murad S. Taqqu

STAMATIS CAMBANIS

A Glimpse of his Life and Work

In the almost three years since the untimely death of Stamatis Cambanis there has been a quite extraordinary outpouring of affection and tribute to this remarkable scientist, scholar, teacher and human being. One can no doubt point to even more gifted researchers in our field or to even more inspirational teachers. But Stamatis Cambanis had few peers in the combination of these abilities mingled with extraordinary human qualities which led to his incomparable impact and influence. Stamatis would have shrunk from eulogies, insisting that the time be "better" spent in furthering the causes which were dear to his heart. It is therefore especially fitting that a tribute should take the form of a volume of this kind. Some of the works will undoubtedly impinge on or even develop topics which he himself influenced. In any case one suspects that all the contributors are conscious as they write of Stamatis' influence (and perhaps even sense his presence), producing a volume of their best scholarship as a lasting tribute to his life and work.

This introductory article consists of those parts, emphasizing (a) the life and career of Stamatis Cambanis, (b) his technical contributions, and (c) some of his special personal qualities. These substantially (and unashamedly) contain material from two earlier accounts, [1, 2] which seemed to capture many of the aspects of Stamatis' character and work that were so evident to his colleagues and friends.

Life and Career

Born on July 8, 1943 in Athens, Greece, Stamatis received his elementary and high school education there, entering the National University in 1961. He graduated in 1966 with a B.S. in electrical and mechanical engineering.

Stamatis never made a hasty decision and his marriage was no exception, having met his future wife Miranda at age 2, captured in a delightful photograph on their living room wall. They were married in 1967 and their strong mutual support through the years was clearly a most significant factor in their remarkable complementary achievements in Science and the Arts. Perhaps even more important was their influence in the raising of their two sons, Alexis and Thanassis, whose spectacular scholastic achievements and sense of real values and service to others, repeatedly taxed their parents' noted modesty.

After receiving his doctorate in electrical engineering from Princeton University in 1969, Stamatis took a postdoctoral position at the University of North Carolina Statistics Department, which had begun a "specialty track" in statistical communication theory. He spent his entire career in the department, taking a permanent faculty appointment in 1971; he was promoted to full professor in 1981, and served as chairman from 1986–1993.

Stamatis was stereotypical of a class of electrical engineers working with basic stochastic methods with no interest in practical electronic systems. His slide rule satisfied his limited needs for numerical calculation until classroom teaching requirements forced his conversion to the hand calculator, his use of which persisted well into the computer age. Television was not admitted to his home until more recent years (though probably as a thoughtful family decision rather than electronic avoidance). He strongly recognized real requirements of colleagues for up-to-date computer facilities, and worked very hard to see them met as a definite priority over his own needs. Indeed it was only perhaps two years before his death that he was persuaded to accept a used computer for his own office – largely for e-mail purposes, to which he became rapidly addicted.

As described below, Stamatis vigorously pursued a variety of research interests including (his much beloved) stable processes, optimal sampling designs, and more recently, properties of wavelets. His publications include some 75 papers and the editorship of three books. He especially loved collaboration and his infectious enthusiasm attracted co–workers from all parts of the world. His work in the development of the "Center for Stochastic Processes" helped bring wide recognition to his department, providing a framework for collaborative research with a broad spectrum of junior and senior visitors. He greatly enjoyed directing students' research, insisting on high standards and bringing out the very best in those whom he advised. His great enthusiasm extended also to the classroom, where he was an outstandingly effective teacher at all levels of undergraduate and graduate instruction over a wide range of probability and statistical theory.

From 1986–1993 Stamatis served as department chairman, devoting, it seemed, more than 24 hours a day to this task. The department ran like clockwork under his infinitely patient attention to detail and thoughtful consideration of every issue. He had hoped to educate his colleagues to accept responsibility for automatic department operation, through an intricate committee structure and was naturally disappointed when their (our) devotion did not match his, and his memoranda accumulated on less efficient desks. As chairman he also developed excellent relationships with university administrators, and took an energetic part in negotiation of federal funding.

The enormous energy devoted to departmental administration and maintaining an active research program clearly took a physical toll and he relinquished the chairmanship in June 1993, perhaps already feeling the onset of the disease which claimed his life less than two years later. Even before and after his own chairmanship he always exerted a strong influence on the operation of the department through gentle but persistent reminders to chairmen and colleagues concerning matters needing attention.

Stamatis also played a full role in service to the scientific profession. He served on the editorial boards of several journals, on professional society committees and in conference organization. As a Trustee and Executive Committee member he was influential in the origination of the National Institute of Statistical Sciences. He was honored by the statistical and engineering professions by his election to Fellowships of the IMS in 1984, the IEEE (1989) and to membership of the ISI in 1991. His native country recognized his signal contributions with the award of an Honorary Doctorate from the University of Athens in 1987.

In December 1993 Stamatis underwent cancer surgery, but for the next year remained active between ensuing treatments, giving full attention to departmental matters. He played a key role in the organization of the IMS–Bernoulli World Congress held in June 1994 and was very glad that the timing of treatments at least allowed him to attend the start of that meeting and greet many friends from all parts of the world.

Even when finally unable to come to the office, Stamatis continued his strong interest in departmental affairs and his home became a gathering place in the late afternoons for colleagues, visitors and students along with his family and friends. At all times he was the dignified and gracious host. He died on April 12, 1995 after his lengthy and extraordinarily courageous battle with the disease that finally claimed his life.

Research Contributions[1]

Stamatis' interest in random processes was ignited during his graduate student days in Princeton, and already in his Ph.D. dissertation he had obtained new orthogonal series and integral representations for harmonizable processes. During the ensuing 27 years he made fundamental contributions to the areas of probability theory, random processes, estimation theory, com-

[1]As noted in the text the contents of this section are taken from Reference 2 with only slight modifications.

munication systems, and signal analysis and processing.

Stamatis is internationally known for his major contributions to the theory of stable (both Gaussian and non-Gaussian) processes. The non-Gaussian stable processes have heavier tails than the Gaussian ones, and are thus more appropriate for models with outliers. In a series of twenty seven papers beginning in 1980, Stamatis thoroughly studied the analytic behavior of sample paths as well as the ergodicity, Wold decomposition, Markovian, and prediction properties of stable processes. Through his individual and joint contributions it was shown that, although non-Gaussian stable processes share some properties with the Gaussian ones, they differ in terms of their ergodicity, their moving average and spectral representations, their prediction, and the detectability of sure signals embedded in them.

Stamatis also made fundamental contributions to the study of stochastic and multiple Wiener integrals. The well known results for the Wiener processes were extended to general Gaussian processes, and the associated stochastic calculus was developed. A stochastic integral representation and an orthogonal series expansion in terms of multiple Wiener integrals were obtained for every L_2-functional of a Gaussian process. Further applications to the identification of nonlinear systems with Gaussian input were also given.

The problem of sampling designs has many applications to the detection of sure signals in noise, filtering, and the estimation of regression coefficients. In these applications, one desires to estimate a weighted integral of a random process over a finite interval from observations at a finite number of properly selected sampling points, which may be deterministic or random. Stamatis obtained asymptotically optimal deterministic designs when the discrete–time estimators use simple non–optimal coefficients. This work is significant in practice because the latter estimators require knowledge of the covariance function of the process and involve covariance matrix inversions that may not be stable for large sample sizes. These sampling designs can dramatically outperform uniformly–spaced designs. He also investigated a variety of Monte Carlo integration schemes, including stratified and stratified/symmetrized sampling, and obtained their convergence properties. Applications to filtering, prediction, and detection problems were given.

In the area of second order random processes and linear functionals, Stamatis obtained new series representations for harmonizable processes, weakly continuous processes, and general second–order processes, valid over finite or infinite intervals, generalizing the classical Karhunen-Loève representation. He then applied the new representations to linear mean-square filtering.

Other areas to which Stamatis made significant contributions include bandlimited processes and sampling expansions, rate distortion theory, quantization and delta modulation, and more recently, wavelet transforms and wavelet representations of stationary and nonstationary random processes. It was typical for him to maintain a strenuous and interactive program in these latter topics even after his surgery and subsequent debilitating treatments. A consummately crafted seminar at this time displayed his obvious authority in the field and the ability of the "master craftsman" to provide an outstanding educational experience for his audience.

Special Personal Qualities

His notable scholarly, professional and administrative achievements would have assured Stamatis Cambanis of a lasting place in the *memories* of the countless number of those with whom he interacted. But it is the additional remarkable personal qualities which assures him of a permanent place in so many *hearts.* His love of professional interaction arose from his love of people. His exceptional impact on his students in class or thesis direction resulted directly from his concerns for them as individual people. He was vitally concerned with the development of others – students, colleagues and friends, seeking recognition of their achievements (never of his own), and the success of those he had influenced brought him obvious joy and satisfaction.

Stamatis was never set back by adversity – regarding it as a challenge for the future. Indeed his uniformly positive personality made it very difficult for him to see the dark side to any situation. He saw not only the "silver lining" but also the potential for good to be extracted from the very cloud itself. He gave of himself beyond measure to his university, profession and to all whom his path crossed.

One expects memorial tributes, like letters of recommendation, to be exaggerated, larger than life descriptions. Those who knew Stamatis personally will attest that even the high admiration we have attempted to convey is inadequate to describe this superb colleague and wonderfully sensitive human being. Those who did not have the good fortune of meeting him will almost surely know some who did and who will forcefully attest to the professional and personal impact of this regrettably short but extraordinarily well lived span of 51 years.

As was well put by a speaker at the Athens memorial conference, Stamatis would have wanted us not to eulogize, but to "roll up our sleeves" in the human and scientific pursuits that were so dear to him. And so we respectfully dedicate this volume in grateful memory of Stamatis Cambanis, a distinguished scholar and inspirational teacher but above all, true friend.

Gopinath Kallianpur and Ross Leadbetter

References

[1] Stamatis Cambanis, 1943-1995, Obituary article in IMS Bulletin **24**, 1995, p. 231-2, by Ross Leadbetter.

[2] Obituary, "Stamatis Cambanis 1943-1995", Information Theory Society Newsletter **45**, No. 2, 1995, p. 1-3, by Elias Masry.

Publications of Stamatis Cambanis

Books Edited

Probability Theory on Vector Spaces IV, with A. Weron, Lecture Notes in Mathematics, No. 1391, Springer, 1989.

Stable Processes and Related Topics, with G. Samorodnitsky and M.S. Taqqu, Birkhäuser, 1991.

Stochastic Processes, A Festshrift in Honour of Gopinath Kallianpur, with J.K. Ghosh, R.L. Karandikar and P.K. Sen, Springer, 1993.

Papers

On harmonizable stochastic processes, with B. Liu, *Information and Control*, **17** (1970), 183-202.

Bases in L_2 spaces with applications to stochastic processes with orthogonal increments, *Proc. Amer. Math. Soc.*, **29** (1971), 284-290.

On the representation of weakly continuous stochastic processes, with E. Masry, *Information Sciences*, **3** (1971), 277-190. Erratum **4** (1972), 289-290.

Gaussian processes and Gaussian measures, with B.S. Rajput, *Ann. Math. Statist.*, **43** (1972), 1944-1952.

A general approach to linear mean square estimation problems, *IEEE Trans. Information Theory*, **IT-19** (1973), 110-114.

On some continuity and differentiability properties of paths of Gaussian processes, *J. Mult. Anal.*, **3** (1973), 420-434.

Representation of stochastic processes of second order and linear operations, *J. Math. Anal. Appl.*, **42** (1973), 603-620.

The representation of stochastic processes without loss of information, with E. Masry, *SIAM J. Appl. Math.*, **24** (1973), 628-633.

Some remarks on a theorem of Kolmogorov, with G.D. Allen, in *Vector and Operator Valued Measures and Applications*, D.H. Tucker and H.B. Maynard, eds., Academic Press, (1973), 1-5.

Some zero-one laws for Gaussian processes, with B.S. Rajput, *Ann. Probab.*, **1** (1973), 304-312.

On the measure induced on L_2 by a stochastic process, *J. Math. Anal. Appl.*, **47** (1974), 111-124.

Delta modulation of the Wiener process, with E. Masry, *IEEE Trans. Communications*, **COM-23** (1975), 1297-1300.

The measurability of a stochastic process of second order and its linear space, *Proc. Amer. Math. Soc.*, **47** (1975), 467-475.

On the expansion of a bivariate distribution and its relationship to the output of a nonlinearity, with B. Liu, *IEEE Trans. Information Theory*, **IT-17** (1971), 17-25. Reprinted in *Nonlinear Systems*, A.H. Haddad, ed., Benchmark Papers in Electrical Engineering and Computer Sciences, vol. **10**, Dowden, Hutchinson and Ross, Stroudsburg, Pa., (1975), 46-54.

On the path absolute continuity of second order processes, *Ann. Probab.*, **3** (1975), 1050-1054.

Some remarks on the equivalence of Gaussian processes, with A.G. Gualtierotti, *J. Math. Anal. Appl.*, **49** (1975), 226-236.

Bandlimited processes and certain nonlinear transformations, with E. Masry, *J. Math. Anal. Appl.*, **53** (1976), 558-568.

Inequalities for $Ek(X,Y)$ when the marginals are fixed, with G. Simons and W. Stout, *Z. Wahr. verw. Geb.*, **36** (1976), 285-294.

The relationship between the detection of sure and stochastic signals in noise, *SIAM J. Appl. Math.*, **31** (1976), 558-568.

Zakai's class of bandlimited functions and processes: Its characterization and properties, with E. Masry, *SIAM J. Appl. Math.*, **30** (1976), 10-21.

On the rate distortion function of a memoryless Gaussian vector source whose components have fixed variances, with H.M. Leung, *Information and Control*, **34** (1977), 198-209.

Some properties and generalizations of multivariate Eyraud-Gumbel-Morgenstern distributions, *J. Mult. Anal.*, **7** (1977), 551-559.

Gaussian processes: Nonlinear analysis and stochastic calculus, with S.T. Huang, in *Measure Theory Applications to Stochastic Analysis*, G. Kallianpur and D. Kolzow, eds., Lecture Notes in Mathematics, No. 695, Springer, (1978), 165-177.

On the rate distortion functions of spherically invariant vectors and sequences, with H.M. Leung, *IEEE Trans. Information Theory*, **IT-24** (1978), 367-373.

On the reconstruction of the covariance of stationary Gaussian processes observed through zero-memory nonlinearities, with E. Masry, *IEEE Trans. Information Theory*, **IT-24** (1978), 485-494.

Stochastic and multiple Wiener integrals, with S.T. Huang, *Ann. Probab.*, **6** (1978), 585-614.

On the representation of nonlinear systems with Gaussian inputs, with S.T. Huang, *Stochastics*, **2** (1979), 173-189.

Spherically invariant processes: Their nonlinear structure, discrimination, and estimation, with S.T. Huang, *J. Mult. Anal.*, **9** (1979), 59-83.

On the rate distortion functions of memoryless sources under a magnitude-error criterion, with H.M. Leung, *Information and Control*, **44** (1980), 116-133.

On the reconstruction of the covariance of stationary Gaussian processes observed through zero-memory nonlinearities - Part II, with E. Masry, *IEEE Trans. Information Theory*, **IT-26** (1980), 503-507.

Signal identification after noisy nonlinear transformations, with E. Masry, *IEEE Trans. Information Theory*, **IT-26** (1980), 50-58.

Some path properties of p-th order and symmetric stable processes, with G. Miller, *Ann. Probab.*, **8** (1980), 1148-1156.

Consistent estimation of continuous-time signals from samples of noisy nonlinear transformations, with E. Masry, *IEEE Trans. Information Theory*, **IT-27** (1981), 84-96.

Dyadic sampling approximations for non-sequency-limited signals, with M.K. Habib, *Information and Control*, **49** (1981), 199-211.

Linear problems in p-th order and stable processes, with G. Miller, *SIAM J. Appl. Math.*, **41** (1981), 43-69.

On the theory of elliptically contoured distributions, with S. Huang and G. Simons, *J. Mult. Anal.*, **11** (1981), 368-385.

Sampling approximations for non-band-limited harmonizable random signals, with M.K. Habib, *Information Sciences*, **23** (1981), 143-152.

Finite sampling approximations for non-band-limited signals, with M.K. Habib, *IEEE Trans. Information Theory*, **IT-28** (1982), 67-73.

Probability and expectation inequalities, with G. Simons, *Z. Wahr. verw. Geb.*, **59** (1982), 1-25.

Random designs for estimating integrals of stochastic processes, with C. Schoenfelder, *Ann. Statist.*, **10** (1982), 526-538.

Truncation error bounds for the cardinal sampling expansion of bandlimited signals, with E. Masry, *IEEE Trans. Information Theory*, **IT-28** (1982), 605-612.

Complex symmetric stable variables and processes, in *Contributions to Statistics: Essays in Honour of Norman L. Johnson*, P.K. Sen, ed., North Holland, New York, (1983), 63-79.

On α-symmetric multivariate distributions, with R. Keener and G. Simons, *J. Mult. Anal.*, **13** (1983), 213-233.

Sampling designs for the detection of signals in noise, with E. Masry, *IEEE Trans. Information Theory*, **IT-29** (1983), 83-104.

A simple class of asymptotically optimal quantizers, with N.L. Gerr, *IEEE Trans. Information Theory*, **IT-29** (1983), 664-676.

Prediction of stable processes: Spectral and moving average representations, with A.R. Soltani, *Z. Wahr. verw. Geb.*, **66** (1984), 593-612.

Similarities and contrasts between Gaussian and other stable signals, *Fifth Aachen Colloquium on Mathematical Methods in Signal Processing*, (1984), 113-120.

Spectral density estimation for stationary stable processes, with E. Masry, *Stoch. Proc. Appl.*, **18** (1984), 1-31.

Convergence of quadratic form in p-stable random variables and θ_p-radonifying operators, with J. Rosinski and W. Woyczynski, *Ann. Probab.*, **13** (1985), 885-897.

Sampling designs for time series, in *Handbook of Statistics, Volume 5: Time Series in Time Domain*, E.J. Hannan, P.R. Krishnaiah and M.M. Rao, eds., Elsevier, (1985), 337-362.

Analysis of a delayed delta modulator, with N.L. Gerr, *IEEE Trans. Information Theory*, **IT-32** (1986), 496-512.

Analysis of adaptive differential PCM of a stationary Gauss-Markov input, with N.L. Gerr, *IEEE Trans. Information Theory*, **IT-33** (1987), 350-359.

Ergodic properties of stationary stable processes, with C.D. Hardin and A. Weron, *Stoch. Proc. Appl.*, **24** (1987), 1-18.

Commentary on I. J. Schoenberg's Work on Metric Geometry, with D. Richards, in *I. J. Schoenberg: Selected Papers*, Vol. **1**, C. de Boor, ed., Birkhäuser, (1988), 189-191.

Estimating random integrals from noisy observations: Sampling designs and their performance, with J.A. Bucklew, *IEEE Trans. Information Theory*, **IT-34** (1988), 111-127.

Innovations and Wold decompositions of stable sequences, with C.D. Hardin and A. Weron, *Probab. Th. Rel. Fields*, **79** (1988), 1-27.

Performance of discrete-time predictors of continuous-time processes, with E. Masry, *IEEE Trans. Information Theory*, **IT-34** (1988), 655-668.

Random filters which preserve the stability of random inputs, *Adv. Appl. Probab.*, **20** (1988), 275-294.

Admissible and singular translates of stable processes, with M. Marques, in *Probability Theory on Vector Spaces* IV, S. Cambanis and A. Weron, eds., Lecture Notes in Mathematics, No. 1391, Springer, (1989), 239-257.

On prediction of harmonizable stable processes, with A.G. Miamee, *Sankhyā* Ser. A, **51** (1989), 269-294.

Two classes of self-similar stable processes with stationary increments, with M. Maejima, *Stoch. Proc. Appl.*, **32** (1989), 329.

Trapezoidal Monte Carlo integration, with E. Masry, *SIAM J. Numer. Anal.*, **27** (1990), 225-246

On stable Markov processes, with R.J. Adler and G. Samorodnitsky, *Stoch. Proc. Appl.*, **34** (1990), 1-17.

On the oscillation of infinitely divisible processes, with J.P. Nolan and J. Rosinski, *Stoch. Proc. Appl.*, **35** (1990), 87-97.

Conditional variance of symmetric stable variables, with W. Wu, in *Stable Processes and Related Topics*, S. Cambanis, G. Samorodnitsky, and M. S. Taqqu, eds., Boston Birkhäuser, (1991), 85-99.

Dichotomies for certain product measures and stable processes, with M. Marques, *Probab. Math. Stat.*, **12** (1991), 271-289.

On Eyraud-Gumbel-Morgenstern random processes, in *Advances in Probability Distributions with Given Marginals*, G. Dall'Aglio et al., eds., Kluwer Academic, (1991), 207-222.

Characterizations of linear and harmonizable fractional stable motions, with M. Maejima and G. Samorodnitsky, *Stoch. Proc. Appl.*, **42** (1992), 91-110.

Multiple regression on stable vectors, with W. Wu, *J. Mult. Anal.*, **41** (1992), 243-272.

On the statistics of the error in predictive coding for stationary Ornstein-Uhlenbeck processes, with T. Koski, *IEEE Trans. Information Theory*, **38** (1992), 1029-1040.

Random filters which preserve the normality of non-stationary random inputs, in *Nonstationary Stochastic Processes and their Applications*, A.G. Miamee, ed., World Scientific, (1992), 219-237.

Sampling designs for estimating integrals of stochastic processes, with K. Benhenni, *Ann. Statist.*, **20** (1992), 161-194.

Sampling designs for estimating integrals of stochastic processes using quadratic mean derivatives, with K. Benhenni, in *Approximation Theory*, G. Anastassiou, ed., Dekker, (1992), 93-123.

Trapezoidal stratified Monte Carlo integration, with E. Masry, *SIAM J. Numer. Anal.*, **29** (1992), 284-301.

Sampling designs for estimation of a random process, with Y.C. Su, *Stoch. Proc. Appl.*, **37** (1993), 47-89.

Stable mixed moving averages, with V. Mandrekar, J. Rosiński, and D. Surgailis, *Probab. Th. Rel. Fields*, **97** (1993), 543-558.

Stable processes: Moving averages versus Fourier transforms, with C. Houdré, *Probab. Th. Rel. Fields*, **95** (1993), 75-85.

Laws of large numbers for periodically and almost periodically correlated processes, with J. Leskow, C. Houdré, and H.L. Hurd, *Stoch. Proc. Appl.*, **53** (1994), 37-54.

Wavelet approximation of deterministic and random signals: Convergence properties and rates, with E. Masry, *IEEE Trans. Information Theory*, **40** (1994), 1013-1029.

On prediction of heavy-tailed autoregressive sequences: Regression versus best linear prediction, with I. Fakhre-Zakeri, *Theor. Probab. Appl.*, **39** (1994), 217-233.

Sampling designs for regression coefficient estimation with correlated errors, with Y.C. Su, *Ann. Inst. Statist. Math.*, **46** (1994), 707-722.

Conditional variance for stable random vectors, with S. Fotopoulos, *Probab. Math. Statist.*, **15** (1995), 195-214.

On the continuous wavelet transform of second order random processes, with C. Houdré, *IEEE Trans. Information Theory*, **41** (1995), 628-642.

Chaotic behavior of infinitely divisible processes, with K. Podgorski and A. Weron, *Studia Mathematica*, **115** (1995), 109-127.

Exact convergence rates of the Euler-Maruyama scheme, with applications to sampling design, with Y. Su, *Stochastics & Stochastic Reports*, **59** (1996), 211-240.

Forward and reversed time prediction of autoregressive sequences, with I. Fakhre-Zakeri, *J. Appl. Probab.* **33** (1996), 1053-1060.

Unpublished manuscripts

Admissible translates of stable processes: A survey and some new results, Proc. Conference on the 150 years of the University of Athens, Greece, and the award of Honorary Doctorates in Mathematics (1987).

Bootstrapping the sample mean for data from general distributions, with W. Wu and E. Carlstein, Center for Stochastic Processes Technical Report No. 296, (1990).

Non–orthogonal wavelet approximation with rates of deterministic signals, with G.A. Anastassiou, University of North Carolina Center for Stochastic Processes Technical Report No. 456, (1995).

On the conditional variance for scale mixtures of normal distributions, with S. Fotopoulos and L. He, University of North Carolina Center for Stochastic Processes Technical Report No. 413, (1996).

Asymptotically optimal quantizers of bivariate random vectors, with Y. Su, University of North Carolina Center for Stochastic Processes Technical Report No. 476, (1996).

The effect of quantization on the performance of sampling designs, with K. Benhenni, University of North Carolina Center for Stochastic Processes Technical Report No. 481, (1996).

Spectral Representation and Structure of Stable Self-Similar Processes

K. Burnecki, J. Rosiński and A. Weron

Abstract

In this paper we establish a spectral representation of any symmetric stable self-similar process in terms of multiplicative flows and cocycles. A structure of this class of self-similar processes is studied. Applying the Lamperti transformation, we obtain a unique decomposition of a symmetric stable self-similar process into three independent parts: mixed fractional motion, harmonizable and evanescent. This decomposition is illustrated by graphical presentation of corresponding kernels of their spectral representations.

1. Introduction

Following the idea of Rosiński [Ros 1] for stationary processes, we obtain a unique in distribution decomposition of a symmetric α-stable self-similar process $\{X_t\}_{t\in\mathbf{R}_+}$ in three independent parts,

$$X \stackrel{d}{=} X^{(1)} + X^{(2)} + X^{(3)}.$$

Here $\{X_t^{(1)}\}_{t\in\mathbf{R}_+}$ corresponds to a superposition of moving averages in the theory of stationary processes (see [SRMC]). We will call it mixed fractional motion (MFM). This class contains the mixed linear fractional α-stable motion in the terminology of Burnecki, Maejima and Weron [BMW]. The second class $\{X_t^{(2)}\}_{t\in\mathbf{R}_+}$ is harmonizable and $\{X_t^{(3)}\}_{t\in\mathbf{R}_+}$ is called evanescent.

Definition 1.1. A stochastic process $\{X_t\}_{t\in T}$ is called symmetric α-stable or Lévy $S\alpha S$ or, shortly, $S\alpha S$ process for $\alpha \in (0,2]$, if for every $n \in N$ and any $a_1, \ldots, a_n$, $t_1, \ldots, t_n \in T$, the random variable $Y = \sum_{i=1}^n a_i X_{t_i}$ has a symmetric stable distribution with index α.

Definition 1.2. A family of functions $\{f_t\}_{t\in T} \subset L^\alpha(S, \mathcal{B}, \mu)$, where $(S, \mathcal{B}, \mu)$ is a standard Lebesgue space, is said to be the kernel of a spectral representation of an $S\alpha S$ process $\{X_t\}_{t\in T}$ if

$$\{X_t\}_{t\in T} \stackrel{d}{=} \left\{ \int_S f_t(s) M(ds) \right\}_{t\in T}, \tag{1}$$

where M is an independently scattered random measure on $\mathcal{B}$ such that

$$E\exp\{iuM(A)\} = \exp\{-|u|^\alpha \mu(A)\}, \quad u \in \mathbf{R},$$

for every $A \in \mathcal{B}$ with $\mu(A) < \infty$. A kernel $\{f_t\}_{t\in T}$ is said to be minimal if $\sigma\{f_t/f_u : t, u \in T\} = \mathcal{B}$ modulo μ.

Every separable in probability $S\alpha S$ process has a minimal representation (see [Har] and [JW]). Note that the definition of minimality given here is equivalent to the original definition but is easier to formulate (see [Ros 2]); the latter work provides several workable tests for the verification of minimality in concrete cases. We will also consider complex stable processes. In the complex case, f_t are complex valued and M is invariant under rotations.

2. General spectral representation

From now on we will consider processes indexed by $T = \mathbf{R}_+ = (0, \infty)$. A stochastic process $\{X_t\}_{t>0}$ is said to be *H-self-similar* $(H - ss)$ if $\{X_{ct}\}_{t>0} =^d \{c^H X_t\}_{t>0}$, for every $c > 0$. In this section we will characterize the kernel of a spectral representation of a self-similar $S\alpha S$ stochastic process. Without loss of generality, we may and do assume that underlying measure space $(S, \mathcal{B}, \mu)$ for the kernel is Borel. A collection $\{\phi_t\}_{t>0}$ of measurable maps from S onto S such that

$$\phi_{t_1 t_2}(s) = \phi_{t_1}(\phi_{t_2}(s)) \tag{2}$$

and $\phi_1(s) = s$ for all $s \in S$ and $t_1, t_2 > 0$ is called a *multiplicative flow*. This flow is said to be measurable if the map $\mathbf{R}_+ \times S \ni (t, s) \mapsto \phi_t(s) \in S$ is measurable. Given a σ-finite measure μ on $(S, \mathcal{B})$, $\{\phi_t\}_{t>0}$ is said to be *nonsingular* if $\mu(\phi_t^{-1}(A)) = 0$ if and only if $\mu(A) = 0$ for every $t > 0$ and $A \in \mathcal{B}$.

Let A be a locally compact second countable group. A measurable map $\mathbf{R}_+ \times S \ni (t, s) \to a_t(s) \in A$ is said to be a cocycle for a measurable flow $\{\phi_t\}_{t>0}$ if for every $t_1, t_2 > 0$

$$a_{t_1 t_2}(s) = a_{t_2}(s) a_{t_1}(\phi_{t_2}(s)) \quad \textit{for all } s \in S. \tag{3}$$

Theorem 2.1. *Let $\{f_t\}_{t>0} \subset L^\alpha(S, \mu)$ be the kernel of a measurable minimal spectral representation of a measurable $H - ss$ $S\alpha S$ process $\{X_t\}_{t>0}$. Then there exists a unique modulo μ nonsingular flow $\{\phi_t\}_{t>0}$ on (S, μ) and a cocycle $\{a_t\}_{t>0}$ taking values in $\{-1, 1\}$ ($\{|z| = 1\}$ in the complex case) such that, for each $t > 0$,*

$$f_t = t^H a_t \Big\{\frac{d\mu \circ \phi_t}{d\mu}\Big\}^{1/\alpha} (f_1 \circ \phi_t) \quad \mu - a.e. \tag{4}$$

Proof. Since $t \to f_t$ is minimal, then for each $c > 0$, $\{1/c^H f_{ct}\}_{t>0}$ and $\{f_t\}_{t>0}$ are kernels of minimal representations of the same $H-ss$ $S\alpha S$ process. Applying Theorem 2.2 in [Ros 1] there exist a one-to-one and onto function $\Phi_c : S \to S$ and a function $h_c : S \to \mathbf{R} - \{0\}$ such that, for each $t > 0$,

$$f_{ct} = (c^H)(h_c)(f_t \circ \Phi_c) \quad \mu - a.e., \tag{5}$$

and

$$\frac{d(\mu \circ \Phi_c)}{d\mu} = |h_c|^\alpha, \quad \mu - a.e. \tag{6}$$

For every $t, c_1, c_2 > 0$, it is true that, $\mu - a.e$

$$f_{c_1c_2t} = (c_2^H)(h_{c_2})(f_{c_1t} \circ \Phi_{c_2}) = (c_2^H c_1^H)(h_{c_2})(h_{c_1} \circ \Phi_{c_2})(f_t \circ \Phi_{c_1} \circ \Phi_{c_2}) \tag{7}$$

and

$$f_{c_1c_2t} = (c_1^H c_2^H)(h_{c_1c_2})(f_t \circ \Phi_{c_1c_2}).$$

We infer from Theorem 2.2 in [Ros 1], that for every $c_1, c_2 > 0$,

$$h_{c_1c_2} = (h_{c_2})(h_{c_1} \circ \Phi_{c_2}), \quad \mu - a.e., \tag{8}$$

and

$$\Phi_{c_1c_2} = \Phi_{c_1} \circ \Phi_{c_2}, \quad \mu - a.e. \tag{9}$$

In order to conclude the proof it is enough to rewrite the arguments of the proof of Theorem 3.1 in [Ros 1] replacing the additive group $\mathbf{R}$ with the multiplicative $\mathbf{R}_+$. Therefore, $\phi_t = \Phi_t$ is the map and putting $a_t = h_t/|h_t|$ ends the proof. ∎

Remark. It is possible to present another proof of the theorem using the Lamperti transformation defined in the following:

Lemma 2.1. [Lam] *If $\{Y_t\}_{t\in\mathbf{R}}$ is a stationary process and if, for some $H > 0$,*

$$X_t = t^H Y_{\log t}, \quad for \ t > 0, \ X_0 = 0, \tag{10}$$

then X_t is H-ss. Conversely, every nontrivial ss-process with $X_0 = 0$ is obtained in this way from some stationary process Y.

First we need to see that the Lamperti transformation leading from self-similar to stationary processes preserves the minimality of the spectral representation. To this end it is enough to verify condition (iii) of Theorem 3.8 in [Ros 2] with $F = \{e^{-tH} f_{e^t}\}_{t\in\mathbf{R}}$. It is trivially satisfied as the condition is fulfilled for $F = \{f_t\}_{t\in\mathbf{R}_+}$. Now, taking $Y_t = e^{-tH} X_{e^t}$, we obtain a stationary process whose minimal representation is defined by Theorem 3.1 in

[Ros 1] in terms of a unique flow and a corresponding cocycle on the additive group **R**. In order to conclude the proof we apply the reciprocal transformation $X_t = t^H Y_{\log t}$ which leads to the minimal spectral representation of the process X, as stated in Theorem 2.1.

Corollary 2.1. *Since there is a correspondence between self-similar and stationary processes through Lamperti transformation, every minimal representation $t \to f_t$ (4) given in terms of a flow ϕ_t and a cocycle a_t defines the kernel of a minimal spectral representation $\{f_t^1\}_{t\in\mathbf{R}}$ of the corresponding stationary process as*

$$f_t^1 = a_t^1 \left\{ \frac{d\mu \circ \phi_t^1}{d\mu} \right\}^{1/\alpha} (f_0 \circ \phi_t^1), \quad \mu - a.e. \tag{11}$$

such that

$$\phi_t^1(s) = \phi_{e^t}(s),\ a_t^1(s) = a_{e^t}(s),\ f_0^1(s) = f_1(s) \quad \textit{for all } s \in S \textit{ and } t \in \mathbf{R}.$$

Conversely if (11) is the kernel of a minimal spectral representation of a stationary process then (4) defines the kernel of a minimal representation of an $H-ss$ process in terms of a pair $\{a_t, \phi_t\}_{t>0}$ such that

$$\phi_t(s) = \phi_{\log t}^1(s),\ a_t(s) = a_{\log t}^1(s),\ f_1(s) = f_0^1(s) \quad \textit{for all } s \in S \textit{ and } t > 0.$$

Remark. Combining results of Theorem 3.1 in [Ros 1] and Theorem 2.1, we try to describe classes of transformations leading from self-similar to stationary processes and conversely in a similar way as in Theorems 3.1 and 3.2 in [BMW], which are the following.

Theorem 2.2. [BMW] *Let $0 < H < \infty$.*

(i) If for some continuous functions $\theta, \psi : (0,\infty) \to \mathbf{R}$ and a nontrivial stationary process $\{Y_t\}_{t\in\mathbf{R}}$,

$$X_t = \begin{cases} \theta(t) Y_{\psi(t)}, & \textit{for } t > 0 \\ 0, & \textit{for } t = 0 \end{cases} \tag{12}$$

is $H-ss$, then $\theta(t) = t^H$ and $\psi(t) = a \log t$ for some $a \in \mathbf{R}$.

(ii) If for some continuous functions $\zeta, \eta : \mathbf{R} \to (0,\infty)$ such that η is invertible and for a nontrivial $H-ss$ process $\{X_t\}_{t\in\mathbf{R}}$,

$$Y_t = \zeta(t) X_{\eta(t)}, \quad t \in \mathbf{R},$$

is stationary, then

$$\zeta(t) = e^{-bHt} \textit{ and } \eta(t) = e^{bt} \textit{ for some } b \in \mathbf{R}.$$

Sketch of the proof. Let us concentrate on (i). We will support the thesis that $\theta = t^H$ and $\psi = a\log t$ using Theorem 3.1 in [Ros 1] and Theorem 2.1 which concern minimal spectral representations of stationary and self-similar processes, respectively. First we notice that any transformation of the form $X_t = \theta(t)Y_{\psi(t)}$ for a nontrivial stationary process Y and functions $\theta,\ \psi:\ (0,\infty)\to\mathbf{R}$ such that ψ is onto preserves minimality of the spectral representation. It is obvious since $F = \{\theta(t)f^1_{\psi(t)}\}_{t>0}$ satisfies condition (iii) of Theorem 3.8 in [Ros 2] as $\{f^1_t\}_{t\in\mathbf{R}}$ (the spectral representation of process Y) is rigid in $L^\alpha(S,\mu)$. Thus X is $H-ss$ with the spectral representation as follows:

$$f_t = \theta(t)a^1_{\psi(t)}\Big\{\frac{d\mu\circ\phi^1_{\psi(t)}}{d\mu}\Big\}^{1/\alpha}(f_0\circ\phi^1_{\psi(t)})\quad \mu - a.e.$$

Now we use the fact that the process X has a spectral representation defined by (4) and compare them. We immediately obtain that $\theta(t) = t^H$. Furthermore, it is easy to see that the spectral representations are equivalent if

$$\phi^1_{\psi(t_1t_2)} = \phi^1_{\psi(t_1)+\psi(t_2)} \text{ and } \psi(1) = 0.$$

This yields either

$$\psi(t_1t_2) = \psi(t_1)+\psi(t_2) \quad \text{for all } t_1,t_2>0 \tag{13}$$

or

$$\psi(t_1t_2) = \psi(t_1)+\psi(t_2)+c \quad \text{for some } t_1,t_2>0 \text{ and } c\neq 0.$$

Since ψ is continuous, the latter implies that Y is trivial. The equivalence (13) leads to the statement $\psi(t) = a\log t$ for some real constant a. ■

3. Mixed fractional motion

The simplest $H-ss$ $S\alpha S$ process is obtained from a kernel of the form

$$f_t(s) = t^{H-\frac{1}{\alpha}}f\left(\frac{s}{t}\right), \qquad t,s>0, \tag{14}$$

considered with Lebesgue control measure on $(0,\infty)$, $f\in L^\alpha((0,\infty),\mathrm{Leb})$. A $S\alpha S$ process with such representation will be called a *fractional motion* (FM). A superposition of independent FM processes of type (14) is called a *mixed fractional motion* (MFM).

Definition 3.1. An $H-ss$ $S\alpha S$ process $\{X_t\}_{t>0}$ is said to be a MFM if it admits a spectral representation with a kernel $\{g_t\}_{t>0}$ defined on

$(W \times (0,\infty), \mathcal{B}_W \otimes \mathcal{B}_{(0,\infty)}, \nu \otimes \text{Leb})$, for some Borel measure space $(W, \mathcal{B}_W, \nu)$, such that

$$g_t(w,u) = t^{H-\frac{1}{\alpha}} g\left(w, \frac{u}{t}\right), \tag{15}$$

$(w,u) \in W \times (0,\infty), \quad t > 0.$

We will give a few examples of FM and MFM processes. We begin with the simplest one.

Example 3.1. Let $0 < \alpha < 2$, $H = \frac{1}{\alpha}$ and $\{X\}_{t>0}$ be a Lévy motion. Then

$$X_t = \int_0^t M(ds) = \int_0^\infty f(s/t) M(ds),$$

where

$$f(s) = I[0 < s < 1]$$

and M is $S\alpha S$ on $(0,\infty)$ with Lebesgue control measure.

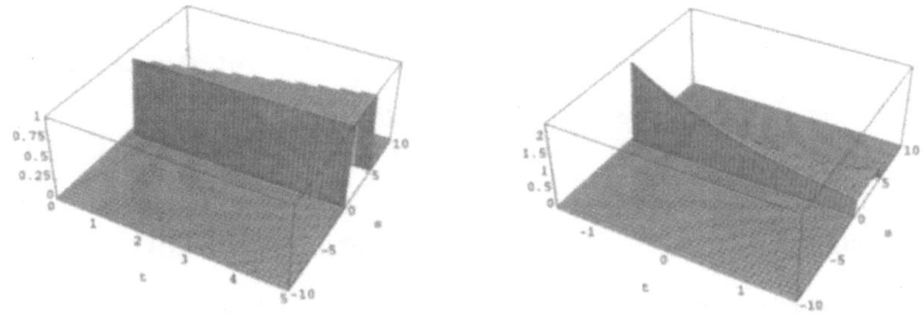

(a) (b)

Figure 1: (a) The kernel of the spectral representation of Lévy motion, (b) the kernel of the corresponding stationary process through the Lamperti transformation for $H = 1/1.8$ (i.e. Ornstein–Uhlenbeck process).

Example 3.2. Let $f \in L^\alpha(\mathbf{R}^d, \text{Leb})$. Let

$$f_t(s) = t^{H-\frac{d}{\alpha}} f\left(\frac{s}{t}\right), \quad s \in \mathbf{R}^d, \ t > 0,$$

and let M be a $S\alpha S$ random measure on $\mathbf{R}^d$ with Lebesgue control measure. It is easy to check that a $S\alpha S$ process $\{X_t\}_{t>0}$ with such spectral representation is H-ss. We will show that $\{X_t\}_{t>0}$ is a MFM. Indeed, let $W = S_d$ be the unit sphere in $\mathbf{R}^d$ equipped with the uniform probability measure ν and let

$$g(w,u) = (c_d u^{d-1})^{1/\alpha} f(uw), \quad (w,u) \in S_d \times (0,\infty),$$

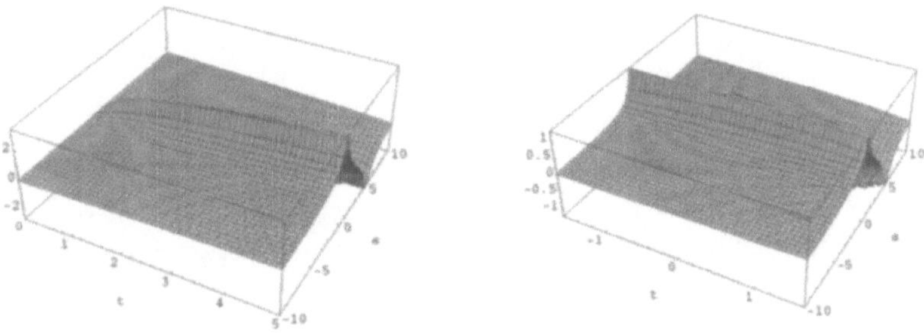

(a) (b)

Figure 2: (a) The kernel of the spectral representation of log-fractional motion, (b) the kernel of the corresponding stationary process for $H = 1/1.8$.

where $c_d = 2\pi^{d/2}/\Gamma(d/2)$ is the surface area of S_d. Using polar coordinates, we get for every $a_1, \ldots, a_n \in \mathbf{R}$, $t_1, \ldots, t_n > 0$,

$$\begin{aligned} \int_{\mathbf{R}^d} |\sum a_j f_{t_j}(s)|^\alpha \, ds &= c_d \int_{S_d} \int_0^\infty |\sum a_j t_j^{H-\frac{d}{\alpha}} f\left(\frac{uw}{t_j}\right) |^\alpha u^{d-1} \, du \nu(dw) \\ &= \int_{S_d} \int_0^\infty |\sum a_j t_j^{H-\frac{1}{\alpha}} g\left(w, \frac{u}{t_j}\right) |^\alpha \, du \nu(dw), \end{aligned}$$

which proves the claim.

Comparing the kernel from the above example with the general form (4) we obtain that $S = \mathbf{R}^d \setminus \{0\}$, $\phi_t(s) = t^{-1}s$, $f_1(s) = f(s)$, and $\frac{d\mu\circ\phi_t}{d\mu} = t^{-d}$. The following well-known $H-ss$ processes are special cases of Example 3.2.

Example 3.3. Let $1 < \alpha < 2$ and $H = \frac{1}{\alpha}$. Then a log-fractional motion (cf. [KMV]) $\{X_t\}_{t>0}$ is defined by

$$X_t = \int_{-\infty}^{\infty} \log\left|\frac{t-s}{s}\right| M(ds) = \int_{-\infty}^{\infty} f(s/t) M(ds),$$

where

$$f(s) = \log|1/s - 1|$$

and M is $S\alpha S$ on $\mathbf{R}$ with Lebesgue control measure.

Example 3.4. Let $0 < H < 1$, $0 < \alpha < 2$, $H \neq \frac{1}{\alpha}$. Put $\beta = H - \frac{1}{\alpha}$. Then a linear fractional stable motion (cf. [CMS]) $\{X_t\}_{t>0}$ is defined by

$$X_t = \int_{-\infty}^{0} p[(t-s)^\beta - (-s)^\beta] \, M(ds)$$

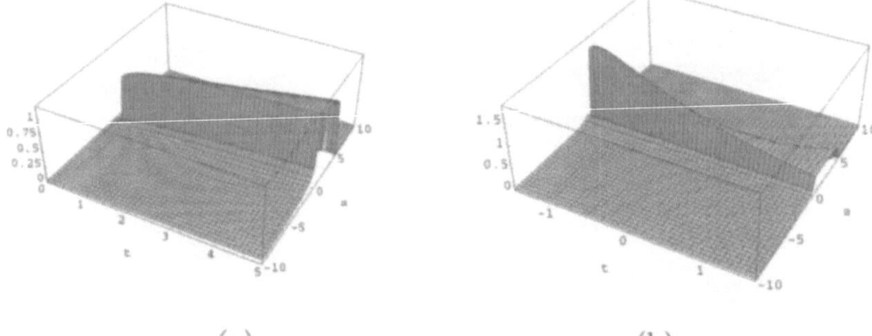

(a) (b)

Figure 3: (a) The kernel of the spectral representation of linear fractional stable motion for $H-1/\alpha = 0.1$, (b) the kernel of the corresponding stationary process for $H = 0.1 + 1/1.8$.

$$
\begin{aligned}
&+ \int_0^\infty \left(I[0 < s < t][p(t-s)^\beta - qs^\beta] + I[t < s]q[(s-t)^\beta - s^\beta] \right) M(ds) \\
=& \int_{-\infty}^\infty t^\beta f(s/t) M(ds),
\end{aligned}
$$

where

$$
\begin{aligned}
f(s) \; = \; & I[s < 0]p[(1-s)^\beta - (-s)^\beta] \\
& + I[0 < s < 1][p(1-s)^\beta - qs^\beta] + I[s > 1]q[(s-1)^\beta - s^\beta],
\end{aligned}
$$

and M is $S\alpha S$ on $\mathbf{R}$ with Lebesgue control measure.

The next theorem shows that the kernel of a spectral representation of any MFM can be defined on $\mathbf{R}^2$ in a canonical way.

Theorem 3.1. (Canonical representation of a MFM). *Let σ be a σ-finite measure on the unit circle S_2 of $\mathbf{R}^2$ and let μ be a measure on $\mathbf{R}^2 \setminus \{0\}$ whose representation in polar coordinates is*

$$\mu(dr, d\theta) = r^{\alpha H - 1} dr\, \sigma(d\theta), \qquad r > 0,\ \theta \in S_2. \tag{16}$$

Let $f : \mathbf{R}^2 \setminus \{0\} \mapsto \mathbf{R}$ (or $\mathbf{C}$) be such that

$$\int_{\mathbf{R}^2 \setminus \{0\}} |f(z)|^\alpha \mu(dz) < \infty.$$

Then the family of functions $\{f_t\}_{t>0} \subset L^\alpha(\mathbf{R}^2 \setminus \{0\}, \mu)$, given by

$$f_t(z) = f(t^{-1} z), \tag{17}$$

is the kernel of a spectral representation of an $S\alpha S$ process, which is $H-ss$ and MFM. Conversely, every MFM admits a (canonical) representation (16)–(17).

Proof. We will show only the converse part. Consider a MFM with a representation (15). Since S is a Borel space, S is measurably isomorphic to a Borel subset of S_2. Let $\Phi : S \mapsto S_2$ denote this isomorphism and let $\sigma = \nu \circ \Phi^{-1}$. Define a function f on $\mathbf{R}^2 \setminus \{0\}$ as

$$f(z) = \begin{cases} g\left(\Phi^{-1}\left(\frac{z}{|z|}\right), |z|\right) |z|^{1/\alpha - H}, & \text{if } \frac{z}{|z|} \in \Phi(S) \\ 0, & \text{otherwise.} \end{cases}$$

Let μ be a measure on $\mathbf{R}^2 \setminus \{0\}$ given by (16). Then

$$\begin{aligned} \int_{\mathbf{R}^2\setminus\{0\}} |\sum a_j f_{t_j}(z)|^\alpha \mu(dz) &= \int_{\mathbf{R}^2\setminus\{0\}} |\sum a_j f(t_j^{-1} z)|^\alpha \mu(dz) \\ &= \int_{S_2} \int_0^\infty |\sum a_j f(t_j^{-1} r\theta)|^\alpha r^{\alpha H - 1} \, dr\sigma(d\theta) \\ &= \int_S \int_0^\infty |\sum a_j f(t_j^{-1} r\Phi(s))|^\alpha r^{\alpha H - 1} \, dr\nu(ds) \\ &= \int_S \int_0^\infty |\sum a_j t_j^{H - 1/\alpha} g(s, t_j^{-1} r)|^\alpha \, dr\nu(ds), \end{aligned}$$

for every $t_1, \ldots, t_n > 0$ and $a_1, \ldots, a_n \in \mathbf{R}\,(\mathbf{C})$. This ends the proof. ∎

Remark. The Lamperti transformation maps FMs onto moving average processes and MFMs onto mixed moving averages (see [SRMC]). Considering above examples it seems that MFMs appear more naturally than FMs. This is quite opposite to the relation between mixed and the usual moving averages.

It is clear that a stable process may have many spectral representations with different kernels defined on various measure spaces. However, we can identify one property, common to all such representations, which characterizes MFMs.

Theorem 3.2. *Let $\{X_t\}_{t>0}$ be a $S\alpha S$ $H-ss$ process with an arbitrary representation (1). Then X is MFM if and only if*

$$\int_0^\infty t^{-\alpha H - 1} |f_t(s)|^\alpha dt < \infty \quad \mu - a.e. \tag{18}$$

Proof. Condition (18) is equivalent to

$$\int_{-\infty}^\infty e^{-\alpha H t} |f_{e^t}(s)|^\alpha dt < \infty \quad \mu - a.e.$$

Using Theorem 2.1 in [Ros 2] in conjunction with (10) concludes the proof. ∎

4. Decomposition of stable self-similar processes

Similarly, as in the case of stationary $S\alpha S$ processes, Theorem 2.1 allows one to use ergodic theory ideas in the study of $S\alpha S$ self-similar processes. In particular, the Hopf decomposition of the underlying space S of the spectral representation (4) into invariant parts C and D, such that the flow ϕ_t is conservative on C and dissipative on D, generates a decomposition of $\{X_t\}_{t>0}$ into two independent $S\alpha S$ $H-ss$ processes $\{X_t^C\}_{t>0}$ and $\{X_t^D\}_{t>0}$. We will characterize the latter process.

Theorem 4.1. *$\{X_t^D\}_{t>0}$ is a MFM and one can choose a minimal representation of $\{X_t^D\}_{t>0}$ of the form (15). Furthermore, $\{X_t^D\}_{t>0}$ is a FM if and only if $\{\phi_t\}_{t>0}$ restricted to D is ergodic.*

Proof. Using Corollary 2.1 we infer that the process $\{X_t^D\}_{t>0}$ corresponds, by Lamperti transformation, to a stationary $S\alpha S$ process $\{Y_t\}_{t\in\mathbf{R}}$ generated by a dissipative flow. From Theorem 4.4 in [Ros 1] we obtain that $\{Y_t\}_{t\in\mathbf{R}}$ is a mixed moving average, implying that $\{X_t^D\}_{t>0}$ is a MFM.

We will now prove the second part of the theorem. Since a moving average representation kernel is minimal (see e.g. [Ros 2]), (14) is minimal as well. Since f_t in (4) is minimal, then also f_t restricted to D is minimal. By Theorem 3.6 in [Ros 1] we infer that the (multiplicative) flow ϕ_t is equivalent to the flow $\psi_t(s) = t^{-1}s,\ \ t,s>0$. Since $\{\psi_t\}$ is ergodic, so is $\{\phi_t\}$. Now suppose that $\{\phi_t\}$ is ergodic. By the first part of this theorem, $\{X_t\}$ admits a minimal representation of the form (15) whose flow is given by $\psi_t(w,u) = (w, t^{-1}u)$. Since the latter flow is equivalent to $\{\phi_t\}$ by the foregoing theorem, it must be ergodic which is only possible when ν is a point-mass measure. Thus (15) reduces to (14). ∎

The class generated by conservative flows consists of harmonizable processes and processes of a third kind (evanescent).

Definition 4.1. An $H-ss$ process $\{X_t\}_{t>0}$ is said to be *harmonizable* if it admits the representation

$$\{X_t\}_{t>0} =_d \left\{\int_{\mathbf{R}} t^{H+is} N(ds)\right\}_{t>0}, \qquad (19)$$

where N is a complex-valued rotationally invariant $S\alpha S$ measure with the finite control measure ν on S.

Note that the representation (19) is minimal and it is generated by an identity flow acting on S with $a_t(s) = t^{is}$ as the corresponding multiplicative cocycle. It is easy to prove the converse.

Proposition 4.1. *Let $\{X_t\}_{t>0}$ be a measurable complex-valued $H-ss$ $S\alpha S$ process generated by an identity flow. Then $\{X_t\}_{t>0}$ is harmonizable.*

Proof. Let

$$S_0 = \{s : a_{t_1t_2}(s) = a_{t_1}(s)a_{t_2}(s) \text{ for Leb} \otimes \text{Leb } a.a.\ (t_1,t_2)\}.$$

Now it is enough to show that, for each $s \in S_0$, there exist a unique $k(s) \in \mathbf{R}$ such that

$$a_t(s) = t^{ik(s)}.$$

To this end we follow the proof of Proposition 5.1 in [Ros 1] and next define a finite measure $\mu_0(ds) = |f(s)|^\alpha \mu(ds)$ on S. Therefore, (19) holds with $\nu = \mu_0 \circ k^{-1}$. ■

Theorem 4.2. *Let $\{f_t\}_{t>0}$ be the kernel of a minimal spectral representation of the form (4) for a complex-valued $S\alpha S$ harmonizable process $\{X_t\}_{t>0}$. Then $\{\phi_t\}_{t>0}$ is the identity flow and (4) reduces to*

$$f_t(s) = t^{H+is} f(s). \tag{20}$$

Proof. Since (20) follows from the proof of the previous proposition, we only need to show that $\{\phi_t\}_{t>0}$ is the identity flow. However, the representation (19) is minimal and is induced by the identity flow $\psi_t(s) = s$, for all t, s, so that by Theorem 3.6 in [Ros 1], ϕ_t, being equivalent to the identity flow, must be the identity. ■

Example 4.1. Let

$$\{X_t\}_{t>0} =_d \left\{ \int_{-\infty}^{\infty} t^{H+is} \frac{e^{is}-1}{is} |s|^{-(H-1/2)} M(ds) \right\}_{t>0},$$

where M is a complex-valued rotationally invariant $S\alpha S$ measure. The process X corresponds via the Lamperti transformation to the harmonizable representation of fractional Gaussian noise (cf. [ST]).

Remark. There cannot be any nonzero real-valued stationary harmonizable process. Using the Lamperti transformation, the same statement is valid about real-valued harmonizable self-similar processes. However, the class of real-valued self-similar processes whose spectral represenation is generated by the identity flow is slightly larger. Any process of this class must be of the form $X_t = t^H X_1$ (cf. Proposition 5.2 in [Ros 1]).

Definition 4.2. A stochastic process whose minimal representation (4) contains a conservative flow without fixed points will be called *evanescent*.

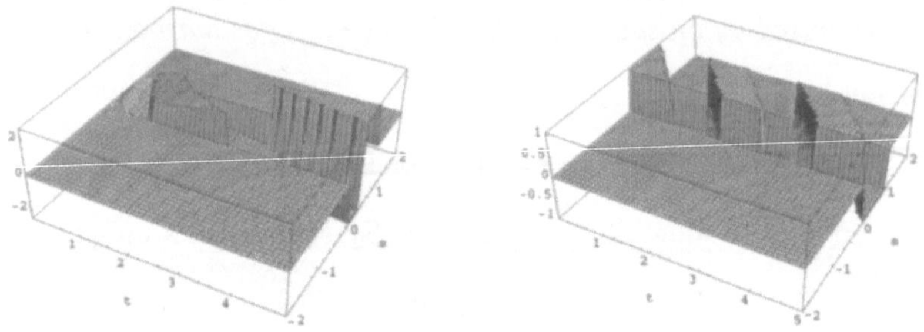

(a) (b)

Figure 4: (a) The kernel of the spectral representation of the evanescent process, (b) the kernel of the corresponding stationary process for $H = 1/1.8$.

This class is not well understood at present. The next theorem is useful to verify whether or not a process is evanescent.

Theorem 4.3. *Let $\{X_t\}_{t>0}$ be a $S\alpha S$ $H-ss$ process with an arbitrary representation (1). Then $\{X_t\}_{t>0}$ is evanescent if and only if*

$$\mu\{s \in S: \int_0^\infty t^{-\alpha H-1}|f_t(s)|^\alpha \, dt < \infty\} = 0$$

and

$$\mu\{s \in S: \; f_{t_1 t_2}(s) f_1(s) = f_{t_1}(s) f_{t_2}(s) \text{ for a.a. } t_1, t_2 > 0\} = 0$$

Proof. It is a direct consequence of the results of Section 6 in [Ros 1] combined with the Lamperti transformation, and Lemma 2.1. ■

We will give two examples of evanescent processes.

Example 4.2. Let

$$\{X_t\}_{t>0} =_d \left\{ \int_0^1 t^H \cos \pi[\log t + s] M(ds) \right\}_{t>0},$$

where $[x]$ denotes the largest integer not exceeding x. Then X does not have a corresponding harmonizable or a mixed moving average component, and therefore it is an example of an evanescent process.

Example 4.3. Let $\{X_t\}_{t>0}$ be the real part of a harmonizable process, i.e.,

$$\{X_t\}_{t>0} \stackrel{d}{=} \left\{ \int_{[0,2\pi)\times\mathbf{R}} t^H \cos(s + w \log t) \, Z(ds, dw) \right\}_{t>0},$$

where Z is a real-valued $S\alpha S$ random measure with control measure Leb $\otimes \nu$ and ν is a finite measure on $\mathbf{R}$ (see [Ros 2], Example 4.9). Here $\phi_t(s, w) = (s +_{2\pi} w \log t, w)$, where "$+_{2\pi}$" denotes addition modulo 2π.

Theorem 4.4. *Every $S\alpha S$ self-similar process $\{X_t\}_{t>0}$ admits a unique decomposition into three independent parts*

$$\{X_t\}_{t>0} \stackrel{d}{=} \{X_t^{(1)}\}_{t>0} + \{X_t^{(2)}\}_{t>0} + \{X_t^{(3)}\}_{t>0},$$

where the first process on the right-hand side is a MFM, the second is harmonizable, and the thrid one is an $H-ss$ evanescent process.

Proof. Since the set D of Hopf decomposition and the set of fixed points for a flow are invariant, we obtain a decomposition of self-similar processes analogous to the decomposition of stationary processes (see Theorem 6.1 in [Ros 1]). ∎

References

[BMW] K. Burnecki, M. Maejima, and A. Weron, *The Lamperti transformation for self-similar processes*, Yokohama Math. J. **44** (1997), 25–42.

[CMS] S. Cambanis, M. Maejima, and G. Samorodnitsky, *Characterization of linear and harmonizable fractional stable motions*, Stoch. Proc. Appl. **42** (1992), 91–110.

[Har] C. D. Hardin, Jr., *On the spectral representation of symmetric stable processes*, J. Multivariate Anal. **12** (1982), 385–401.

[JW] A. Janicki and A. Weron, *Simulation and Chaotic Behavior of α−Stable Stochastic Processes*, Marcel Dekker Inc., New York, 1994.

[KMV] Y. Kasahara, M. Maejima, and W. Vervaat, *Log-fractional stable processes*, Stoch. Proc. Appl. **36** (1988), 329–339.

[Lam] J. W. Lamperti, *Semi-stable stochastic processes*, Trans. Amer. Math. Soc. **104** (1962), 62–78.

[Ros 1] J. Rosiński, *On the structure of stationary stable processes*, Ann. Probab. **23** (1995), 1163–1187.

[Ros 2] J. Rosiński, *Minimal integral representations of stable processes*, Studia Math., to appear.

[ST] G. Samorodnitsky and M. S. Taqqu, *Stable Non-Gaussian Random Processes: Stochastic Models with Infinite Variance*, Chapman & Hall, London, 1994.

[SRMC] D. Surgailis, J. Rosiński, V. Mandrekar, and S. Cambanis, *Stable mixed moving averages*, Probab. Th. Rel. Fields **97** (1993), 543–558 .

Krzysztof Burnecki
Hugo Steinhaus Center for Stochastic Methods
Technical University, 50-370 Wrocław, Poland
email:burnecki@im.pwr.wroc.pl

Jan Rosiński
Mathematics Department
University of Tennessee, Knoxville, TN 37996-1300, USA
email:rosinski@math.utk.edu

Aleksander Weron
Hugo Steinhaus Center for Stochastic Methods
Technical University, 50-370 Wrocław, Poland
email:weron@im.pwr.wroc.pl

Three Elementary Proofs of the Central Limit Theorem with Applications to Random Sums*

T. Cacoullos, N. Papadatos and V. Papathanasiou

Abstract

Three simple proofs of the classical CLT are presented. The proofs are based on some basic properties of *covariance kernels* or w-functions in conjunction with bounds for the total variation distance. Applications to random sum CLT's are also given.

AMS 1991 *subject classifications.* Primary 60F15; secondary 60F05.
Key words and phrases: covariance kernels, CLT, random sums.

1. Introduction

Consider a (real) absolutely continuous r.v. X with d.f. F, density f, mean μ, and finite variance σ^2. The (characterizing) covariance kernel $w(\cdot)$ is defined for every x in the interval support of X by the relation

$$\sigma^2 w(x) f(x) = \int_{-\infty}^{x} (\mu - t) f(t)\, dt = \int_{x}^{\infty} (t - \mu) f(t)\, dt; \qquad (1.1)$$

w appears in the basic *covariance identity* ([4], Lemma 3.1)

$$\text{Cov}[X, g(X)] = \sigma^2 E[w(X) g'(X)], \qquad (1.2)$$

provided that the (otherwise arbitrary) absolutely continuous function g satisfies $E|w(X)g'(X)| < \infty$. Interestingly enough, the same w-function appears both in the upper and lower bounds on the variance of $g(X)$.

The purpose of the paper is to give elementary proofs of the CLT in the i.i.d. (univariate) case, by only using properties of the covariance kernel w. It should be noted that these proofs are heavily dependent on some previous

*Work partially supported by the Greek Ministry of Industry, Energy and Technology, under Grant 1369/95.

results ([4],[6], [7] and [3]). Furthermore, multivariate extensions have been given by [9] and [5].

The first proof shows the weak convergence to the standard normal d.f. Φ of the d.f. F_n of the standardized partial sums

$$S_n = \frac{X_1 + \cdots + X_n - n\mu}{\sqrt{n}\sigma}.$$

The assumption used here is that the covariance kernel w of X (where X has the same d.f. as $X_1, X_2, \ldots$) has finite variance:

$$\mathrm{Var}[w(X)] < \infty. \tag{1.3}$$

The second proof is stronger, since assumption (1.3) is unnecessary and, at the same time, the conclusion is strengthened to

$$\varrho(F_n, \Phi) \to 0, \text{ as } n \to \infty. \tag{1.4}$$

In this paper, $\varrho(F, G)$ denotes the total variation distance between the d.f.'s F and G (or the corresponding r.v.'s X and Y), defined by

$$\varrho(F, G) = \sup_A\{|F(A) - G(A)|, A \text{ Borel}\}, \tag{1.5}$$

where $F(A) = P[X \in A]$, $G(A) = P[Y \in A]$.

In the third proof, the stronger assumption (1.3) is used in order to obtain the bound

$$\varrho(F_n, \Phi) \leq c/\sqrt{n}, \tag{1.6}$$

for some constant c (see (2.10)), depending only upon the d.f. of X. Obviously (1.6) implies (1.4), and furthermore it gives a bound on the rate of convergence.

It is worth noting that the last technique is also applicable to sums of independent (not necessarily identically distributed) r.v.'s, as well as random sums. The first case has been treated in [3], while the random sums are discussed in Section 3.

2. Three elementary proofs of CLT

Without loss of generality, take $E[X_j] = 0$ and $\mathrm{Var}[X_j] = 1$, $j = 1, 2, \ldots$, and let F_n be the d.f. of the standardized sum $S_n = (X_1 + \cdots + X_n)/\sqrt{n}$. Suppose also that w_n is the w-function (covariance kernel) of S_n. Obviously, w_1 is the w-function of $S_1 = X_1$.

In Theorem 2 of [6], it was shown that for any r.v. Y with w-function w_Y,

$$E[w_Y^2(Y)] \geq 1, \tag{2.1}$$

and equality characterizes the normal distribution. The stability of the preceding characterization was also proved in Theorem 3, namely,

$$E[w_{Y_n}^2(Y_n)] \to 1 \text{ implies } \frac{Y_n - E[Y_n]}{\sqrt{\mathrm{Var}[Y_n]}} \xrightarrow{\text{d}} N(0,1) \text{ as } n \to \infty.$$

It should be noted that the condition $E[w_{Y_n}^2(Y_n)] \to 1$ is equivalent to $Var[w_{Y_n}(Y_n)] \to 0$, since $E[w_Y(Y)] = 1$ for any r.v. Y. Thus, the proof of the CLT requires showing that

$$\mathrm{Var}[w_n(S_n)] \to 0, \text{ as } n \to \infty. \tag{2.2}$$

A proof of (2.2) under the assumption (1.3) was given in [6], Theorem 4, by employing certain properties of the w-functions. A simpler proof is given below.

Theorem 2.1 (c.f. [6]). *Let $X, X_1, X_2, \ldots, X_n, \ldots$ be i.i.d. and absolutely continuous r.v.'s with $E[X] = 0$, $E[X^2] = 1$ and $E[w^2(X)] < \infty$, where w is the covariance kernel of X. Then, (2.2) holds.*

For the proof we make use of the following Lemma (c.f. relation (3.3) of [5]).

Lemma 2.1. *Under the notations of Theorem 2.1, for $j \le n$ we have*

$$E[w_j(S_j)w_n(S_n)] = E[w_n^2(S_n)].$$

Proof. Set $A_j = E[w_j(S_j)w_n(S_n)]$. From the basic covariance identity (1.2) we get for arbitrary g

$$E[S_j g(S_j)] = E[w_j(S_j)g'(S_j)].$$

On the other hand, using the same identity,

$$\begin{aligned}
E[S_j g(S_j)] &= E\left[\left(\sqrt{\frac{j-1}{j}}S_{j-1} + \sqrt{\frac{1}{j}}X_j\right) g(S_j)\right] \\
&= \sqrt{\frac{j-1}{j}}E\left\{E\left[S_{j-1}\, g(S_j)\,|X_j\right]\right\} + \sqrt{\frac{1}{j}}E\left\{E\left[X_j\, g(S_j)\,|S_{j-1}\right]\right\} \\
&= \frac{j-1}{j}E\left[w_{j-1}(S_{j-1})g'(S_j)\right] + \frac{1}{j}E\left[w(X_j)g'(S_j)\right].
\end{aligned}$$

Hence,

$$E[w_j(S_j)g'(S_j)] = \frac{j-1}{j}E\left[w_{j-1}(S_{j-1})g'(S_j)\right] + \frac{1}{j}E\left[w(X_j)g'(S_j)\right]. \tag{2.3}$$

Applying (2.3) to $g'(x) = w_n\left(\sqrt{j/n}\,x + \sqrt{(n-j)/n}\,S^*_{n-j}\right)$, where $S^*_{n-j} = (X_{j+1} + \cdots + X_n)/\sqrt{n-j}$ for $j < n$ and $S^*_0 \equiv 0$, we obtain

$$\begin{aligned} E&\left[w_j(S_j)w_n(S_n)\,\middle|S^*_{n-j}\right] \\ &= \frac{j-1}{j}E\left[w_{j-1}(S_{j-1})w_n(S_n)\,\middle|S^*_{n-j}\right] + \frac{1}{j}E\left[w(X_j)w_n(S_n)\,\middle|S^*_{n-j}\right]. \end{aligned}$$

Thus, taking expectations with respect to S^*_{n-j} and using the fact that

$$E[w(X_j)w_n(S_n)] = E[w_1(S_1)w_n(S_n)]$$

for each j, we conclude that $jA_j = (j-1)A_{j-1} + A_1$. This implies that $A_1 = \cdots = A_n$ and the proof is complete. ■

Proof of Theorem 2.1. Set $\sigma_n = \text{Var}\left[w_n(S_n)\right]$. It follows from Lemma 2.1 that

$$\sigma_n - \sigma_{n+1} = E[w_n^2(S_n)] - E[w_{n+1}^2(S_{n+1})] = E[w_n(S_n) - w_{n+1}(S_{n+1})]^2.$$

Therefore, σ_n decreases and hence it converges since $\sigma_1 < \infty$ by the assumptions. Thus,

$$\sigma_n - \sigma_{2n} = E[w_{2n}(S_{2n}) - w_n(S_n)]^2 \to 0.$$

Consequently,

$$\begin{aligned} E[w_{2n}(S_{2n}) - w_n(S_n)]^2 &\geq E[w_{2n}(S_{2n}) - E[w_{2n}(S_{2n})|S_n]]^2 \\ &= E\left\{\text{Var}[w_{2n}(S_{2n})|S_n]\right\} \to 0. \end{aligned}$$

Furthermore, if $S^*_n = (X_{n+1} + \cdots + X_{2n})/\sqrt{n}$, we have

$$\begin{aligned} E&[w_{2n}(S_{2n}) - E[w_{2n}(S_{2n})|S_n]]^2 \\ &= E\left\{E\left([w_{2n}(S_{2n}) - E[w_{2n}(S_{2n})|S_n]]^2\,|S^*_n\right)\right\} \\ &\geq E\left\{E^2\left([w_{2n}(S_{2n}) - E[w_{2n}(S_{2n})|S_n]]\,|S^*_n\right)\right\} \\ &= E\left[E[w_{2n}(S_{2n})|S^*_n] - 1\right]^2 \\ &= E\left[E[w_{2n}(S_{2n})|S_n] - 1\right]^2 \\ &= \text{Var}\left\{E[w_{2n}(S_{2n})|S_n]\right\} \to 0. \end{aligned}$$

Hence, $\sigma_{2n} = \text{Var}\left\{E[w_{2n}(S_{2n})|S_n]\right\} + E\left\{\text{Var}[w_{2n}(S_{2n})|S_n]\right\} \to 0$ by the above arguments, which completes the proof. ■

The second proof of CLT is based on the bound (see [7], Theorem 1.1)

$$\varrho(F, \Phi) \leq 2E|w_X(X) - 1| \leq 2\sqrt{\text{Var}[w_X(X)]}, \tag{2.4}$$

where X is any standardized r.v. with d.f. F. In fact, the factor 2 in (2.4) can be replaced by 3/2, as shown in the following

Lemma 2.2. *If X has mean zero, variance one, density f and d.f. F,*

$$\varrho(F, \Phi) \leq (3/2)E|w_X(X) - 1| \leq (3/2)\sqrt{\mathrm{Var}[w_X(X)]}. \tag{2.5}$$

Proof. For an arbitrary Borel set A, consider the function $\psi_A(x)$, $x \in R$, defined by (see [7], [2], and references therein)

$$\psi_A(x) = \exp(x^2/2) \int_{-\infty}^{x} (I(t \in A) - \Phi(A)) \exp(-t^2/2)\, dt.$$

Applying the basic covariance identity (1.2) with $g = \psi_A$ and using relation (1.3) of [7], we get

$$\begin{aligned} F(A) - \Phi(A) &= E\{\psi_A'(X) - X\psi_A(X)\} \\ &= E\{\psi_A'(X)[1 - w_X(X)]\} \\ &= E\{I(X \in A)[1 - w_X(X)]\} + E\{X\psi_A(X)[1 - w_X(X)]\}. \end{aligned}$$

Observe that for $A_0 = \{x \in R : w_X(x) \leq 1\}$,

$$\begin{aligned} E\{I(X \in A)[1 - w_X(X)]\} &= \int_A [1 - w_X(x)]\, dF(x) \\ &\leq \int_{A_0} [1 - w_X(x)]\, dF(x) = \frac{1}{2} E|w_X(X) - 1|. \end{aligned} \tag{2.6}$$

Moreover,

$$\begin{aligned} E\{X\psi_A(X)[1 - w_X(X)]\} &\leq E|X\psi_A(X)||1 - w_X(X)| \\ &\leq E|1 - w_X(X)|, \end{aligned} \tag{2.7}$$

since for each x and A,

$$|x\psi_A(x)| \leq (|x|/\varphi(x)) \min\{\Phi(x), 1 - \Phi(x)\} \leq 1.$$

Now, the first inequality in (2.5) follows from (2.7) and (2.7), in view of the fact that

$$\sup_A |F(A) - \Phi(A)| = \sup_A [F(A) - \Phi(A)],$$

and the second one is obvious since $E[w_X(X)] = 1$. ■

We now state the following CLT (for a proof see [7], Theorem 5.1).

Theorem 2.2 (CPU [7]). *Let $X, X_1, X_2, \ldots, X_n, \ldots$ be i.i.d. and absolutely continuous r.v.'s with mean zero and variance one. Then, under the notations of Theorem 2.1,*

$$E|w_n(S_n) - 1| \to 0. \tag{2.8}$$

Note that (2.8), combined with (2.5), implies a strengthened CLT (L_1 convergence of the densities). However, $w(X)$ need not have finite variance, since (2.8) is an immediate consequence of the (weak) law of large numbers applied to $\overline{w_n} = (w(X_1) + \cdots + w(X_n))/n$ (see CPU [7] for more details).

The third elementary proof of the CLT is based on the following lemma, which may be of some interest in itself.

Lemma 2.3 (CPP [3], Lemma 2.1.) *Let X, Y be two independent and standardized absolutely continuous r.v.'s and consider the r.v. $S = aX+bY$, where a, b are real constants such that $a^2 + b^2 = 1$. Then*

$$\mathrm{Var}[w_S(S)] \le a^4 \mathrm{Var}[w_X(X)] + b^4 \mathrm{Var}[w_Y(Y)],$$

where w_X, w_Y, w_S, are the w-functions of X, Y, S, respectively.

As an immediate consequence of the above lemma we have the following

Theorem 2.3. *If $X, X_1, X_2, \ldots, X_n, \ldots$ are as in Theorem 2.1, then*

$$\varrho(F_n, \Phi) \le c/\sqrt{n}, \tag{2.9}$$

where the constant c can be taken as

$$c = (3/2)\sqrt{Var[w(X)]}. \tag{2.10}$$

Proof. If we apply Lemma 2.3 to $S_n = \sqrt{(n-1)/n}\, S_{n-1} + \sqrt{1/n}\, X_n$, we get

$$\sigma_n \le \left(\frac{n-1}{n}\right)^2 \sigma_{n-1} + \frac{1}{n^2}\sigma_1,$$

where $\sigma_n = \mathrm{Var}[w_n(S_n)]$. Thus, by induction on n, we conclude that $\sigma_n \le \sigma_1/n$, and (2.9) follows from (2.5). ∎

3. Applications to Random Sums

In this section, we give some applications of the above results to random sums (i.e., when the sample size n is no longer constant, but is a discrete nonnegative integer-valued random variable N, independent of the sequence

$\{X_i\}$). It is proved that under appropriate conditions (in fact, when N is likely to be large), the CLT continues to hold for the standardized sums of N i.i.d. standardized r.v.'s. It should be noted however, that here the term 'standardized' sums has a somewhat different meaning in the sense that

$$S_N = \frac{1}{\sqrt{N}}(X_1 + \cdots + X_N) \,, \tag{3.1}$$

(where $S_0 \equiv 0$ by definition) need not have variance one.

We first prove the following auxiliary result:

Lemma 3.1. *Suppose $X, X_1, \ldots, X_i, \ldots$ are i.i.d. r.v.'s as in Theorem 2.1. Then, for any nonnegative integer-valued r.v. N, independent of $\{X_i\}$,*

$$\varrho(F_N, \Phi) \leq P[N \leq m] + c/\sqrt{m+1}, \; m = 0, 1, \ldots \,, \tag{3.2}$$

where F_N is the d.f. of S_N and c can be taken as in Theorem 2.3.

Proof. For an arbitrary Borel set A we have, by (2.9),

$$\begin{aligned} |P[S_N \in A] - \Phi(A)| &\leq \sum_{n=0}^{\infty} P[N = n] \, |P[S_N \in A | N = n] - \Phi(A)| \\ &\leq \sum_{n=0}^{m} P[N = n] + \sum_{n=m+1}^{\infty} cP[N = n]/\sqrt{n}, \end{aligned}$$

and the proof is complete. ■

Using the above Lemma, we can easily establish the following:

Theorem 3.1. *Let $\{X_i\}$ satisfy the assumptions of Lemma 3.1 and suppose that the nonnegative integer-valued r.v.'s N_n in probability tend to infinity, in the sense that for any $m > 0$, $P[N_n > m] \to 1$ as $n \to \infty$. Then,*

$$\varrho(F_n, \Phi) \to 0, \; \text{as } n \to \infty, \tag{3.3}$$

where F_n is the d.f. of S_{N_n}.

Proof. It follows from Lemma 3.1 that for any $\epsilon > 0$ (arbitrarily small) and $m > 0$ (arbitrarily large),

$$\varrho(F_n, \Phi) \leq \epsilon + c/\sqrt{m+1} \quad \text{when } n > n_0(\epsilon, m)$$

and the assertion follows from the arbitrariness of m and ϵ. ■

Corollary 3.1. *If $N_n/a_n \to \Theta$ in probability, where Θ is an arbitrary positive r.v. and $a_n \to \infty$,*

$$\varrho(F_n, \Phi) \to 0, \quad \text{as } n \to \infty. \tag{3.4}$$

Proof. It is easy to verify the conditions of Theorem 3.1. ∎

Remark 3.1. Similar conditions imposed on $\{N_n\}$ can also be found in [1]; Theorem 17.2, p. 147 and [8], p. 258, where the sums are scaled by $\sqrt{a_n} = \sqrt{n}$ instead of $\sqrt{N_n}$ and Θ equals one. Billingsley considers the general case of $\{N_n\}$ not necessarily independent of $\{X_i\}$. He notes that the assumption that N_n in probability tends to infinity *does not suffice alone* for the convergence in distribution of F_n to Φ. For a counterexample see [1], pp. 143–144. It is therefore a crucial assumption that the sequences $\{N_n\}$ and $\{X_i\}$ are independent.

Remark 3.2. The restriction here to r.v.'s having w-functions with finite variance makes it possible not only to relax the condition on $\{N_n\}$, but also to obtain estimates for the rate of convergence. In fact, the following theorem provides uniform estimates for the rate of convergence, under somewhat stronger conditions imposed on $\{N_n\}$.

Theorem 3.2. *Let the sequences $\{X_i\}$, $\{a_n\}$ and $\{N_n\}$ be as in Corollary 3.1, and suppose that* [1]

$$\sup_n \left\{ \sqrt{a_n}\, E\left|\frac{N_n}{a_n} - \Theta\right| \right\} = C < \infty,$$

where Θ is a positive r.v. such that $P[\Theta \ge \delta] = 1$ for some $\delta > 0$. Then,

$$\varrho(F_n, \Phi) \le O(a_n^{-1/2}), \quad \text{as } n \to \infty. \tag{3.5}$$

Proof. By using Markov's inequality, it is easily verified that

$$P[2N_n \le \delta a_n] \le P\left[\left|\frac{N_n}{a_n} - \Theta\right| \ge \frac{\delta}{2}\right] \le \frac{2C}{\delta\sqrt{a_n}}.$$

Applying Lemma 3.1 with $m = [\delta a_n/2]$, we get

$$\varrho(F_n, \Phi) \le \frac{2C/\delta + c\sqrt{2}/\sqrt{\delta}}{\sqrt{a_n}}$$

and the proof is complete. ∎

---

[1]Thanks are due to the referee for pointing out the insufficiency of an earlier condition for concluding (3.5); also that the proof of the CLT via w-functions, in the case of a noninterval support, is not straightforward, and hence omitted.

References

[1] Billingsley, P. (1968). *Convergence of Probability Measures.* Wiley, New York.

[2] Bolthausen, E. (1984). An estimate of the remainder in a combinatorial central limit theorem. *Z. Wahrsch. Verw. Gebiete* **66**, 379–386.

[3] Cacoullos, T., Papadatos, N., and Papathanasiou, V. (1996). Variance inequalities for covariance kernels and applications to central limit theorems. *Theory Probab. Appl.* **71**, 195–201.

[4] Cacoullos, T. and Papathanasiou, V. (1989). Characterizations of distributions by variance bounds. *Statist. Probab. Lett.* **7**, 351–356.

[5] Cacoullos, T. and Papathanasiou, V. (1992). Lower variance bounds and a new proof of the central limit theorem. *J. Multivariate Anal.* **43**, 173–184.

[6] Cacoullos, T., Papathanasiou, V., and Utev, S. (1992). Another characterization of the normal law and a proof of the central limit theorem connected with it. *Theory Probab. Appl.* **37**, 648–657 (in Russian).

[7] Cacoullos, T., Papathanasiou, V., and Utev, S. (1994). Variational inequalities with examples and an application to the central limit theorem. *Ann. Probab.* **22**, 1607–1618.

[8] Feller, W. (1966). *An introduction to Probability Theory and its Applications.* Vol. 2, Wiley, New York.

[9] Papathanasiou, V. (1996). Multivariate variational inequalities and the central limit theorem. *J. Multivariate Anal.* **58**, 189–196.

T. Cacoullos and V. Papathanasiou
University of Athens
Department of Mathematics
Panepistemiopolis, 157 84 Athens
Greece

N. Papadatos
University of Cyprus
Department of Mathematics and Statistics
P.O. Box 537, Nicosia 1678
Cyprus

Almost Everywhere Convergence and SLLN Under Rearrangements

Sergei Chobanyan and V. Mandrekar*

Introduction

The almost everywhere (a.e.) convergence of trigonometric Fourier series for $L_2(0,1)$ functions was conjectured by Luzin (1922) and was partially solved by Kolmogorov and Silvestrov in (1925). The full solution was given by Carleson (1966). In the work of Garsia (1964), (see Garsia (1970)) almost everywhere convergence of a rearrangement of series of orthogonal functions was initiated using the so-called Garsia Inequality (GI). His convergence result was generalized by Nikishin (1967) who removed the assumption of orthogonality. We prove an inequality which generalizes GI using the technique introduced in Chobanyan (1990) (for other references see Chobanyan (1994)). As a consequence of this inequality we derive GI and a generalization of Nikishin's result to a series of Banach space valued random variables.

It is well-known (Doob (1953)) that the Kolmogorov strong law of large numbers (SLLN) for independent (not necessarily identically distributed) summands is derived as a consequence of a.e. convergence of a series. A natural question arises as to the validity of SLLN for orthogonal random variables under a rearrangement. The first SLLN for orthogonal random variables with bounded second moments can be found in Doob (1953). This was derived as a consequence of the Rademacher–Men'shov inequality using series convergence. In Beck et al. (1975) a SLLN was proved for Banach space valued random variables satisfying the weak orthogonality condition, essentially using the Rademacher–Men'shov inequality on the real line. We prove the Rademacher–Men'shov inequality under natural orthogonality of Hilbert space valued random variables and give a SLLN in Hilbert space. Since weak orthogonality implies orthogonality used here for Hilbert-space valued random variables, our result strengthens that in the work of Beck et al. (1975) for Hilbert-space valued random variables. In fact, one obtains for orthogonal random variables $\{\xi_n\}$ with $\sigma_n^2 = E\|\xi_n\|^2$ bounded,

$$\frac{1}{N}\sum_{k=1}^{N} \xi_{\pi(k)} \to 0 \ \ a.e. \tag{1}$$

*This work is partly supported by the Georgian Academy of Sciences Grant 1.16

for every permutation $\pi: \mathbf{N} \to \mathbf{N}$. However, if $\{\sigma_n^2\}$ is an unbounded sequence of non-negative reals, then there exists orthogonal random variables $\{\xi_n\}$ with $E\xi_n = 0, E\xi_n^2 = \sigma_n^2$ and a permutation π so that

$$\frac{1}{N}\sum_{k=1}^{N} \xi_{\pi(k)} \to \infty \quad a.e.$$

(see Tandori (1982)). Now we can show that under the condition $\sum \sigma_n^2 \log n (\log\log n)^{1+\epsilon}/n^2 < \infty$ for some $\epsilon > 0$, there exists a permutation $\pi: \mathbf{N} \to \mathbf{N}$ so that (1) holds. This is done by using an inequality due to Maurey and Pisier (1975). We note that even for pairwise independent random variables, $\sum_{m=1}^{\infty} \sigma_m^2/m^2 < \infty$ is not enough to guarantee SLLN (Csorgo et al. (1983)).

1. Generalized Garsia inequality and its applications

Let X be a Banach space with norm $\|\cdot\|$ and let $(\Omega, \mathcal{F}, \mathcal{P})$ be a probability space. We denote by $\xi: \Omega \to X$ a Bochner measurable function and call it an X-valued random variable. We first state a generalization of Garsia Inequality (Garsia (1970)).

Theorem 1. *Let $\Phi: \mathbf{R} \to \mathbf{R}^+$ be a convex increasing function and $\xi_1, \cdots, \xi_n$ be X-valued random variables such that $E\Phi(\|\xi_i\|) < \infty, i = 1, 2, \cdots, n$.*

(a) If $\xi_1 + \xi_2 + \cdots + \xi_n = 0$. Then there exists a permutation $\pi: \{1, 2, \cdots, n\} \to \{1, 2, \cdots, n\}$ so that for any collection of signs $\theta = (\theta_1, \theta_2, \cdots, \theta_n)$

$$E\Phi(\max_{1\le k\le n} \|\xi_{\pi(1)} + \cdots + \xi_{\pi(k)}\|) \le E\Phi(\max_{1\le k\le n} \|\xi_{\pi(1)}\theta_1 + \cdots + \xi_{\pi(k)}\theta_k\|).$$

(b) Let $\Phi(t) = t^p, p \ge 1$, then there exists a permutation $\pi: \{1, \cdots, n\} \to \{1, \cdots, n\}$ so that for any collection of signs $\theta = (\theta_1, \theta_2, \cdots, \theta_n)$,

$$(E\max_{1\le k\le n} \|\xi_{\pi(1)} + \cdots + \xi_{\pi(k)}\|^p)^{\frac{1}{p}}$$
$$\le (E\max_{1\le k\le n} \|\xi_{\pi(1)}\theta_1 + \cdots + \xi_{\pi(k)}\theta_k\|^p)^{\frac{1}{p}} + 2E(\|\sum_{i=1}^{n} \xi_i\|^p)^{\frac{1}{p}},$$

or

$$E\max_{1\le k\le n} \|\xi_{\pi(1)} + \cdots + \xi_{\pi(n)}\|^p$$
$$\le 2^p(E\max_{1\le k\le n} \|\xi_{\pi(1)}\theta_1 + \cdots + \xi_{\pi(k)}\theta_k\|^p + E(\|\sum_{i=1}^{n} \xi_i\|^p).$$

For the proof we need the following

Lemma. *If* $\{a_1, \cdots, a_n\} \in X$ *with* $\sum_{i=1}^n a_i = 0$, *then for any signs* $\theta = (\theta_1, \cdots, \theta_n)$ *there exists a permutation* $\lambda: \{1, \cdots, n\} \to \{1, \cdots, n\}$ *satisfying*

$$\max_{1\le k\le n} \|a_1 + \cdots + a_k\| + \max_{1\le k\le n} \|a_1\theta_1 + \cdots + a_k\theta_k\| \ge 2 \max_{1\le k\le n} \|a_{\lambda(1)} + \cdots + a_{\lambda(k)}\|.$$

Proof. Denote $|(a)| = \max_{1\le k\le n} \|a_1 + \cdots + a_k\|$. Then $|.|$ is a norm on $(a_1, \cdots, a_n)$'s. Then using the triangle inequality with $a\theta = (a_1\theta_1, \cdots, a_n\theta_n)$,

$$|(a)| + |(a\theta)| \ge |(a + a\theta)| = 2|(a^+)|.$$

Here (a^+) is the ordered subcollection of (a) corresponding to $\theta_i = +1$. We also have

$$|(a)| + |(a\theta)| = |(-a)| + |(a\theta)| \ge 2|(a^-)|,$$

where a^- corresponds to $\theta_i = -1$. Hence

$$|(a)| + |(a\theta)| \ge 2\max(|(a^+)|, |(a^-)|).$$

In view of the fact that $\sum_{i=1}^n a_i = 0$, we can find a permutation λ so that

$$\max(|(a^+), (a^-)|) = |(a_\lambda)|.$$

In fact λ is as follows: first go over indices of (a^+) followed by those of (a^-) in reverse order. ■

Proof of Theorem 1. Let $\sigma: \{1, \cdots, n\} \to \{1, \cdots, n\}$ be a permutation and $\theta = (\theta_1, \cdots, \theta_n)$ be a collection of signs. We have by the properties of Φ using Lemma 1,

$$E\Phi(|(\xi_\sigma)|) + E\Phi(|(\xi_\sigma\theta)|) \ge 2E\Phi(\frac{1}{2}(|(\xi_\sigma)| + |(\xi_\sigma\theta)|) \ge 2E\Phi(|(\xi_\lambda)|)).$$

Let π be a permutation minimizing $E\Phi(|(\xi_\sigma)|)$. Then from the above inequality we get

$$E\Phi(|(\xi_\pi)|) \le E\Phi(|(\xi_\pi\theta)|),$$

which proves part (a). To prove (b) we consider the sequence $a_1, \cdots, a_n, -s$ with $s = \sum_{i=1}^n a_i$ and apply part (a) and triangle inequality. ■

Corollary 1 (Garsia inequality). *For any finite orthonormal system* $\{\phi_1, \cdots, , \phi_n\}$ *and* $\{\alpha_1, \cdots, \alpha_n\} \subset \mathbf{R}$, *there exists a permutation* $\pi: \{1, \cdots, n\} \to \{1, \cdots, n\}$ *so that*

$$E \max_{1\le k\le n} |\alpha_{\pi(1)}\phi_{\pi(1)} + \cdots + \alpha_{\pi(k)}\varphi_{\pi(k)}|^2 \le C\sum_{i=1}^n \alpha_i^2,$$

where C *is an absolute constant.*

Proof. In Theorem 1(b), set $X = \mathbf{R}$, $\Phi(t) = t^2$, $\xi_i = \alpha_i\phi_i$. We obtain that there exists a permutation π such that

$$E\max_{1\le k\le n}\Big(\sum_1^k \alpha_{\pi(i)}\phi_{\pi(i)}\Big) \le 4\Big(E\max_{1\le k\le n}\Big(\sum_1^k \alpha_{\pi(i)}\phi_{\pi(i)}\theta_i\Big)^2 + E\Big(\sum_1^n \alpha_i\phi_i\Big)^2\Big).$$

Taking the average with respect to 2^n signs, we get

$$\frac{1}{2^n}\sum_\theta E\max_{1\le k\le n}\Big(\sum_1^k \alpha_{\pi(i)}\phi_{\pi(i)}\theta_i\Big)^2 = EE_r\max_{1\le k\le n}\Big(\sum_1^k \alpha_{\pi(i)}\phi_{\pi(i)}r_i\Big)^2$$

$$\le 2EE_r\Big(\sum_{i=1}^n \alpha_i\phi_i r_i\Big)^2 = 2E\Big(\sum_{i=1}^n \alpha_i^2\phi_i^2\Big) = 2\sum_{i=1}^n \alpha_i^2 .$$

Here $\{r_1, \cdots, r_n\}$ are Rademacher functions and the inequality follows from the Levy inequality. Thus we get the Garsia inequality with $C = 9$. ■

Now we prove a generalization of the Nikishin theorem.

Theorem 2. *Let $\sum \xi_k$ be a series with ξ_k an X-valued random variable for each k. Assume that there exists a subsequence (S_{n_k}) of $S_n = \sum_{k=1}^n \xi_k$ that converges a.e. to an X-valued random variable s. Then the series $\sum \xi_{\sigma(k)}$ converges to s a.e. for some permutation σ of $\mathbf{N}$ if the following condition holds: For any permutation π of $\mathbf{N}$ there exists a sequence of signs (θ_k) such that the series*

$$\sum \xi_{\pi(k)}\theta_k$$

converges a.e. and

$$\sup_n \Big\|\sum_1^n \xi_{\pi(k)}\theta_k\Big\| \le g\, a.e.,$$

where g is some measurable function from $L_0(\Omega, \mathcal{F}, \mathcal{P})$.

Proof. We consider the series $\sum \eta_k$, where $\eta_k = \xi_k/(f+g+1)$, where $f = \sup_k \|S_{n_k}\|$. By the condition, $S_{n_k} \to s/(f+g+1)$ a.s. We introduce

$$Q_N = \sup_{q_1\cdots q_l > N}\ \sup_\sigma \inf_\theta E\max_{1\le k\le l}\|\eta_{\sigma(1)}\theta_1 + \cdots + \eta_{\sigma(k)}\theta_k\|,$$

$$L_N = \sup_{n_u, n_v > N} E\|S^\eta_{n_u} - S^\eta_{n_v}\|,$$

where in Q_N, the first sup is taken over all finite indices $q_1, \cdots, q_l$, that exceed N and the second sup is over all permutations σ: $\{q_1, \cdots, q_l\} \to \{q_1, \cdots, q_l\}$. In L_N, $S^\eta_{n_u} = \sum_1^{n_u} \eta_k$. Under the condition of the theorem and the Lebesgue dominated convergence theorem, one can show that Q_N and L_N converge to zero as $N \to \infty$. Now choose a sequence N_k so that $Q_{N_k} < 1/k$ and $L_{N_k} < 1/k$. Then we consider the blocks $\{\eta_{N_k}, \cdots, \eta_{N_{k+1}}\}$ and choose π_k as in Theorem 1(b). The rearrangement π defined by π_k's is the desired permutation.

We have by Theorem 1(b) that for any $\theta = (\theta_1, \cdots, \theta_{N_{k+1}-N_{n_k}})$,

$$E \max_{1 \le m \le N_{k+1}-N_k} \|\eta_{\pi_k(N_k+1)} + \eta_{\pi_k(N_k+2)} + \cdots + \eta_{\pi_k(N_k+m)}\|$$

$$\le E \max_{1 \le m \le N_{k+1}-N_k} \|\eta_{\pi_k(N_k+1)}\theta_1 + \cdots + \eta_{\pi_k(N_{k+1})}\theta_k\|$$

$$+ E\|S_{N_{k+1}} - S_{N_k}\| \le \frac{2}{k}.$$

Let $\zeta_k = \eta_{\pi(k)}$ and take $N_u < p \le N_{u+1}$, $N_v < q \le N_{v+1}$ $(N_u < N_v)$. Then

$$E \max_{1 \le m \le q-p} \|\zeta_p + \zeta_{p+1} + \cdots + \zeta_{p+m}\|$$

$$\le E \max_{1 \le m \le p-N_u} \|\zeta_{N_u+1} + \cdots + \zeta_{N_u+m}\| + E\|S_{N_v} - S_{N_u}\|$$

$$+ E \max_{1 \le m \le q-N_v} \|\zeta_{N_v+1} + \cdots + \zeta_{N_v+m}\|.$$

This is $\le 2/u + 1/u + 2/v \to 0$ as $u, v \to \infty$. Thus we get a.s convergence of $\sum \eta_{\pi(k)}$ to $s/(f+g+1)$ which gives that $\sum \xi_{\pi(k)} \to s$ a.e. ■

Corollary 2.

(a) *Let for each $k = 1, 2, \cdots$, ξ_k be X-valued random variables such that $S_n = \sum_{k=1}^n \xi_k$ converges in probability to s and $\sum_k \xi_k r_k$ converges $P \times \lambda$ - a.s. with $\{r_k\}$ Rademacher functions and λ the Lebesgue measure. Then there exists a permutation π of* **N** *such that $\sum_k \xi_{\pi(k)}$ converges $P -$ a.e. to s.*

(b) *Let X be a type 2 space and $\{\xi_k\}$ be as above satisfying $S_n \to s$ in probability and $\sum \|\xi_k\|^2 < \infty$ a.e. Then there exists a permutation π of* **N** *so that $\sum \xi_{\pi(k)}$ converges a.e. to s.*

Proof. It suffices to prove (a), as in type 2 space X, a.e. convergence of $\sum \|\xi_k^2\|$ implies a.e. convergence of $\sum \xi_k r_k$. Let σ be any permutation of **N**. Then by the Levy inequality,

$$E_t \max_n \|\sum_{k=1}^n \xi_{\sigma(k)}(\omega) r_{\sigma(k)}(t)\| \le 2E_t \|\sum_{k=1}^\infty \xi_{\sigma(k)}(\omega) r_{\sigma(k)}(t)\|$$

$$= 2E_t \|\sum_{k=1}^\infty \xi_k(\omega) r_k(t)\| = g(\omega) \text{ a.e.}$$

This implies the existence of signs $\theta = (\theta_k)$ satisfying the condition of the theorem. ∎

Remark. Corollary 2(b) gives the Nikshin Theorem if $X = \mathbb{R}$.

2. SLLN for Hilbert-space valued orthogonal random variables

In this section we consider $X = H$ a real Hilbert space. Two H-valued random variables ξ, η with $E\|\xi\|^2$, $E\|\eta\|^2$ finite are said to be orthogonal if $E(\xi, \eta)_H = 0$. We note that orthogonality is weaker condition than weak orthogonality considered in Beck and Warren (1975). We start by considering the following extension of Rademacher–Men'shov inequality.

Theorem 3. *Let $\xi_1, \cdots, \xi_n$ be Hilbert-space valued orthonormal random variables, i.e. $E(\xi_i, \xi_j) = \delta_{ij}$. Then for any reals $\alpha_1, \cdots, \alpha_n$*

$$E \max_{1 \le k \le n} \|\alpha_1 \xi_1 + \cdots + \alpha_k \xi_k\|^2 \le (\log_2 n + 2)^2 \sum_{k=1}^{n} \alpha_k^2 .$$

Proof. Let

$$D_n = \sup_{\xi_1 \cdots \xi_n} \sup_{\alpha_1^2 + \cdots + \alpha_n^2 \le 1} E \max_{1 \le k \le n} \|\sum_{i=1}^{k} \alpha_i \xi_i\|^2,$$

where the first supremum is over all orthonormal $\xi_1, \cdots, \xi_n$'s. Define for $\{b_1, \cdots, b_n\} \in H$,

$$|(b_1, \cdots, b_n)| = \max_{1 \le k \le n} \|\sum_{i=1}^{k} b_i\|.$$

Then we have for any $n \in \mathbf{N}$, collection of H-valued random variables $\xi_1, \cdots, \xi_{2n}$ and reals $\alpha_1, \cdots, \alpha_{2n}$

$$\begin{aligned} |(\alpha_1 \xi_1, \cdots, \alpha_{2n} \xi_{2n})|^2 &= |(\alpha_1 \xi_1, \cdots, \alpha_{2n} \xi_{2n}, -\sum_{i=1}^{2n} \alpha_i \xi_i)|^2 \\ &= \max\{|(\alpha_1 \xi_1, \cdots, \alpha_n \xi_n)|, |(-s, \alpha_{2n} \xi_{2n}, \cdots, \alpha_{n+1} \xi_{n+1})|)^2, \\ &\le |(\alpha_1, \xi_1, \cdots, \alpha_n \xi_n)|^2 |(-s, \alpha_{2n} \xi_{2n}, \cdots, \alpha_{n+1} \xi_{n+1})|^2 \end{aligned} \tag{1}$$

where $s = -\sum_{i=1}^{2n} \alpha_i \xi_i$. If ξ_i's are orthonormal and $(\alpha_1, \cdots, \alpha_{2n})$ satisfy $\sum_{i=1}^{2n} \alpha_i^2 = 1$, then denoting by $p = \sum_{i=1}^{n} \alpha_i^2$, $q = \sum_{n+1}^{2n} \alpha_i^2$ we get from the above inequality

$$E \max_{1 \le k \le 2n} \|\sum_{i=1}^{k} \alpha_i \xi_i\|^2 \le pD_n + (1 + q^{1/2} D_n^{1/2})^2$$

$$\leq (1 + D_n^{1/2})^2 . \tag{2}$$

Since (2) is true for any collection of $(\xi_1, \cdots, \xi_{2n})$ orthonormal and $\alpha_1, \cdots, \alpha_{2n}$ reals $(\sum_{i=1}^{2n} \alpha_i^2 = 1)$, we get

$$D_{2n} \leq (1 + D_n^{1/2})^2.$$

As $D_1 = 1$, we get $D_{2^k} \leq (k+1)^2$. For any $n \in \mathbf{N}$, $2^k < n \leq 2^{k+1}$ for some k. So

$$D_n \leq D_{2^{k+1}} \leq (k+2)^2 \leq (\log_2 n + 2)^2.$$

■

From the above inequality, following the proof in Doob (1953) (Theorem 4.2, p. 157), we get

Corollary 3. *Let $\{\xi_k\}$ be a sequence of orthogonal H-valued random variables with $E\|\xi_k\|^2 = \sigma_k^2$. If $\sum_1^\infty \sigma_k^2 (\log k)^2 < \infty$, then the series $\sum_{k=1}^\infty \xi_k$ converges a.e.*

If we apply this corollary to $\{\xi_k/k\}$ and use the Kronecker lemma, we have

Corollary 4 (SLLN). *Let $\{\xi_k\}$ be a sequence of orthogonal random variables with $\sum_{k=1}^\infty \sigma_k^2 (\log k)^2/k^2 < \infty$. Then*

$$\lim_{n\to\infty} \frac{1}{n} \sum_{k=1}^n \xi_k \to 0 \ \ a.e.$$

In particular, if $\sup_k \sigma_k^2 < \infty$, then $\sum_{k=1}^\infty \sigma_{\pi(k)}^2 (\log k)^2/k^2 < \infty$ for every rearrangement $\{\xi_{\pi(k)}\}$ ensures a.e. convergence of the series $\sum_{k=1}^\infty \xi_{\pi(k)}/k$, and hence we get by the Kronecker lemma

$$\frac{1}{n} \sum_{k=1}^n \xi_{\pi(k)} \ \to 0 \ \ \text{a.e.} \tag{3}$$

Tandori (1982), has shown that if $\limsup_k \sigma_k^2 = \infty$, then there exists a sequence $\{\xi_k\}$ of orthogonal random variables ($H = \mathbf{R}$) with $E\xi_k = 0$, $E\xi_k^2 = \sigma_k^2$ so that

$$\frac{1}{n} \sum_{k=1}^n \xi_{\pi(k)} \ \to \infty \ \ a.e.$$

We now want to obtain a SLLN of the form (3) for a rearrangement which will correspond to permutations over blocks of integers. We first give a lemma which gives convergence in (3) over a subsequence $\{n_k\}$ $(n_k \to \infty)$ under the usual condition of Kolmogorov SLLN (i.e. $\sum_k \sigma_k^2/k^2 < \infty$).

Lemma 1. *Let $\{\xi_k\}$ be a sequence of orthogonal H-valued random variables satisfying $\sum_{k=1}^{\infty} \sigma_k^2/k^2 < \infty$ with $\sigma_k^2 = E\|\xi_k\|^2$. Then for any sequence $\{n_k\}$ ($n_k \to \infty$)so that $\sup_k (k \log k)^2/n_k^2 < \infty$, we have*

$$\frac{1}{n_k} \sum_{i=1}^{n_k} \xi_j \to 0 \quad a.e.$$

Proof. In view of the Kronecker lemma, it suffices to prove that$\sum_k \xi_k/n_k$ converges a.e. By Theorem 3, this happens if $\sum_k \sigma_k^2 (\log k)^2/n_k^2 < \infty$. Now write $\sigma_k^2(\log k)^2/n_k^2 = \sigma_k^2(k \log k)^2/(n_k^2 k^2)$. So under the conditions of Lemma 2 we get the result. In particular, $n_k = 2^k$ gives a result of Moricz (1983). ∎

Proposition 1. *Let $\{\xi_k\}$ be a sequence of Hilbert space-valued orthogonal random variables with $\sigma_k^2 = E\|\xi_k\|^2$ satisfying*

$$\sum_l \left(\frac{1}{n_l n_{l+1}} \sum_{k=n_l}^{n_{l+1}} \sigma_i^2\right)^{1/2} < \infty , \tag{4}$$

where $\{n_l\} \in \mathbf{N}$ is an increasing sequence tending to infinity. Then there exists a rearrangement of $\{\xi_{\pi(k)}\}$ so that

$$\lim_{n\to\infty} \frac{1}{n} \sum_{k=1}^{n} \xi_{\pi(k)} \to 0 \quad a.e.$$

We need the following inequality due to Maurey and Pisier (1975) (see also Chobanyan (1994)).

If $\{a_1, \cdots, a_n\} \subset H$ (more generally in type 2 Banach space X) and $\lambda_1, \cdots, \lambda_n$ are positive real numbers, then

$$\frac{1}{n!} \sum_\sigma \max_{1\le k\le n} \|\lambda_1 a_{\sigma(1)} + \cdots + \lambda_k a_{\sigma(k)}\|^2 \le \frac{C}{n} \left(\sum_{i=1}^{n} \lambda_k^2\right)\left(\|\sum_{j=1}^{n} a_j\|^2 + \sum_{j=1}^{n} \|a_j\|^2\right) , \tag{5}$$

where σ denotes permutations over $\{1, \cdots, n\}$ and C is an absolute constant.

Proof of Proposition 1. It suffices to prove that $\sum_{k=1}^{\infty} \xi_{\pi(k)}/k$ converges a.e. for some permutation π. Let $\{n_l\}$ be as above, then for $l \in \mathbf{N}$ we get by (5)

$$\begin{aligned} &\frac{1}{(n_{l+1} - n_l)} \sum_{\pi_l} E \max_{1\le k\le n_{l+1}-n_l} \|\sum_{i=1}^{k} \frac{1}{n_l + i} \xi_{\pi_l(n_l+i)}\|^2 \\ &\le \frac{C}{n_{l+1} - n_l} \sum_{i=n_l+1}^{n_{i+1}} \frac{1}{i^2} E\left(\|\sum_{j=n_l+1}^{n_{l+1}} \xi_j\|^2 + \sum_{j=n_l+1}^{n_{l+1}} \|\xi_j\|^2\right) . \end{aligned}$$

By orthogonality,

$$\leq \frac{2C}{n_{l+1}-n_l}\left(\frac{1}{n_l}-\frac{1}{n_{l+1}}\right)\sum_{i=n_l+1}^{n_{l+1}} \sigma_i^2 = \frac{2C}{n_l n_{l+1}}\sum_{i=n_l}^{n_{l+1}} \sigma_i^2 ,$$

where π_l are permutations of $(n_l+1,\cdots,n_{l+1})$. So for some π_l

$$E \max_{1\leq k\leq n_{l+1}-n_l} \|\sum_{i=1}^{k} \frac{1}{n_l+i}\xi_{\pi_l(n_l+i)}\|^2 \leq \frac{2C}{n_l n_{l+1}}\sum_{i=n_l+1}^{n_{l+1}} \sigma_i^2 .$$

Let π be a permutation of $\mathbf{N}$ determined by permutations $\{\pi_l\}$. Then

$$E\sum_{l=1}^{\infty} \max_{1\leq k\leq n_{l+1}-n_l} \|\sum_{i=1}^{k} \frac{1}{n_l+i}\xi_{\pi(n_l+i)}\|$$

$$\leq \sum_{l=1}^{\infty}(E \max_{1\leq k\leq n_{l+1}-n_l} \|\sum_{l=1}^{k} \frac{1}{n_{l+1}}\xi_{\pi(n_l+i)}\|^2)^{1/2}$$

$$\leq (2C)^{1/2}\sum_{l=1}^{\infty}(\frac{1}{n_l n_{l+1}}\sum_{l=n_l+1}^{n_{l+1}} \sigma_i^2)^{1/2} < \infty .$$

This implies a.e. convergence of $\sum_{k=1}^{\infty} \xi_{\pi(k)}/k$. ■

Corollary 5. *Let $\{\xi_k\}$ be a sequence of H-valued orthogonal random variables with $\sigma_k^2 = E|\xi_k|^2$ and let $\{a_k\}$ be an increasing sequence of non-negative real numbers such that $\sum 1/a_{2^k}^2 < \infty$. Then $\sum_{k=1}^{\infty} \frac{a_k\sigma_k^2}{k^2} < \infty$ implies that there exists a rearrangement $\{\xi_\pi\}$ such that $1/n\sum_{j=1}^{n} \xi_{\pi(j)} \to 0$ a.e. In particular, such a rearrangement exists if $\sum_k \frac{\sigma_k^2}{k^2}(\log k)(\log\log k)^{1+\xi} < \infty$.*

Proof. Choose $n_k = 2^k$ in Proposition 1 and let $\beta_k = a_k\sigma_k^2/k^2$. Then

$$\sum\left(\frac{1}{2^{2k+1}}\sum_{2^k}^{2^{k+1}} \frac{i^2\beta_i}{a_i}\right)^{1/2} \leq \sum_k\left(\frac{2^{2k+2}}{2^{k+1}a_{2^k}}\sum_{2^k}^{2^{k+1}} \beta_i\right)^{1/2}$$

$$\leq C\left(\sum_k \frac{1}{a_{2^k}^2}\right)^{1/2}\left(\sum_k\left(\sum_{2^k}^{2^{k+1}} \beta_i\right)\right),$$

where the last inequality follows from the Schwartz inequality. Thus, by Proposition 1, we obtain the result. In particular, we can take $a_k = \log k\,(\log\log k)^{1+\epsilon}$ for some $\epsilon > 0$. ■

References

[1] Beck A., Giesy D., Warren P. (1975). Recent developments in the theory of strong law of large numbers for vector valued random variables. *Th. Prob. Appl.* **20** 106–133.

[2] Beck A., Warren P. (1972). Weak orthogonality. *Pac. J. Math.* **41**, 1–11.

[3] Carleson L. (1966). On convergence and growth of partial sums of Fourier series. *Acta Math.* (Uppsala). **116**, 135–157.

[4] Chobanyan S. (1990). On some inequalities related to permutations of summands in a normed space, Preprint, Muskhelishvili Inst. Comp. Math., Georgian Acad. Sc. 1–21.

[5] Chobanyan S. (1994). Convergence a.s. of rearranged series in Banach spaces and associated inequalities. Prob. in Banach spaces 9 (eds. J. Hoffman-Jorgensen, J. Kuelbs, and M.B. Marcus), Birkhäuser Boston.

[6] Csorgo S., Tandori K., Totek W. (1983). On strong law of large numbers for pairwise independent random variables. *Acta Math. Hung.* **42**, 319–330.

[7] Garsia A. (1964). Existence of almost everywhere convergent rearrangement for Fourier series of L_2 functions. *Ann. Math.* **79**, 623–629.

[8] Garsia A. (1970). Topics in almost everywhere convergence, Markham Publishing Company, Chicago.

[9] Kolmogorov A.N., Silvestrov G.S. (1925). Sur la convergence des séries de Fourier. *Comptes Rend. Acad. Sci. Paris.* **178**, 303–305.

[10] Luzin N.N. (1951). Integral and Trigonometric Series. 2nd ed. Gostekhizdat. Moscow (Russian).

[11] Moricz F.(1983). On the Cesaro means of orthogonal sequences of random variables. *Ann. Prob.* **11**, 827–832.

[12] Nikishin E.M.(1967). On convergent rearrangements of functional series. *Matem. Zametki* **1**, 126–136. (Russian).

[13] Tandori K. (1982). Bemerkung zum Gesetzt der grossen Zahlen. *Acta Math. Hung.* **39** 361–362.

S. Chobanyan and V. Mandrekar
Department of Statistics and Probability
Michigan State University
Wells Hall
East Lansing, MI 48824-1027

Sufficient Conditions for the Existence of Conditional Moments of Stable Random Variables *

Renata Cioczek-Georges and Murad S. Taqqu

Abstract

Conditional moments $E[|X_2|^p|X_1 = x]$ of an α-stable random vector (X_1, X_2) may exist even if $p \geq \alpha$. The precise conditions are stated in Cioczek-Georges and Taqqu [4] and Samorodnitsky and Taqqu [12]. This paper provides the proof for the most delicate cases, namely $1 < \alpha < 2$ and $p < 2\alpha + 1$, which is the maximal range of possible p's when the vector (X_1, X_2) is nondegenerate.

AMS 1980 subject classification. 60E07, 62J02.

Keywords and phrases: stable distributions, stable random vectors, symmetric α-stable, conditional moments.

1. Introduction

Stable random vectors are parametrized by α, $0 < \alpha < 2$, and have finite moments of order only less than α and hence, in particular, they always have infinite variance. But because conditional distributions of stable vectors are typically non-stable, conditional moments of order higher than α may exist. Since conditional moments are often used in probability and statistics, it is important to determine sufficient conditions for their existence, and to obtain, in particular, functional forms for the regression and the conditional variance. These questions have been explored in Samorodnitsky and Taqqu [11]; Cambanis and Wu [3], [13]; Cioczek-Georges and Taqqu [4], [6], [5]; and the results are presented in Samorodnitsky and Taqqu [12]. In addition, Cambanis and Wu [3], and also Cambanis and Fotopoulos [1], consider conditioning on more than one variable. While it is difficult to give a meaningful functional form of the conditional variance-covariance matrix for general stable random vectors, Cambanis and Fotopoulos [1], Cambanis and Fotopoulos and He [2], and Fotopoulos and He [7] do so in the special sub-Gaussian case and also provide its asymptotic behavior.

*This research was supported by the ONR grant N00014-90-J-1287 and NSF grants DMS-9404093 and NCR-9404931 at Boston University.

When proving the existence of conditional moments for stable random variables, real difficulties occur for $\alpha > 1$ and large p's. Although the corresponding results in these cases have been quoted in Cioczek-Georges and Taqqu [4] and Samorodnitsky and Taqqu [12], the proofs have not appeared anywhere before. This paper fills this gap and provides the proofs. The cases considered here are the most delicate and new techniques are developed for their solutions.

Recall that a random vector (X_1, X_2) is symmetric α-stable ($S\alpha S$) if its characteristic function is of the form

$$\phi_{X_1,X_2}(t,r) := \phi(t,r) := E\exp(i(tX_1 + rX_2)) = \exp\Big(-\int_{S_2} |ts_1 + rs_2|^\alpha \Gamma(d\mathbf{s})\Big),$$

where Γ, called the spectral measure, is a finite symmetric measure on the Borel sets of the unit circle S_2 in $\mathbf{R}^2$. For more information about stable random vectors, see [12]. Our main result is:

Theorem 1.1. *Suppose that the spectral measure* Γ *satisfies*

$$\int_{S_2} |s_1|^{-\nu}\Gamma(d\mathbf{s}) < \infty \tag{1.1}$$

with $\nu > 0$. *Then* $E[|X_2|^p | X_1 = x] < \infty$ *a.e. for every*

$$p \le \alpha + \nu \;\text{ and }\; p < 2\alpha + 1.$$

The proof of this theorem for $0 < \alpha \le 1$ can be found in Cioczek-Georges and Taqqu [4]. This paper contains the proof of the harder case $1 < \alpha < 2$.

To understand the reason for the condition $p < 2\alpha + 1$, note that if $X_2 = aX_1$, then $E(|X_2|^p | X_1 = x) < \infty$ for all p, but Example 4.5 in [11] shows that (1.1), applied to a *nondegenerate* $S\alpha S$ random vector (X_1, X_2), cannot be expected to guarantee the existence of conditional moments of order higher or equal to $2\alpha + 1$. The intuition behind the proof of Theorem 1.1, described in [4], is included here as well in order to provide a guide to the reader.

The existence of conditional moments $E[|X_2|^p | X_1 = x]$ is related to the behavior at the origin of the characteristic function

$$\phi_{X_2|x}(r) = \frac{1}{2\pi f(x)} \int_{-\infty}^{\infty} e^{-itx}\phi(t,r)dt, \tag{1.2}$$

of X_2 given $X_1 = x$, where f denotes the density function of the (stable) random variable X_1 (it is shown in [11] that this characteristic function always exists if $X_1 \not\equiv 0$). Moreover, if $E[|X_2|^{2n} | X_1 = x] < \infty$, $n \ge 0$, then a necessary and sufficient condition for the existence of the moment of order $p = 2n + \lambda$, where $0 < \lambda < 2$, is for some $c > 0$

$$\int_0^c r^{-(1+\lambda)}(|\phi^{(2n)}_{X_2|x}(0)| - |\phi^{(2n)}_{X_2|x}(r)|)dr < \infty, \tag{1.3}$$

where $\phi^{(0)}_{X_2|x} = \phi_{X_2|x}$. Also, the moment of order $p = 2n + 2$ exists if and only if $r^{-2}(|\phi^{(2n)}(0)| - |\phi^{(2n)}(r)|)$ is bounded for $0 < r \leq c$. These results are due to Ramachandran (c.f. Theorem 5 in [9]) and their proof can also be found in [12].

For $2 < p \leq 4$ and $4 < p < 5$, we must consider the second and the fourth derivative of $\phi_{X_2|x}$, respectively. Since formal computation of the derivatives can yield divergent terms, special manipulations are needed. We use a number of different techniques in order to obtain an adequate representation of the derivatives. An important one can be written formally as follows:

$$\begin{aligned} \frac{d}{ds}\int_{-\infty}^{\infty} f(t)g(t+cs)dt &= \lim_{h\to 0}\frac{1}{h}\int_{-\infty}^{\infty} f(t)[g(t+c(s+h)) - g(t+cs)]dt \\ &= \lim_{h\to 0}\frac{1}{h}\int_{-\infty}^{\infty}[f(t-ch) - f(t)]g(t+cs)dt \\ &= -c\int_{-\infty}^{\infty} f'(t)g(t+cs)dt. \end{aligned}$$

It is used when $\int_{-\infty}^{\infty} f(t)g'(t+cs)dt$ does not exist but the above equalities hold. If $d/ds \int_{-\infty}^{\infty} f(t)g(t+cs)dt = c\int_{-\infty}^{\infty} f(t)g'(t+cs)dt$, then the above formula is nothing more than the ordinary integration by parts

$$\int_{-\infty}^{\infty} f(t)g'(t+cs)dt = -\int_{-\infty}^{\infty} f'(t)g(t+cs)dt.$$

This important tool enables us to calculate higher order derivatives, albeit in a more complex setting.

To establish the existence of the conditional moment of order $p = \alpha + \nu < 2\alpha + 1$, one needs to bound the differences of the characteristic function or its derivative between the points 0 and r so that condition (1.3) is satisfied. We represent these differences as a sum of a number of terms. We prove that in the cases $p = 2$ and $p = 4$, all these terms are bounded by const. r^2 for $0 < r \leq 1$. For the other values of p, we try to bound (at least some of) them by const. r^γ with $\gamma > p \bmod 2n$, $n = 0, 1$ or 2, so that they become integrable w.r.t. $r^{-1-p \bmod 2n}dr$. Since r is in the neighborhood of zero, the higher the exponent γ the more delicate the bound. In the "high p" cases, namely $2 < p < 2\alpha + 1$, $1/2 < \alpha < 1$, and $4 < p < 2\alpha + 1$, $3/2 < \alpha < 2$, one has $p \bmod 2n < 1$, so a bound of the form const. r is sufficient. But in the cases $p < 2$, $0 < \alpha < 2$, and $2 < p < 4$, $1 < \alpha < 2$, the value of $p \bmod 2n$

is in the interval $(0,2)$, a wider range requiring more delicate estimations. Thus, the case $p > 4$ and $3/2 < \alpha < 2$ is "hard" on the one hand because it requires the fourth derivative of the conditional characteristic function (which is a sum of more than 20 terms!), but "easy" as far as estimation of several of the terms, because $p \bmod 4 < 1$.

Only some of the terms, however, can be bounded by the adequate power of r, and they are, in fact, not the most important ones. It is the remaining terms which are crucial. (Had we attempted to bound them by a power of r, as in Samorodnitsky and Taqqu [11], the highest obtainable exponent γ would equal $p \bmod 2n$ which is sufficient only for the existence of all moments less than p, but not equal to p.) These remaining terms appear as integrals with respect to additional variables. To show that they are integrable with respect to $r^{-1-p \bmod 2n}dr$, we change order of integration and integrate with respect to r using the Gamma integral. The lemmas used in the "high p" cases $2 < p < 2\alpha+1,\ 1/2 < \alpha < 1$, and $4 < p < 2\alpha+1,\ 3/2 < \alpha < 2$, are different from those used in the cases $p < 2,\ 0 < \alpha < 2$, and $2 < p < 4,\ 1 < \alpha < 2$.

Finally, let us observe that the assumption of symmetry on the vector (X_1, X_2) can be dropped in the statement of Theorem 1.1. In [8], Hardin, Samorodnitsky and Taqqu have shown that Theorem 3.1 of [11] can be extended to an arbitrary α-stable vector (X_1, X_2). Their proof can also be applied to the case of Theorem 1.1.

The paper is organized as follows. Section 2 presents the proof of Theorem 1.1 for $1 < \alpha < 2$. Section 3 covers the auxiliary technical results, and the form of the fourth derivative of the conditional characteristic function is given in the appendix.

2. Proof of Theorem 1.1 in the case $1 < \alpha < 2$

We consider here a $S\alpha S$ random vector (X_1, X_2), $1 < \alpha < 2$, satisfying (1.1) with $\nu > 0$ and we let $\sigma_i = (\int_{S_2} |s_i|^\alpha \Gamma(d\mathbf{s}))^{1/\alpha}$, $i = 1, 2$, denote the scale parameter of the random variable X_i. The proof of Theorem 1.1 is divided into four propositions. The first one states the existence of conditional moments of order $p \le 2$. The second one establishes a representation for the real part of $\phi''_{X_2|x}$ (it is different from the one for $0 < \alpha \le 1$). The third one uses that representation to prove the existence of conditional moments of order $p,\ 2 < p \le 4$. Finally, the last proposition improves the result in the case $3/2 < \alpha < 1$, i.e., shows the existence of moments of order higher than 4 if $\nu > 4 - \alpha$. Its proof justifies the representation for Re $\phi^{(4)}_{X_2|x}$ given in the appendix.

Proposition 2.1. *Suppose* $1 < \alpha < 2$ *and* $\nu > 0$ *in* (1.1). *Then* $E[|X_2|^p|X_1 = x] < \infty$ *a.e. for*

$$p \leq \min\{\alpha + \nu, 2\}.$$

Proof. Follow the lines of the proof of Proposition 2.1 in Cioczek-Georges and Taqqu [4]. First, in order to apply Ramachandran's condition (1.3), use the bound

$$0 \leq 1 - |\phi_{X_2|x}(r)| \leq |I_1| + |I_2|,$$

where

$$I_1 = \frac{1}{2\pi f(x)} \int_{-\infty}^{\infty} \cos tx \; e^{-|t|^\alpha \sigma_1^\alpha} \int_{S_2} (|ts_1 + rs_2|^\alpha - |ts_1|^\alpha)\Gamma(ds)dt,$$

and

$$\begin{aligned} I_2 &= \frac{-1}{2\pi f(x)} \int_{-\infty}^{\infty} \cos tx \Big[\exp\Big(-\int_{S_2} |ts_1 + rs_2|^\alpha \Gamma(ds)\Big) \\ &\quad - \exp\Big(-\int_{S_2} |ts_1|^\alpha \Gamma(ds)\Big) \\ &\quad + \exp\Big(-\int_{S_2} |ts_1|^\alpha \Gamma(ds)\Big) \int_{S_2} (|ts_1 + rs_2|^\alpha - |ts_1|^\alpha)\Gamma(ds)\Big] dt. \end{aligned}$$

Then, application of Lemma 3.1 to relation (2.1) of [4] gives, for $0 < r \leq 1$,

$$\begin{aligned} |I_2| &\leq \text{const.} \int_{-\infty}^{\infty} \exp\{-|t|^\alpha \sigma_1^\alpha + \alpha\Gamma(S_2)(r^\alpha + r|t|^{\alpha-1})\}(r^\alpha + r|t|^{\alpha-1})^2 dt \\ &\leq \text{const.}\ r^2, \end{aligned}$$

which implies

$$\int_0^1 \frac{|I_2|}{r^{1+\alpha+\nu}} dr < \infty$$

for $0 < \nu < 2 - \alpha$, and $|I_2|/r^2 \leq$ const. for $0 < r \leq 1$ when $\nu \geq 2 - \alpha$.

Unless stated explicitly, here and in the future const. denotes a finite positive constant, which may change from one expression to another and is independent from r (it depends however on Γ, α or x).

Similarly, all the statements made for I_1 in the case $1/2 < \alpha \leq 1$ hold for $1 < \alpha < 2$, and the proposition is proved. ■

Proposition 2.2. *Suppose* $1 < \alpha < 2$ *and* $\nu > 2 - \alpha$ *in* (1.1). *Then the second derivative of the characteristic function* $\phi_{X_2|x}$ *exists and its real part equals*

$$\text{Re}\phi''_{X_2|x}(r) = \frac{\alpha^2}{2\pi f(x)} \int_{-\infty}^{\infty} \cos tx \exp\Big(-\int_{S_2} |ts_1 + rs_2|^\alpha \Gamma(ds)\Big) \tag{2.1}$$

$$\times\Big(\int_{S_2}(ts_1+rs_2)^{<\alpha-1>}s_2\Gamma(d\mathbf{s})\Big)^2dt$$
$$-\frac{\alpha(\alpha-1)}{2\pi f(x)}\int_{-\infty}^{\infty}\cos tx\exp\Big(-\int_{S_2}|ts_1+rs_2|^\alpha\Gamma(d\mathbf{s})\Big)$$
$$\times\int_{S_2}|ts_1+rs_2|^{\alpha-2}s_2^2\Gamma(d\mathbf{s})dt$$

Proof. Since the second moment $E[X_2^2|X_1=x]$ exists a.e. by Proposition 2.1, the characteristic function $\phi_{X_2|x}$ has a second derivative. We want to show that its real part is given by (2.1). First, note that Re $\phi'_{X_2|x}(r)$ is given by

$$\begin{aligned}(\mathrm{Re}\phi_{X_2|x}(r))' &= \frac{1}{2\pi f(x)}\int_{-\infty}^{\infty}\cos tx\frac{\partial}{\partial r}\phi(t,r)dt\\ &= \frac{1}{2\pi f(x)}\int_{-\infty}^{\infty}\cos tx\exp\Big(-\int_{S_2}|ts_1+rs_2|^\alpha\Gamma(d\mathbf{s})\Big)\\ &\quad\times\Big(-\int_{S_2}\frac{\partial}{\partial r}|ts_1+rs_2|^\alpha\Gamma(d\mathbf{s})\Big)dt\\ &= \frac{-\alpha}{2\pi f(x)}\int_{-\infty}^{\infty}\cos tx\exp\Big(-\int_{S_2}|ts_1+rs_2|^\alpha\Gamma(d\mathbf{s})\Big)\\ &\quad\times\Big(\int_{S_2}(ts_1+rs_2)^{<\alpha-1>}s_2\Gamma(d\mathbf{s})\Big)dt. \qquad (2.2)\end{aligned}$$

The change of order of integration and differentiation needed to justify (2.2) follows from the Lebesgue Dominated Convergence Theorem and the following bounds:

$$\begin{aligned}\left|\frac{|ts_1+(r+h)s_2|^\alpha-|ts_1+rs_2|^\alpha}{h}\right| &\le \alpha(|t|^{\alpha-1}+|r|^{\alpha-1}+|h|^{\alpha-1})\\ &\le \alpha(|t|^{\alpha-1}+2|r|^{\alpha-1})\end{aligned}$$

if $|h|<|r|$, and

$$\begin{aligned}\left|\frac{\phi(t,r+h)-\phi(t,r)}{h}\right| &\le \max(\phi(t,r),\phi(t,r+h)) \qquad (2.3)\\ &\quad\times\int_{S_2}\left|\frac{|ts_1+(r+h)s_2|^\alpha-|ts_1+rs_2|^\alpha}{h}\right|\Gamma(d\mathbf{s})\\ &\le \exp(|2r|^\alpha\sigma_2^\alpha)\exp(-2^{1-\alpha}\sigma_1^\alpha|t|^\alpha)\alpha(|t|^{\alpha-1}+2|r|^{\alpha-1})\end{aligned}$$

by (3.1) and Lemma 3.3.

To establish the desired representation for Re $\phi''_{X_2|x}(r)$ note that

$$\frac{1}{|h|}\Big|\phi(t,r+h)\Big(-\alpha\int_{S_2}(ts_1+(r+h)s_2)^{<\alpha-1>}s_2\Gamma(d\mathbf{s})\Big)$$

$$
\begin{aligned}
&\quad - \phi(t,r)\Big(-\alpha \int_{S_2} (ts_1 + rs_2)^{<\alpha-1>} s_2 \Gamma(d\mathbf{s})\Big)\Big| \\
&\leq \frac{1}{|h|} |\phi(t, r+h) - \phi(t,r)| \, \Big|\alpha \int_{S_2} (ts_1 + (r+h)s_2)^{<\alpha-1>} s_2 \Gamma(d\mathbf{s})\Big| \\
&\quad + \frac{\phi(t,r)}{|h|} \alpha \int_{S_2} |(ts_1 + (r+h)s_2)^{<\alpha-1>} s_2 - (ts_1 + rs_2)^{<\alpha-1>} s_2| \Gamma(d\mathbf{s}).
\end{aligned}
$$

The first term can be bounded in a way similar to the case of the first derivative, namely, for fixed r and small $|h|$ (say $|h| < |r|$), that term is not greater than const. $\exp(-2^{1-\alpha}\sigma_1^\alpha |t|^\alpha)$ $(|t|^{\alpha-1} + \text{const.})^2$, an integrable function of t.

For the second term, assume $ts_1 + rs_2 \neq 0$ in the integrand. By Lemma 3.4 (iv) with $\beta = \alpha - 1$, we have

$$
\begin{aligned}
&\frac{\phi(t,r)}{|h|} |(ts_1 + (r+h)s_2)^{<\alpha-1>} s_2 - (ts_1 + rs_2)^{<\alpha-1>} s_2| \\
&\quad \leq \frac{\phi(t,r)}{|h|} |ts_1 + rs_2|^{\alpha-1} \frac{2|hs_2|}{|ts_1 + rs_2|} |s_2| = 2|s_1|^{\alpha-2} s_2^2 \phi(t,r) \Big| t + \frac{rs_2}{s_1}\Big|^{\alpha-2},
\end{aligned}
$$

which is integrable with respect to $\Gamma(d\mathbf{s}) \times dt$ over $S_2 \times \mathbf{R}$. Indeed, by Lemma 3.3 $\phi(t,r)$ is bounded by a constant (depending on r) times $\exp(-2^{1-\alpha}\sigma_1^\alpha |t|^\alpha)$. The integral

$$
\int_{-\infty}^{\infty} \exp(-2^{1-\alpha}\sigma_1^\alpha |t|^\alpha) \Big| t + \frac{rs_2}{s_1}\Big|^{\alpha-2} dt
$$

can be bounded by a constant using Lemma 3.10 with $\beta = 0$, $p = 0$ and $\eta = \alpha - 2$, and, finally, the integral $\int_{S_2} |s_1|^{\alpha-2} s_2^2 \Gamma(d\mathbf{s})$ is finite by (1.1).

Again by the Lebesgue Dominated Convergence Theorem, we can differentiate under the integral signs to obtain the desired form of Re $\phi''_{X_2|x}(r)$. ∎

Proposition 2.3. *Suppose* $1 < \alpha < 2$ *and* $\nu > 2 - \alpha$ *in* (1.1). *Then* $E[|X_2|^p | X_1 = x] < \infty$ *a.e. for*

$$
p \leq \min\{\alpha + \nu, 4\} \quad \text{and} \quad p < 2\alpha + 1.
$$

Proof. The existence of conditional moments of order $p \leq 2$ is shown in Proposition 2.1. To prove that, under the assumptions, higher conditional moments exist we use condition (1.3) with $n = 1$. Since

$$
\begin{aligned}
0 \leq |\phi''_{X_2|x}(0)| - |\phi''_{X_2|x}(r)| &= |\text{Re}\phi''_{X_2|x}(0)| - |\phi''_{X_2|x}(r)| \\
&\leq |\text{Re}\phi''_{X_2|x}(0)| - |\text{Re}\phi''_{X_2|x}(r)| \\
&\leq |\text{Re}\phi''_{X_2|x}(0) - \text{Re}\phi''_{X_2|x}(r)|, \qquad (2.4)
\end{aligned}
$$

we want to find a bound for (2.4), for $0 < r \le 1$, using the form of Re $\phi''_{X_2|X}(r)$ established in Proposition 2.2. Note that

$$\begin{aligned}
&\text{Re } \phi''_{X_2|x}(0) - \text{Re } \phi''_{X_2|x}(r)\\
&= \frac{1}{2\pi f(x)} \int_{-\infty}^{\infty} \cos tx(\phi(t,0) - \phi(t,r))\Big[\Big(\alpha \int_{S_2} (ts_1 + rs_2)^{<\alpha-1>} s_2 \Gamma(ds)\Big)^2\\
&\quad - \alpha(\alpha-1) \int_{S_2} |ts_1 + rs_2|^{\alpha-2} s_2^2 \Gamma(ds)\Big] dt - \frac{\alpha^2}{2\pi f(x)} \int_{-\infty}^{\infty} \cos tx \phi(t,0)\\
&\quad \times \Big[\Big(\int_{S_2} (ts_1 + rs_2)^{<\alpha-1>} s_2 \Gamma(ds)\Big)^2 - \Big(\int_{S_2} (ts_1)^{<\alpha-1>} s_2 \Gamma(ds)\Big)^2\Big] dt\\
&\quad + \frac{\alpha(\alpha-1)}{2\pi f(x)} \int_{-\infty}^{\infty} \cos tx \phi(t,0) \int_{S_2} (|ts_1 + rs_2|^{\alpha-2} - |ts_1|^{\alpha-2}) s_2^2 \Gamma(ds) dt\\
&= : I_1 + I_2 + I_3, \qquad (2.5)
\end{aligned}$$

and, as we will now show, the integrals I_1, I_2, I_3 are bounded in a suitable way. I_3 is the most delicate one.

We start with I_1 and write it as a sum of the five following integrals:

$$\begin{aligned}
I_{11}: &= \frac{\alpha^2}{2\pi f(x)} \int_{-\infty}^{\infty} \cos tx \Big[\exp(-|t|^\alpha \sigma_1^\alpha) - \exp\Big(-\int_{S_2} |ts_1 + rs_2|^\alpha \Gamma(ds)\Big)\\
&\qquad - \exp(-|t|^\alpha \sigma_1^\alpha) \int_{S_2} (|ts_1 + rs_2|^\alpha - |ts_1|^\alpha) \Gamma(ds)\Big]\\
&\quad \times \Big(\int_{S_2} (ts_1 + rs_2)^{<\alpha-1>} s_2 \Gamma(ds)\Big)^2 dt,\\
I_{12}: &= -\frac{\alpha(\alpha-1)}{2\pi f(x)} \int_{-\infty}^{\infty} \cos tx \Big[\exp(-|t|^\alpha \sigma_1^\alpha) - \exp\Big(-\int_{S_2} |ts_1 + rs_2|^\alpha \Gamma(ds)\Big)\\
&\qquad - \exp(-|t|^\alpha \sigma_1^\alpha) \int_{S_2} (|ts_1 + rs_2|^\alpha - |ts_1|^\alpha) \Gamma(ds)\Big]\\
&\quad \times \int_{S_2} |ts_1 + rs_2|^{\alpha-2} s_2^2 \Gamma(ds) dt,\\
I_{13}: &= \frac{1}{2\pi f(x)} \int_{-\infty}^{\infty} \cos tx \exp(-|t|^\alpha \sigma_1^\alpha) \int_{S_2} (|ts_1 + rs_2|^\alpha - |ts_1|^\alpha) \Gamma(ds)\\
&\quad \times \Big[\Big(\alpha \int_{S_2} (ts_1)^{<\alpha-1>} s_2 \Gamma(ds)\Big)^2 - \alpha(\alpha-1) \int_{S_2} |ts_1|^{\alpha-2} s_2^2 \Gamma(ds)\Big] dt,\\
I_{14}: &= \frac{\alpha^2}{2\pi f(x)} \int_{-\infty}^{\infty} \cos tx \exp(-|t|^\alpha \sigma_1^\alpha) \int_{S_2} (|ts_1 + rs_2|^\alpha - |ts_1|^\alpha) \Gamma(ds)\\
&\quad \times \Big[\Big(\int_{S_2} (ts_1 + rs_2)^{<\alpha-1>} s_2 \Gamma(ds)\Big)^2 - \Big(\int_{S_2} (ts_1)^{<\alpha-1>} s_2 \Gamma(ds)\Big)^2\Big] dt,\\
I_{15}: &= -\frac{\alpha(\alpha-1)}{2\pi f(x)} \int_{-\infty}^{\infty} \cos tx \exp(-|t|^\alpha \sigma_1^\alpha) \int_{S_2} (|ts_1 + rs_2|^\alpha - |ts_1|^\alpha) \Gamma(ds)\\
&\quad \times \int_{S_2} (|ts_1 + rs_2|^{\alpha-2} - |ts_1|^{\alpha-2}) s_2^2 \Gamma(ds) dt,
\end{aligned}$$

provided all of them are finite.

Using (3.3), Lemmas 3.3 and 3.1, the expression in square brackets in both I_{11} and I_{12} can be bounded by const. $\exp(-2^{1-\alpha}|t|^\alpha\sigma_1^\alpha)(r^{2\alpha}+r^2|t|^{2\alpha-2})$. (Recall that we always take $0<r\le 1$.) Now the triangle inequality in the case of I_{11} and Lemma 3.10 with $\eta=\alpha-2$, $\beta=0$ and $\beta=2\alpha-2$, $p=0$, in the case of I_{12}, imply that

$$|I_{11}|+|I_{12}|\le \text{const. } r^2.$$

One can show similarly (i.e. using the first inequality of Lemma 3.1) that the integrand of I_{13} is absolutely integrable, but the bound we would obtain is only of order r. Since we need a better bound, we proceed in a more delicate manner and use the second inequality of Lemma 3.1 as well.

For $\alpha>3/2$, we get

$$\begin{aligned}|I_{13}| &\le \frac{1}{\pi f(x)}\int_0^\infty \exp(-t^\alpha\sigma_1^\alpha)\int_{S_2}(|ts_1+rs_2|^\alpha \\ &\quad +|-ts_1+rs_2|^\alpha-2|ts_1|^\alpha)\Gamma(ds)\text{ const. }(t^{2\alpha-2}+t^{\alpha-2})dt \\ &\le \text{const. } r^2\int_0^\infty \exp(-t^\alpha\sigma_1^\alpha)t^{\alpha-2}(t^{2\alpha-2}+t^{\alpha-2})dt \\ &\le \text{const. } r^2.\end{aligned}$$

For $\alpha<3/2$,

$$\begin{aligned}|I_{13}| &\le \text{const. }\Big\{r^2\int_0^\infty \exp(-t^\alpha\sigma_1^\alpha)t^{\alpha-2}t^{2\alpha-2}dt \\ &\quad +\int_0^r (r^\alpha+rt^{\alpha-1})t^{\alpha-2}dt+\int_r^\infty r^2t^{\alpha-2}t^{\alpha-2}dt\Big\} \\ &\le \text{const. } r^{2\alpha-1}.\end{aligned}$$

For $\alpha=3/2$ and $0<\epsilon<2$,

$$\begin{aligned}|I_{13}| &\le \text{const. }\Big\{r^2\int_0^\infty \exp(-t^\alpha\sigma_1^\alpha)t^{\alpha-2}t^{2\alpha-2}dt \\ &\quad +\int_0^r (r^\alpha+rt^{\alpha-1})t^{\alpha-2}dt+\int_r^\infty \exp(-t^\alpha\sigma_1^\alpha)r^2t^{\alpha-2}t^{\alpha-2}dt\Big\} \\ &\le \text{const. }\Big\{r^2+\int_r^\infty \exp(-t^\alpha\sigma_1^\alpha)r^{2-\epsilon}t^{-1+\epsilon}dt\Big\} \\ &\le \text{const. }\Big\{r^2+r^{2-\epsilon}\int_0^\infty \exp(-t^\alpha\sigma_1^\alpha)t^{-1+\epsilon}dt\Big\} \\ &\le \text{const. } r^{2-\epsilon}.\end{aligned}$$

I_{14} can be bounded using $|a^2-b^2|\le(|a|+|b|)|a-b|$, Lemmas 3.4 (iv) and also 3.1, in the following way:

$$|I_{14}| \le \text{const. }\int_{-\infty}^\infty \exp(-|t|^\alpha\sigma_1^\alpha)(r^\alpha+r|t|^{\alpha-1})(2|t|^{\alpha-1}+r^{\alpha-1})$$

$$
\begin{aligned}
&\times \int_{S_2} |(ts_1+rs_2)^{<\alpha-1>} - (ts_1)^{<\alpha-1>}||s_2|\Gamma(d\mathbf{s})dt \\
\le\ & \text{const. } r \int_{-\infty}^{\infty} \exp(-|t|^\alpha \sigma_1^\alpha)(1+|t|^{\alpha-1}+|t|^{2\alpha-2}) \\
&\times \int_{S_2} |ts_1|^{\alpha-2} r s_2^2 \Gamma(d\mathbf{s})dt \\
\le\ & \text{const. } r^2.
\end{aligned}
$$

To bound I_{15} note that

$$
\begin{aligned}
|I_{15}| \ \le\ & \text{const. } \int_{S_2} |s_1|^{\alpha-2} s_2^2 \int_{-\infty}^{\infty} \exp(-|t|^\alpha \sigma_1^\alpha)(r^\alpha + r|t|^{\alpha-1}) \\
& \times \left| \left| t + \frac{rs_2}{s_1} \right|^{\alpha-2} - |t|^{\alpha-2} \right| dt \Gamma(d\mathbf{s}).
\end{aligned}
$$

We will use Lemma 3.10 twice, once with $\beta = 0$ and $p := p_1$, $0 \le p_1 < \alpha - 1$, and once with $\beta = \alpha - 1$; and $p := p_2$, $0 \le p_2 < 2\alpha - 2$, $p_2 \le 1$, depending on the values of ν in (1.1) and α. Consider

(i) $p_1 = p_2 = 0$ if $\nu < \alpha$ and $\alpha \le 3/2$, to get $|I_{15}| \le$ const. $r \int_{S_2} |s_1|^{\alpha-2}\Gamma(d\mathbf{s}) =$ const. r,

(ii) $p_1 = \alpha - 1 - \epsilon$ and $p_2 = 2\alpha - 2 - \epsilon$, $0 < \epsilon < \alpha - 1$ if $\nu \ge \alpha$ and $\alpha \le 3/2$, to get $|I_{15}| \le$ const. $r^{2\alpha-1-\epsilon} \int_{S_2} |s_1|^{\epsilon-\alpha}\Gamma(d\mathbf{s}) =$ const. $r^{2\alpha-1-\epsilon}$,

(iii) $p_1 = p_2 = 0$ if $\nu \le 1$ and $\alpha > 3/2$, to get $|I_{15}| \le$ const. r,

(iv) $p_1 = 0$, $p_2 = \alpha - 1$ if $1 < \nu < \alpha$ and $\alpha > 3/2$, to get $|I_{15}| \le$ const. $r^\alpha \int_{S_2} |s_1|^{-1}\Gamma(d\mathbf{s}) =$ const. r^α,

(v) $p_1 = 2 - \alpha$, $p_2 = 1$ if $\alpha \le \nu$ and $\alpha > 3/2$, to get $|I_{15}| \le$ const. $r^2 \int_{S_2} |s_1|^{\alpha-3}\Gamma(d\mathbf{s}) =$ const. r^2.

Gathering together the bounds obtained for I_{1i}, $i = 1, \ldots, 5$, we obtain

$$
|I_1| \le \text{const.} \begin{cases} r & \text{for } 2-\alpha < \nu < \alpha \ , \ \ \alpha \le 3/2, \\ r^{2\alpha-1-\epsilon} & \text{for } \alpha \le \nu \ , \ \ \alpha \le 3/2, \\ r & \text{for } 2-\alpha < \nu \le 1 \ , \ \ \alpha > 3/2, \\ r^\alpha & \text{for } 1 < \nu < \alpha \ , \ \ \alpha > 3/2, \\ r^2 & \text{for } \alpha \le \nu \ , \ \ \alpha > 3/2. \end{cases} \tag{2.6}
$$

It is easy to bound I_2 by const. r, using only $|a^2 - b^2| \le (|a| + |b|)\ |a - b|$ and Lemma 3.4 as for I_{14}. However, we need better bounds. Note that

$$
I_2 \ =\ \frac{\alpha^2}{2\pi f(x)} \int_0^\infty \cos tx \exp(-|t|^\alpha \sigma_1^\alpha) \Big[\Big(\int_{S_2} (ts_1 + rs_2)^{<\alpha-1>} s_2 \Gamma(d\mathbf{s}) \Big)^2
$$

$$
\begin{aligned}
&+\Big(\int_{S_2}(ts_1-rs_2)^{<\alpha-1>}s_2\Gamma(ds)\Big)^2-2\Big(\int_{S_2}(ts_1)^{<\alpha-1>}s_2\Gamma(ds)\Big)^2\Big]dt\\
=\ & \frac{\alpha^2}{2\pi f(x)}\int_0^\infty \cos tx\exp(-|t|^\alpha\sigma_1^\alpha)\\
&\times\int_{S_2}\int_{S_2}[(ts_1+rs_2)^{<\alpha-1>}(ts_1'+rs_2')^{<\alpha-1>}s_2s_2'\\
&\times(ts_1-rs_2)^{<\alpha-1>}(ts_1'-rs_2')^{<\alpha-1>}s_2s_2'\\
&-2(ts_1)^{<\alpha-1>}(ts_1')^{<\alpha-1>}s_2s_2']\Gamma(ds)\Gamma(ds')dt\\
=\ & \frac{\alpha^2}{2\pi f(x)}\int_0^\infty \cos tx\exp(-|t|^\alpha\sigma_1^\alpha)t^{2\alpha-2}\int_{S_2}\int_{S_2}s_1^{<\alpha-1>}s_1'^{<\alpha-1>}s_2s_2'\\
&\times\Big[\Big(\Big(1+\frac{rs_2}{ts_1}\Big)^{<\alpha-1>}\Big(1+\frac{rs_2'}{ts_1'}\Big)^{<\alpha-1>}\\
&+\Big(1-\frac{rs_2}{ts_1}\Big)^{<\alpha-1>}\Big(1-\frac{rs_2'}{ts_1'}\Big)^{<\alpha-1>}-2\Big)\Big]\Gamma(ds)\Gamma(ds')dt
\end{aligned}
$$

To bound the inner double integral (over $S_2\times S_2$) use Lemma 3.6.

(i) For either $\alpha\le 3/2,\ \nu<\alpha$ or for $\alpha>3/2,\ \alpha-1<\nu<3-\alpha$, take $p=\nu$ and $\beta=\alpha-1$. Then

$$
\begin{aligned}
|I_2|\ \le\ & \text{const.}\int_0^\infty \exp(-|t|^\alpha\sigma_1^\alpha)t^{2\alpha-2}\int_{S_2}\int_{S_2}|s_1|^{<\alpha-1>}|s_1'|^{<\alpha-1>}|s_2||s_2'|\\
&\times\Big[\Big|\frac{rs_2}{ts_1}\Big|^{\alpha-1+\nu}+\Big|\frac{rs_2'}{ts_1'}\Big|^{\alpha-1+\nu}+\Big|\frac{rs_2rs_2'}{ts_1ts_1'}\Big|^{\frac{\alpha-1+\nu}{2}}\Big]\Gamma(ds)\Gamma(ds')dt\\
\le\ & \text{const.}\int_0^\infty \exp(-|t|^\alpha\sigma_1^\alpha)t^{2\alpha-2-(\alpha-1+\nu)}|r|^{\alpha-1+\nu}\\
&\times\int_{S_2}\int_{S_2}[|s_1|^{-\nu}|s_1'|^{\alpha-1}+|s_1'|^{-\nu}|s_1|^{\alpha-1}\\
&+|s_1'|^{\frac{\alpha-1-\nu}{2}}|s_1'|^{\frac{\alpha-1-\nu}{2}}]\Gamma(ds)\Gamma(ds')dt\\
=\ & \text{const.}\int_0^\infty \exp(-|t|^\alpha\sigma_1^\alpha)t^{\alpha-1-\nu}|r|^{\alpha-1+\nu}\\
&\times\Big[2\int_{S_2}|s_1|^{-\nu}\Gamma(ds)+\Big(\int_{S_2}|s_1|^{-\frac{(\nu-\alpha+1)}{2}}\Gamma(ds)\Big)^2\Big]dt\\
\le\ & \text{const.}\ |r|^{\alpha-1+\nu},
\end{aligned}
$$

since $(\nu-\alpha+1)/2\le\nu$.

(ii) For $\alpha\le 3/2$ and $\nu\ge\alpha$ take $p=\alpha-\epsilon,\ 0<\epsilon<1$ and obtain $|I_2|\le\text{const.}\ |r|^{\alpha-1+\alpha-\epsilon}=\text{const.}\ |r|^{2\alpha-1-\epsilon}$.

(iii) For $\alpha>3/2$ and $\nu\ge 3-\alpha$, take $p=3-\alpha$ and obtain $|I_2|\le\text{const.}\ r^2$.

Thus,

$$I_2 \leq \text{const.} \begin{cases} r^{\alpha-1+\nu} & \text{for } 2-\alpha<\nu<\alpha \ , \ \alpha \leq 3/2, \\ r^{2\alpha-1-\epsilon} & \text{for } \alpha \leq \nu \ , \ \alpha \leq 3/2, \\ r & \text{for } 2-\alpha<\nu\leq\alpha-1 \ , \ \alpha>3/2, \\ r^{\alpha-1+\nu} & \text{for } \alpha-1<\nu<3-\alpha \ , \ \alpha>3/2, \\ r^2 & \text{for } 3-\alpha\leq\nu \ , \ \alpha>3/2. \end{cases} \tag{2.7}$$

At this moment using (2.6) and (2.7) we may conclude that

$$\int_0^1 \frac{|I_1|+|I_2|}{r^{1+\alpha+\nu-2}}dr < \infty$$

for $2-\alpha<\nu<\min(4-\alpha,\alpha+1)$. Moreover, if $\nu \geq 4-\alpha$ and $\alpha>3/2$ then $(|I_1|+|I_2|)/r^2 \leq$ const. for $0<r\leq 1$. To prove similar statements for I_3, we use the technique developed for the term I_1 in Proposition 2.1 (c.f. [4]). First note that

$$\begin{aligned} I_3 &= \frac{\alpha(\alpha-1)}{2\pi f(x)}\int_0^\infty \cos tx \exp(-t^\alpha\sigma_1^\alpha) \\ &\times \int_{S_2}[|ts_1+rs_2|^{\alpha-2}+|ts_1-rs_2|^{\alpha-2}-2|ts_1|^{\alpha-2}]s_2^2\Gamma(ds)dt \\ &= \frac{\alpha(\alpha-1)}{2\pi f(x)}\int_{S_2} s_2^2|s_1|^{\alpha-2}\int_0^\infty \cos tx \exp(-t^\alpha\sigma_1^\alpha) \\ &\times\Big[\Big|t+\frac{rs_2}{s_1}\Big|^{\alpha-2}+\Big|t-\frac{rs_2}{s_1}\Big|^{\alpha-2}-2|t|^{\alpha-2}\Big]dt\Gamma(ds). \end{aligned}$$

Since the exponent $\alpha-2<0$, we integrate by parts:

$$\begin{aligned} L: &= \int_0^\infty \cos tx\exp(-t^\alpha\sigma_1^\alpha)\Big[\Big(\Big|t+\frac{rs_2}{s_1}\Big|^{\alpha-2}+\Big|t-\frac{rs_2}{s_1}\Big|^{\alpha-2}-2t^{\alpha-2}\Big)\Big]dt \\ &= 0-\int_0^\infty(-x\sin tx\exp(-t^\alpha\sigma_1^\alpha)-\cos tx\exp(-t^\alpha\sigma_1^\alpha)\alpha t^{\alpha-1}\sigma_1^\alpha) \\ &\times\frac{1}{(\alpha-1)}\Big(\Big(t+\frac{rs_2}{s_1}\Big)^{<\alpha-1>}+\Big(t-\frac{rs_2}{s_1}\Big)^{<\alpha-1>}-2t^{\alpha-1}\Big)dt \qquad (2.8) \\ &= \frac{1}{\alpha-1}\int_0^\infty e^{-r^\alpha t^\alpha\sigma_1^\alpha}r^\alpha(x\sin rtx+\alpha\sigma_1^\alpha\cos rtx\ r^{\alpha-1}t^{\alpha-1}) \\ &\times\Big(\Big(t+\frac{s_2}{s_1}\Big)^{<\alpha-1>}+\Big(t-\frac{s_2}{s_1}\Big)^{<\alpha-1>}-2t^{\alpha-1}\Big)dt. \end{aligned}$$

Using $|\sin y|\leq|y|$ and the following identity

$$\int_0^\infty e^{-c^\alpha r^\alpha}r^\beta dr = \Gamma\Big(\frac{\beta+1}{\alpha}\Big)/(\alpha c^{\beta+1}) \text{ for } c>0,\alpha>0,\beta>-1, \tag{2.9}$$

we obtain, for $2-\alpha<\nu<\min(4-\alpha,\alpha+1)$,

$$\begin{aligned}\int_0^1 \frac{|I_3|}{r^{1+\alpha+\nu-2}}dr &\le \text{const.} \int_{S_2} s_2^2|s_1|^{\alpha-2} \\ &\quad \times \int_0^\infty \left|\left(t+\frac{s_2}{s_1}\right)^{<\alpha-1>} + \left(t-\frac{s_2}{s_1}\right)^{<\alpha-1>} - 2t^{\alpha-1}\right| \\ &\quad \times \int_0^\infty e^{-r^\alpha t^\alpha \sigma_1^\alpha} r^{1-\nu}(rt+r^{\alpha-1}t^{\alpha-1})dr\, dt\Gamma(ds) \\ &= \text{const.} \int_{S_2} s_2^2|s_1|^{\alpha-1} \int_0^\infty \left|\left(t+\frac{s_2}{s_1}\right)^{<\alpha-1>}\right. \\ &\quad \left. + \left(t-\frac{s_2}{s_1}\right)^{<\alpha-1>} - 2t^{\alpha-1}\right| t^{\nu-2} dt\Gamma(ds) \\ &= \text{const.} \int_{S_2} |s_2|^{\alpha+\nu}|s_1|^{-\nu}\Gamma(ds) \\ &\quad \times \int_0^\infty |(t+1)^{\alpha-1} + (t-1)^{<\alpha-1>} - 2t^{\alpha-1}| t^{\nu-2} dt.\end{aligned}$$

The last expression is finite because of (1.1) and Lemma 3.5, which proves that the conditional moments of order $\alpha+\nu$ exist for $2-\alpha<\nu<\min(4-\alpha,\alpha+1)$. In the case $\nu\ge\alpha+1$ and $\alpha\le 3/2$, the proposition holds since (1.1) is then satisfied with any $\nu'<\alpha+1$. If $\nu\ge 4-\alpha$ and $\alpha>3/2$ then, applying Lemma 3.5 to (2.8), we get

$$|L| \le \text{const.}\ \left|\frac{rs_2}{s_1}\right|^2 \int_0^\infty e^{-t^\alpha\sigma_1^\alpha}(t+t^{\alpha-1})t^{\alpha-3}dt \le \text{const.}\ r^2\left|\frac{s_2}{s_1}\right|^2$$

and $|I_3|/r^2 \le \text{const.} \int_{S_2} s_2^4|s_1|^{\alpha-4}\Gamma(ds) \le \text{const.}$ for $0<r\le 1$. The proposition has been proved. ■

Proposition 2.4. *Suppose* $3/2<\alpha<2$ *and* $\nu>4-\alpha$ *in* (1.1). *Then* $E[|X_2|^p|X_1=x]<\infty$ *a.e. for*

$$p\le\alpha+\nu \quad \text{and} \quad p<2\alpha+1.$$

Proof. Proposition 2.3 implies that $E[|X_2|^4|X_1=x]<\infty$ a.e. and, hence, the fourth derivative of $\phi_{X_2|x}(r)$ exists a.e. We want to find a bound for

$$0\le|\phi^{(4)}_{X_2|x}(0)|-|\phi^{(4)}_{X_2|x}(r)|\le|\text{Re}\phi^{(4)}_{X_2|x}(0)-\text{Re}\phi^{(4)}_{X_2|x}(r)| \tag{2.10}$$

for $0<r\le 1$ and use condition (1.3) with $n=2$ in order to establish the existence of conditional moments $E[|X_2|^p|X_1=x]$ with $p>4$. First, we need to find a form of Re $\phi^{(4)}_{X_2|x}(r)$. Proposition 2.2 gives a form of Re $\phi''_{X_2|x}$. It is easy to see that the (formal) differentiation of Re $\phi''_{X_2|x}$ under the integral sign gives rise to $|t|^{\alpha-3}$ (for $r=0$) which is not integrable around 0. Thus we need to use a different form of Re $\phi''_{X_2|x}(r)$. Integrating the second term in (2.1) by parts, we obtain the same form as for $\alpha \le 1$ (c.f. [4]), namely

$$\begin{aligned}
\mathrm{Re}\phi''_{X_2|x}(r) &= \frac{\alpha^2}{2\pi f(x)} \int_{-\infty}^{\infty} \cos tx \exp\Big(-\int_{S_2} |ts_1 + rs_2|^\alpha \Gamma(ds)\Big) \\
&\times \Big(\int_{S_2} (ts_1 + rs_2)^{<\alpha-1>} s_2 \Gamma(ds)\Big)^2 dt \\
&- \frac{\alpha x}{2\pi f(x)} \int_{-\infty}^{\infty} \sin tx \exp\Big(-\int_{S_2} |ts_1 + rs_2|^\alpha \Gamma(ds)\Big) \\
&\times \Big(\int_{S_2} (ts_1 + rs_2)^{<\alpha-1>} s_2^2 s_1^{-1} \Gamma(ds)\Big) dt \\
&- \frac{\alpha^2}{2\pi f(x)} \int_{-\infty}^{\infty} \cos tx \exp\Big(-\int_{S_2} |ts_1 + rs_2|^\alpha \Gamma(ds)\Big) \\
&\times \Big(\int_{S_2} (ts_1 + rs_2)^{<\alpha-1>} s_1 \Gamma(ds)\Big) \\
&\times \Big(\int_{S_2} (ts_1 + rs_2)^{<\alpha-1>} s_2^2 s_1^{-1} \Gamma(ds)\Big) dt. \qquad (2.11)
\end{aligned}$$

To justify that integration by parts, we use Lemma 3.4 (iv) and $\nu > 1$, to get

$$\frac{\partial}{\partial t}\Big(\int_{S_2} (ts_1 + rs_2)^{<\alpha-1>} s_2^2 s_1^{-1} \Gamma(ds)\Big) = (\alpha - 1) \int_{S_2} |ts_1 + rs_2|^{\alpha-2} s_2^2 \Gamma(ds),$$

which is finite for a.a. t, and Lemma 3.3 with the triangle inequality to get

$$\lim_{|t|\to\infty} \exp\Big(-\int_{S_2} |ts_1 + rs_2|^\alpha \Gamma(ds)\Big) \int_{S_2} (ts_1 + rs_2)^{<\alpha-1>} s_2^2 s_1^{-1} \Gamma(ds) = 0.$$

Similar considerations as in Proposition 2.2 (in particular, (2.3), the triangle inequality, Lemmas 3.3, 3.4 (iv), and 3.10 with $p = 0$, $\eta = \alpha - 2$ and $\beta = 0, \alpha - 1$, and (1.1) with $\nu > \alpha - 3$) allow us to differentiate Re $\phi''_{X_2|x}(r)$, now given in the form (2.11), under the integral signs. Then, using $\nu > 2$, we integrate by parts the terms containing $\int_{S_2} |ts_1 + rs_2|^{\alpha-2} s_2^2 \Gamma(ds)$ and $\int_{S_2} |ts_1 + rs_2|^{\alpha-2} s_2^3 s_1^{-1} \Gamma(ds)$ to obtain:

$$\begin{aligned}
&2\pi f(x) \mathrm{Re}\, \phi^{(3)}_{X_2|x}(r) \\
&= -\alpha^3 \int_{-\infty}^{\infty} \cos tx \exp\Big(-\int_{S_2} |ts_1 + rs_2|^\alpha \Gamma(ds)\Big) \\
&\quad \times \Big(\int_{S_2} (ts_1 + rs_2)^{<\alpha-1>} s_2 \Gamma(ds)\Big)^3 dt \\
&\quad + 3\alpha^2 x \int_{-\infty}^{\infty} \sin tx \exp\Big(-\int_{S_2} |ts_1 + rs_2|^\alpha \Gamma(ds)\Big) \\
&\quad \times \Big(\int_{S_2} (ts_1 + rs_2)^{<\alpha-1>} s_2 \Gamma(ds)\Big)\Big(\int_{S_2} (ts_1 + rs_2)^{<\alpha-1>} s_2^2 s_1^{-1} \Gamma(ds)\Big) dt
\end{aligned}$$

$$
\begin{aligned}
&+ 3\alpha^3 \int_{-\infty}^{\infty} \cos tx \exp\Big(-\int_{S_2} |ts_1 + rs_2|^{\alpha}\Gamma(ds)\Big) \\
&\times\Big(\int_{S_2} (ts_1 + rs_2)^{<\alpha-1>} s_1\Gamma(ds)\Big)\Big(\int_{S_2} (ts_1 + rs_2)^{<\alpha-1>} s_2\Gamma(ds)\Big) \\
&\times\Big(\int_{S_2} (ts_1 + rs_2)^{<\alpha-1>} s_2^2 s_1^{-1}\Gamma(ds)\Big)dt \\
&- 3\alpha^2(\alpha-1) \int_{-\infty}^{\infty} \cos tx \exp\Big(-\int_{S_2} |ts_1 + rs_2|^{\alpha}\Gamma(ds)\Big) \\
&\times\Big(\int_{S_2} (ts_1 + rs_2)^{<\alpha-1>} s_2^2 s_1^{-1}\Gamma(ds)\Big)\Big(\int_{S_2} |ts_1 + rs_2|^{\alpha-2} s_2 s_1\Gamma(ds)\Big)dt \\
&+ \alpha x^2 \int_{-\infty}^{\infty} \cos tx \exp\Big(-\int_{S_2} |ts_1 + rs_2|^{\alpha}\Gamma(ds)\Big) \\
&\times\Big(\int_{S_2} (ts_1 + rs_2)^{<\alpha-1>} s_2^3 s_1^{-2}\Gamma(ds)\Big)dt \\
&- 2\alpha^2 x \int_{-\infty}^{\infty} \sin tx \exp\Big(-\int_{S_2} |ts_1 + rs_2|^{\alpha}\Gamma(ds)\Big) \\
&\times\Big(\int_{S_2} (ts_1 + rs_2)^{<\alpha-1>} s_1\Gamma(ds)\Big)\Big(\int_{S_2} (ts_1 + rs_2)^{<\alpha-1>} s_2^3 s_1^{-2}\Gamma(ds)\Big)dt \\
&- \alpha^3 \int_{-\infty}^{\infty} \cos tx \exp\Big(-\int_{S_2} |ts_1 + rs_2|^{\alpha}\Gamma(ds)\Big) \\
&\times\Big(\int_{S_2} (ts_1 + rs_2)^{\alpha-1>} s_1\Gamma(ds)\Big)^2\Big(\int_{S_2} (ts_1 + rs_2)^{<\alpha-1>} s_2^3 s_1^{-2}\Gamma(ds)\Big)dt \\
&+ \alpha^2(\alpha-1) \int_{-\infty}^{\infty} \cos tx \exp\Big(-\int_{S_2} |ts_1 + rs_2|^{\alpha}\Gamma(ds)\Big) \\
&\times\Big(\int_{S_2} (ts_1 + rs_2)^{<\alpha-1>} s_2^3 s_1^{-2}\Gamma(ds)\Big)\Big(\int_{S_2} |ts_1 + rs_2|^{\alpha-2} s_1^2\Gamma(ds)\Big)dt.
\end{aligned}
$$

We will now obtain the fourth derivative. It is easy to show that all the terms of the third derivative, except the fourth and the last one (which involve the power $\alpha - 2$), can be differentiated under the integral signs. The assumption $\nu \geq 4 - \alpha$ plays a crucial role in the differentiation of the fifth, sixth and seventh terms.

We now focus on the fourth term (the last one can be dealt with in a similar fashion). We want to find the limit

$$
\begin{aligned}
\lim_{h\to 0} \frac{1}{h}\Big[&\int_{-\infty}^{\infty} \cos tx \exp\Big(-\int_{S_2} |ts_1 + (r+h)s_2|^{\alpha}\Gamma(ds)\Big) \\
&\times\Big(\int_{S_2} (ts_1 + (r+h)s_2)^{<\alpha-1>} s_2^2 s_1^{-1}\Gamma(ds)\Big) \\
&\times\Big(\int_{S_2} |ts_1 + (r+h)s_2|^{\alpha-2} s_2 s_1\Gamma(ds)\Big) \\
&- \int_{-\infty}^{\infty} \cos tx \exp\Big(-\int_{S_2} |ts_1 + rs_2|^{\alpha}\Gamma(ds)\Big) \\
&\times\Big(\int_{S_2} (ts_1 + rs_2)^{<\alpha-1>} s_2^2 s_1^{-1}\Gamma(ds)\Big)\Big(\int_{S_2} |ts_1 + rs_2|^{\alpha-2} s_2 s_1\Gamma(ds)\Big)dt\Big]
\end{aligned}
$$

$$
\begin{aligned}
= & \lim_{h\to 0}\frac{1}{h}\int_{-\infty}^{\infty}\cos tx\Big[\exp\Big(-\int_{S_2}|ts_1+(r+h)s_2|^{\alpha}\Gamma(d\mathbf{s})\Big) \\
& -\exp\Big(-\int_{S_2}|ts_1+rs_2|^{\alpha}\Gamma(d\mathbf{s})\Big)\Big] \\
& \times\Big(\int_{S_2}(ts_1+rs_2)^{<\alpha-1>}s_2^2s_1^{-1}\Gamma(d\mathbf{s})\Big)\Big(\int_{S_2}|ts_1+rs_2|^{\alpha-2}s_2s_1\Gamma(d\mathbf{s})\Big)dt \\
& +\lim_{h\to 0}\frac{1}{h}\int_{-\infty}^{\infty}\cos tx\exp\Big(-\int_{S_2}|ts_1+(r+h)s_2|^{\alpha}\Gamma(d\mathbf{s})\Big) \\
& \times\int_{S_2}\Big[(ts_1+(r+h)s_2)^{<\alpha-1>}-(ts_1+rs_2)^{<\alpha-1>}\Big]s_2^2s_1^{-1}\Gamma(d\mathbf{s}) \\
& \times\Big(\int_{S_2}|ts_1+rs_2|^{\alpha-2}s_2s_1\Gamma(d\mathbf{s})\Big)dt \\
& +\lim_{h\to 0}\frac{1}{h}\int_{-\infty}^{\infty}\cos tx\exp\Big(-\int_{S_2}|ts_1+(r+h)s_2|^{\alpha}\Gamma(d\mathbf{s})\Big) \\
& \times\Big(\int_{S_2}(ts_1+(r+h)s_2)^{<\alpha-1>}s_2^2s_1^{-1}\Gamma(d\mathbf{s})\Big) \\
& \times\Big[\int_{S_2}|ts_1+(r+h)s_2|^{\alpha-2}s_2s_1\Gamma(d\mathbf{s})-\int_{S_2}|ts_1+rs_2|^{\alpha-2}s_2s_1\Gamma(d\mathbf{s})\Big]dt.
\end{aligned}
$$

We can change the order of integration and differentiation in the first two limits by using the usual bounds, and Lemma 3.10 and Corollary 3.1 with $p=0$. The third term is the most delicate. This is the case where the idea of appropriate "integration by parts," as explained in the introduction, is implemented. We first make a change of variables $t := t - hs_2/s_1$. After obvious manipulations we get three limits:

$$
\begin{aligned}
& \lim_{h\to 0}\int_{S_2}\int_{-\infty}^{\infty}\frac{1}{\frac{hs_2}{s_1}}\Big[\cos(t-\frac{hs_2}{s_1})x-\cos tx\Big]\exp\Big(-\int_{S_2}|ts_1'+(r+h)s_2'|^{\alpha}\Gamma(d\mathbf{s}')\Big) \\
& \quad\times\Big(\int_{S_2}(ts_1''+(r+h)s_2'')^{<\alpha-1>}(s_2'')^2(s_1'')^{-1}\Gamma(d\mathbf{s}'')\Big)\Big|t+\frac{rs_2}{s_1}\Big|^{\alpha-2}s_2^2s_1^{\alpha-2}dt\Gamma(d\mathbf{s}) \\
& +\lim_{h\to 0}\int_{S_2}\int_{-\infty}^{\infty}\frac{1}{\frac{hs_2}{s_1}}\cos\Big(t-\frac{hs_2}{s_1}\Big)x \\
& \quad\times\Big[\int_{S_2}\Big(\Big(t-\frac{hs_2}{s_1}\Big)s_1''+(r+h)s_2''\Big)^{<\alpha-1>}(s_2'')^2(s_1'')^{-1}\Gamma(d\mathbf{s}'') \\
& \quad-\int_{S_2}(ts_1''+(r+h)s_2'')^{<\alpha-1>}(s_2'')^2(s_1'')^{-1}\Gamma(d\mathbf{s}'')\Big] \\
& \quad\times\exp\Big(-\int_{S_2}|ts_1'+(r+h)s_2'|^{\alpha}\Gamma(d\mathbf{s}')\Big)\Big|t+\frac{rs_2}{s_1}\Big|^{\alpha-2}s_2^2s_1^{\alpha-2}dt\Gamma(d\mathbf{s}) \\
& +\lim_{h\to 0}\int_{S_2}\int_{-\infty}^{\infty}\frac{1}{\frac{hs_2}{s_1}}\cos(t-\frac{hs_2}{s_1})x \\
& \quad\times\Big(\int_{S_2}\Big(\Big(t-\frac{hs_2}{s_1}\Big)s_1''+(r+h)s_2''\Big)^{<\alpha-1>}(s_2'')^2(s_1'')^{-1}\Gamma(d\mathbf{s}'')\Big)
\end{aligned}
$$

$$\times\Big[\exp\Big(-\int_{S_2}|(t-\frac{hs_2}{s_1})s_1'+(r+h)s_2'|^\alpha\Gamma(ds')\Big)$$
$$-\exp\Big(-\int_{S_2}|ts_1'+(r+h)s_2'|^\alpha\Gamma(ds')\Big)\Big]\Big|t+\frac{rs_2}{s_1}\Big|^{\alpha-2}s_2^2s_1^{\alpha-2}dt\Gamma(ds). \quad (2.12)$$

The integrand of $\int_{S_2}\int_{-\infty}^{\infty}$ in the first limit can be bounded uniformly in $|h| < |r|$ by an integrable function (Lemma 3.10). The integrand in the second limit is bounded by

$$\begin{aligned} T(h) &= \text{const. } \exp(-2^{1-\alpha}\sigma_1^\alpha|t|^\alpha) \\ &\times\Big[\int_{S_2}\Big|t+\frac{rs_2''}{s_1''}+\frac{hs_2''}{s_1''}\Big|^{\alpha-2}|s_2''|^2|s_1''|^{\alpha-2}\Gamma(ds'')\Big]\Big|t+\frac{rs_2'}{s_1}\Big|^{\alpha-2}s_2^2|s_1|^{\alpha-2}. \end{aligned}$$

Write $T(h) = (T(h)-T(0))+T(0)$. The term $T(0)$ is integrable by Corollary 3.1 and $\lim_{h\to 0}\int T(h) = \int T(0)$ by Lemma 3.11, with $0 < p < 2(\alpha-2)+1$. Thus, by the generalization of the Lebesgue Dominated Convergence Theorem (c.f. Theorem 17, p. 32 in Royden [10]) the first two limits in (2.12) can be taken under the integral signs.

The integrand in the third limit in (2.12) is the most delicate to bound. Using (3.1) and Lemma 3.3, we first bound it (up to a multiplicative constant) by

$$\begin{aligned} U(h) &= \exp\Big\{-2^{1-\alpha}\sigma_1^\alpha\min\Big(\Big|t-\frac{hs_2}{s_1}\Big|^\alpha,|t|^\alpha\Big) \\ &\times\Big(|t|^{2(\alpha-1)}+1+\Big|\frac{s_2}{s_1}\Big|^{2(\alpha-1)}\Big)\Big\}\Big|t+\frac{rs_2}{s_1}\Big|^{\alpha-2}s_2^2|s_1|^{\alpha-2}. \end{aligned}$$

Since the bound $U(h)$ depends on h, we need to show that the generalized Lebesgue Dominated Convergence Theorem applies here as well. We focus on the case $hs_2/s_1 > 0$ (the case $hs_2/s_1 < 0$ is analogous), for which

$$\min\Big(\Big|t-\frac{hs_2'}{s_1'}\Big|,|t|\Big) = \begin{cases} \Big|t-\frac{hs_2'}{s_1'}\Big| & \text{for } t \geq \frac{hs_2'}{2s_1'}, \\ |t| & \text{for } t < \frac{hs_2'}{2s_1'}. \end{cases} \quad (2.13)$$

Then

$$\begin{aligned} \int_{-\infty}^{\infty}U(h)dt &= \int_{\frac{hs_2}{2s_1}}^{\infty}\exp\Big(-2^{1-\alpha}\sigma_1^\alpha\Big|t-\frac{hs_2}{s_1}\Big|^\alpha\Big)\Big(|t|^{2\alpha-2}+1+\Big|\frac{s_2}{s_1}\Big|^{2\alpha-2}\Big) \\ &\times\Big|t+\frac{rs_2}{s_1}\Big|^{\alpha-2}s_2^2|s_1|^{\alpha-2}dt+\int_{-\infty}^{\frac{hs_2}{2s_1}}\exp(-2^{1-\alpha}\sigma_1^\alpha|t|^\alpha)\Big|t+\frac{rs_2}{s_1}\Big|^{\alpha-2} \\ &\times\Big(|t|^{2\alpha-2}+1+\Big|\frac{s_2}{s_1}\Big|^{2\alpha-2}\Big)\Big|t+\frac{rs_2}{s_1}\Big|^{\alpha-2}s_2^2|s_1|^{\alpha-2}dt \end{aligned}$$

$$= \int_{-\infty}^{\infty} \exp(-2^{1-\alpha}\sigma_1^\alpha|t|^\alpha)\Big[\Big(\Big|t+\frac{hs_2}{s_1}\Big|^{2\alpha-2}+1+\Big|\frac{s_2}{s_1}\Big|^{2\alpha-2}\Big)$$
$$\times\Big|t+\frac{hs_2}{s_1}+\frac{rs_2}{s_1}\Big|^{\alpha-2} I_{(-\frac{hs_2}{2s_1},\infty)}(t)$$
$$+\Big(|t|^{2\alpha-2}+1+\Big|\frac{s_2}{s_1}\Big|^{2\alpha-2}\Big)\Big|t+\frac{rs_2}{s_1}\Big|^{\alpha-2} I_{(-\infty,\frac{hs_2}{2s_1})}(t)\Big] s_2^2|s_1|^{\alpha-2}dt.$$

Since $2\alpha-2>0$, use the triangle inequality to show that the integrand of the last integral is bounded by

$$V(h) = \text{const.}\exp(-2^{1-\alpha}\sigma_1^\alpha|t|^\alpha)(|t|^{2\alpha-2}+1+\Big|\frac{s_2}{s_1}Big|^{2\alpha-2})$$
$$\times\Big[\Big|t+\frac{hs_2}{s_1}+\frac{rs_2}{s_1}\Big|^{\alpha-2}+\Big|t+\frac{rs_2}{s_1}\Big|^{\alpha-2}\Big]|s_1|^{\alpha-2}s_2^2.$$

$V(0)$ is integrable[1] by Lemma 3.10. Moreover, we get

$$\lim_{h\to 0}\int_{-\infty}^{\infty}\int_{S_2} V(h)\Gamma(ds)dt = \int_{-\infty}^{\infty}\int_{S_2} V(0)\Gamma(ds)dt$$

by using Lemma 3.12 with a $0<p<\alpha-1$ and such that $\int_{S_2}|s_1|^{\alpha-2-2(\alpha-1)-p}\Gamma(ds)<\infty$. Hence also,

$$\lim_{h\to 0}\int_{-\infty}^{\infty}\int_{S_2} U(h)\Gamma(ds)dt = \int_{-\infty}^{\infty}\int_{S_2} U(0)\Gamma(ds)dt,$$

so that the third limit in (2.12) can also be taken under the integral sign.

Then the expression in (2.12) equals

$$x\int_{-\infty}^{\infty}\sin tx\exp\Big(-\int_{S_2}|ts_1+rs_2|^\alpha\Gamma(ds)\Big)$$
$$\times\Big(\int_{S_2}(ts_1+rs_2)^{<\alpha-1>}s_2^2s_1^{-1}\Gamma(ds)\Big)\Big(\int_{S_2}|ts_1+rs_2|^{\alpha-2}s_2^2\Gamma(ds)\Big)dt$$
$$-(\alpha-1)\int_{-\infty}^{\infty}\cos tx\exp\Big(-\int_{S_2}|ts_1+rs_2|^\alpha\Gamma(ds)\Big)$$
$$\times\Big(\int_{S_2}|ts_1+rs_2|^{\alpha-2}s_2^2\Gamma(ds)\Big)^2dt$$
$$+\alpha\int_{-\infty}^{\infty}\cos tx\exp\Big(-\int_{S_2}|ts_1+rs_2|^\alpha\Gamma(ds)\Big)$$
$$\times\Big(\int_{S_2}(ts_1+rs_2)^{<\alpha-1>}s_1\Gamma(ds)\Big)$$
$$\times\Big(\int_{S_2}(ts_1+rs_2)^{<\alpha-1>}s_2^2s_1^{-1}\Gamma(ds)\Big)\Big(\int_{S_2}|ts_1+rs_2|^{\alpha-2}s_2^2\Gamma(ds)\Big)dt.$$

[1] s_1, in $V(0)$, appears as $|s_1|^{-(2\alpha-2)+(\alpha-2)}=|s_1|^{-\alpha}$ which is always integrable with respect to $\Gamma(ds)$ since $\nu>4-\alpha>\alpha$. If in the third derivative of the characteristic function we had not integrated $\int_{S_2}|ts_1+rs_2|^{\alpha-2}s_2^2\Gamma(ds)$ and similar terms by parts, we would have obtained here $|s_1|$ to a power too negative, namely, $|s_1|^{-(2\alpha-2)+(\alpha-3)}=|s_1|^{-(\alpha+1)}$, which is not necessarily integrable with respect to $\Gamma(ds)$ (note that $4-\alpha<1+\alpha$).

We can now write down a representation for Re $\phi^{(4)}_{X_2|x}$ (see Appendix). However the expression is quite complicated (it has 27 terms!). We state only what types of terms appear. All the terms are of the form

$$\text{const.} \int_{-\infty}^{\infty} \text{trig}(tx) \exp\Big(-\int_{S_2} |ts_1 + rs_2|^\alpha \Gamma(ds)\Big) g(t,r) dt$$

where *trig* is the cosine or sine function and where $g(t,r)$ can be of the following type:

(i) a product of 2,3 or 4 terms of the form

$$\text{(A)} \quad \Big(\int_{S_2} (ts_1 + rs_2)^{<\alpha-1>} s_1^\kappa s_2^\lambda \Gamma(ds)\Big) \text{ with } (\kappa, \lambda) = (1,0), (0,1), (-1,2), (-2,3),$$

(ii) a product of 2 terms of the form

$$\text{(B)} \quad \Big(\int_{S_2} |ts_1 + rs_2|^{\alpha-2} s_1^\kappa s_2^\lambda \Gamma(ds)\Big) \text{ with } (\kappa, \lambda) = (2,0), (1,1), (0,2), (-1,3), (-2,4),$$

(iii) a product of zero, one or two terms (A) and exactly one term (B).

We will now show that the representation that we have obtained for Re $\phi^{(4)}_{X_2|x}(r)$ also yields suitable bounds for $|\text{Re}\phi^{(4)}_{X_2|x}(0) - \text{Re}\phi^{(4)}_{X_2|x}(r)|$, $0 < r \leq 1$, namely

$$\int_0^1 \frac{|\text{Re } \phi^{(4)}_{X_2|x}(0) - \text{Re } \phi^{(4)}_{X_2|x} x(r)|}{r^{1+\alpha+\nu-4}} dr < \infty \tag{2.14}$$

for $4 - \alpha < \nu < \alpha + 1$.

We express the differences in the corresponding terms of Re $\phi^{(4)}_{X_2|x}(r)$ using the identities

$$\begin{aligned} h(r)k(r) - h(0)k(0) &= (h(r) - h(0))k(0) - h(r)(k(r) - k(0)), \\ h(r)k(r)l(r) - h(0)k(0)l(0) &= (h(r) - h(0))k(0)l(0) + h(r)(k(r) \\ &\quad - k(0))l(0) + h(r)k(r)(l(r) - k(0)) \end{aligned}$$

and their obvious generalization to product of more than 3 terms, where h is always the exponential and $k \cdot l \cdot \ldots$ is g with terms (A) preceding terms (B), if any. Then we use the following bounds:

(a) $|\exp(-\int_{S_2} |ts_1 + rs_2|^\alpha \Gamma(ds)) - \exp(-\int_{S_2} |ts_1|^\alpha \Gamma(ds))|$
$\leq$ const. $\exp(-2^{1-\alpha}\sigma_1^\alpha |t|^\alpha)(|t|^{\alpha-1} + 1)r$, obtained by (3.1), Lemmas 3.3 and 3.1;

(b) $\exp(-\int_{S_2}|ts_1+rs_2|^\alpha\Gamma(ds)) \le \exp(|r|^\alpha\sigma_2^\alpha)\exp(-2^{1-\alpha}\sigma_1^\alpha|t|^\alpha)$ (Lemma 3.3);

(c) $|\int_{S_2}((ts_1+rs_2)^{<\alpha-1>}-(ts_1)^{<\alpha-1>})s_1^\kappa s_2^\lambda\Gamma(ds)| \le$ const. $r|t|^{\alpha-2}$, obtained by Lemma 3.4 (iv) and (1.1) with $\nu \ge 4-\alpha \ge 2-\alpha-\kappa$;

(d) $|\int_{S_2}(ts_1+rs_2)^{<\alpha-1>}s_1^\kappa s_2^\lambda\Gamma(ds)| \le$ const. $(|t|^{\alpha-1}+1)$, obtained by the triangle inequality and (1.1) with $\nu > 2 \ge -\kappa$.

Terms (B), i.e. those involving the exponent $\alpha-2$, are the nasty ones. Their differences will have to be treated separately.

Since $\alpha > 1$, bounds of the form const. $(|t|^{\alpha-1})^k$, $k \ge 0$, are integrable w.r.t. t around $t=0$. Therefore, using (a), (b), (c) and (d), differences of terms with g of the type (i) can be bounded by const. r, which is integrable w.r.t. $r^{-\alpha-\nu+3}dr$ over $(0,1]$.

Similarly, if g is of type (ii) or (iii), the integrals (w.r.t. t) which involve differences of either exponentials or differences of terms (A), can also be bounded by const. r, using (a) (type (ii)) or (a),(b),(c) and (d) (type (iii)). We also use here the fact that const. $|t|^{2\alpha-4}$ is integrable w.r.t. t around $t=0$.

We now have to deal with expressions involving differences of terms (B). When g is a product of two terms (B) (that is, g is of type (ii)) then we also have to show integrability w.r.t. $r^{-\alpha-\nu-3}dr$ of the following integral

$$\begin{aligned} I &= \int_{-\infty}^{\infty}\exp\Big(-\int_{S_2}|ts_1+rs_2|^\alpha\Gamma(ds)\Big)\Big|\int_{S_2}|ts_1+rs_2|^{\alpha-2}s_1^\kappa s_2^\lambda\Gamma(ds) \\ &\quad\times\int_{S_2}|ts_1'+rs_2'|^{\alpha-2}s_1'^{\kappa'}s_2'^{\lambda'}\Gamma(ds') \\ &\quad-\int_{S_2}|ts_1|^{\alpha-2}s_1^\kappa s_2^\lambda\Gamma(ds)\int_{S_2}|ts_1'|^{\alpha-2}s_1'^{\kappa'}s_2'^{\lambda'}\Gamma(ds')\Big|dt. \end{aligned}$$

Using (b), (2.9), and Lemma 3.9 we get, for $4-\alpha<\nu<\alpha+1$,

$$\begin{aligned} \int_0^1\frac{I}{r^{\alpha+\nu-3}}dr &\le \text{const.}\int_{S_2}|s_1|^{\kappa+\alpha-2}|s_2|^\lambda\int_{S_2}|s_1'|^{\kappa'+\alpha-2}|s_2'|^{\lambda'} \\ &\quad\times\int_{-\infty}^{\infty}\left|\left|t+\frac{s_2}{s_1}\right|^{\alpha-2}\left|t+\frac{s_2'}{s_1'}\right|^{\alpha-2}-|t|^{2\alpha-4}\right| \\ &\quad\times\int_0^\infty e^{-2^{1-\alpha}r^\alpha|t|^\alpha\sigma_1^\alpha}r^{\alpha-\nu}dr\,dt\,\Gamma(ds')\Gamma(ds) \\ &= \text{const.}\int_{S_2}|s_1|^{\kappa+\alpha-2}|s_2|^\lambda\int_{S_2}|s_1'|^{\kappa'+\alpha-2}|s_2'|^{\lambda'}\int_{-\infty}^{\infty} \\ &\quad\times\left|\left|t+\frac{s_2}{s_1}\right|^{\alpha-2}\left|t+\frac{s_2'}{s_1'}\right|^{\alpha-2}-|t|^{2\alpha-4}\right|t^{\nu-\alpha-1}dt\,\Gamma(ds')\Gamma(ds) \\ &\le \text{const.}\Big(\int_{S_2}|s_1|^{\kappa+2-\nu}|s_2|^{\lambda+\alpha-4+\nu}\Gamma(ds) \\ &\quad\times\int_{S_2}|s_1'|^{\kappa'+\alpha-2}|s_2'|^{\lambda'}\Gamma(ds')\Big) \end{aligned}$$

$$+\int_{S_2} |s_1|^{\kappa+\alpha-2}|s_2|^{\lambda}\Gamma(ds)\int_{S_2} |s_1'|^{\kappa'+2-\nu}|s_2'|^{\lambda'+\alpha-4+\nu}\Gamma(ds'),$$

which is finite by (1.1).

When g contains only one term (B) (that is, g is of type (iii)), then we have to show integrability w.r.t. $r^{-\alpha-\nu+3}dr$ of the following integral

$$J=\int_{-\infty}^{\infty} e^{-2^{1-\alpha}|t|^{\alpha}\sigma_1^{\alpha}}|t|^{\gamma}\int_{S_2} \left||ts_1+rs_2|^{\alpha-2}-|ts_1|^{\alpha-2}\right||s_1|^{\kappa}|s_2|^{\lambda}\Gamma(ds)dt,$$

which was obtained from the respective difference by using bounds (b) and possibly (d). Hence $0\le\gamma\le 2\alpha-2$. Again

$$\int_0^1 \frac{J}{r^{\alpha+\nu-3}}dr \le \int_0^{\infty} \frac{J}{r^{\alpha+\nu-3}}dr \le \text{const.}\int_{S_2} |s_1|^{\kappa+2-\nu}|s_2|^{\lambda+\alpha-4+\nu}\Gamma(ds)<\infty$$

by (2.9), Lemma 3.8, and (1.1). This yields (2.14) and hence ends the proof. ■

3. Inequalities

We establish here several inequalities that are used in the proof of the preceding results. It is the sharpness of these inequalities that allows us to obtain such a general result as Theorem 1.1. We divide them in lemmas according to the form they possess.

The first inequalities, whose proofs we omit, are easy consequences of Lemma 3.1 of Samorodnitsky and Taqqu [11], Cambanis and Wu [3] and applications of the triangle inequality and the Mean Value Theorem.

Lemma 3.1. *Suppose that $\int_{S_2}|s_1|^{-\nu}\Gamma(ds)<\infty$ and $1<\alpha<2$. Then*

$$\left|\int_{S_2}(|ts_1+rs_2|^{\alpha}-|ts_1|^{\alpha})\Gamma(ds)\right|\le \alpha\Gamma(S_2)(|r|^{\alpha}+|r||t|^{\alpha-1})$$

and, if $\nu\ge 2-\alpha$,

$$\left|\int_{S_2}(|ts_1+rs_2|^{\alpha}+|ts_1-rs_2|^{\alpha}-2|ts_1|^{\alpha})\Gamma(ds)\right|\le \text{ const. } r^2|t|^{\alpha-2}.$$

The constant depends on α and Γ.

Lemma 3.2. *For $x,y\in\mathbf{R}$,*

$$|e^{-x}-e^{-y}| \le e^{-\min(x,y)}|x-y|, \tag{3.1}$$

$$|e^{-x}-e^{-y}+e^{-y}(x-y)| \le \frac{1}{2}e^{-y}e^{|x-y|}(x-y)^2, \tag{3.2}$$

$$|e^{-x}-e^{-y}+e^{-y}(x-y)| \le \frac{1}{2}e^{-\min(x,y)}(x-y)^2. \tag{3.3}$$

Lemma 3.3. *For $\alpha > 1$ and $t, r \in \mathbf{R}$,*

$$\exp\{-\int_{S_2} |ts_1 + rs_2|^\alpha \Gamma(ds)\} \leq \exp(|r|^\alpha \sigma_2^\alpha) \exp(-2^{1-\alpha}\sigma_1^\alpha |t|^\alpha).$$

Lemma 3.4. *For $z \in \mathbf{R}$ and $0 < \beta \leq 1$,*

$$(i) \quad Bigl||1+z|^\beta - 1\Big| \leq |z|^\beta, \qquad (ii) \quad \Big||1+z|^\beta - 1\Big| \leq |z|,$$
$$(iii) \quad |(1+z)^{<\beta>} - 1| \leq 2|z|^\beta, \qquad (iv) \quad |(1+z)^{<\beta>} - 1| \leq 2|z|.$$

Lemma 3.5. *For $z \in \mathbf{R}$ and $0 < \beta \leq 1$,*

$$|(z+1)^{<\beta>} + (z-1)^{<\beta>} - 2z^{<\beta>}| \leq \text{const.} \min(|z|^\beta, |z|^{\beta-2}),$$

and for $1 < \beta \leq 3$,

$$|(z+1)^{<\beta>} + (z-1)^{<\beta>} - 2z^{<\beta>}| \leq \text{const.} \min(|z|, |z|^{\beta-2}),$$

where constants depend on β alone.

Proof. Note that for $\beta = 1$ the lemma is trivial. For other cases first consider $|z| \leq 1$. Then

$$\begin{aligned} & |(z+1)^{<\beta>} + (z-1)^{<\beta>} - 2z^{<\beta>}| \\ = & \ |((z+1)^\beta - 1) - ((1-z)^\beta - 1) - 2z^{<\beta>}| \\ \leq & \ |(1+z)^\beta - 1| + |(1-z)^\beta - 1| + 2|z|^\beta, \end{aligned}$$

which is bounded by const. $|z|^\beta$ if $0 < \beta < 1$ by Lemma 3.4 and by const. $|z|$ if $\beta > 1$ by the mean value theorem. If $1 < |z| \leq 2$, then

$$|(z+1)^{<\beta>} + (z-1)^{<\beta>} - 2z^{<\beta>}| \leq |z|^\beta (|1 + \frac{1}{z})^{<\beta>} - 1| + |(1 - \frac{1}{z})^{<\beta>} - 1|,$$

which is bounded by a constant if $0 < \beta < 1$ by Lemma 3.4 and by const. $|z|^{\beta-1}$ if $\beta > 1$ by the mean value theorem. In both cases we can bound it further by const. $|z|^{\beta-2}$ since $1 \leq 2/|z|$.

Now consider $f(w) = (1+w)^{<\beta>} + (1+w)^{<\beta>} - 2$ for $|w| < 1/2$. Then $f'(w) = \beta|1+w|^{\beta-1} - \beta|1-w|^{\beta-1}$ and $f''(w) = \beta(\beta-1)(1+w)^{<\beta-2>} + \beta(\beta-1)(1-w)^{<\beta-2>}$. Since $f(0) = f'(0) = 0$, there exists $0 < \theta < 1$ such that $|(1+w)^{<\beta>} + (1-w)^{<\beta>} - 2| = |f(w)| = \frac{1}{2}|f''(\theta w)|w^2 \leq$ const. w^2. Using this inequality with $w = 1/z$, we obtain for $|z| > 2$, $|(z+1)^{<\beta>} + (z-1)^{<\beta>} - 2z^{<\beta>}| = |z|^\beta |f(\frac{1}{z})| \leq$ const. $|z|^{\beta-2}$, and the lemma is proved. ∎

Lemma 3.6. *For $x, y \in \mathbf{R}$ and $0 < \beta < 1$,*

$$\begin{aligned} & |(1+x)^{<\beta>}(1+y)^{<\beta>} + (1-x)^{<\beta>}(1-y)^{<\beta>} - 2| \\ \leq & \ \text{const.} \ (|x|^{\beta+p} + |y|^{\beta+p} + |xy|^{(\beta+p)/2}), \end{aligned}$$

where $\beta \leq p \leq 2 - \beta$ and const. *depends on p and β alone.*

Proof. Note that if $|y| > 1/2$ and $|x| \leq 1/2$, then by Lemma 3.4 (iii) and (iv)

$$\begin{aligned}
&|(1+x)^{<\beta>}(1+y)^{<\beta>} - 1| \\
&= |((1+x)^{<\beta>} - 1)(1+y)^{<\beta>} + ((1+y)^{<\beta>} - 1)| \\
&\leq 2|x|^\beta + 2|x||y|^\beta + 2|y|^\beta \leq (2+1+2)|y|^\beta \leq 5\, 2^p |y|^{\beta+p},
\end{aligned}$$

with $p > 0$. An analogous inequality holds for $|y| \leq 1/2$ and $|x| > 1/2$ by symmetry. If both $|x| > 1/2$ and $|y| > 1/2$ then, as above,

$$\begin{aligned}
&|(1+x)^{<\beta>}(1+y)^{<\beta>} - 1| \leq 2|x|^\beta(1+|y|^\beta) + 2|y|^\beta \\
&\leq 2(|x|^\beta + |y|^\beta + |x|^\beta|y|^\beta) \leq 2^{1+p}(|x|^{\beta+p} + |y|^{\beta+p} + |x|^{\frac{\beta+p}{2}}|y|^{\frac{\beta+p}{2}}),
\end{aligned}$$

since $\beta \leq p$. If either $|y| > 1/2$ or $|x| > 1/2$ we get, by combining the previous inequalities,

$$|(1+x)^{<\beta>}(1+y)^{<\beta>} - 1| \leq \text{ const. } (|x|^{\beta+p} + |y|^{\beta+p} + |x|^{\frac{\beta+p}{2}}|y|^{\frac{\beta+p}{2}}).$$

Using the triangle inequality we conclude that if $|y| > 1/2$ or $|x| > 1/2$, then

$$\begin{aligned}
&|(1+x)^{<\beta>}(1+y)^{<\beta>} + (1-x)^{<\beta>}(1-y)^{<\beta>} - 2| \\
&\leq \text{ const. } (|x|^{\beta+p} + |y|^{\beta+p} + (|x||y|)^{\frac{\beta+p}{2}}),
\end{aligned}$$

where const. depends only on p. We want to show that the last inequality (with possibly different constant) also holds if $|y| \leq 1/2$ and $|x| \leq 1/2$. Define $g(x,y) = (1+x)^{<\beta>}(1+y)^{<\beta>} + (1-x)^{<\beta>}(1-y)^{<\beta>} - 2$. Then

$$\begin{aligned}
g(0,0) &= 0,\ g_x(x,y) = \beta|1+x|^{\beta-1}(1+y)^{<\beta>} - \beta|1-x|^{\beta-1}(1-y)^{<\beta>} \\
&= g_x(y,x), \\
g_x(0,0) &= 0 = g_y(0,0), \\
g_{xx}(x,y) &= \beta(\beta-1)(1+x)^{<\beta-2>}(1+y)^{<\beta>} \\
&\quad + \beta(\beta-1)(1-x)^{<\beta-2>}(1-y)^{<\beta>} \\
&= g_{yy}(y,x), \\
g_{xy}(x,y) &= \beta^2|1+x|^{\beta-1}|1+y|^{\beta-1} + \beta^2|1-x|^{\beta-1}||1-y|^{\beta-1}.
\end{aligned}$$

For $|x|, |y| \leq \frac{1}{2}$ there exist $\theta_1 = \theta_1(x,y)$ and $\theta_2 = \theta_2(x,y)$, $|\theta_1|, |\theta_2| < 1$ such that

$$\begin{aligned}
|g(x,y)| &= |g(x,y) - g(0,0)| \\
&= \frac{1}{2}|g_{xx}(\theta_1 x, \theta_2 y)x^2 + 2g_{xx}(\theta_1 x, \theta_2 y)xy + g_{yy}(\theta_1 x, \theta_2 y)y^2| \\
&\leq \frac{1}{2}\Big[2\beta(1-\beta)\Big(\frac{1}{2}\Big)^{\beta-2}\Big(\frac{3}{2}\Big)^\beta(x^2+y^2) + 4\beta^2\Big(\frac{1}{2}\Big)^{2\beta-2}|xy|\Big].
\end{aligned}$$

Thus $|g(x,y)| \leq$ const. $(x^2+y^2+|xy|) \leq$ const. $(|x|^{\beta+p}+|y|^{\beta+p}+|xy|^{\frac{\beta+p}{2}})$ since $\beta+p\leq 2$. ■

Lemma 3.7. *For $z \leq -2$ or $z \geq -1/2$ and $-1<\eta<0$, $0\leq p\leq 1$,*

(i) $||1+z|^{\eta}-1| \leq c_1|z|^p$,

(ii) $|(1+z)^{<\eta>}-1| \leq c_2|z|^p$,

where $c_1=\max((-\eta)2^{1-\eta},2^{-\eta})$ *and* $c_2=2^{1-\eta}$.

Proof. To prove (i) for $-1/2\leq z\leq 1$, note that by the mean value theorem, we obtain $||1+z|^{\eta}-1| = (-\eta)(1+\theta z)^{\eta-1}|z| \leq (-\eta)2^{1-\eta}|z|^p$, where $0<\theta<1$. For either $z>1$ or $z\leq -2$, we have $||1+z|^{\eta}-1| = \{||1+z|^{-\eta}-1|\}/\{|1+z|^{-\eta}\} \leq |z|^{-\eta}/|z/2|^{-\eta} = 2^{-\eta} \leq 2^{-\eta}|z|^p$, which establishes (i).

The proof of (ii) is similar with Lemma 3.4 (iii) used in the case $z>1$ or $z\leq -2$ instead of the triangle inequality. ■

Lemma 3.8. *The following inequalities hold for $-1<\eta<0$ and $-\eta-1<\beta<-\eta$:*

$$\int_{\mathbf{R}} ||t+z|^{\eta}-|t|^{\eta}|\,|t|^{\beta}dt \;\leq\; \text{const. } |z|^{\eta+\beta+1},$$

$$\int_{\mathbf{R}} |(t+z)^{<\eta>}-t^{<\eta>}|\,|t|^{\beta}dt \;\leq\; \text{const. } |z|^{\eta+\beta+1},$$

where const. *depends only on η and β.*

Proof. Without loss of generality, let us assume that $z>0$. In the regions $\{t\leq -2z\}\cup\{t\geq z/2\} = \{-1/2\leq z/t<0\}\cup\{0<z/t\leq 2\}$ and $\{-z/2\leq t<z/2\} = \{2<z/t\}\cup\{z/t\leq -2\}$ we can apply Lemma 3.4 (i) with $p_1=1$ and $p_2=0$, respectively, to get

$$\begin{aligned}
&\int_{\mathbf{R}} ||t+z|^{\eta}-|t|^{\eta}||t|^{\beta}dt \\
&\leq \Big(\int_{-\infty}^{-2z}+\int_{z/2}^{\infty}\Big)c_1\Big|\frac{z}{t}\Big|t^{\beta+\eta}dt+\int_{-z/2}^{z/2}c_1|t|^{\beta+\eta}dt \\
&\quad+\int_{-2z}^{-z/2}(|t+z|^{\eta}|t|^{\beta}+|t|^{\beta+\eta})dt
\end{aligned}$$

which is bounded by const. $|z|^{\beta+\eta+1}$. This proves the first inequality. (For $-2z<t<-z/2$ we have used the fact that $|t|^{\beta}\leq 2^{|\beta|}|z|^{\beta}$.) The second inequality can be obtained in an analogous way using Lemma 3.4 (ii). ■

Lemma 3.9. *The following inequalities hold for $-1/2<\eta<0$ and $-2\eta-1<\beta<0$:*

$$\int_{\mathbf{R}} \big||t+z_1|^{\eta}|t+z_2|^{\eta}-|t|^{2\eta}\big||t|^{\beta}dt \;\leq\; \text{const. } (|z_1|^{2\eta+\beta+1}+|z_2|^{2\eta+\beta+1}),$$

$$\int_{\mathbf{R}} \big|(t+z_1)^{<\eta>}(t+z_2)^{<\eta>}-|t|^{2\eta}\big||t|^{\beta}dt \;\leq\; \text{const. } (|z_1|^{2\eta+\beta+1}+|z_2|^{2\eta+\beta+1}),$$

where const. *depends only on η and β.*

Proof. Without loss of generality, let us assume that $z_1 > 0$ and $|z_2| > z_1$. (The case $z_1 = z_2$ is covered by Lemma 3.8.) Note that

$$\int_{\mathbf{R}} \Big||t+z_1|^\eta |t+z_2|^\eta - |t|^{2\eta}\Big| |t|^\beta dt$$
$$\leq \int_{\mathbf{R}} \Big||t+z_1|^\eta - |t|^\eta\Big| |t+z_2|^\eta |t|^\beta dt + \int_{\mathbf{R}} \Big||t+z_2|^\eta - |t|^\eta\Big| |t|^{\eta+\beta} dt.$$

The second integral above can be bounded by const. $|z_2|^{2\eta+\beta+1}$ using Lemma 3.8. We now focus on the first integral. As in the proof of Lemma 3.8, divide the domain of integration into the regions where the Lemma 3.4 can be used with $p = 0$ or $p = 1$. Let $A = (-2z_2, -z_2/2)$ or $A = (z_2/2, 2z_2)$ depending on the sign of z_2. We have

$$\Big(\int_{-\infty}^{-2z_1} + \int_{-z_1/2}^{\infty}\Big) \Big||t+z_1|^\eta - |t|^\eta\Big| |t+z_2|^\eta |t|^\beta dt$$
$$\leq \Big(\int_{-\infty}^{-2z_1} + \int_{z_1/2}^{\infty}\Big) c_1 \Big|\frac{z_1}{t}\Big| |t+z_2|^\eta |t|^{\eta+\beta} dt$$
$$+ \int_{-z_1/2}^{z_1/2} c_1 |t+z_2|^\eta |t|^{\eta+\beta} dt \leq c_1 |z_1|$$
$$\times \int_{\{|t|>z_1/2\}\cap A^c} \Big||t+z_2|^\eta - |t|^\eta\Big| |t|^{\eta+\beta-1} dt$$
$$+ c_1|z_1| \int_{\{|t|>z_1/2\}\cap A^c} |t|^{2\eta+\beta-1} dt + c_1 |z_1| \int_A |t+z_2|^\eta |t|^{\eta+\beta-1} dt$$
$$+ c_1 \int_{-z_1/2}^{z_1/2} \Big||t+z_2|^\eta - |t|^\eta\Big| |t|^{\eta+\beta} dt$$
$$+ c_1 \int_{-z_1/2}^{z_1/2} |t|^{2\eta+\beta} dt \leq c_1^2 |z_1| \int_{|t|>z_1/2} |t|^{2\eta+\beta-1} dt$$
$$+ c_1 |z_1| \int_{|t|>z_1/2} |t|^{2\eta+\beta-1} dt$$
$$+ \text{const. } |z_1| |z_2|^{\eta+\beta-1} \int_A |t+z_2|^\eta dt$$
$$+ c_1^2 \int_{-z_1/2}^{z_1/2} |t|^{2\eta+\beta} dt + c_1 \int_{-z/2}^{z_1/2} |t|^{2\eta+\beta} dt$$
$$\leq \text{const. } |z_1|^{2\eta+\beta+1} + \text{const. } |z_1||z_2|^{2\eta+\beta}$$
$$\leq \text{const. } (|z_1|^{2\eta+\beta+1} + |z_2|^{2\eta+\beta+1}).$$

Now, consider the other part of the main integral, i.e.

$$\int_{-2z_1}^{-z_1/2} \Big||t+z_1|^\eta - |t|^\eta\Big| |t+z_2|^\eta |t|^\beta dt$$
$$\leq \text{const. } |z_1|^\beta \int_{-z_1}^{z_1/2} \Big||t|^\eta - |t-z_1|^\eta\Big| |t+z_2-z_1|^\eta dt$$

$$\leq \text{ const. } |z_1|^\beta c_1 \int_{-z_1}^{z_1/2} |t|^\eta |t+z_2-z_1|^\eta dt,$$

where we again applied Lemma 3.4.

Recall that we assume $|z_2| > z_1 > 0$ and consider two cases. First, if $z_2 < z_1$, i.e. $z_2 < 0$ and $z_2 < -z_1$, then $|t+z_2-z_1| \geq z_1/2 - z_2 \geq (3/2)z_1$ for $-z_1 < t < z_1/2$ and the above integral can be easily bounded by const. $z_1^{2\eta+\beta+1}$. Second (this is the crucial case), if $z_2 > z_1 > 0$, put $B = (-2(z_2-z_1), -(z_2-z_1)/2)$ and note that $B \cap (-z_1, z_1/2) \neq \emptyset$ iff $z_2 < 3z_1$. Then

$$\begin{aligned}
&\text{const. } |z_1|^\beta \int_{-z_1}^{z_1/2} |t+z_2-z_1|^\eta |t|^\eta dt \\
&\leq \text{ const.} |z_1|^\beta \int_{(-z_1,z_1/2)\cap B^c} \Big| |t+z_2-z_1|^\eta - |t|^\eta \Big| |t|^\eta dt \\
&\quad + \text{const. } |z_1|^\beta \int_{-z_1}^{z_1/2} |t|^{2\eta} dt + \text{const.} |z_1|^\beta \\
&\quad \times \int_{(-z_1,z_1/2)\cap B} |t+z_2-z_1|^\eta |t|^\eta dt \\
&\leq \text{ const. } |z_1|^\beta \int_{-z_1}^{z_1/2} |t|^{2\eta} dt + \text{const. } |z_1|^\beta |z_2-z_1|^\eta \\
&\quad \times \int_{(-z_1,z_1/2)\cap B} |t+z_2-z_1|^\eta dt \\
&\leq \text{ const. } |z_1|^{\beta+2\eta+1} + \text{const. } |z_1|^\beta |z_2-z_1|^{2\eta+1} I[z_2 < 3z_1] \\
&\leq \text{ const. } |z_1|^{\beta+2\eta+1} + \text{const. } |z_1|^\beta (|z_2|^{2\eta+1} I[z_2 < 3z_1] + |z_1|^{2\eta+1}) \\
&\leq \text{ const. } |z_1|^{\beta+2\eta+1}.
\end{aligned}$$

Hence, the first inequality of the lemma is proved. The second inequality can be proved in the similar fashion using part (ii) of Lemma 3.4 instead of part (i). ■

Lemma 3.10. *The following inequality holds for* $c > 0$, $0 < \alpha < 2$, $-1 < \eta < 0$ *and* $-1-\eta < \beta$:

$$\int_{\mathbf{R}} \exp(-c|t|^\alpha) \Big| |t+z|^\eta - |t|^\eta \Big| |t|^\beta dt \leq \text{ const. } |z|^p$$

with

$$0 \leq p < \beta + \eta + 1 \quad \text{for} \quad -1-\eta < \beta < 0,$$

and

$$0 \leq p < \eta + 1 \quad \text{or} \quad \beta \leq p < \beta + \eta + 1, \; p \leq 1 \quad \text{for} \quad 0 \leq \beta.$$

const. *depends only on* c, α, η, β *and* p.

Proof. Let $0 \le p_1 \le 1$, $p_1 < \beta + \eta + 1$, and assume, without loss of generality, that $z > 0$. Since $\{t \ge -z/2\} = \{-2 \ge z/t\} \cup \{z/t > 0\}$ and $\{t \le -2z\} = \{-1/2 \le z/t < 0\}$ we can apply Lemma 3.7 to get

$$\begin{aligned} J_1 : &= \int_{\{t \ge -z/2\} \cup \{t \le -2z\}} \exp(-c|t|^\alpha) \big| |t+z|^\eta - |t|^\eta \big| |t|^\beta dt \\ &\le \int_{\{t \ge -z/2\} \cup \{t \le -2z\}} \exp(-c|t|^\alpha) c_1 \Big| \frac{z}{t} \Big|^{p_1} |t|^{\beta+\eta} dt \\ &\le c_1 |z|^{p_1} \int_{\mathbf{R}} \exp(-c|t|^\alpha) |t|^{\beta+\eta-p_1} dt \\ &= \text{const. } |z|^{p_1}. \end{aligned}$$

The last integral converges because $p_1 < \beta + \eta + 1$. Now

$$\begin{aligned} J_2 : &= \int_{-2z}^{-z/2} \exp(-c|t|^\alpha) \ \big| |t+z|^\eta - |t|^\eta \big| \ |t|^\beta dt \\ &= \int_{-z}^{z/2} \exp(-c|t-z|^\alpha) \ \big| |t|^\eta - |t-z|^\eta \big| \ |t-z|^\beta dt. \end{aligned}$$

For $-z < t < z/2$ (and $t \ne 0$) we have $|t-z|^\beta \le 2^{|\beta|} z^\beta$, $|t-z| \ge |t|$ and $\{-z \le t < 0\} = \{1 \le -z/t\}$, $\{0 < t \le z/2\} = \{-z/t \le -2\}$. We can therefore apply Lemma 3.7 with $0 \le p_2 < \eta + 1$ to $|1 - |1 - z/t|^\eta|$ to obtain

$$\begin{aligned} J_2 &\le 2^{|\beta|} z^\beta \int_{-z}^{z/2} \exp(-c|t|^\alpha) c_1 \Big| \frac{z}{t} \Big|^{p_2} |t|^\eta dt \\ &\le \Big(2^{|\beta|} c_1 \int_{-\infty}^{\infty} \exp(-c|t|^\alpha) |t|^{\eta - p_2} dt \Big) \ |z|^{\beta + p_2} \\ &= \text{const. } |z|^{\beta+p_2}. \end{aligned}$$

For $\beta \ge 0$, note that there exists a constant depending on c, α and β such that

$$\exp(-c|t|^\alpha)|t|^\beta \le \text{const. } \exp(-c|t|^\alpha/2). \tag{3.4}$$

Thus, in the case $\beta \ge 0$, J_2 can also be bounded by const. $|z|^{p_2}$ with $0 \le p_2 < \eta + 1$.

The statement of the above lemma holds with either $p = p_1 = \beta + p_2$ or $p = p_1 = p_2$. ∎

The following lemma considers the case $\beta = \eta$ but allows for the presence of other z's.

Lemma 3.11. *The following inequality holds for* $c > 0$, $0 < \alpha < 2$, $-1/2 < \eta < 0$ *and* $0 \le p < 2\eta + 1$*:*

$$\int_{\mathbf{R}} \exp(-c|t|^\alpha) \big| |t+z_1|^\eta - |t+z_2|^\eta \big| \ |t+z_3|^\eta dt \le \text{const. } |z_1 - z_2|^p,$$

where const. *depends only on* c, α, η *and* p.

Proof. The main tools are "translation" and use of Lemmas 3.7 and 3.10. Translating, we get

$$\int_{\mathbf{R}} \exp(-c|t|^\alpha)\big||t+z_1|^\eta - |t+z_2|^\eta\big||t+z_3|^\eta dt$$
$$= \int_{\mathbf{R}} \exp(-c|t-z_2|^\alpha)\big||t+z_1-z_2|^\eta - |t|^\eta\big||t+z_3-z_2|^\eta dt.$$

As in the proof of Lemma 3.10, we assume, without loss of generality, that $z_1 - z_2 > 0$. Putting $A:=\{t\colon t \geq -(z_1-z_2)/2\} \cup \{t\colon t \leq -2(z_1-z_2)\}$ and using Lemma 3.7 with $0 \leq p < 2\eta+1$, we obtain

$$\int_A \exp(-c|t-z_2|^\alpha)\left|\left|1+\frac{z_1-z_2}{t}\right|^\eta - 1\right||t|^\eta|t+z_3-z_2|^\eta dt \tag{3.5}$$
$$\leq \text{const. } |z_1-z_2|^p \int_{\mathbf{R}} \exp(-c|t-z_2|^\alpha)|t|^{\eta-p}|t+z_3-z_2|^\eta dt$$
$$\leq \text{const. } |z_1-z_2|^p \int_{\mathbf{R}} \exp(-c|t-z_2|^\alpha)\big||t+z_3-z_2|^\eta - |t|^\eta\big||t|^{\eta-p} dt$$
$$+ \text{const. } |z_1-z_2|^p \int_{\mathbf{R}} \exp(-c|t-z_2|^\alpha)|t|^{2\eta-p} dt.$$

The second term can be bounded by const.$|z_1-z_2|^p$ after changing variables $t - z_2 \to t$ in the integral and using Lemma 3.10 with $2\eta - p$ instead of η, and $\beta = p = 0$. For the first term, under the assumption $z_3 - z_2 > 0$, we split the integral over $\mathbf{R}$ into two integrals, one over $A_1 = \{t\colon t \geq -(z_3-z_2)/2\} \cup \{t\colon t \leq -2(z_3-z_2)\}$, the other over A_1^c. The integral over A_1 is bounded by const.$|z_3-z_2|^0 \int_{\mathbf{R}} \exp(-c|t-z_2|^\alpha)|t|^{2\eta-p} \leq$ const., where as above the const. does not depend on z_2. Using Lemma 3.7 with $p-\eta$ instead of p, we get, as in the argument involving J_2 in Lemma 3.10,

$$\int_{A_1^c} \exp(-c|t-z_2|^\alpha)\big||t+z_3-z_2|^\eta - |t|^\eta\big||t|^{\eta-p} dt$$
$$= \int_{-(z_3-z_2)}^{(z_3-z_2)/2} \exp(-c|t-z_3|^\alpha)\big||t|^\eta - |t-z_3+z_2|^\eta\big||t-z_3+z_2|^{\eta-p} dt$$
$$\leq \text{const. } |z_3-z_2|^{\eta-p} \int_{\mathbf{R}} \exp(-c|t-z_3|^\alpha)|t|^{\eta-(p-\eta)}|z_3-z_2|^{p-\eta} dt$$
$$= \text{const.} \int_{\mathbf{R}} \exp(-c|t-z_3|^\alpha)|t|^{2\eta-p} dt \leq \text{const.}$$

Thus, we have shown that (3.5) is bounded by const.$|z_1-z_2|^p$.

Now, consider the integral over A^c:

$$\int_{-2(z_1-z_2)}^{-(z_1-z_2)/2} \exp(-c|t-z_2|^\alpha)\big||t+z_1-z_2|^\eta - |t|^\eta\big||t+z_3-z_2|^\eta dt$$
$$= \int_{-(z_1-z_2)}^{(z_1-z_2)/2} \exp(-c|t-z_1|^\alpha)\big||t|^\eta - |t-z_1+z_2|^\eta\big||t+z_3-z_1|^\eta dt$$
$$\leq \text{const. } |z_1-z_2|^p \int_{\mathbf{R}} \exp(-c|t-z_1|^\alpha)|t|^{\eta-p}|t+z_3-z_1|^\eta dt$$

with $0 \leq p < 2\eta + 1$. The integral on the right hand side is of the same type as the one obtained in the first inequality of (3.5), with z_2 replaced by z_1. Thus, it can be bounded by a constant and the lemma is proved. ■

Corollary 3.1. *For* $c > 0$, $0 < \alpha < 2$, $-1/2 < \eta < 0$ *and* $0 \leq p < 2\eta + 1$

$$\int_{\mathbf{R}} \exp(-c|t|^\alpha)\Big||t+z_1|^\eta|t+z_3|^\eta - |t+z_2|^\eta|t+z_4|^\eta\Big|dt$$
$$\leq \text{ const. } (|z_1 - z_2|^p + |z_3 - z_4|^p),$$

where const. *depends only on* c, α, η *and* p.

Proof. Writing $u_i = |t + z_i|^\eta$, $i = 1, \ldots, 4$, we have $u_1u_3 - u_2u_4 = (u_1 - u_2)u_3 + (u_3 - u_4)u_2$, and so the result follows from Lemma 3.11. ■

Lemma 3.12. *The following inequality holds for* $c > 0$, $0 < \alpha < 2$, $-1 < \eta < 0$, $\beta \geq 0$ *and* $0 \leq p < \eta + 1$*:*

$$\int_{\mathbf{R}} \exp(-c|t|^\alpha)\Big||t+z_1|^\eta - |t+z_2|^\eta\Big||t|^\beta dt \leq \text{const. } |z_1 - z_2|^p,$$

where const. *depends only on* c, α, η, β *and* p.

Proof. By (3.4) it is enough to show the above inequality in the case $\beta = 0$. Now, follow the lines of the proof of Lemma 3.11, use the definition of A in that lemma and note that

$$\int_A \exp(-c|t - z_2|^\alpha)\Big||t + z_1 - z_2|^\eta - |t|^\eta\Big|dt$$
$$\leq \text{ const. } |z_1 - z_2|^p \int_{\mathbf{R}} \exp(-c|t - z_2|^\alpha)|t|^{\eta - p}dt$$
$$\leq \text{ const. } |z_1 - z_2|^p$$

with $0 \leq p < \eta + 1$. Similar bounds (with z_1 replaced by z_2 and vice versa) hold for $\int_{A^c}$, which establishes the lemma. ■

Appendix. The fourth derivative of the conditional characteristic function

The following table contains the terms of $2\pi f(x)\ \text{Re}\phi^{(4)}_{X_2|x}(r)$. The right column indicates the type of term—(i), (ii) or (iii)—according to the classification of Proposition 2.4. The terms are ordered in the same way as the corresponding terms in $\text{Re}\phi^{(3)}_{X_2|x}(r)$, and the groups of terms coming from the same term in $\text{Re}\phi^{(3)}_{X_2|x}(r)$ are separated by a line. Observe that since some of the terms (like the fourth and the twelfth, the eighth and the thirteenth, the eleventh and the last one) differ only by a constant, $\text{Re}\phi^{(4)}_{X_2|x}(r)$ can be expressed as a sum of 22 terms instead of 27.

No.	Term	Type
(1)	$\alpha^4 \int_{-\infty}^{\infty} \cos tx\, \phi(t,r)(\int_{S_2}(ts_1+rs_2)^{<\alpha-1>} s_2\Gamma(d\mathbf{s}))^4 dt$	(i)
(2)	$-3\alpha^3(\alpha-1)\int_{-\infty}^{\infty} \cos tx\, \phi(t,r)(\int_{S_2}(ts_1+rs_2)^{<\alpha-1>} s_2\Gamma(d\mathbf{s}))^2$ $\int_{S_2}\lvert ts_1+rs_2\rvert^{\alpha-2} s_2^2\Gamma(d\mathbf{s})dt$	(iii)
(3)	$-3\alpha^3 x\int_{-\infty}^{\infty} \sin tx\, \phi(t,r)(\int_{S_2}(ts_1+rs_2)^{<\alpha-1>} s_2\Gamma(d\mathbf{s}))^2$ $(\int_{S_2}(ts_1+rs_2)^{<\alpha-1>} s_2^2 s_1^{-1}\Gamma(d\mathbf{s}))dt$	(i)
(4)	$3\alpha^2(\alpha-1)x\int_{-\infty}^{\infty} \sin tx\, \phi(t,r)(\int_{S_2}\lvert ts_1+rs_2\rvert^{\alpha-2} s_2^2\Gamma(d\mathbf{s}))$ $(\int_{S_2}(ts_1+rs_2)^{<\alpha-1>} s_2^2 s_1^{-1}\Gamma(d\mathbf{s}))dt$	(iii)
(5)	$3\alpha^2(\alpha-1)x\int_{-\infty}^{\infty} \sin tx\, \phi(t,r)(\int_{S_2}(ts_1+rs_2)^{<\alpha-1>} s_2\Gamma(d\mathbf{s}))$ $(\int_{S_2}\lvert ts_1+rs_2\rvert^{\alpha-2} s_2^3 s_1^{-1}\Gamma(d\mathbf{s}))dt$	(iii)
(6)	$-3\alpha^4\int_{-\infty}^{\infty} \cos tx\, \phi(t,r)(\int_{S_2}(ts_1+rs_2)^{<\alpha-1>} s_2\Gamma(d\mathbf{s}))^2$ $(\int_{S_2}(ts_1+rs_2)^{<\alpha-1>} s_1\Gamma(d\mathbf{s}))(\int_{S_2}(ts_1+rs_2)^{<\alpha-1>} s_2^2 s_1^{-1}\Gamma(d\mathbf{s}))dt$	(i)
(7)	$3\alpha^3(\alpha-1)\int_{-\infty}^{\infty} \cos tx\, \phi(t,r)(\int_{S_2}\lvert ts_1+rs_2\rvert^{\alpha-2} s_1 s_2\Gamma(d\mathbf{s}))$ $(\int_{S_2}(ts_1+rs_2)^{<\alpha-1>} s_2\Gamma(d\mathbf{s}))(\int_{S_2}(ts_1+rs_2)^{<\alpha-1>} s_2 s_1^{-1}\Gamma(d\mathbf{s}))dt$	(iii)
(8)	$3\alpha^3(\alpha-1)\cos tx\, \phi(t,r)(\int_{S_2}(ts_1+rs_2)^{<\alpha-1>} s_1\Gamma(d\mathbf{s}))$ $(\int_{S_2}\lvert ts_1+rs_2\rvert^{\alpha-2} s_2^2\Gamma(d\mathbf{s}))(\int_{S_2}(ts_1+rs_2)^{<\alpha-1>} s_2 s_1^{-1}\Gamma(d\mathbf{s}))dt$	(iii)
(9)	$3\alpha^3(\alpha-1)\int_{-\infty}^{\infty} \cos tx\, \phi(t,r)(\int_{S_2}(ts_1+rs_2)^{\alpha-1>} s_1\Gamma(d\mathbf{s}))$ $(\int_{S_2}(ts_1+rs_2)^{<\alpha-1>} s_2\Gamma(d\mathbf{s}))(\int_{S_2}\lvert ts_1+rs_2\rvert^{\alpha-2} s_2^3 s_1^{-1}\Gamma(d\mathbf{s}))dt$	(iii)
(10)	$3\alpha^3(\alpha-1)\int_{-\infty}^{\infty} \cos tx\, \phi(t,r)(\int_{S_2}(ts_1+rs_2)^{<\alpha-1>} s_2\Gamma(d\mathbf{s}))$ $(\int_{S_2}(ts_1+rs_2)^{<\alpha-1>} s_2^2 s_1^{-1}\Gamma(d\mathbf{s}))(\int_{S_2}\lvert ts_1+rs_2\rvert^{\alpha-2} s_2 s_1\Gamma(d\mathbf{s}))dt$	(iii)
(11)	$-3\alpha^2(\alpha-1)^2\int_{-\infty}^{\infty} \cos tx\, \phi(t,r)(\int_{S_2}\lvert ts_1+rs_2\rvert^{\alpha-2} s_2^3 s_1^{-1}\Gamma(d\mathbf{s}))$ $(\int_{S_2}\lvert ts_1+rs_2\rvert^{\alpha-2} s_2 s_1\Gamma(d\mathbf{s}))dt$	(ii)
(12)	$-3\alpha^2(\alpha-1)x\int_{-\infty}^{\infty} \sin tx\, \phi(t,r)(\int_{S_2}(ts_1+rs_2)^{<\alpha-1>} s_2^2 s_1^{-1}\Gamma(d\mathbf{s}))$ $(\int_{S_2}\lvert ts_1+rs_2\rvert^{\alpha-2} s_2^2\Gamma(d\mathbf{s}))dt$	(iii)
(13)	$-3\alpha^3(\alpha-1)\int_{-\infty}^{\infty} \cos tx\, \phi(t,r)(\int_{S_2}(ts_1+rs_2)^{<\alpha-1>} s_1\Gamma(d\mathbf{s}))$ $(\int_{S_2}(ts_1+rs_2)^{<\alpha-1>} s_2^2 s_1^{-1}\Gamma(d\mathbf{s})(\int_{S_2}\lvert ts_1+rs_2\rvert^{\alpha-2} s_2^2\Gamma(d\mathbf{s})))dt$	(iii)
(14)	$3\alpha^2(\alpha-1)^2\int_{-\infty}^{\infty} \cos tx\, \phi(t,r)(\int_{S_2}\lvert ts_1+rs_2\rvert^{\alpha-2} s_2^2\Gamma(d\mathbf{s}))^2 dt$	(ii)

No.	Term	Type
(15)	$-\alpha^2 x^2 \int_{-\infty}^{\infty} \cos tx\ \phi(t,r)(\int_{S_2}(ts_1+rs_2)^{<\alpha-1>} s_2\Gamma(ds))$ $(\int_{S_2}(ts_1+rs_2)^{<\alpha-1>} s_2^3 s_1^{-2}\Gamma(ds))dt$	(i)
(16)	$\alpha(\alpha-1)x^2 \int_{-\infty}^{\infty} \cos tx\ \phi(t,r)(\int_{S_2}\|ts_1+rs_2\|^{\alpha-2} s_2^4 s_1^{-2}\Gamma(ds))dt$	(iii)
(17)	$2\alpha^3 x \int_{-\infty}^{\infty} \sin tx\ \phi(t,r)(\int_{S_2}(ts_1+rs_2)^{<\alpha-1>} s_2\Gamma(ds))$ $(\int_{S_2}(ts_1+rs_2)^{<\alpha-1>} s_1\Gamma(ds))(\int_{S_2}(ts_1+rs_2)^{<\alpha-1>} s_2^3 s_1^{-2}\Gamma(ds))dt$	(i)
(18)	$-2\alpha^2(\alpha-1)x \int_{-\infty}^{\infty} \sin tx\ \phi(t,r)(\int_{S_2}\|ts_1+rs_2\|^{\alpha-2} s_2 s_1\Gamma(ds))$ $(\int_{S_2}(ts_1+rs_2)^{<\alpha-1>} s_2^3 s_1^{-2}\Gamma(ds))dt$	(iii)
(19)	$-2\alpha^2(\alpha-1)x \int_{-\infty}^{\infty} \sin tx\ \phi(t,r)(\int_{S_2}(ts_1+rs_2)^{<\alpha-1>} s_1\Gamma(ds))$ $(\int_{S_2}\|ts_1+rs_2\|^{\alpha-2} s_2^4 s_1^{-2}\Gamma(ds))dt$	(iii)
(20)	$\alpha^4 \int_{-\infty}^{\infty} \cos tx\ \phi(t,r)(\int_{S_2}(ts_1+rs_2)^{<\alpha-1>} s_2\Gamma(ds))$ $(\int_{S_2}(ts_1+rs_2)^{<\alpha-1>} s_1\Gamma(ds))^2(\int_{S_2}(ts_1+rs_2)^{<\alpha-1>} s_2^3 s_1^{-2}\Gamma(ds))dt$	(i)
(21)	$-2\alpha^3(\alpha-1) \int_{-\infty}^{\infty} \cos tx\ \phi(t,r)(\int_{S_2}(ts_1+rs_2)^{<\alpha-1>} s_1\Gamma(ds))$ $(\int_{S_2}\|ts_1+rs_2\|^{\alpha-2} s_2 s_1\Gamma(ds))(\int_{S_2}(ts_1+rs_2)^{<\alpha-1>} s_2^3 s_1^{-2}\Gamma(ds))dt$	(iii)
(22)	$-\alpha^3(\alpha-1) \int_{-\infty}^{\infty} \cos tx\ \phi(t,r)(\int_{S_2}(ts_1+rs_2)^{<\alpha-1>} s_1\Gamma(ds))^2$ $(\int_{S_2}\|ts_1+rs_2\|^{\alpha-2} s_2^4 s_1^{-2}\Gamma(ds))dt$	(iii)
(23)	$-\alpha^3(\alpha-1) \int_{-\infty}^{\infty} \cos tx\ \phi(t,r)(\int_{S_2}(ts_1+rs_2)^{<\alpha-1>} s_2\Gamma(ds))$ $(\int_{S_2}(ts_1+rs_2)^{<\alpha-1>} s_2^3 s_1^{-2}\Gamma(ds))(\int_{S_2}\|ts_1+rs_2\|^{\alpha-2} s_1^2\Gamma(ds))dt$	(iii)
(24)	$\alpha^2(\alpha-1)^2 \int_{-\infty}^{\infty} \cos tx\ \phi(t,r)(\int_{S_2}\|ts_1+rs_2\|^{\alpha-2} s_2^4 s_1^{-2}\Gamma(ds))$ $(\int_{S_2}\|ts_1+rs_2\|^{\alpha-2} s_1^2\Gamma(ds))dt$	(ii)
(25)	$\alpha^2(\alpha-1)x \int_{-\infty}^{\infty} \sin tx\ \phi(t,r)(\int_{S_2}(ts_1+rs_2)^{<\alpha-1>} s_2^3 s_1^{-2}\Gamma(ds))$ $(\int_{S_2}\|ts_1+rs_2\|^{\alpha-2} s_2 s_1\Gamma(ds))dt$	(iii)
(26)	$\alpha^3(\alpha-1) \int_{-\infty}^{\infty} \cos tx\ \phi(t,r)(\int_{S_2}(ts_1+rs_2)^{<\alpha-1>} s_1\Gamma(ds))$ $(\int_{S_2}(ts_1+rs_2)^{<\alpha-1>} s_2^3 s_1^{-2}\Gamma(ds))(\int_{S_2}\|ts_1+rs_2\|^{\alpha-2} s_2 s_1\Gamma(ds))dt$	(iii)
(27)	$-\alpha^2(\alpha-1)^2 \int_{-\infty}^{\infty} \cos tx\ \phi(t,r)(\int_{S_2}\|ts_1+rs_2\|^{\alpha-2} s_2^3 s_1^{-1}\Gamma(ds))$ $(\int_{S_2}\|ts_1+rs_2\|^{\alpha-2} s_2 s_1\Gamma(ds))dt$	(ii)

References

[1] S. Cambanis and S. B. Fotopoulos. Conditional variance for stable random vectors. *Probability and Mathematical Statistics*, **15**:195–214, 1995.

[2] S. Cambanis, S. B. Fotopoulos, and L. He. On the conditional variance for scale mixtures of normal distributions. Preprint, 1996.

[3] S. Cambanis and W. Wu. Multiple regression on stable vectors. *Journal of Multivariate Analysis*, **41**:243–272, 1992.

[4] R. Cioczek-Georges and M. S. Taqqu. How do conditional moments of stable vectors depend on the spectral measure? *Stochastic Processes and their Applications*, **54**:95–111, 1994.

[5] R. Cioczek-Georges and M. S. Taqqu. Form of the conditional variance for symmetric stable random variables. *Statistica Sinica*, **5**:351–361, 1995.

[6] R. Cioczek-Georges and M. S. Taqqu. Necessary conditions for the existence of conditional moments of stable random variables. *Stochastic Processes and their Applications*, **56**:233–246, 1995.

[7] S. B. Fotopoulos and L. He. Form of the conditional variance-covariance matrix for α-stable scale mixtures of normal distributions. Preprint, 1996.

[8] C. D. Hardin Jr., G. Samorodnitsky, and M. S. Taqqu. Non-linear regression of stable random variables. *The Annals of Applied Probability*, **1**:582–612, 1991.

[9] B. Ramachandran. On characteristic functions and moments. *Sankhya*, **31**(Series A):1–12, 1969.

[10] H. L. Royden. *Real Analysis*. Macmillan, third edition, 1988.

[11] G. Samorodnitsky and M. S. Taqqu. Conditional moments and linear regression for stable random variables. *Stochastic Processes and their Applications*, **39**:183–199, 1991.

[12] G. Samorodnitsky and M. S. Taqqu. *Stable Non-Gaussian Processes: Stochastic Models with Infinite Variance*. Chapman and Hall, New York, London, 1994.

[13] W. Wu and S. Cambanis. Conditional variance of symmetric stable variables. In S. Cambanis, G. Samorodnitsky, and M. S. Taqqu, editors, *Stable Processes and Related Topics*, volume 25 of *Progress in Probability*, pp. 85–99, Birkhäuser, Boston, 1991.

Renata Cioczek-Georges
Department of Statistics
University of North Carolina at Chapel Hill
Chapel Hill, NC 27599-3260, USA
Email: renata@stat.unc.edu

Murad S. Taqqu
Boston University
Department of Mathematics
111 Cummington Street
Boston, MA 02215-2411, USA
Email: murad@math.bu.edu

How Heavy are the Tails of a Stationary HARCH(k) Process? A Study of the Moments*†

Paul Embrechts, Gennady Samorodnitsky
Michel M. Dacorogna, and Ulrich A. Müller

Abstract

Probabilistic properties of HARCH(k) processes as special stochastic volatility models are investigated. We present necessary and sufficient conditions for the existence of a strongly stationary version of a HARCH(k) process with finite $(2m)$th moments, $m \geq 1$. Our approach is based on the general Markov chain techniques of (Meyn and Tweedie, 1993). The conditions are explicit in the case of second moments, and also in the case of 4th moments of the HARCH(2) process. We also deduce explicit *necessary* and explicit *sufficient* conditions for higher order moments of general HARCH(k) models. We start by studying the HARCH(2) process (in which case our results are the most explicit) and then generalize the results to a general HARCH(k) process.

1. Introduction

In [Müller et al., 1997], the Heterogeneous Auto-Regressive Conditional Heteroskedastic process (HARCH) has been introduced. This HARCH process has been developed as an improvement of traditional ARCH-type models in order to describe the behavior of financial time series, including some newly detected properties such as the long memory of volatility and the asymmetry between volatilities with different degrees of resolution in time.

The HARCH process is useful in practice only if its basic properties are known. In Section 2, we give some motivation for using HARCH and determining its stationarity and moment conditions.

The HARCH process is a random recursion, whose equations are presented in Section 3. In order to understand how heavy are the tails of a stationary

---

**AMS 1991 subject classification.* Primary 60K30, 90A09. Secondary 60H25

†*Keywords and phrases:* HARCH processes, heavy tails, Markov chains, heteroscedasticity, Harris recurrence, random recursion, stochastic volatility.

HARCH(k) process, we study its moments. In Section 4, we start by presenting in more detail how a necessary condition for the existence of the 2nd moment can be derived and to show, using a Markov chain approach, that the necessary condition presented in [Müller et al., 1997] is also sufficient for the existence of a stationary HARCH(k) process with a finite second moment. This is done first for the HARCH(2) process and then generalized to HARCH(k). As a rule, we first analyze the HARCH(2) process, and then consider the general HARCH(k) process. The reason for this is two-fold. First, the results are often most explicit for the HARCH(2) model. Second, the arguments and main ideas are the easiest to follow in this case.

With the above in mind, we next present the derivation of an explicit necessary and sufficient condition for the existence of the 4th moment of a HARCH(2) model. The following step is to derive a general theorem giving necessary and sufficient conditions for the existence of a stationary HARCH(2) process with finite $(2m)$th moments. We are then able to derive an explicit necessary condition for a finite $(2m)$th moment of an HARCH(2) and derive an explicit sufficient condition for the existence of the $(2m)$th moment.

The second part of the paper contains a generalization to HARCH(k) models. As mentioned above, we start by proving an explicit necessary and sufficient condition for a finite 2nd moment. We further give a general theorem for stationarity with finite moments, where we prove that the condition is both necessary and sufficient. The condition is in terms of existence of a nonnegative solution of a certain system of linear equations. For higher order moments of HARCH(k) models we give an explicit necessary condition and an explicit sufficient condition for a finite $(2m)$th moment.

2. Motivation

The HARCH process has been developed for describing the behavior of financial time series, with price quotes from the foreign exchange market being the best-studied example; see [Müller et al., 1997]. Financial time series exhibit clusters of high and low volatility, i.e. autoregressive conditional heteroskedasticity.

Additional facts have been found in recent empirical research of high-frequency data in finance: (1) a long memory in the volatility (positive autocorrelation of absolute price changes decaying slower than exponentially) and (2) asymmetry between volatilities observed with different time resolutions. The HARCH process precisely reproduces these empirical facts.

The asymmetry between different volatilities deserves some special attention here as it is a quite newly detected effect. Figure 1 shows that the

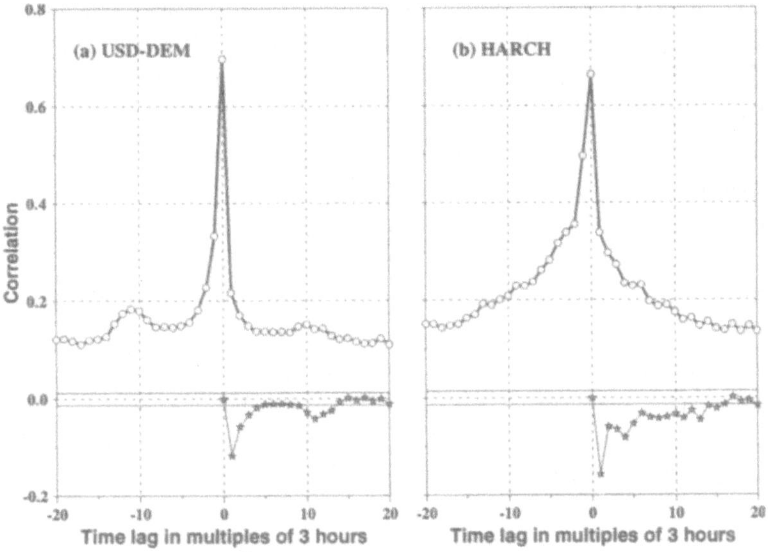

Figure 1: *Asymmetric lead-lag correlation of fine and coarse volatilities of a half-hourly USD/DEM series (a) compared to a similar time series generated by Monte-Carlo simulation using the HARCH model (b). Lead-lag correlations of fine and coarse volatilities are computed with a half-hourly grid in ϑ-time as defined in [Guillaume et al., 1997]. The fine volatility is defined as the mean absolute half-hourly price change within 3 hours (ϑ-time); the coarse volatility is the absolute price change over a whole 3 hour interval (ϑ-time). The thin curve indicates the asymmetry: the difference between correlations at positive and corresponding negative lags. Sampling period: 8 years, from 1 Jan 1987 00:00 to 1 Jan 1995 00:00 (GMT). The confidence limits represent the 95% confidence interval of a Gaussian random walk.*

lead-lag correlation between a high-resolution volatility and another volatility measured with lower resolution in time is asymmetric: low-resolution (coarse) volatility predicts high-resolution (fine) volatility better than the other way around. In Figure 1 (a), the case of the US Dollar expressed in German Marks (USD/DEM) is shown, but the effect is found also for all other foreign exchange rates we studied.

Figure 1 (b) shows the case of a time series synthetically generated from a HARCH process whose 8 parameters were fitted on a long USD/DEM sample. This synthetic data exhibits the same asymmetric behavior of lead-lag correlation as the empirical USD/DEM data in Figure 1 (a), whereas the other ARCH-type processes lead to symmetric lead-lag correlograms. More details on this study can be found in [Müller et al., 1997].

The correct reproduction of many empirical properties makes HARCH an attractive choice for modeling the behavior of financial time series. In order to use a HARCH model in practice, however, some basic theoretical properties of HARCH should be known; among these certainly the conditions for the stationarity and the existence of finite higher moments of the unconditional distribution function are crucial.

There is yet another motivation for studying the moment conditions. Financial risk analysis is based upon assumptions on the probability of extreme price movements. Extreme events in the tails of the distribution function are strongly related to the existence of the higher moments. If the $(2m)$th moment of a distribution is finite and the $(2(m+1))$th moment is not, the tail index of the distribution, if this distribution has a regularly varying tail, must lie between $2m$ and $2(m+1)$. The knowledge of the moment conditions might help to use HARCH in financial risk management.

3. The HARCH process

The HARCH(k) process equations as first presented in [Müller et al., 1997] are: specify $r_0, \ldots, r_{k-1}$ and

$$\begin{aligned} r_n &= \sigma_n \, \varepsilon_n \;, \; n = 0, 1, 2, \ldots \;, \\ \sigma_n^2 &= c_0 + \sum_{j=1}^{k} c_j \left(\sum_{i=1}^{j} r_{n-i} \right)^2 , \\ c_0 > 0 \;, \; c_k &> 0 \;, \; c_j \geq 0 \text{ for } j = 1 \ldots k-1 \;, \end{aligned} \tag{3.1}$$

where r_n is the return of a financial asset time series as defined in [Guillaume et al., 1997]. The ϵ_n's are independent identically distributed (iid) symmetric random variables, whose distribution is assumed to have a non-zero absolutely continuous component, which has a positive density on an interval $(-a, a)$ for some $a > 0$. Typical examples are a normal distribution $N(0, \sigma^2)$ or a Student-t distribution with zero expectation. Denote $a_m = \mathrm{E}(|\epsilon|^{2m})$, $m \geq 1$. For example, if the ϵ_n's have the standard normal distribution, then it is well known that $a_m = \prod_{j=1}^{m}(2j-1)$, $m \geq 1$. We will always assume that ϵ_n has a finite absolute moment of at least the same order as we want r_n to have. (That is, if we are discussing conditions for existence of the finite 6th moment of r_n, we will assume that $\mathrm{E}(|\epsilon_n|^6) < \infty$.) In

[Müller et al., 1997] the moment conditions deduced in this paper were first announced.

In Figure 2, we present a realization of a HARCH process fitted to the returns of the USD/DEM foreign exchange rate measured over 30 minute time intervals. The figure displays 1 month of 30 minutes re-

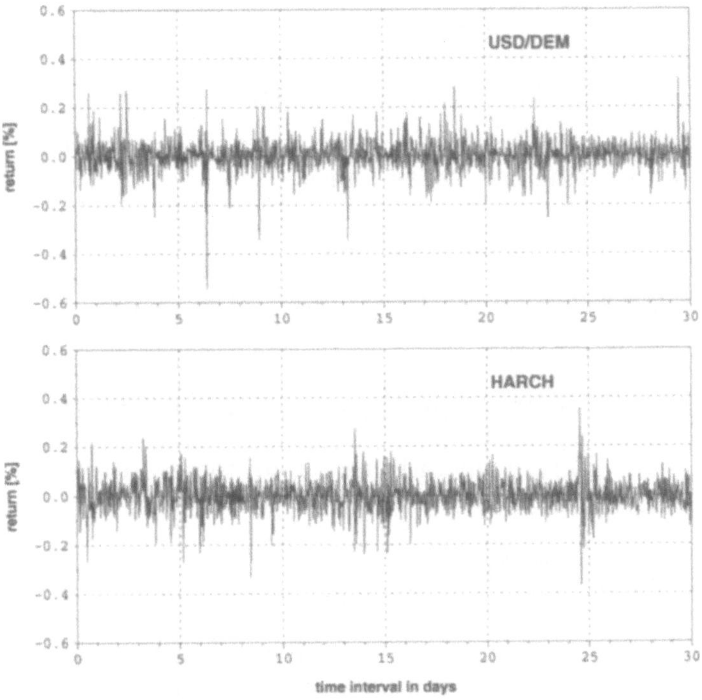

Figure 2: *A comparison between 1 month of 30 minutes (in business time scale) returns of the USD/DEM foreign exchange rate and a Monte-Carlo realization of a HARCH process fitted to the same rate over 30 minute time intervals.*

turns for the real FX rate (measured on the business time scale described in [Dacorogna et al., 1993]) and a Monte-Carlo realization of the fitted HARCH process. Both graphs present fat-tails and heteroskedasticity although not exactly at the same place since the realization of the HARCH process is drawn with a random number generator for the ε_n. The similarity in behavior between the two curves is striking and is confirmed by other statistical analyses see [Müller et al., 1997].

4. Condition for stationarity with a finite 2nd moment for HARCH(2)

4.1. Necessity

Let us first recall the way to get to the condition for the existence of a stationary distribution with a finite 2nd moment for a HARCH(2) process. The equation of such a process is

$$\begin{aligned} r_n &= \sigma_n \varepsilon_n, \; n = 0, 1, 2, \dots \; , \\ \sigma_n^2 &= c_0 \; + \; c_1 r_{n-1}^2 \; + \; c_2 (r_{n-1} + r_{n-2})^2 \; . \end{aligned} \tag{4.1}$$

The scaled returns ε_n satisfy the assumptions of the previous section. If r_n is stationary with a finite 2nd moment, then

$$\mathrm{E}(r_n^2) \quad = \quad \mathrm{E}(r_{n-i}^2) \quad , \quad i \geq 1 \, . \tag{4.2}$$

Recall that $a_1 = \mathrm{E}(\epsilon_n)^2$. If we now use (4.1), we can write

$$\begin{aligned} \mathrm{E}(r_n^2) &= a_1 \; \mathrm{E}(\sigma_n^2) \qquad\qquad (4.3) \\ &= a_1 \left(c_0 \; + \; (c_1 + c_2) \mathrm{E}(r_{n-1}^2) \; + \; c_2 \mathrm{E}(r_{n-2}^2) \; + \; 2c_2 \mathrm{E}(r_{n-1} r_{n-2}) \right) \; . \end{aligned}$$

Since we know that ε_n is iid with mean zero, the expectation of the cross product is zero and using (4.2) we obtain

$$\mathrm{E}(r_n^2) \quad = \quad \frac{c_0}{\frac{1}{a_1} - [(c_1 + c_2) + c_2]} \; . \tag{4.4}$$

The reason why we write the denominator this way will become clear later. A necessary condition for stationarity with a finite 2nd moment for a HARCH(2) process then becomes

$$(c_1 + c_2) + c_2 \quad < \quad \frac{1}{a_1} \tag{4.5}$$

which is equivalent to (3.6) in [Müller et al., 1997] when $k = 2$ (see condition (6.10) for the reason for writing the left hand side above in this form).

4.2. Sufficiency of the condition $(c_1 + c_2) + c_2 < 1/a_1$

The derivation of the expression of $\mathrm{E}(r_n^2)$ in (4.4) as a function of the coefficients c_1 and c_2 is not sufficient to prove the existence of a stationary version of the HARCH(2) process with finite second moments. In order to do this, we need to reformulate the problem in terms of a Markov chain. We do not follow here the path chosen by [Engle, 1982] and [Bollerslev, 1986] because of the cross term $r_{n-1} r_{n-2}$ which makes the matrix formulation of the problem difficult.

With the HARCH(2) process (r_n) we can associate a two-dimensional Markov chain

$$X_n = (r_{n-1}, r_n),\ n \geq 1\,. \tag{4.6}$$

The various properties of (r_n) can now readily be derived through standard results on Markov chains with state space $\mathbb{R}^2$. We base our analysis strongly on the excellent book by Meyn and Tweedie (1993), where one can find many results that give sufficiency of various recurrence conditions and existence of stationary distributions with finite moments. The latter, in our language, means existence of a stationary solution to (4.1) with appropriate finite moments.

Let us summarize the main results. Let $\{X_n,\, n \geq 0\}$ be a Markov chain, with values in an Euclidean space E and transition probabilities $P(x, A)$, $x \in E$ and A a measurable set. For such Markov chains, the key notion is that of an *irreducible T-chain.* Because of the conditions on the density of the noise variables ε_n's, the Markov chain we are working with (in the case of the HARCH(2) process this is the Markov chain in (4.6)) are irreducible, with an irreducibility measure being the Lebesgue measure in a sufficiently small neighborhood of the origin. For example, for the Markov chain corresponding to the HARCH(2) process, such a neighborhood can be $c_0(-a, a)^2$. Therefore, by part (iii) of Theorem 6.0.1 in [Engle, 1982] the property of being an irreducible T-chain is equivalent to showing the weak Feller property, which is equivalent to the following weak continuity property:

(C) The conditional distribution of X_n given $X_{n-1} = y_k$ converges weakly to that of X_n given $X_{n-1} = y$ if $y_k \to y$.

See also the relevant definitions on irreducible T-chains (pp. 87, 127) and the result that for a T-chain; every compact set is *petite* (p. 121).

There are three levels of theorems in [Meyn and Tweedie, 1993] that will bring us to the desired result. We combine these results in the following lemma.

Lemma 1 *(i) Suppose that there exists a real $V \geq 0$ with $V \not\equiv 0$ and a compact set C such that*

$$V(X_n) - PV(X_n) \geq 0 \text{ for } X_n \notin C, \tag{4.7}$$

where P expresses the expectation value at the following step of the Markov chain, i.e. $PV(x) = \mathrm{E}(V(X_n)|X_{n-1} = x) = \int_E V(y)P(x, dy)$. Then there exists an invariant measure and the Markov chain is Harris recurrent.

(ii) Suppose that there exists a real $V \geq 0$ and $V \not\equiv 0$ and a compact set C such that

$$V(X_n) - PV(X_n) \geq -b\, 1_C(X_n) + 1 \tag{4.8}$$

where 1_C is the indicator function of the set C (1 if in C, 0 if outside), and b a finite positive number. Then there exists a unique stationary distribution π and (X_n) is positive Harris recurrent.

(iii) Suppose that X_n is positive Harris recurrent and there exist non-negative measurable functions f, V so that

$$V(X_n) - PV(X_n) \;\geq\; f(X_n) \;-\; b \tag{4.9}$$

with a finite $b \;\geq\; 0$. Then the π-expectation of f, $\pi f \;=\; \int f(x)\,\pi(dx)$, is finite, specifically: $\pi f \;\leq\; b < \infty$.

Part (*i*) of the lemma is, essentially Theorem 12.3.3 of [Meyn and Tweedie, 1993]. Part (*ii*) of the lemma is essentially Theorem 11.3.4 or 13.0.1(iv) of [Meyn and Tweedie, 1993], and it gives a sufficient condition for existence of a stationary version of a Markov chain. Part (*iii*) of the lemma is essentially Theorem 14.3.7 of [Meyn and Tweedie, 1993], and it gives a sufficient condition for the existence of the appropriate moments of the stationary distribution.

To prove the ergodicity and existence of moments of HARCH(2), or, indeed, for any HARCH(k), we thus need to prove the weak continuity (C) of the distribution of X_n given $X_{n-1} = y$ in y, and then to find a condition on the coefficients of the model so that parts (*ii*) and (*iii*) hold. The main idea behind the above approach is to prove that the Markov chain tends to drift back to a neighbourhood of its initial position from any position far out.

4.2.1. Continuity condition (C) on X

We consider a Markov chain (X_n) as defined in (4.6). To prove that it is a T-chain, we need to examine the continuity condition (C). Because of (4.1) and (4.6) every new state is described by analytic functions of the previous state and a new random variable. Thus the weak continuity of the Markov chain holds by the continuous mapping theorem (see [Billingsley, 1968], Theorem 5.1, p. 30).

4.2.2. Drift from the tails to the center

We now turn to show that our Markov chain eventually tends to drift back to the center if the coefficients c_1, c_2 satisfy condition (4.5), $c_1 + 2c_2 < 1/a_1$. We start by choosing a number $0 < \alpha < 1$ such that

(i) $\alpha \;>\; c_2\, a_1$,

(ii) $c_1 + c_2 + \alpha/a_1 < 1/a_1$.

We then define a function V as

$$V(x,y) \;\equiv\; \alpha x^2 \;+\; y^2 \;+\; 2c_2 a_1 xy\;, \tag{4.10}$$

where, because of condition (i) above, $V \geq 0$. Indeed the discriminant satisfies

$$c_2^2 a_1^2 \;-\; \alpha \;<\; c_2 a_1 \;-\; \alpha \;<\; 0\,. \tag{4.11}$$

By inserting (4.10) and (4.1),

$$\begin{aligned} V(r_{n-1}, r_n) - PV(r_{n-1}, r_n) \;&=\; \alpha r_{n-1}^2 \;+\; 2c_2 a_1 r_{n-1} r_n \;+\; r_n^2 \\ &-\; E\Big[\, \alpha r_n^2 + 2c_2 a_1 r_n \varepsilon_{n+1}(c_0 + c_1 r_n^2 + c_2(r_n + r_{n-1})^2)^{1/2} \\ &+\; \varepsilon_{n+1}^2(c_0 + c_1 r_n^2 + c_2(r_n + r_{n-1})^2) \,\Big]\,. \end{aligned} \tag{4.12}$$

Using $\mathrm{E}(\varepsilon_n) = 0$ and $\mathrm{E}(\varepsilon_n^2) = a_1$, we have

$$V(r_{n-1}, r_n) - PV(r_{n-1}, r_n) \;=\; (\alpha - c_2 a_1) r_{n-1}^2 + (1 - \alpha - (c_1 + c_2) a_1) r_n^2 - c_0\, a_1\,. \tag{4.13}$$

The two coefficients $\alpha - c_2 a_1$ and $1 - \alpha - (c_1 + c_2) a_1$ are positive, and the last expression can be made as large as we wish if (r_{n-1}, r_n) is outside of a compact set.

Now, under condition (4.5), we have found a function V that satisfies the requirements (ii) and (iii) in Lemma 1 with $f(x_1, x_2) = \theta(x_1^2 + x_2^2)$, with $0 < \theta < (\alpha - c_2 a_1) \wedge (1 - \alpha - (c_1 + c_2))$. Thus the stationarity and the existence of the second moment are proven if (4.5) holds.

5. Existence of the 4th moment of a HARCH(2)

5.1. A necessary condition

Since $r_n = \sigma_n \varepsilon_n$, we deduce that

$$\mathrm{E}(r_n^4) \;=\; \mathrm{E}(\sigma_n^4)\, \mathrm{E}(\varepsilon_n^4) \;=\; a_2\, \mathrm{E}(\sigma_n^4). \tag{5.1}$$

Therefore,

$$\mathrm{E}(\sigma_n^4) \;=\; \frac{1}{a_2}\, \mathrm{E}(r_n^4). \tag{5.2}$$

With the help of (4.1), we obtain

$$\begin{aligned} \mathrm{E}(\sigma_n^4) \;&=\; \mathrm{E}([c_0 + c_1 r_{n-1}^2 + c_2(r_{n-1} + r_{n-2})^2]^2) \\ &=\; c_0^2 \;+\; 2c_0(c_1 + c_2)\mathrm{E}(r_{n-1}^2) \;+\; (c_1 + c_2)^2 \mathrm{E}(r_{n-1}^4) \;+\; 2c_0 c_2 \mathrm{E}(r_{n-2}^2) \\ &+\; 2c_2(c_1 + 3c_2)\mathrm{E}(r_{n-1}^2 r_{n-2}^2) \;+\; c_2^2 \mathrm{E}(r_{n-2}^4)\,, \end{aligned} \tag{5.3}$$

where we have used

$$\mathrm{E}(r_{n-1} r_{n-2}) \;=\; \mathrm{E}(r_{n-1}^3 r_{n-2}) \;=\; \mathrm{E}(r_{n-1} r_{n-2}^3) \;=\; 0, \tag{5.4}$$

deduced from the iid nature of ε_n's and their symmetric distribution. For ease of notation, we define

$$L = \mathrm{E}(r_n^2)\,, \quad M_0 = \mathrm{E}(r_n^4) \quad \text{and} \quad M_2 = \mathrm{E}(r_{n-1}^2 r_{n-2}^2)\,, \tag{5.5}$$

which do not depend on n if the process is stationary. Rewrite (5.3) as

$$\frac{1}{a_2} M_0 = c_0^2 + 2c_0(c_1 + 2c_2)\, L + (c_2^2 + (c_1 + c_2)^2)\, M_0 + 2c_2(c_1 + 3c_2)\, M_2$$

and compute, in a similar way, an expression for M_2:

$$\begin{aligned} M_2 &= \mathrm{E}(r_n^2 r_{n-1}^2) = \mathrm{E}(\sigma_n^2 r_{n-1}^2) E(\varepsilon_n^2) = a_1\, \mathrm{E}(\sigma_n^2 r_{n-1}^2) \\ &= a_1\, \mathrm{E}(r_{n-1}^2 [c_0 + c_1 r_{n-1}^2 + c_2(r_{n-1} + r_{n-2})^2]) \\ &= a_1 \Big(c_0\, L + (c_1 + c_2)\, M_0 + c_2\, M_2 \Big)\,. \end{aligned} \tag{5.6}$$

We are now left with a system of two equations with two variables M_0 and M_2 (the quantity L is known from the stationarity condition (4.4)):

$$\begin{cases} [1 - a_2(c_2^2 + (c_1 + c_2)^2)] M_0 - 2a_2 c_2(c_1 + 3c_2) M_2 \\ \qquad\qquad = a_2 c_0^2 + 2a_2 c_0(c_1 + 2c_2) L, \\ -(c_1 + c_2) M_0 + (\frac{1}{a_1} - c_2) M_2 = c_0 L. \end{cases} \tag{5.7}$$

This linear system can be solved, yielding

$$M_0 = \frac{a_2 c_0^2 (1 - a_1 c_2) + 2a_2 c_0 (c_1 + 2c_2 + a_1 c_2^2)\, L}{(1 - a_1 c_2) - a_2[c_2^2 + (c_1 + c_2)^2] - a_1 a_2 c_2 (c_1^2 + 6c_1 c_2 + 4c_2^2)}\,. \tag{5.8}$$

Because of the condition (4.5), the numerator is positive and hence a necessary condition for the existence of a stationary HARCH(2) process with finite 4th moment becomes

$$1 - a_2[c_2^2 + (c_1 + c_2)^2] - a_1 c_2 [1 + a_2(c_1^2 + 6c_1 c_2 + 4c_2^2)] > 0\,. \tag{5.9}$$

This condition is more stringent than (4.5). The solution for M_2 is

$$M_2 = \frac{a_1 a_2 c_0^2 (c_1 + c_2)}{D} + \frac{2a_1 a_2 c_0 (c_1 + c_2)(c_1 + 2c_2 + a_1 c_2^2)}{(1 - a_1 c_2)\, D} L + \frac{a_1 c_0}{1 - a_1 c_2} L\,, \tag{5.10}$$

where D is the denominator of the right-hand side of (5.8).

An easy check of this condition is to compare it to the condition for an ARCH(1) process. We know that HARCH(2) becomes ARCH(1) if $c_2 = 0$. In (Engle, 1982; p. 992), the condition for the existence of the $2m$th moment is given. Assume that the ϵ_n's have the standard normal distribution. In that case, translated into our notation that condition becomes

$$c_1^m \prod_{j=1}^{m} (2j-1) \ < \ 1. \tag{5.11}$$

In the case of interest, the 4th moment, we get

$$3c_1^2 \ < \ 1 \quad \Rightarrow \quad c_1^2 \ < \ \frac{1}{3},$$

the same condition as in (5.9) if $c_2 = 0$ (and $a_2 = 3$).

The fact that (5.9) is also sufficient for the existence of the 4th moment will follow from the general theory for HARCH(2) models to which we now turn.

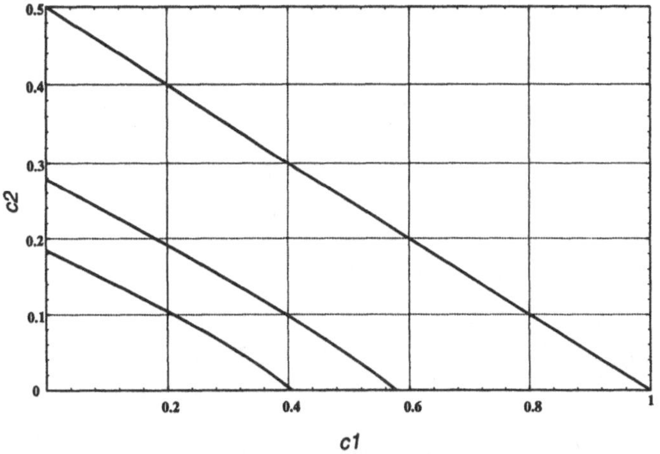

Figure 3: *Moment conditions for HARCH(2) processes. The straight line on the right represents the boundary for the 2nd moment condition. The middle curve is the boundary for the existence of a finite 4th moment. The left curve represents the boundary for the existence condition of the 6th moment.*

6. (2m)th moment of HARCH(2)

6.1. The system of equations

To get a necessary condition for the existence of a stationary distribution of the HARCH(2) process with a finite (2m)th moment, $m \geq 2$, we denote

$$\begin{aligned} M_0 &= \mathrm{E}(r_n^{2m}), \\ M_i &= \mathrm{E}(r_n^{2(m-i)} r_{n-1}^{2i}), \quad i = 1, \ldots, m-1 \end{aligned} \tag{6.1}$$

under stationarity conditions, in which case there is no dependence on n. If the (2m)th moment is finite, then all the moments of smaller orders are finite, $a_m < \infty$, and $M_0, M_1, \ldots, M_{m-1}$ above are all finite. Observe that

$$\mathrm{E}(r_n^{2m}) = \mathrm{E}(\sigma_n^{2m})\, \mathrm{E}(\varepsilon_n^{2m}), \tag{6.2}$$

and so using (6.2) and $a_m = \mathrm{E}(\varepsilon_n^{2m})$ we have that

$$\begin{aligned} M_0 &= \mathrm{E}(r_n^{2m}) = a_m\, \mathrm{E}\,[\, (c_0 + c_1 r_{n-1}^2 + c_2(r_{n-1} + r_{n-2})^2)^m \,] \\ &= a_m\, \mathrm{E}\,[\, (c_0 + (c_1 + c_2) r_{n-1}^2 + c_2 r_{n-2}^2 + 2c_2 r_{n-1} r_{n-2})^m \,] \\ &= a_m \sum_{j=1}^{m} \binom{m}{j} c_0^j\, \mathrm{E}\,[\, ((c_1 + c_2) r_{n-1}^2 + c_2 r_{n-2}^2 + 2c_2 r_{n-1} r_{n-2})^{m-j} \,] \\ &\quad + a_m\, \mathrm{E}\,[\, ((c_1 + c_2) r_{n-1}^2 + c_2 r_{n-2}^2 + 2c_2 r_{n-1} r_{n-2})^m \,] \\ &\equiv a_m \vartheta_{0,m} + a_m\, \mathrm{E}\,[\, ((c_1 + c_2) r_{n-1}^2 + c_2 r_{n-2}^2 + 2c_2 r_{n-1} r_{n-2})^m \,], \end{aligned} \tag{6.3}$$

where $\vartheta_{0,m}$ involves only moments of order less than $2m$. Moreover, we have

$$\begin{aligned} &\mathrm{E}[\, ((c_1 + c_2) r_{n-1}^2 + c_2 r_{n-2}^2 + 2c_2 r_{n-1} r_{n-2})^m \,] \\ &= \sum_{j=0}^{[m/2]} \binom{m}{2j} (2c_2)^{2j}\, \mathrm{E}\,[\, (r_{n-1}^{2j} r_{n-2}^{2j}) \left((c_1 + c_2) r_{n-1}^2 + c_2 r_{n-2}^2 \right)^{m-2j} \,] \\ &= \sum_{j=0}^{[m/2]} \binom{m}{2j} (2c_2)^{2j}\, \mathrm{E}\,[\, (r_{n-1}^{2j} r_{n-2}^{2j}) \\ &\quad \times \sum_{i=0}^{m-2j} \binom{m-2j}{i} (c_1 + c_2)^i\, c_2^{m-2j-i}\, r_{n-1}^{2i}\, r_{n-2}^{2(m-2j-i)} \,] \\ &= \sum_{j=0}^{[m/2]} \sum_{i=0}^{m-2j} \frac{m!}{(2j)!\, i!\, (m-2j-i)!}\, 2^{2j}\, (c_1 + c_2)^i\, c_2^{m-i}\, \mathrm{E}(r_{n-1}^{2(i+j)}\, r_{n-2}^{2(m-j-i)}) \end{aligned}$$

$$= \sum_{j=0}^{[m/2]} \sum_{i=0}^{m-2j} \frac{m!}{(2j)!\, i!\, (m-2j-i)!} 2^{2j} \, (c_1+c_2)^i \, c_2^{m-i} \, M_{m-(i+j)} \,, \tag{6.4}$$

where $M_m \equiv M_0$, and $[\cdot]$ is the entire function. By replacing (6.4) in (6.3), we obtain

$$M_0 = a_m \vartheta_{0,m} + a_m \sum_{j=0}^{[m/2]} \sum_{i=0}^{m-2j} \frac{m!}{(2j)!\, i!\, (m-2j-i)!} 2^{2j} (c_1+c_2)^i c_2^{m-i} M_{m-(i+j)}. \tag{6.5}$$

Furthermore, for every $i = 1, ..., m-1$,

$$\begin{aligned} M_i &= \mathrm{E}(r_n^{2(m-i)} r_{n-1}^{2i}) = \mathrm{E}(\varepsilon_n^{2(m-i)}) \, \mathrm{E}(r_{n-1}^{2i} \sigma_n^{2(m-i)}) \\ &= a_{m-i} \, \mathrm{E} \left[\, r_{n-1}^{2i} (c_0 + (c_1+c_2) r_{n-1}^2 + c_2 r_{n-2}^2 + 2c_2 r_{n-1} r_{n-2})^{m-i} \, \right] \\ &= a_{m-i} \sum_{j=1}^{m-i} \binom{m-i}{j} \\ &\quad \times c_0^j \, \mathrm{E} \left[\, r_{n-1}^{2i} ((c_1+c_2) r_{n-1}^2 + c_2 r_{n-2}^2 + 2c_2 r_{n-1} r_{n-2})^{m-i-j} \, \right] \\ &\quad + \; a_{m-i} \, \mathrm{E} \left[\, r_{n-1}^{2i} ((c_1+c_2) r_{n-1}^2 + c_2 r_{n-2}^2 + 2c_2 r_{n-1} r_{n-2})^{m-i} \, \right] \\ &= a_{m-i} \, \vartheta_{i,m} \\ &\quad + \; a_{m-i} \, \mathrm{E} \left[\, r_{n-1}^{2i} ((c_1+c_2) r_{n-1}^2 + c_2 r_{n-2}^2 + 2c_2 r_{n-1} r_{n-2})^{m-i} \, \right] , \end{aligned} \tag{6.6}$$

where $\vartheta_{i,m}$ involves only moments of orders less than $2m$. Now,

$$\begin{aligned} &\mathrm{E} \left[\, r_{n-1}^{2i} ((c_1+c_2) r_{n-1}^2 + c_2 r_{n-2}^2 + 2c_2 r_{n-1} r_{n-2})^{m-i} \, \right] \\ &= \sum_{j=0}^{[\frac{m-i}{2}]} (2c_2)^{2j} \binom{m-i}{2j} \mathrm{E} \left[\, r_{n-1}^{2(i+j)} r_{n-2}^{2j} ((c_1+c_2) r_{n-1}^2 + c_2 r_{n-2}^2)^{m-i-2j} \right] \\ &= \sum_{j=0}^{[\frac{m-i}{2}]} (2c_2)^{2j} \binom{m-i}{2j} \sum_{d=0}^{m-i-2j} \binom{m-i-2j}{d} (c_1+c_2)^d \, c_2^{m-i-2j-d} \\ &\quad \times \mathrm{E} \left[\, r_{n-1}^{2(i+j)} r_{n-2}^{2j} r_{n-1}^{2d} r_{n-2}^{2(m-i-2j-d)} \, \right] \\ &= \sum_{j=0}^{[\frac{m-i}{2}]} \sum_{d=0}^{m-i-2j} \frac{(m-i)!}{(2j)!\, d!\, (m-i-2j-d)!} 2^{2j} \, (c_1+c_2)^d \, c_2^{m-i-d} \, M_{m-(i+j+d)} \,. \end{aligned} \tag{6.7}$$

Therefore,

$$M_i = a_{m-i}\,\vartheta_{i,m} + a_{m-i} \sum_{j=0}^{[\frac{m-i}{2}]} \sum_{d=0}^{m-i-2j} \frac{(m-i)!}{(2j)!\,d!\,(m-i-2j-d)!} \times 2^{2j}\,(c_1+c_2)^d\,c_2^{m-i-d}\,M_{m-(i+j+d)}\,, \quad i = 1,\ldots, m-1\,. \tag{6.8}$$

Equations (6.5) and (6.8) form a system of m equations with m unknowns $(M_0, M_1, \ldots, M_{m-1})$. Let us call this system ϱ_m.

6.2. The main theorem: necessary and sufficient moment conditions for HARCH(2)

The following theorem presents a general result on the existence of a stationary version of the HARCH(2) process with a finite $(2m)$th moment.

Theorem 1. *The HARCH(2) process has a stationary version with a finite $(2m)$th moment, if and only if, for every $j = 1, \ldots, m$, the system ϱ_j has a positive solution, with $\vartheta_{i,j}$, $j = 2, \ldots, m$, $i = 0, \ldots, j-1$, determined by solving $\varrho_1, \ldots, \varrho_{j-1}$, $j = 2, \ldots, m$. If this is the case, then ϱ_m has a unique solution given by (6.1).*

Proof. The proof is by induction on m. For $m = 1$, we have already proved this statement. Assume now that this statement is also true for $m \geq 1$, and let us prove it for $m + 1$.

Assume first that the the process is stationary, and the $(2(m+1))$th moment is finite. Then the $(2m)$th moment is finite as well, and so by the assumption of the induction, $\varrho_1, \ldots, \varrho_m$ all have a unique (positive) solution given by (6.1), and we use these solutions to compute $\vartheta_{i,m+1}$ for $i = 0, 1, \ldots, m$. Since $M_0, M_1, \ldots, M_m$, defined in (6.1), obviously yield a positive solution to ϱ_{m+1}, we only need to prove that this solution is unique.

First of all, suppose that there is a nonnegative solution $\underline{y}$ to ϱ_{m+1}, such that $y_i < M_i$ for at least one $i = 0, \ldots, m$. We now proceed as follows: take a nonstationary HARCH(2) with the same coefficients, beginning with $r_0 = r_1 = 0$. Observe that all the moments of the type $\mathrm{E}(r_n^{2j} r_{n-1}^{2k})$, $0 \leq j + k \leq m$, are for $n = 1$ less than or equal to their corresponding stationary values (this is just the positivity of the moments), and

$$\mathrm{E}(r_n^{2(m+1-k)} r_{n-1}^{2k}) \leq y_k, \quad \text{with } k = 0, 1, \ldots, m\,, \tag{6.9}$$

(once again this is just the nonnegativity of $\underline{y}$). Since $\varrho_1, \ldots, \varrho_{k+1}$ are systems of equations with nonnegative coefficients, we conclude that the above is true for all $n \geq 1$.

Since we have assumed that the HARCH(2) process (r_n) has a stationary version, it follows immediately that our Markov chain (r_{n-1}, r_n) has a stationary distribution. Moreover, we have proved that the existence of a finite 2nd moment implies (4.5), which yields that the Markov chain (r_{n-1}, r_n) is positive Harris recurrent. Therefore by Theorem 13.0.1 of [Meyn and Tweedie, 1993] we conclude that $(r_{n-1},\ r_n)$ converges weakly to its stationary version. Therefore, so do the products $r_n^{2(m+1-k)} r_{n-1}^{2k}$, $k = 0, 1, \ldots, m$. Hence, by Fatou's lemma, each M_k does not exceed the lowest subsequential limit of $E(r_n^{2(m+1-k)} r_{n-1}^{2k})$, and so $M_k \leq y_k$, $k = 0,\ 1, \ldots,\ m$, which contradicts our assumption on $\underline{y}$.

Therefore $\underline{m}$, given by (6.1), is the smallest nonnegative solution of ϱ_{m+1} and, if ϱ_{m+1} has another nonnegative solution $\underline{y}$, we must have $y_i \geq M_i$ for all $i = 0, 1, \ldots,\ m$, and $y_i > M_i$ for at least one i. Thus for any $\alpha > 0$, $\underline{y}_\alpha = (1+\alpha)\underline{m} - \alpha\underline{y}$ is yet another solution to ϱ_{m+1}. However, if α is small enough, we have $\underline{y}_\alpha \geq 0$, and some components of $\underline{y}_\alpha$ will be less then those of $\underline{m}$. This contradicts the already established fact that $\underline{m}$ is the smallest nonnegative solution to ϱ_{m+1}. Therefore, $\underline{m}$ is the only nonnegative solution of ϱ_{m+1}.

Suppose now that ϱ_{m+1} has another, not necessary nonnegative, solution $\underline{y}$. Consider $\underline{y}_\alpha$ above. For all $|\alpha|$ small enough, $\underline{y}_\alpha \geq 0$, and for no $\alpha \neq 0$ it is equal to $\underline{m}$. Therefore, $\underline{m}$ is the *only* solution of ϱ_{m+1}.

In the opposite direction, assume that the system ϱ_j has a positive solution for all $j = 1, \ldots,\ m+1$. By the assumption of the induction, we know that the model has a stationary distribution with finite $(2m)$th moment, and as above, is positive Harris recurrent. Once again, set $r_0 = r_1 = 0$, and observe that all the moments of the type $\mathrm{E}(r_n^{2i} r_{n-1}^{2k})$, $0 \leq i+k \leq m+1$, do not exceed the corresponding solution of the systems $\varrho_1, \ldots,\ \varrho_{k+1}$ for $n = 1$, and so, by the nonnegativity of the coefficients of these systems of equations, for each $n \geq 1$. By Fatou's lemma, we conclude that the stationary version of $\mathrm{E}(r_n^{2(m+1)})$ does not exceed the corresponding solution of ϱ_{m+1}, and so is finite.

This completes the proof of Theorem 1. ∎

6.3. An explicit necessary and an explicit sufficient condition

We now move to establish some more explicit necessary conditions for the existence of the $(2m)$th moment of the HARCH(2) models. First of all, it follows from (6.5) $(i = j = 0,\ i = m,\ j = 0)$ that

$$M_0 > a_m\,(c_2^m \ +\ (c_1 + c_2)^m) M_0,$$

implying that

$$(c_1 + c_2)^m + c_2^m < \frac{1}{a_m} \tag{6.10}$$

is a necessary condition for the existence of a finite $(2m)$th moment, $m = 1, 2, \ldots$. We can get a stricter necessary condition. It follows from (6.8) with $j = 0$, $d = m - i$ that

$$M_i > a_{m-i}\,(c_1 + c_2)^{m-i}\,M_0, \quad i = 1, \ldots, m-1\,. \tag{6.11}$$

We rewrite (6.5) in the form

$$M_0 = a_m \vartheta_{0,m} + a_m \sum_{d=0}^{m} M_d$$

$$\times \sum_{j=0}^{\mathrm{Min}[d,(m-d)]} 2^{2j}\,(c_1 + c_2)^{m-d-j}\,c_2^{d+j} \frac{m!}{(2j)!\,(m-d-j)!(d-j)!}, \tag{6.12}$$

and substituting (6.11) into (6.12) we obtain

$$\begin{aligned} M_0 &> a_m M_0[(c_1 + c_2)^m + c_2^m] \\ &\quad + a_m \sum_{d=1}^{m-1} M_d \sum_{j=0}^{\mathrm{Min}[d,(m-d)]} 2^{2j}\,(c_1 + c_2)^{m-d-j}\,c_2^{d+j} \\ &\quad \times \frac{m!}{(2j)!\,(m-d-j)!(d-j)!} \\ &> a_m M_0\,[\,((c_1 + c_2)^m + c_2^m) + \sum_{d=1}^{m-1} a_{m-d}\,(c_1 + c_2)^{m-d} \\ &\quad \times \sum_{j=0}^{\mathrm{Min}[d,(m-d)]} 2^{2j}\,(c_1 + c_2)^{m-d-j}\,c_2^{d+j} \frac{m!}{(2j)!\,(m-d-j)!(d-j)!}\,]\,. \end{aligned} \tag{6.13}$$

Then, a necessary condition for existence of a finite $(2m)$th moment is

$$\begin{aligned} &(c_1 + c_2)^m + c_2^m + \sum_{d=1}^{m-1} a_{m-d} \\ &\quad \times \sum_{j=0}^{\mathrm{Min}[d,(m-d)]} 2^{2j}\,(c_1 + c_2)^{2(m-d)-j}\,c_2^{d+j} \frac{m!}{(2j)!\,(m-d-j)!(d-j)!} \\ &< \frac{1}{a_m}, \quad m = 1, 2, \ldots\,. \end{aligned} \tag{6.14}$$

Now we switch to computing a more explicit sufficient condition for the existence of a $(2m)$th moment of the HARCH(2) models. We claim that

$$(c_1 + 4c_2)^m \quad < \quad \frac{1}{a_m} \tag{6.15}$$

is such a sufficient condition. Indeed, define

$$V(x,y) \quad = \quad \alpha x^{2m} + y^{2m}, \quad 0 < \alpha < 1\,. \tag{6.16}$$

It is enough to prove that, under (6.15), with a suitable choice of α we have

$$V(x,y) \; - \; PV(x,y) \; \geq \; \theta(x^{2m} + y^{2m}), \quad \text{with } \theta > 0 \tag{6.17}$$

outside of a compact set, as in Section 4.2. Observe that

$$\begin{aligned} V(x,y) - PV(x,y) \quad &= \quad (\alpha x^{2m} + y^{2m}) - (\alpha y^{2m} + a_m(c_0 + c_1 y^2 + c_2(x+y)^2)^m) \\ &= \quad \alpha x^{2m} + (1-\alpha) y^{2m} - a_m(c_1 y^2 + c_2(x+y)^2)^m - \gamma(x,y)\,, \end{aligned} \tag{6.18}$$

where $\gamma(x,y)$ is a polynomial of a lower order. It is, therefore, enough to prove that there is an α such that for all x, y not both equal to zero,

$$\alpha x^{2m} + (1-\alpha) y^{2m} - a_m(c_1 y^2 + c_2(x+y)^2)^m \; > \; \theta(x^{2m} + y^{2m}), \quad \text{with } \theta > 0\,. \tag{6.19}$$

By the homogeneity of the terms in the inequality (6.19), it suffices to show that for all $(x,y) \neq (0,0)$,

$$\alpha x^{2m} \; + \; (1-\alpha) y^{2m} \; > \; a_m(c_1 y^2 + c_2(x+y)^2)^m\,. \tag{6.20}$$

We are finally ready to specify α. Let

$$\alpha \quad = \quad \frac{1 - a_m^{1/m}\, c_1}{2}\,. \tag{6.21}$$

We have, by convexity of the function $f(t) = t^m$,

$$\alpha x^{2m} \; + \; (1-\alpha) y^{2m} \; \geq \; (\alpha x^2 \; + \; (1-\alpha) y^2)^m\,, \tag{6.22}$$

and so inequality (6.20) will follow once we prove that

$$\alpha x^2 \; + \; (1-\alpha) y^2 \; > \; a_m^{1/m}\,(c_1 y^2 \; + \; c_2(x+y)^2)\,. \tag{6.23}$$

We have

$$\begin{aligned} &\alpha x^2 + (1-\alpha) y^2 \; - \; a_m^{1/m}\,(c_1 y^2 + c_2(x+y)^2) \\ &\quad = \; x^2(\alpha - a_m^{1/m} c_2) - 2a_m^{1/m} c_2 xy + (1 - \alpha - (c_1 + c_2) a_m^{1/m}) y^2\,. \end{aligned} \tag{6.24}$$

Using (6.15) and (6.21), we can write

$$\alpha - a_m^{1/m} c_2 = \frac{1 - a_m^{1/m} (c_1 + 2c_2)}{2} > 0 . \tag{6.25}$$

Moreover,

$$(1-\alpha) - (c_1 + c_2)\, a_m^{1/m} = \frac{1 + a_m^{1/m} c_1}{2} - a_m^{1/m} (c_1 + c_2)$$

$$= \frac{1 - a_m^{1/m} (c_1 + 2c_2)}{2} = \alpha - a_m^{1/m} c_2 > 0 . \tag{6.26}$$

Finally, by (6.25) and (6.26),

$$(\alpha - a_m^{1/m} c_2)\left((1-\alpha) - (c_1 + c_2)\, a_m^{1/m}\right) - a_m^{2/m} c_2^2$$

$$= \left(\frac{1 - a_m^{1/m} (c_1 + 2c_2)}{2}\right)^2 - a_m^{2/m} c_2^2 > 0$$

because, by (6.15),

$$\frac{1 - a_m^{1/m} (c_1 + 4c_2)}{2} > 0 .$$

This proves the inequality (6.23), and so the condition (6.15) is a sufficient condition for the existence of a $(2m)$th moment.

7. Stationarity condition with finite 2nd moment for HARCH(k)

We now discuss the more difficult case: conditions for stationarity and finite moments of the general HARCH(k) process, and we start with the easiest and most explicit part: finiteness of the second moment. The following condition has been shown in [Müller et al., 1997] (relation (3.6)) to be necessary for existence of a finite second moment

$$\sum_{j=1}^{k} j\, c_j < \frac{1}{a_1} . \tag{7.1}$$

We will prove now that this condition is also sufficient.

Let $\alpha_1 = 1$. A simple inductive argument establishes that we can choose $\alpha_2, \ldots, \alpha_k$ in such a way that

$$a_1(S_i + \ldots + S_k) < \alpha_i < \alpha_{i-1} - a_1 S_{i-1}, \quad i = 2, \ldots, k , \tag{7.2}$$

where $S_i = c_i + ... + c_k$ with $i = 1, ..., k$.

Indeed, a condition equivalent to (7.1) is

$$\sum_{j=1}^{k} S_j < \frac{1}{a_1}, \tag{7.3}$$

and so $a_1(S_2 + ... + S_k) < 1 - a_1 S_1$, implying that there is a number strictly between the two, which we declare to be α_2. Assuming that we have chosen $\alpha_1, \ldots, \alpha_i$, $i < k$, we have by (7.2),

$$a_1(S_{i+1} + ... + S_k) < \alpha_i - a_1 S_i , \tag{7.4}$$

allowing us to choose α_{i+1} in the interior of the above interval. Having chosen $\alpha_1, ..., \alpha_k$ as above, we define a function $V : \mathbb{R}^k \longrightarrow \mathbb{R}$ by

$$V(x_1, ..., x_k) = \sum_{i=1}^{k} \alpha_i x_i^2 + 2a_1 \sum_{i=1}^{k} \sum_{j=i+1}^{k} \beta_j x_i x_j, \tag{7.5}$$

where

$$\beta_j = S_j + ... + S_k, \quad j = 2, \ldots, k . \tag{7.6}$$

We claim that $V(x_1, ..., x_k) \geq 0$. To this end we need to show that the matrix

$$A = \begin{pmatrix} \alpha_1 & a_1\beta_2 & a_1\beta_3 & \cdots\cdots & a_1\beta_k \\ a_1\beta_2 & \alpha_2 & a_1\beta_3 & \cdots\cdots & a_1\beta_k \\ \cdot & & & & \\ \cdot & & & & \\ \cdot & & & & \\ a_1\beta_k & a_1\beta_k & a_1\beta_k & \cdots\cdots & \alpha_k \end{pmatrix} \tag{7.7}$$

is non-negatively definite. For this, it suffices to note that A is the covariance matrix of the following Gaussian random vector

$$X_i = a_1^{1/2} \sum_{j=i}^{k} S_j^{1/2} G_j + (\alpha_i - a_1(S_i + ... + S_k))^{1/2} U_i, \quad i = 1, \ldots, k, \tag{7.8}$$

where $G_1, ..., G_k$, $U_1, ..., U_k$ are iid $N(0,1)$ random variables.

Now, we turn to the quantity needed for the different levels described in Subsection 4.2.2

$$V(r_{n-k+1}, ..., r_n) - PV(r_{n-k+1}, ..., r_n)$$

$$= \sum_{i=1}^{k} \alpha_i \, r_{n+1-i}^2 \; + \; 2a_1 \sum_{i=1}^{k} \sum_{j=i+1}^{k} \beta_j \, r_{n+1-i} \, r_{n+1-j}$$

$$- E \left[\sum_{i=2}^{k} \alpha_i \, r_{n+2-i}^2 \; + \; \alpha_1 \, \varepsilon_{n+1}^2 \, (c_0 + \sum_{i=1}^{k} c_i \, (\sum_{j=1}^{i} r_{n+1-j})^2 \,) \right.$$

$$+ 2a_1 \sum_{i=2}^{k} \sum_{j=i+1}^{k} \beta_j \, r_{n+2-i} \, r_{n+2-j}$$

$$\left. + \, 2\varepsilon_{n+1} \, (c_0 + \sum_{i=1}^{k} c_i \, (\sum_{m=1}^{i} r_{n+1-m})^2 \,)^{1/2} \sum_{j=2}^{k} \beta_j r_{n+2-j} \right] . \tag{7.9}$$

Using similar arguments as in the case $k = 2$, we can simplify the above as

$$V(r_{n-k+1}, ..., r_n) - PV(r_{n-k+1}, ..., r_n)$$

$$= \sum_{i=1}^{k} \alpha_i \, r_{n+1-i}^2 \; + \; 2a_1 \sum_{i=1}^{k} \sum_{j=i+1}^{k} \beta_j \, r_{n+1-i} \, r_{n+1-j}$$

$$- \sum_{i=2}^{k} \alpha_i \, r_{n+2-i}^2 \; - \; a_1(c_0 \; - \; \sum_{i=1}^{k} c_i \, (\sum_{j=1}^{i} r_{n+1-j})^2)$$

$$- \, 2a_1 \sum_{i=2}^{k} \sum_{j=i+1}^{k} \beta_j \, r_{n+2-i} \, r_{n+2-j} \, . \tag{7.10}$$

Setting $\alpha_{k+1} = \beta_{k+1} = 0$, we can rewrite the expression in terms of the S_i and we obtain

$$V(r_{n-k+1}, ..., r_n) - PV(r_{n-k+1}, ..., r_n)$$

$$= \sum_{i=1}^{k} (\alpha_i - \alpha_{i+1} - a_1 S_i) \, r_{n+1-i}^2$$

$$- \, 2a_1 \sum_{i=1}^{k} \sum_{j=i+1}^{k} (\beta_j - \beta_{j+1} - S_j) \, r_{n+1-i} \, r_{n+1-j} \; - \; c_0 \, . \tag{7.11}$$

By (7.6), we conclude that

$$V(r_{n-k+1}, ..., r_n) - PV(r_{n-k+1}, ..., r_n) = \sum_{i=1}^{k} (\alpha_i - \alpha_{i+1} - a_1 S_i) \, r_{n+1-i}^2 - c_0 \, . \tag{7.12}$$

Let

$$\vartheta_i = \alpha_i \; - \; \alpha_{i+1} \; - \; a_1 S_i, \quad i \; = \; 1, \ldots, \; k \, . \tag{7.13}$$

It follows from (7.2) that $\vartheta_i > 0$, $i = 1, ..., k$, and so

$$\Theta = \min_{i=1,..,k} \vartheta_i > 0 . \tag{7.14}$$

Letting

$$f(x_1, ..., x_k) = x_1^2 + \cdots + x_k^2 \tag{7.15}$$

and denoting by B_a $(a > 0)$ the sphere centered at 0 with radius a, we conclude from (7.12) that

$$\begin{aligned} &V(r_{n-k+1}, ..., r_n) - PV(r_{n-k+1}, ..., r_n) \\ &\quad \geq \Theta f(x_1, ..., x_k)/2 - c_0 1_{B(2c_0/\Theta)}(x_1, \ldots, x_k) \end{aligned} \tag{7.16}$$

for all $(x_1, \ldots, x_k)$. Therefore, all levels of Section 4.2 (Lemma 1) are satisfied for this Markov chain. That is, the stationary HARCH(k) process has, under the condition (7.1), finite second moments.

8. Existence of the 4th moment of a general HARCH(k)

8.1. The system of equations

Assume first, that the 2nd moments are finite. That is, we assume that the condition (7.1) holds and denote

$$\begin{aligned} L &= \mathrm{E}(r_n^2), \\ M_0 &= \mathrm{E}(r_n^4), \\ M_i &= \mathrm{E}(r_n^2 r_{n-i}^2), \qquad i = 1, \ldots, k-1 , \\ n_{ij} &= \mathrm{E}(r_n^2 r_{n-i} r_{n-j}), \quad i = 1, \ldots, k-2, \quad j = i+1, \ldots, k-1 . \end{aligned}$$

We have

$$n_{ij} = \mathrm{E}(r_n^2 r_{n-i} r_{n-j}) = \mathrm{E}(\varepsilon_n^2)\, \mathrm{E}(\sigma_n^2 r_{n-i} r_{n-j}) .$$

Recalling that $\mathrm{E}(\varepsilon_n^2) = a_1$, we can write the expression as

$$\begin{aligned} n_{ij} &= a_1\, \mathrm{E}(\sigma_n^2 r_{n-i} r_{n-j}) = a_1\, \mathrm{E}\, [\, r_{n-i} r_{n-j}\, (c_0 + \sum_{l=1}^{k} c_l\, (\sum_{d=1}^{l} r_{n-d})^2\,]] \\ &= a_1 \sum_{l=1}^{k} c_l\, \mathrm{E}\, [\, r_{n-i} r_{n-j} (\sum_{d=1}^{l} r_{n-d})^2\,] . \end{aligned} \tag{8.1}$$

We evaluate the expectation under the sum in (8.1) by examining different cases for the index l.

Case 1. $1 \le l < i$.

$$\begin{aligned}
&\mathrm{E}\,[\, r_{n-i} r_{n-j} (\sum_{d=1}^{l} r_{n-d})^2 \,] \\
&= \mathrm{E}\,[\, r_{n-i} r_{n-j} \sum_{d=1}^{l} r_{n-d}^2 \,] + 2 \sum_{d_1=1}^{l} \sum_{d_2=d_1+1}^{l} \mathrm{E}\,[\, r_{n-i} r_{n-j} r_{n-d_1} r_{n-d_2} \,] \\
&= \sum_{d=1}^{l} \mathrm{E}\,[\, r_{n-i} r_{n-j} r_{n-d}^2 \,] \\
&= \sum_{d=1}^{l} n_{i-d,j-d} \,, \qquad (8.2)
\end{aligned}$$

since the expectation of a product of only odd powers of r_n is zero.

Case 2. $i \le l < j$.
Here we have

$$\begin{aligned}
\mathrm{E}[r_{n-i} r_{n-j} (\sum_{d=1}^{l} r_{n-d})^2] &= \sum_{d=1}^{i-1} \mathrm{E}[r_{n-d}^2 r_{n-i} r_{n-j}] + 2 \sum_{d_2=i+1}^{l} \mathrm{E}[r_{n-i}^2 r_{n-j} r_{n-d_2}] \\
&= \sum_{d=1}^{i-1} n_{i-d,j-d} + 2 \sum_{d=i+1}^{l} n_{d-i,j-i} \,. \qquad (8.3)
\end{aligned}$$

Case 3. $j \le l \le k$.
Similar to the above we have

$$\begin{aligned}
\mathrm{E}[r_{n-i} r_{n-j} (\sum_{d=1}^{l} r_{n-d})^2] &= \sum_{d=1}^{i-1} \mathrm{E}[r_{n-d}^2 r_{n-i} r_{n-j}] + 2 \sum_{d=i+1}^{l} \mathrm{E}[r_{n-i}^2 r_{n-j} r_{n-d}] \\
&= \sum_{d=1}^{i-1} n_{i-d,j-d} + 2 \sum_{d=i+1}^{j-1} n_{d-i,j-i} + 2M_{j-i} + 2 \sum_{d=j+1}^{l} n_{j-i,d-i} \,. \qquad (8.4)
\end{aligned}$$

Putting together all three cases and using (8.1), we obtain an equation for the n_{ij}'s:

$$\begin{aligned}
\frac{1}{a_1} n_{ij} &= \sum_{l=1}^{i-1} c_l \sum_{d=1}^{l} n_{i-d,j-d} + \sum_{l=i}^{j-1} c_l \left(\sum_{d=1}^{i-1} n_{i-d,j-d} + 2 \sum_{d=i+1}^{l} n_{d-i,j-i} \right) \\
&+ \sum_{l=j}^{k} c_l \left(\sum_{d=1}^{i-1} n_{i-d,j-d} + 2 \sum_{d=i+1}^{j-1} n_{d-i,j-i} + 2M_{j-i} + 2 \sum_{d=j+1}^{l} n_{j-i,d-i} \right) , \qquad (8.5)
\end{aligned}$$

with $1 \le i < j \le k-1$.

We now turn to finding an equation for M_0:

$$\begin{aligned}
M_0 &= \mathrm{E}(r_n^4) = a_2\,\mathrm{E}(\sigma_n^4) = a_2\,\mathrm{E}\,[\,(c_0 + \sum_{l=1}^{k} c_l\,(\sum_{d=1}^{l} r_{n-d})^2)^2\,] \\
&= a_2 \left[c_0^2 + 2c_0 \sum_{l=1}^{k} c_l\,\mathrm{E}\,[\,(\sum_{d=1}^{l} r_{n-d})^2\,]\mathrm{E}\,[\,(\sum_{l=1}^{k} c_l\,(\sum_{d=1}^{l} r_{n-d})^2)^2\,] \right] \\
&= a_2 \left[c_0^2 + 2c_0 \sum_{l=1}^{k} c_l\, l\, L + \mathrm{E}\,[\,(\sum_{l=1}^{k} c_l \sum_{d=1}^{l} r_{n-d}^2 \right. \\
&\qquad \left. + 2\sum_{l=1}^{k} c_l \sum_{d_1=1}^{l} \sum_{d_2=d_1+1}^{l} r_{n-d_1} r_{n-d_2})^2\,] \right].
\end{aligned}$$

We continue to develop the right hand side of the equation by introducing more partial sums,

$$\begin{aligned}
\frac{M_0}{a_2} &= c_0^2 + 2c_0\,L \sum_{l=1}^{k} l\,c_l + \mathrm{E}\,[\,(\sum_{l=1}^{k} c_l \sum_{d=1}^{l} r_{n-d}^2)^2\,] \\
&\quad + 4\sum_{l_1=1}^{k}\sum_{l_2=1}^{k} c_{l_1} c_{l_2} \sum_{d_3=1}^{l_1}\sum_{d_1=1}^{l_2}\sum_{d_2=d_1+1}^{l_2} \mathrm{E}(r_{n-d_3}^2 r_{n-d_1} r_{n-d_2}) \\
&\quad + 4\sum_{l_1=1}^{k}\sum_{l_2=1}^{k} c_{l_1} c_{l_2} \sum_{d_1=1}^{l_1}\sum_{d_2=d_1+1}^{l_1}\sum_{d_3=1}^{l_2}\sum_{d_4=d_3+1}^{l_2} \mathrm{E}(r_{n-d_1} r_{n-d_2} r_{n-d_3} r_{n-d_4}) \\
&= c_0^2 + 2c_0\,L \sum_{l=1}^{k} l\,c_l + \mathrm{E}\,[\,(\sum_{d=1}^{k} r_{n-d}^2 \sum_{l=d}^{k} c_l)^2\,] \\
&\quad + 4\sum_{l_1=1}^{k}\sum_{l_2=1}^{k} c_{l_1} c_{l_2} \sum_{d_1=1}^{l_1}\sum_{d_2=d_1+1}^{l_2}\sum_{d_3=d_2+1}^{l_2} n_{d_2-d_1,d_3-d_1} \\
&\quad + 4\sum_{l_1=1}^{k}\sum_{l_2=1}^{k} c_{l_1} c_{l_2} \sum_{d_1=1}^{\mathrm{Min}[l_1,l_2]}\sum_{d_2=d_1+1}^{l_1}\sum_{d_3=d_1+1}^{l_2} \mathrm{E}(r_{n-d_1}^2 r_{n-d_2} r_{n-d_3}).
\end{aligned}$$

Using now the definitions we set at the beginning of this section, we get

$$\begin{aligned}
\frac{M_0}{a_2} &= c_0^2 + 2c_0\,L \sum_{l=1}^{k} l\,c_l + M_0 \sum_{d=1}^{k}(\sum_{l=d}^{k} c_l)^2 \\
&\quad + 2\sum_{l_1=1}^{k}\sum_{l_2=1}^{k} c_{l_1} c_{l_2} \sum_{d_1=1}^{l_1}\sum_{d_2=d_1+1}^{l_2} M_{d_2-d_1} \\
&\quad + 4\sum_{l_1=1}^{k}\sum_{l_2=1}^{k} c_{l_1} c_{l_2} \sum_{d_1=1}^{l_1}\sum_{d_2=d_1+1}^{l_2}\sum_{d_3=d_2+1}^{l_2} n_{d_2-d_1,d_3-d_1}
\end{aligned}$$

$$+4\sum_{l_1=1}^{k}\sum_{l_2=1}^{k} c_{l_1} c_{l_2} \sum_{d_1=1}^{\mathrm{Min}[l_1,l_2]} \sum_{d_2=d_1+1}^{l_1} \sum_{d_3=d_2+1}^{l_2} n_{d_2-d_1,d_3-d_1}$$

$$+4\sum_{l_1=1}^{k}\sum_{l_2=1}^{k} c_{l_1} c_{l_2} \sum_{d_1=1}^{\mathrm{Min}[l_1,l_2]} \sum_{d_2=d_1+1}^{l_2} \sum_{d_3=d_2+1}^{l_1} n_{d_2-d_1,d_3-d_1}\,. \qquad (8.6)$$

This is an equation for the variable M_0.

Finally, we study the equations for the variables M_i's with the index $i = 1, \ldots, k-1$:

$$\begin{aligned} M_i &= \mathrm{E}(r_n^2 r_{n-i}^2) = \mathrm{E}(\sigma_n^2 r_{n-i}^2)\ \mathrm{E}(\varepsilon_n^2) = a_1\ \mathrm{E}(\sigma_n^2 r_{n-i}^2) \\ &= a_1\ \mathrm{E}\left[\ r_{n-i}^2\ (c_0 + \sum_{l=1}^{k} c_l\ (\sum_{d=1}^{l} r_{n-d})^2)\ \right] \\ &= a_1 c_0\ L\ +\ a_1 \sum_{l=1}^{k} c_l\ \mathrm{E}\left[\ r_{n-i}^2 (\sum_{d=1}^{l} r_{n-d})^2\ \right] \\ &= a_1 c_0\ L\ +\ a_1 \sum_{l=1}^{i-1} c_l\ \mathrm{E}\left[\ r_{n-i}^2\ \sum_{d=1}^{l} r_{n-d}^2\ \right]\ +\ a_1 \sum_{l=i}^{k} c_l\ \mathrm{E}\left[\ r_{n-i}^2 \sum_{d=1}^{l} r_{n-d}^2\ \right] \\ &\quad + 2a_1 \sum_{l=i}^{k} c_l\ \mathrm{E}\left[\ r_{n-i}^2 \sum_{d_1=1}^{l} r_{n-d_1} \sum_{d_2=d_1+1}^{l} r_{n-d_2}\ \right]. \end{aligned}$$

Using again the definitions at the beginning of the section, we obtain the equations for M_i:

$$\begin{aligned} \frac{M_i}{a_1} &= c_0\ L\ +\ \sum_{l=1}^{i-1} c_l \sum_{d=1}^{l} M_{i-d} + \sum_{l=i}^{k} c_l \left(M_0 + \sum_{d=1}^{i-1} M_{i-d} + \sum_{d=i+1}^{l} M_{d-i} \right) \\ &\quad + 2\sum_{l=i}^{k} c_l \sum_{d_1=i+1}^{l} \sum_{d_2=d_1+1}^{l} n_{d_1-i,d_2-i}\,. \end{aligned} \qquad (8.7)$$

Overall, equations (8.5), (8.6) and (8.7) give a system of $\frac{k^2-k+2}{2}$ equations with as many unknowns. We also mention that it follows from Theorem 2 below that existence of a positive solution to this system of equations, constitutes, together with (4.5), a necessary and sufficient condition for the existence of a stationary version of the HARCH(k) process with a finite 4th moment. The following two subsections give more explicit conditions.

8.2. An explicit necessary condition for the existence of 4th moments of HARCH(k)

We start with showing that, if the 4th moment is finite, then we must have

$$n_{ij}\ \geq\ 0, \quad \text{with}\ \ i = 1, \ldots, k-2\ \ \text{and}\ \ j = i+1,\ ..\ , k-1\,. \qquad (8.8)$$

To this end, observe that

$$P(r_n \geq 0, r_{n-1} \geq 0, \ldots, r_{n-k+1} \geq 0) \;\geq\; 2^{-k} \;>\; 0 \tag{8.9}$$

for every $n \geq k$.

Define a function

$$f(x_1, \ldots, x_k) \;=\; x_1^4 + x_2^4 + \ldots + x_k^4 \,. \tag{8.10}$$

If the 4th moment of HARCH(k) is finite, we conclude from Theorem 14.3.3 of [Meyn and Tweedie, 1993] that the set $\mathcal{S}_f$ of f-regular points has probability 1 (under the steady state), and so by (8.9), it follows that there is an f-regular point $(x_1, \ldots, x_k)$ with $x_j \geq 0$, $j = 1, \ldots, k$. We then set

$$r_n \;=\; x_n \,, \quad n \;=\; 1, \ldots, k \;, \tag{8.11}$$

and we observe that

$$n_{ij}^{(n)} \;=\; \mathrm{E}(r_n^2 r_{n-i} r_{n-j}) \geq 0 \,, \quad i = 1, \cdots k{-}2, \; j = i{+}1, \cdots, k{-}1 \tag{8.12}$$

for $n = k$. We claim that (8.12) holds for all $n \geq k$. The proof is by induction on n. We have seen that (8.12) holds for $n = k$. Assume that it holds for all $k \leq l \leq n$ and let us prove it for $n+1$. As in equation 8.5, we have,

$$\begin{aligned}\frac{n_{ij}^{(n+1)}}{a_1} \;=\; & \sum_{l=1}^{i-1} c_l \sum_{d=1}^{l} n_{i-d,j-d}^{(k_1(i,j,d))} + \sum_{l=i}^{j-1} c_l \left(\sum_{d=1}^{i-1} n_{i-d,j-d}^{(k_2(i,j,d))} + 2 \sum_{d=i+1}^{l} n_{d-i,j-i}^{(k_3(i,j,d))} \right) \\ & + \sum_{l=j}^{k} c_l \left(\sum_{d=1}^{j-1} n_{i-d,j-d}^{(k_4(i,j,d))} + 2 \sum_{d=i+1}^{j-1} n_{d-i,j-i}^{(k_5(i,j,d))} \right. \\ & \left. + 2\mathrm{E}(r_{n+1-i}^2 r_{n+1-j}^2) + 2 \sum_{d=j+1}^{l} n_{i-d,j-d}^{(k_5(i,j,d))} \right) ,\end{aligned} \tag{8.13}$$

where

$$k_p(i,j,d) \;\leq\; n, \quad p \;=\; 1, \ldots, 5 \,. \tag{8.14}$$

Therefore, by the assumption of the induction, $n_{ij}^{(n+1)} \geq 0$, and (8.12) has been proved. Let

$$g(x_1, \ldots, x_k) \;=\; x_k^2 \, x_{k-i} \, x_{k-j}, \quad i \;=\; 1, \ldots, k-2, \; j \;=\; i+1, \ldots, k+1 \,. \tag{8.15}$$

Observe that

$$|g(x_1, \ldots, x_k)| \;\leq\; \max_{j \leq k} |x_j|^4 \;\leq\; f(x_1, \ldots, x_k) \,.$$

Therefore, by Theorem 14.3.3 of [Meyn and Tweedie, 1993], $\mathrm{E}(r_n^2 r_{n-i} r_{n-j})$ converges to its stationary version, and so it follows by (8.12) that this stationary expectation is nonnegative. This proves (8.8).

The discussion above allows us to get explicit necessary conditions for the existence of the 4th moment of the HARCH(k) models. First of all, it follows from (8.6) that

$$M_0 \geq a_2[c_0^2+2c_0\, M\sum_{l=1}^{k} l\, c_l+M_0\sum_{d=1}^{k}(\sum_{l=d}^{k} c_l)^2+2\sum_{l_1=1}^{k}\sum_{l_2=1}^{k} c_{l_1}\, c_{l_2}\sum_{d_1=1}^{l_1}\sum_{d_2=d_1+1}^{l_2} M_{d_2-d_1}]$$

and so

$$\begin{aligned} M_0 &\geq a_2[c_0^2 + 2c_0\, M\sum_{l=1}^{k} l\, c_l + M_0\sum_{d=1}^{k}(\sum_{l=d}^{k} c_l)^2 \\ &\quad + a_1\sum_{l_1=1}^{k}\sum_{l_2=1}^{k} c_{l_1}c_{l_2}\sum_{j=1}^{l_2-1}\mathrm{Min}[l_1,(l_2-j)]M_j] \\ &\geq a_2[c_0^2 + 2c_0 M\sum_{l=1}^{k} l\, c_l + M_0\sum_{d=1}^{k}(\sum_{l=d}^{k} c_l)^2 \\ &\quad + 2\sum_{j=1}^{k-1} M_j\sum_{l=1}^{k-j} l(2k-2l+1-j)]\,. \end{aligned} \tag{8.16}$$

Moreover, it follows from (8.7) that

$$\begin{aligned} \frac{M_i}{a_1} &\geq c_0\, M + \sum_{l=1}^{i-1} c_l \sum_{d=1}^{l} M_{i-d} \\ &\quad + \sum_{l=i}^{k} c_l\left(\sum_{d=1}^{i-1} M_{i-d} + M_0 + \sum_{d=i+1}^{l} M_{d-i}\right)\,, \quad i = 1,\ldots,k-1\,. \end{aligned} \tag{8.17}$$

We immediately see from (8.16) that

$$M_0 \geq a_2\,[\, c_0^2 + 2c_0\, M\sum_{l=1}^{k} l\, c_l + M_0\sum_{d=1}^{k}(\sum_{l=d}^{k} c_l)^2\,]\,,$$

and so

$$\sum_{d=1}^{k}\Big(\sum_{l=d}^{k} c_l\Big)^2 < \frac{1}{a_2} \tag{8.18}$$

is a necessary condition for the existence of a finite 4th moment; but it is, clearly, insufficient. We can get a stricter necessary condition as follows. From (8.17), we know that for every $i = 1,\ldots,k-1$,

$$M_i \geq M_0 a_1 \sum_{l=i}^{k} c_l\,. \tag{8.19}$$

Substituting (8.19) into (8.16), we obtain

$$M_0 \geq a_2 \Bigg[c_0^2 + 2c_0\, M \sum_{l=1}^{k} l c_l + M_0 \sum_{d=1}^{k} (\sum_{l=d}^{k} c_l)^2 + 2a_1 M_0 \sum_{j=1}^{k-1} \sum_{l=j}^{k} c_j \sum_{l=1}^{k-j} l(2k-2l+1-j) \Bigg], \tag{8.20}$$

and so a necessary condition for the existence of a finite 4th moment is

$$\sum_{d=1}^{k} (\sum_{l=d}^{k} c_l)^2 \; + \; 2a_1 \sum_{j=1}^{k-1} \sum_{l=j}^{k} c_j \sum_{l=1}^{k-j} l(2k-2l+1-j) \; < \; \frac{1}{a_2}\,. \tag{8.21}$$

8.3. A sufficient condition for the existence of the 4th moment of HARCH(k)

We now move to derive some explicit sufficient conditions for the existence of a finite 4th moment of a HARCH(k) model. Specifically, we claim that if

$$\left(\sum_{j=1}^{k} j^2\, c_j \right)^2 \; < \; \frac{1}{a_2}\,, \tag{8.22}$$

then a finite 4th moment exists.

Indeed, if (8.22) holds, one can choose positive numbers $\alpha_1, \ldots, \alpha_{k-1}$ such that

$$\alpha_{k-1} \; > \; a_2(kc_k) \sum_{j=1}^{k} j^2\, c_j\,, \tag{8.23}$$

$$\alpha_{i-1} \; - \; \alpha_i \; > \; a_2 \left(\sum_{l=i}^{k} l\, c_l \right) \sum_{j=1}^{k} j^2\, c_j\,, \quad i \; = \; 2, \ldots, \; k-1\,, \tag{8.24}$$

$$1 \; - \; \alpha_1 \; > \; a_2 \left(\sum_{l=1}^{k} l\, c_l \right) \sum_{j=1}^{k} j^2\, c_j\,. \tag{8.25}$$

Using these numbers define a function g

$$g(x_0, x_1, \ldots, x_{k-1}) \; = \; \sum_{j=0}^{k-1} \alpha_j\, x_j^4\,, \quad \alpha_0 \; = \; 1\,. \tag{8.26}$$

As before, we need to show that the difference $g - Pg$ can be made as large as we wish, away from a compact set. We have

$$\begin{aligned} g(r_n, r_{n-1}, \cdots, r_{n-k+1}) &- Pg(r_n, r_{n-1}, \cdots, r_{n-k+1}) \\ &= \sum_{j=0}^{k-1} \alpha_j\, r_{n-j}^4 - \sum_{j=1}^{k-1} \alpha_j\, r_{n-j+1}^4 - a_2\, [\, c_0 + \sum_{l=1}^{k} c_l\, (\sum_{d=1}^{l} r_{n-d+1})^2\,]^2 \\ &= \sum_{j=0}^{k-1} \alpha_j\, r_{n-j}^4 - \sum_{j=1}^{k-1} \alpha_j\, r_{n-j+1}^4 - a_2\, [\, \sum_{l=1}^{k} c_l\, (\sum_{d=1}^{l} r_{n-d+1})^2\,]^2 - \Theta\,, \end{aligned} \tag{8.27}$$

with Θ being a polynomial of order 2. Thus, it is enough to prove that, away from the origin,

$$\sum_{j=0}^{k-1} \alpha_j\, r_{n-j}^4 \;-\; \sum_{j=1}^{k-1} \alpha_j\, r_{n-j+1}^4 \;-\; a_2 \left(\sum_{l=1}^{k} c_l (\sum_{d=1}^{l} r_{n-d+1})^2 \right)^2 \;>\; 0\,. \tag{8.28}$$

With the help of (8.23)-(8.25), we estimate the left hand side of (8.28) as

$$\begin{aligned} &\geq \sum_{j=0}^{k-1} \alpha_j\, r_{n-j}^4 - \sum_{j=1}^{k-1} \alpha_j\, r_{n-j+1}^4 - a_2 \left(\sum_{l=1}^{k} l\, c_l \sum_{d=1}^{l} r_{n-d+1}^2 \right)^2 \\ &= \sum_{j=0}^{k-1} \alpha_j\, r_{n-j}^4 - \sum_{j=1}^{k-1} \alpha_j\, r_{n-j+1}^4 - a_2 \left(\sum_{d=1}^{k} r_{n-d+1}^2 \sum_{l=d}^{k} l\, c_l \right)^2 \\ &\geq \sum_{j=0}^{k-1} \alpha_j\, r_{n-j}^4 - \sum_{j=1}^{k-1} \alpha_j\, r_{n-j+1}^4 - a_2\, (\sum_{d=1}^{k} \sum_{l=d}^{k} l\, c_l\,) \sum_{d=1}^{k} r_{n-d+1}^4 \sum_{l=d}^{k} l\, c_l \\ &= \sum_{j=0}^{k-1} \alpha_j\, r_{n-j}^4 - \sum_{j=1}^{k-1} \alpha_j\, r_{n-j+1}^4 - a_2\, (\sum_{i=1}^{k} i^2\, c_i\,) \sum_{j=0}^{k-1} r_{n-j}^4 \sum_{l=j+1}^{k} l\, c_l \\ &= \sum_{j=0}^{k-2} r_{n-j}^4 \left(\alpha_j - \alpha_{j+1} - a_2\, (\sum_{i=1}^{k} i^2\, c_i\,) \sum_{l=j+1}^{k} l\, c_l \right) \\ &\quad + r_{n-k+1}^4 \left(\alpha_{k-1} - a_2\, (k c_k) \sum_{i=1}^{k} i^2\, c_i \right) > 0\,. \end{aligned}$$

This proves (8.28), and so (8.22) is a sufficient condition for the existence of a finite 4th moment for the HARCH(k) models.

9. General moments of HARCH(k)

Let $m \geq 2$. As always, if the (2m)th moments are finite, there is a system of equations these moments must satisfy. Define, under the assumption of stationarity,

$$M_0 = \mathrm{E}(r_n^{2m}) , \tag{9.1}$$

$$n_{i_0,i_1,\ldots,i_{k-1}} = \mathrm{E}(r_n^{2i_0} r_{n-1}^{i_1} \cdots r_{n-k+1}^{i_{k-1}}) , \tag{9.2}$$

$$1 \leq i_0 \leq m-1 , \quad 0 \leq i_j \leq 2m-2 , \quad j = 1,\ldots,k-1 ,$$

$$2i_0 + i_1 + \ldots + i_{k-1} = 2m .$$

If the process has a finite (2m)th moment, then we have

$$\begin{aligned} M_0 &= \mathrm{E}(r_n^{2m}) = \mathrm{E}(\varepsilon_n^{2m})\, \mathrm{E}(\sigma_n^{2m}) = a_m\, \mathrm{E}(\sigma_n^{2m}) \\ &= a_m\, \mathrm{E}\Big[(c_0 + \sum_{j=1}^{k} c_j (\sum_{d=1}^{j} r_{n-d})^2)^m \Big] a_m\, \vartheta_{0,m} \\ &\quad + a_m\, \mathrm{E}\Big[\sum_{j=1}^{k} c_j (\sum_{d=1}^{j} r_{n-d})^2)^m \Big] , \end{aligned} \tag{9.3}$$

where $\vartheta_{0,m}$ involves only moments of orders less than $2m$. We further have

$$\mathrm{E}[(\sum_{j=1}^{k} c_j (\sum_{d=1}^{j} r_{n-d})^2)^m] = \sum_{p_1,\ldots,p_k}^{m} \frac{m!}{p_1! \cdots p_k!} \prod_{i=1}^{k} c_i^{p_i}\, \mathrm{E}\, [\prod_{j=1}^{k} (\sum_{d=1}^{j} r_{n-d})^{2p_j}] . \tag{9.4}$$

with the following relations

$$0 \leq p_i \leq m , \quad \text{and} \ \sum_{i=1}^{k} p_i = m , \quad i = 1,\ldots, k . \tag{9.5}$$

Concentrating on the expectation in (9.4), we further obtain

$$\begin{aligned} &\mathrm{E}\, [\prod_{j=1}^{k} (\sum_{d=1}^{j} r_{n-d})^{2p_j}] \\ &= \sum_{l \in \mathcal{L}(p_1,\ldots,p_k)} \prod_{i=0}^{k-1} \binom{2(p_{k-i} + \cdots + p_k) - (l_{k-i+1} + \cdots + l_k)}{l_{k-i}} \\ &\quad \times \mathrm{E}\, [\prod_{j=1}^{k} r_{n-j}^{l_j}] , \end{aligned} \tag{9.6}$$

where for a vector $\underline{p} = (p_1, \ldots, p_k)$,

$$\begin{aligned} \mathcal{L}(\underline{p}) &= \mathcal{L}(p_1, \ldots, p_k) \\ &= \{ \underline{l} = (l_1, \ldots, l_k) \mid 0 \leq l_k \leq 2p_k,\ 0 \leq l_{k-1} + l_k \leq 2(p_{k-1} + p_k), \ldots, \\ &\quad 0 \leq l_2 + \ldots + l_k \leq 2(p_2 + \ldots + p_k), l_1 + l_2 + \ldots + l_k \\ &= 2(p_1 + p_2 + \ldots + p_k) \} \end{aligned} \tag{9.7}$$

Define a function $b : \mathbb{R}^k \longrightarrow \{1, \ldots, k\}$ by

$$b(x_1, ..., x_k) = \min\{j, x_j \neq 0\} . \tag{9.8}$$

Observe that for every $\underline{l} \in \mathcal{L}(\underline{p})$ we have

$$b(\underline{l}) \leq b(\underline{p}) . \tag{9.9}$$

Indeed, if $b(\underline{l}) > j$, then $l_1 = l_2 = \ldots = l_j = 0$, and so we have

$$2(p_1 + \cdots + p_k) = l_1 + \ldots + l_k = l_{j+1} + \cdots + l_k \leq 2(p_{j+1} + \cdots + p_k) ,$$

implying that $p_1 = p_2 = \ldots = p_j = 0$, and so $b(\underline{p}) > j$. It is clear that for every $\underline{l} \in \mathcal{L}(\underline{p})$ such that $l_{b(\underline{l})}$ is odd, we have

$$\mathrm{E} [\prod_{j=1}^{k} r_{n-j}^{l_j}] = 0 .$$

Therefore, by (9.9), we have

$$\begin{aligned} &\mathrm{E} [\prod_{j=1}^{k} (\sum_{d=1}^{j} r_{n-d})^{2p_j}] \\ &= \sum_{i=1}^{b(\underline{p})} \left[\sum_{\underline{l} \in \mathcal{L}(\underline{p})} \prod_{j=0}^{k-1} \binom{2(p_{k-j} + \cdots + p_k) - (l_{k-j+1} + \cdots + l_k)}{l_{k-j}} \right. \\ &\quad \left. \times\ n_{l_1/2, l_2, \ldots, l_k} + i\, M_0 \right] , \end{aligned} \tag{9.10}$$

with the following condition on the indices of the second summation

$$b(\underline{l}) = i , \quad l_i \text{ even}, \quad l_i \neq 2m . \tag{9.11}$$

Substituting (9.10) into (9.4), we obtain

$$\mathrm{E} [(\sum_{j=1}^{k} c_j (\sum_{d=1}^{j} r_{n-d})^2)^m]$$

$$= \sum_{i=1}^{k} \sum_{p_i,\dots,p_k} \frac{l!}{p_i!\cdots p_k!} \prod_{j=i}^{k} c_j^{p_j}$$
$$\times \sum_{d=1}^{i} \Bigg[\sum_{\underline{l}\in\mathcal{L}(\underline{p})} \prod_{j=0}^{k-d} \binom{2(p_{k-j}+\cdots+p_k)-(l_{k-j+1}+\cdots+l_k)}{l_{k-j}}$$
$$\times \; n_{l_d/2,l_{d+1},l_k,0,\dots,0} \; + \; d\, M_0 \Bigg] ,$$

where the indices $\underline{p}$ and $\underline{l}$ follow (9.5) and(9.11) respectively, with $p_1 = \ldots = p_{i-1} = 0$. That is, we have

$$\mathrm{E}\,[\,(\sum_{j=1}^{k} c_j\,(\sum_{d=1}^{j} r_{n-d})^2)^m\,] \;=\; M_0 \sum_{j=1}^{k}(\sum_{i=j}^{k} c_j)^m \;+\; \sum_{i=1}^{k} \sum_{p_i,\dots,p_k} \frac{l!}{p_i!\cdots p_k!} \prod_{j=i}^{k} c_j^{p_j}$$
$$\times \sum_{d=1}^{i} \sum_{\underline{l}\in\mathcal{L}(\underline{p})} n_{l_d/2,l_{d+1},\dots,l_k,0,\dots,0} \prod_{j=0}^{k-d} \binom{2(p_{k-j}+\cdots+p_k)-(l_{k-j+1}+\cdots+l_k)}{l_{k-j}} ,$$

where the subscript of n above is of length k. Therefore, we obtain the following equation

$$M_0 \;=\; M_0\, a_m \sum_{j=1}^{k}(\sum_{i=j}^{k} c_j)^m \;+\; a_m \sum_{i=1}^{k} \sum_{p_i,\dots,p_k} \frac{l!}{p_i!\cdots p_k!} \prod_{j=i}^{k} c_j^{p_j}$$
$$\times \sum_{d=1}^{i} \sum_{\underline{l}\in\mathcal{L}(\underline{p})} n_{l_d/2,l_{d+1},\dots,l_k,0,\dots,0} \prod_{j=0}^{k-d}$$
$$\times \binom{2(p_{k-j}+\cdots+p_k)-(l_{k-j+1}+\cdots+l_k)}{l_{k-j}} + a_m\, \vartheta_{0,m}\,. \quad (9.12)$$

In a similar manner, it is possible to obtain a system of equations for $n_{i_0,i_1,\dots,i_{k-1}}$. This gives us a system of

$$\sum_{i=0}^{m-1} \binom{2i+k-2}{k-2}$$

linear equations with as many unknowns. We call this system ϱ_m.

Combining the ideas we used in analyzing the cases of $k = 2$ and $m = 1$ (Section 6.2), we obtain in the same way, the following theorem.

Theorem 2. *A HARCH(k) model has a stationary version with a finite (2m)th moment if and only if, for every $j = 1,\dots,m$, the system ϱ_j has a positive solution. In this case, ϱ_k has a unique solution, given by (9.1) and (9.2).*

9.1. An explicit necessary condition

Since by Theorem 2, we have

$$n_{i_0,i_1,\dots,i_k} \;\geq\; 0 \quad \text{for all } i_0, i_1, \dots, i_k,$$

it follows from (9.12) that

$$M_0 \;>\; a_m\, M_0 \sum_{j=1}^{k} \Big(\sum_{i=j}^{k} c_j\Big)^m ,$$

and so

$$\sum_{j=1}^{k} \Big(\sum_{i=j}^{k} c_j\Big)^m \;<\; \frac{1}{a_m} \tag{9.13}$$

is a necessary condition for the existence of a finite $(2m)$th moment.

9.2. An explicit sufficient condition

We claim that if

$$\left(\sum_{j=1}^{k} j^2\, c_j \right)^m \;<\; \frac{1}{a_m} , \tag{9.14}$$

then the HARCH(k) model has a finite $(2m)$th moment.

The proof proceeds exactly like in the case $m = 2$ (Section 8.3). We take

$$g(x_0, x_1, \dots, x_{k-1}) \;=\; \sum_{j=0}^{k-1} \alpha_j\, x_j^{2m}, \qquad \alpha_0 \;=\; 1 , \tag{9.15}$$

and note that under (9.14) we may choose $\alpha_1, \dots, \alpha_{k-1}$ in such a way that

$$\alpha_{k-1} \;>\; a_m(k c_k) \left(\sum_{j=1}^{k} j^2\, c_j \right)^{m-1} , \tag{9.16}$$

$$\alpha_{i-1} - \alpha_i \;>\; a_m \left(\sum_{l=i}^{k} l\, c_l \right) \left(\sum_{j=1}^{k} j^2\, c_j \right)^{m-1} , \quad i \;=\; 2, \dots, k-1 , \tag{9.17}$$

$$1 \;-\; \alpha_1 \;>\; a_m \left(\sum_{l=1}^{k} l\, c_l \right) \left(\sum_{j=1}^{k} j^2\, c_j \right)^{m-1} . \tag{9.18}$$

Then proceed exactly as in Section 8.3.

10. Summary

In this paper we give a necessary and sufficient condition for the existence of moments of a general stationary HARCH(k) process using a Markov chain approach. Our interest in the tail behavior of this process leads us to study the convergence of the higher order moments of HARCH processes. We give a theorem that governs the stationarity with finite moments and we prove that the condition is both necessary and sufficient. The condition is in terms of the existence of a non-negative solution of a system of linear equations. Unfortunately, the general condition we obtain is not explicit. For every HARCH(k) process one has explicit necessary and sufficient conditions for finiteness of the second moment. In the case of the HARCH(2) process, we can give the explicit necessary and sufficient condition for the existence of the 4th moment. In all other cases, we give explicit necessary conditions and explicit sufficient conditions which will allow one to study the tail behavior of HARCH processes.

References

[Billingsley, 1968] Billingsley, P., *Convergence of Probability Measures*, Wiley, New York, 1968.

[Bollerslev, 1986] Bollerslev, T., Generalized autoregressive conditional heteroskedasticity, *Jrnl. of Econometrics*, **31**, 307–327 (1986).

[Dacorogna et al., 1993] Dacorogna, M.M., Müller, U.A., Nagler, R.J., Olsen, R.B. and Pictet, O.V., A geographical model for the daily and weekly seasonal volatility in the *FX* market, *Jrnl. of Int'l. Money and Finance*, **12** (4), 413–438 (1993).

[Engle, 1982] Engle, R.E., Autoregressive conditional heteroskedasticity with estimate of the variance of U.K. inflation, *Econometrica*, **50**, 984–1008 (1982).

[Guillaume et al., 1997] Guillaume, D.M., Dacorogna, M.M., Davé, R.D., Müller, U.A., Olsen, R.B. and Pictet, O.V., From the bird's eye to the microscope: A survey of new stylized facts of the intra-daily foreign exchange markets, *Finance and Stochastics*, **7**(2), 95–129 (1997).

[Meyn and Tweedie, 1993] Meyn, S.P. and Tweedie, R.L., *Markov Chains and Stochastic Stability*, Springer-Verlag, Heidelberg, 1993.

[Müller et al., 1997] Müller, U.A., Dacorogna, M.M., Davé, R.D., Olsen, R.B., Pictet, O.V., and von Weizsäcker, J.E., Volatilities of different time resolutions—analyzing the dynamics of market components, forthcoming in the *Jrnl. of Empirical Finance*, **4**(2–3), 213–239 (1997).

Paul Embrechts
Department of Mathematics
ETH Zentrum
CH-8092 Zürich
Switzerland
telephone 41-1-632-3419
e-mail: embrechts@math.ethz.ch

Gennady Samorodnitsky
School of Operations Research and Industrial Engineering
Cornell University
Ithaca, NY 14853
U.S.A.
telephone 1-607-255-9141
e-mail: gennady@orie.cornell.edu

Michel M. Dacorogna
Olsen & Associates
Research Institute for Applied Economics
Seefeldstrasse 233
CH–8008 Zürich
Switzerland
telephone 41-1-386-4848
e-mail: daco@olsen.ch

Ulrich A. Müller
Olsen & Associates
Research Institute for Applied Economics
Seefeldstrasse 233
CH–8008 Zürich
Switzerland
telephone 41-1-386-4848
e-mail: ulrichm@olsen.ch

Use of Stochastic Comparisons in Communication Networks

A. Ephremides

Abstract

Many problems of optimization in the field of communication networks are difficult to track. However, use of sample-path comparison methods can occasionally be of help. In this paper we review a few instances where such comparisons have led to helpful results.

I. Introduction

Queuing models have been widely used in the study of communication networks. Most of the successful applications of these models have been in the area of performance evaluation. However, some have occurred in the area of network control as well.

Early work on simple control problems of queuing networks has led to optimal policies that are of the so-called "threshold" or "switch" form; that is, the exercise of a fixed control option depends on whether the stochastic state of the system is in a certain subregion of its range or not. Such results are very useful because they are easy to implement. Typical examples include the "join-the-shortest-queue" [1] and "slow-server-activation" [2], [3] problems. In the first example, an arriving customer to a service system that consists of two equal-rate exponential servers with separate queues should join the shortest queue in order to minimize average waiting time. In the second, the slower of two exponential servers servicing a single queue should be activated if and only if the queue size exceeds a certain threshold (again, in order to minimize average total delay). Similar results were obtained in a variety of other systems of a more complicated nature. For example in [4], a model was considered that is useful in integrated-service, high-speed networks. Voice-call requests are arriving at a node along with data packets. The call requests are either accepted (in which case a fixed amount of bandwidth is allocated to them for their duration) or rejected. The data packets, on the other hand, are stored in a queue and served at the rate of the residual capacity of the node (which is equal to a fixed total less the amount allocated to the accepted calls). If the objective is to minimize the weighted sum of call-blocking probability and average packet delay, then the policy for accepting or rejecting a call should depend on whether the two-dimensional state vector (i, j), where $i =$ the number of packets in the queue, and $j =$ the number of on-going, accepted calls, lies in a region of the first quadrant of a two-dimensional integer grid that is defined simply by a step-wise constant boundary curve.

In all of these problems, an essential assumption has been the assumption of a certain degree of Markovian structure of the system (i.e., Poisson arrivals and/or exponential service times, and/or exponential call-holding times).

In this paper we demonstrate how the use of stochastic comparisons permits the derivation of optimal policy results in two other problems. The first concerns the scheduling of service in a queue if every arriving customer has a deadline by which service must be initiated (or completed). If the deadline is not met, either the customer is lost or a tardiness penalty is paid. In this model the Markovian assumptions are still essential for the service times, but not for the arrival process or the deadline times.

The second problem revisits the case of slow-server activation. However, this time the Markovian assumptions are relaxed. Instead, the packets are considered to be of fixed length and the two servers model outgoing links of different speeds. Thus, the service time of each packet takes one of two fixed values depending on whether the packet is served by the high- or the low-speed link. The Poisson assumption on the arrivals, however, is retained. This set of assumptions models more accurately the situation in packet switched networks. Although the result is again of the same threshold character as in the all-exponential case, it is not contrary to intuition. The constant (and known) values of the service times permit a different way of time-delay accounting and lead to the same result.

In general, the aforementioned results are of "existence" type. The actual computation of the thresholds and switch curves is in fact very difficult and thus the applicability to real systems has reduced value. However, in the second problem that is examined here, we compute bounds for the values of the optimal threshold (both lower and upper) that render the result more useful.

II. Scheduling with deadlines

This work was reported in detail in [5], [6] and is thus only briefly summarized here.

We consider the problem of scheduling the transmission of messages over a single communication link when each message has constraints on its waiting time or complete transmission time. This problem arises in applications that involve time-critical message contents. We wish to model situations in which the penalty incurred when deadlines are not met implies either the complete loss of the message or another form of tardiness cost. We are interested in characterizing the scheduling strategy which minimizes a cost function that reflects the nature of the penalty incurred. The models and the results of the paper apply equally well to numerous other applications that involve service stations, queues, and deadlines. Thus, a much more general terminology could be used. We choose to stick with the message transmission application in order to focus attention on the important problem of real time communication.

Here we consider only the case in which the messages have constraints on their waiting times. Each message upon its arrival at time t_i "announces" a deadline d_i, so that if by time $e_i = t_i + d_i$ (called its "extinction time"), transmission does not *commence*, the message is considered lost and never scheduled for service. The objective is to find a scheduling policy which minimizes the average number of lost messages over any time interval. We show that under nonexplosive, but otherwise arbitrary, arrival and arbitrary deadline processes, and for exponential service (i.e., transmission) times that are independent of each other and of the arrival and deadline processes, the policy of scheduling the eligible customer with the shortest time to extinction (denoted by STE) is optimal among all nonpreemptive and nonidling policies. In fact, we show the optimality in the sense of stochastic order. When considered over the broader class of only nonpreemptive policies, the optimal policy, if it exists, can be found in the class of STEI policies, namely those that are allowed to idle, but schedule according to the STE rule when they do not idle. As a special case, in the situations in which deadlines are deterministic and identical for all messages, the pure STE policy is optimal within the class of nonpreemptive policies.

These results do not seem to be easily extendable beyond the class of a single link. However, under a slightly different set of assumptions and operating conditions, some results can be obtained for a tandem network of links. Specifically, we may assume that no messages are discarded or lost and instead, all messages are scheduled, regardless of whether their deadlines have expired or not. The penalty is, however, incurred when a deadline is missed. A penalty function is of the form $h(c_i - e_i)$ where c_i is the time at which the message arriving at t_i with deadline d_i (and extinction time e_i, which, in this case, is also called due time) completes transmission in the network, and h is a real valued convex function with $h(x) = 0 \ \forall x \leq 0$. For this system, we consider a finite operating time horizon and wish to obtain a scheduling policy that minimizes the total penalty function (usually called tardiness). Under nonexplosive, but otherwise arbitrary, arrival and arbitrary deadline processes, and independent identically distributed service times that are also independent of the arrival and deadline processes, we show that the policy which schedules, at each node, the message with the earliest due time is optimal among all nonpreemptive and nonidling scheduling policies.

II.1. Constraints on waiting times

Let us consider a single server queuing system with unlimited buffer space size that represents a single link of a communication system. Let t_i be the arrival instant of the ith message whose deadline is d_i. We define by $e_i = t_i + d_i$ the extinction time of that message. That is, if by time e_i the transmission does not commence, the message is considered lost and never scheduled for service. At any instant t, a message with extinction time e_i is termed eligible for transmission if $e_i - t > 0$. Let $\{T_i = t_i - t_{i-1}\}_{i=1}^{\infty}$ (with $t_0 = 0$) be the sequence of interarrival times and let S_i be the duration of the ith service time in the system. By service time, we mean the transmission time of a message which may include processing and propagation times as well.

We make the following assumption throughout the paper.

A1: $\{S_i\}_{i=1}^{\infty}$ is a sequence of independent identically distributed RV's (random variables) which are independent of $\{T_i\}_{i=1}^{\infty}$ and $\{d_i\}_{i=1}^{\infty}$. Also, the arrival process is nonexplosive, that is, $\lim_{i\uparrow\infty} t_i = \infty$ with probability 1.

Let $E(t)$ denote the (increasing) ordered set of extinction times of eligible messages at time t. Let $H_a(t)$ denote the set of all arrival instants by time t and $H_d(t)$ the set of corresponding deadlines. Also, let $C_s(t)$ be the expended portion of the service in progress at time t and let $i(t)$ be the condition of the server at time t (1 is busy, 0 if idle). Then, under assumption A1, $z(t) = (E(t), H_a(t), H_d(t), C_s(t), i(t))$ is a useful description of the system. Let Z denote the allowable range of values of $z(t)$.

The control action is to decide at appropriate decision instants, whether to transmit and, if so, which message out of the currently eligible pool of messages. We restrict attention to nonanticipative policies throughout this paper. This means that the control action has to be based only on the past evolution of the system, specifically, the knowledge of the service times of the messages waiting in the queue is not available. Let Γ_0 be the class of nonpreemptive and nonidling policies and let Γ_1 be the class of nonpreemptive policies, while Γ is the global class of all possible scheduling policies.

For every policy in Γ_0, the decision instants are the instants of service completion (provided that $E(t)$ at these instants is nonempty) or of arrivals to an empty queue. Denote by STE, the policy in Γ_0 which at every decision instant schedules the eligible message with the shortest time to extinction. Let STEI denote the class of policies in Γ, which are allowed possibly to idle when messages are waiting in the queue, but which schedule according to the STE mechanism when they choose not to idle.

Following the standard notation, we say that a real-valued random variable X is *stochastically smaller* than a random variable Y, and write $X \leq_{st} Y$, if $P(X > z) \leq P(Y > z)\ \forall z \in R$. Order relations for stochastic processes can be considered as an extension of the definitions for vector-valued random variables. Let $X = \{X(t), t \in \Lambda\}$ and $Y = \{Y(t), t \in \Lambda\}$ be two real-valued processes, where $\Lambda \subset R$. Let $D \stackrel{\text{def}}{=} D_R[0,\infty)$, the space of right continuous functions from R_+ to R with left limits at all $t \in [0,\infty)$, be the space of their sample paths. We say that the process X is *stochastically smaller* than the process Y, and write $X \leq_{st} Y$, if $P\{f(X) > z\} \leq P\{f(Y) > z\}\ \forall z \in R$, where $f : D \to R$ is measurable and $f(x) \leq f(y)$ whenever $x, y \in D$ and $x(t) \leq y(t)$, $\forall t \in \Lambda$. The following equivalence [7], [8] often provides an easy way to prove stochastic order relations without explicit computation of distributions.

1) $X \leq_{st} Y$.
2) $P(g[X(t_1),\dots,X(t_n)] > z) \leq P(g[Y(t_1),\dots,Y(t_n)] > z)$ for all $(t_1,\dots,t_n)$, $\varepsilon\Lambda^n$, all $z \in R$, all $n \in N$, and for all $g : R^n \to R$ measurable and such that $x_j \leq y_j$, $1 \leq j \leq n$ implies $g(x_1,\dots,x_n) \leq g(y_1,\dots,y_n)$.
3) There exists two stochastic processes $\tilde{X} = \{\tilde{X}(t), t \in \Lambda\}$ and $\tilde{Y} = \{Y(t), t \in \Lambda\}$ on a common probability space such that $\mathcal{L}(X) = \mathcal{L}(\tilde{X})$, $\mathcal{L}(Y) = \mathcal{L}(\tilde{Y})$, and $\tilde{X}(t) \leq \tilde{Y}(t)$, $\forall t \in \Lambda$ a.s. Here $\mathcal{L}(\cdot)$ denotes the law of a process on the space of its sample paths.

Returning to our problem, let $L^{\pi}(z)$ denote the process $\{L_1^{\pi}(z), t \geq 0\}$, where $L_t^{\pi}(z)$ is the number of messages lost by time t when starting from state z at time 0 and applying the scheduling policy π.

We first consider optimality within Γ_0.

Theorem II.1. *Consider a single server queue under assumption* A1. *Assume further that the common distribution of the service times is exponential. Then, the STE policy minimizes, in the sense of stochastic order, the number of messages lost by any time among all policies in the class* Γ_0, *that is,*

$$L^{ste}(z) \leq_{st} L^{\pi}(z) \quad \forall \pi \in \Gamma_0,\ \forall z \in Z.$$

To prove this theorem we need first the following lemma which we prove in detail in order to illustrate how the stochastic comparisons are used.

Lemma II.1. *Consider a single server queue as in Theorem* II.1. *Let an arbitrary policy $\pi \in \Gamma_0$ act on the system in $[t_0, \infty)$, where t_0 is an arbitrary decision instant. There exists a policy $\hat{\pi} \in \Gamma_0$ that schedules the customer with the shortest time to extinction at time t_0 (and is appropriately defined in $[t_0, \infty)$) and satisfies*

$$L^{\hat{\pi}}(z) \leq_{st} L^{\pi}(z) \quad \forall z \in Z.$$

Proof. Assume that π does not schedule the customer with the shortest time to extinction at time t_0. (If it does, the result follows trivially by letting $\hat{\pi}$ be the same as π.) We drop z from $L_t^{\pi}(z)$ and $L_t^{\hat{\pi}}(z)$ for notational convenience.

The idea of the proof is to define $\hat{\pi}$ appropriately in $[t_0, \infty)$ and to construct two coupled processes $(L_t^{\hat{\pi}}, \tilde{L}_t^{\pi})$ on the same (given) probability space so that $\tilde{L}_t^{\pi}$ and L_t^{π} have the same distribution and $L_t^{\hat{\pi}} \leq \tilde{L}_t^{\pi}$ a.s. $\forall t \geq t_0$.

Suppose $E(t_0) = \{e_1, \ldots, e_n\}$ with $n \geq 2$. We agree to denote by e_i either the extinction time or the message with that extinction time. Let π schedule e_k $(2 \leq k \leq n)$ at time t_0. We will construct $\tilde{\pi} \in \Gamma_0$ which schedules e_{k-1} at time t_0 (and is appropriately defined in $[t_0, \infty)$) and satisfies the assertion of the lemma. The required policy $\hat{\pi}$ can then be generated by induction on k.

Consider the system evolving under policies $\tilde{\pi}$ and π. Couple the realizations by giving them the same arrival and deadline processes in $[t_0, \infty)$. Let σ be the completion instant of the service which begins at time t_0 under $\tilde{\pi}$. Take the service under π to end at σ as well. This is permissible since the service times are independent and identically distributed. Three cases exhaust the possibilities.

Case 1. $\sigma \geq e_k$: In this case, under both policies, all messages eligible at time t_0, except e_k for π and e_{k-1} for $\tilde{\pi}$ that have extinction times less than or equal to σ are lost, and so are all arrivals in $(t_0, \sigma]$ whose extinction times are similarly less than or equal to σ. The states under π and $\tilde{\pi}$ are therefore matched at time σ. In $[\sigma, \infty)$, define $\tilde{\pi}$ to be identical to π; this is possible since we may take corresponding service times to be equal under π and $\tilde{\pi}$. Thus,

$$\tilde{L}_t^{\pi} = \begin{cases} L_t^{\tilde{\pi}} & \text{if } t \in [t_0, e_{k-1}) \bigcup [e_k, \infty); \\ L_t^{\tilde{\pi}} + 1 & \text{if } t \in [e_{k-1}, e_k). \end{cases}$$

Case 2. $\sigma < e_{k-1}$: First, it is clear that $\tilde{L}_t^{\pi} = L_t^{\tilde{\pi}}$ $\forall t \in [t_0, \sigma)$, and at time σ, the sets of extinction times under π and $\tilde{\pi}$ differ only in that e_{k-1} is included in that set under π as compared to e_k under $\tilde{\pi}$. Let $\tilde{\pi}$ follow π for $t \geq \sigma$ except that it schedules e_k when (and if) π schedules e_{k-1}. Thus, $\tilde{\pi}$ is well defined in $[\sigma, \tau)$, where τ is the end of the current busy period under π.

Suppose π eventually schedules e_{k-1}, that is, suppose that e_{k-1} meets its deadline under π. Since $e_{k-1} < e_k$, the message with extinction time e_k will also meet its deadline under $\tilde{\pi}$; and the states under π and $\tilde{\pi}$ are matched at time τ. Letting $\tilde{\pi}$ follow π in $[\tau, \infty)$, one thus obtains $\tilde{L}_t^\pi = L_t^{\tilde{\pi}}$ $\forall t \in [\sigma, \infty)$.

Suppose now that π does not manage eventually to schedule e_{k-1} before its expiration. Since π is nonidling, we have necessarily that $\tau \leq e_{k-1}$. If $\tau \geq e_k$, then e_k is lost under $\tilde{\pi}$. Letting $\tilde{\pi}$ follow π in $[\tau, \infty)$, one obtains

$$\tilde{L}_t^\pi = \begin{cases} L_1^{\tilde{\pi}} & \text{if } t \in [\sigma, e_{k-1}) \bigcup [e_k, \infty); \\ L_t^{\tilde{\pi}} + 1 & \text{if } t \in [e_{k-1}, e_k). \end{cases}$$

If, however, $\tau < e_k$, then at time τ, the queue is empty under π and e_k is the only eligible message under $\tilde{\pi}$. Let $\tilde{\pi}$ begin serving e_k at time τ. We must now consider the following two subcases.

a) Suppose there are no arrivals while e_k is in service under $\tilde{\pi}$. The states are then matched from the instant message e_k finishes service under $\tilde{\pi}$, thus,

$$\tilde{L}_t^\pi = \begin{cases} L_t^{\tilde{\pi}} & \text{if } t \in [\sigma, e_{k-1}); \\ L_t^{\tilde{\pi}} + 1 & \text{if } t \in [e_{k-1}, \infty). \end{cases}$$

b) Suppose there is at least one arrival while e_k is in service under $\tilde{\pi}$. Let the arrival which begins service first under π have extinction time e_b. Take the service time of e_b under π to be equal to the residual service time of e_k under $\tilde{\pi}$. This is possible because of assumption A1 and the memoryless property of the exponentially distributed service times. In this way, we ensure that e_b and e_k will finish service at the same time instant σ_1. Suppose that $\sigma_1 \geq e_b$. Then e_b is lost under $\tilde{\pi}$ and the states under the two policies are matched at time σ_1. One thus concludes that

$$\tilde{L}_t^\pi = \begin{cases} L_t^{\tilde{\pi}} & \text{if } t \in [\sigma, e_{k-1}) \bigcup [e_b, \infty); \\ L_t^{\tilde{\pi}} + 1 & \text{if } t \in [e_{k-1}, e_b). \end{cases}$$

Suppose now that $\sigma_1 < e_b$. Let $\tilde{\pi}$ follow π for $t \leq \sigma_1$. Take the corresponding service times under π and $\tilde{\pi}$ to be equal. Thus, $\tilde{\pi}$ is well defined in $[\sigma_1, \tau_1)$ where τ_1 is the end of this busy period under π. If $\tau_1 \geq e_b$, we let $\tilde{\pi}$ follow π from time τ_1 onwards, and thus have the same relations between $\tilde{L}_t^{\pi}$ and $L_t^{\tilde{\pi}}$ as in the situation just described. If instead $\tau_1 < e_b$, let $\tilde{\pi}$ begin serving e_b at time τ_1. We are now back to a situation previously described. One easily repeats the arguments to obtain

$$\tilde{L}_t^{\pi} = \begin{cases} L_t^{\tilde{\pi}}, & \text{if } t \in [\sigma, e_{k-1}); \\ L_t^{\tilde{\pi}} + 1 & \text{if } t \in [e_{k-1}, \tau); \end{cases}$$

and $\tilde{L}_t^{\pi} \geq L_t^{\tilde{\pi}}, \forall t \in [\tau_1, \infty)$.

Case 3. $e_{k-1} \leq \sigma < e_k$: It is clear that

$$\tilde{L}_t^{\pi} = \begin{cases} L_t^{\tilde{\pi}}, & \text{if } t \in [t_0, e_{k-1}); \\ L_t^{\tilde{\pi}} + 1 & \text{if } t \in [e_{k-1}, \sigma) \end{cases}$$

and at time σ, under $\tilde{\pi}$, message e_k is eligible for service in addition to all messages that are eligible under π. Consequently, we can proceed as in Case 2. We thus conclude that

$$\tilde{L}_t^{\pi} \geq L_t^{\tilde{\pi}} \quad \forall t \in [\sigma, \infty).$$

The observation that the processes $\tilde{L}^{\pi}$ and L^{π} have the same law now completes the proof.

We can now prove the original theorem.

Proof of Theorem II.1. Start with an arbitrary policy $\pi \in \Gamma_0$ acting on the system in an initial state z. By Lemma II.1, we can construct an alternative policy $\pi_1 \in \Gamma_0$ which schedules according to the STE-rule at the first decision instant along its trajectory, and which satisfies the relation $L^{\pi_1} \leq_{st} L^{\pi}$. We proceed inductively; that is, by repeating the same construction n times we can define a policy $\pi_n \in \Gamma_0$ which schedules according to the STE-rule at least at the first n decision points along its trajectory, and satisfies

$$L^{\pi_n} \leq_{st} L^{\pi_{n-1}} \leq_{st} L^{\pi_1} \leq L^{\pi}.$$

Fix $x \in R$, a positive integer k and $t_i \in [0, \infty)$, $1 \leq i \leq k$ and pick $g : R^k \to R$. Let A^{γ} denote the event $\{g(L_{t_1}^{\gamma}, \ldots, L_{t_k}^{\gamma}) > x\}$ for a policy $\gamma \in \Gamma_0$. Let $t_j = \max_{1 \leq i \leq k} t_j$ and take $\{S_n\}_1^{\infty}$ to be the service times of the messages in the system. Since the policies STE and π_n agree on their first n decisions, one has

$$
\begin{aligned}
P(A^{ste}) &= P\left(A^{ste} \bigcap \left\{\sum_{i=1}^{n} S_i \geq t_j\right\}\right) \\
&\quad + P\left(A^{ste} \bigcap \left\{\sum_{i=1}^{n} S_i < t_j\right\}\right) \\
&\leq P\left(A^{\pi_n} \bigcap \left\{\sum_{i=1}^{n} S_i \geq t_j\right\}\right) + P\left(\sum_{i=1}^{n} S_i < t_j\right) \\
&\leq P(A^{\pi_n}) + P\left(\sum_{i=1}^{n} S_i < t_j\right) \\
&\leq P(A^{\pi}) + P\left(\sum_{i=1}^{n} S_i < t_j\right), \quad \forall n \in N.
\end{aligned}
$$

Taking the limit as $n \nearrow \infty$, it now follows that $P(A^{ste}) \leq P(A^{\pi})$, that is $L^{ste} \leq_{st} L^{\pi}$. ■

Next, we consider optimality within Γ_1, the broader class of only non-preemptive policies. Now, the idling of the server is allowed. In this case, examples can be easily constructed to show that the STE policy is no longer optimal; the basic intuition being that when all the messages awaiting service have large extinction times, it pays to idle in expectation of a message with a very short deadline. However, the philosophy of STE-scheduling still plays an important role, as the following result demonstrates.

Proposition II.1. *Consider a single server queue under assumption* A1. *Then, for every policy* $\pi \in \Gamma_1$, *there exists a policy* $\tilde{\pi}$, *such that*

$$L^{\tilde{\pi}}(z) \leq_{st} L^{\pi}(z), \quad \forall z \in Z.$$

As in the previous theorem, the proof can be worked out in two steps. In the first step, assuming that t_0 is an arbitrary decision instant at which π schedules a message that is not the one with the smallest time to extinction, we construct, using the knowledge of π, a policy $\tilde{\pi}$, which schedules the message with the smallest time to extinction and satisfies $L^{\tilde{\pi}} \leq_{st} L^{\pi}$. This then can be used recursively to improve upon any given policy in Γ_1 until the improving policy belongs to the class of STEI policies. The key observation which in fact facilitates the arguments in the first step is that when idling is permitted, the policy $\tilde{\pi}$, which we construct, can be chosen to follow π at all times beyond t_0. Exponentiality of the service times is therefore not needed. We omit the details.

Corollary II.1. *Consider a single server queuing system under assumption* A1 *and* A2. *Then for every initial state $z \in Z$, we have*

$$L^{ste}(z) \leq_{st} L^{\pi}(z) \quad \forall \pi \in \Gamma_1.$$

Proof. First, let us note that by Proposition II.1, it suffices to prove the claim for $\pi \in$ STEI. We proceed as be fore. Let t_0 be an instant at which a policy $\pi \in$ STEI chooses to idle. It now suffices to construct $\tilde{\pi} \in \Gamma_1$ which schedules the customer with the shortest time to extinction at time t_0 and satisfies $L^{\tilde{\pi}} \leq_{st} L^{\pi}$.

Let $E(t_0) = \{e_1 \cdots e_n\}$ with $n \geq 1$. Because of Assumption A2, it is clear that the arrivals in (t_0, ∞) have extinction times no smaller than e_n. Let τ be the first instant after t_0 at which π decides to schedule a message. Consider the following two cases.

1) Suppose $\tau < e_1$. Since $\pi \in$ STEI, π schedules e_1 at time τ. Take the first service time under π (which begins at time τ) to be the same as the service time of e_1 under $\tilde{\pi}$. Following its own trajectory, therefore, $\tilde{\pi}$ can determine the instant of completion of service of e_1 under π. Let $\tilde{\pi}$ idle from the time the service of e_1 is finished under $\tilde{\pi}$ to the time of completion of service of e_1 under π. Letting $\tilde{\pi}$ follow π from this time onwards, one obtains $\tilde{L}_t^{\pi} = L_t^{\tilde{\pi}}, \forall t \geq t_0$.

2) Suppose $\tau \geq e_1$. Let π schedule E^* at time τ. Take the service time of e^* under π to be equal to that of e_1 under $\tilde{\pi}$. Let $\tilde{\pi}$ never schedule e^* and instead idle, if necessary, as in the previous case. It is then clear that

$$\tilde{L}_t^{\pi} = \begin{cases} L_t^{\tilde{\pi}} & \text{if } t \in [t_0, e_1) \bigcup [e^*, \infty); \\ L_t^{\tilde{\pi}} + 1 & \text{if } t \in [e_1, e^*). \end{cases}$$

Because of assumption A1, L^{π} and $\tilde{L}^{\pi}$ have the same distribution, and so, $L^{\tilde{\pi}} \leq_{st} L^{\pi}$. ∎

There are many extensions and modifications to these results that are possible (see e.g., [5], [6]) and which attempt to model more closely realistic communication network systems. They all follow the same basic technique that was displayed above and thus we leave it to the interested reader to explore them further.

III. Threshold control of deterministic servers

An important problem in network control concerns bandwidth allocation. Although this problem can be complicated and environment-specific (e.g., bandwidth allocation means different things in different network systems), the designer may obtain useful insights from the rigorous consideration of basic abstractions of this problem in terms of simple queuing models.

In this paper we consider a single-queue system of two servers. Under the assumption of Poisson arrivals and exponential service times, if server one is faster than server two, it is known [2], [3] that the activation policy that minimizes average delay always activates the faster server when the queue is non-empty and activates the slower server when the queue size exceeds a threshold that can be precomputed (although the calculation of the threshold is rather complicated). This problem models the situation in which a second (slower) channel is available to be allocated to a data link with a primary (faster) channel. Here we consider the case of deterministic service times. This assumption destroys the full Markovian structure of the system, but permits an alternative approach of pathwise (or stochastic) comparisons that generalizes the threshold result and, more importantly, permits the calculation of good upper and lower bounds on the threshold value.

Specifically, we consider a queuing system composed of an infinite-size buffer and two servers, S_1 and S_2, with constant, but different, service times T_1 and T_2 respectively. We assume $T_2 > T_1$. The arrival process into the queue is Poisson with rate $\lambda < T_1^{-1} + T_2^{-1}$. We wish to find the optimal server activation policy which minimizes the mean sojourn time in the system.

As mentioned earlier, for such a queuing system a threshold control policy is optimal for the case of exponential servers with service rates μ_1 and μ_2. A policy π_{m_0} is a threshold policy of threshold level m_0 if:

1) the fast server is kept busy whenever possible (i.e., when there are packets in the queue), and
2) the slow server is utilized only when the queue size is larger than a threshold m_0.

The threshold m_0 depends, in a complicated manner, upon λ, μ_1, and μ_2. For its derivation the sojourn time J_m is computed for $m = 0, 1, 2, 3, \ldots$ and for given λ, μ_1, and μ_2. The threshold m_0 is that value of m which minimizes J_m. Rubinovich in [9] computed a number of such thresholds for a set of values of λ, μ_1, and μ_2. We will make use of these results to evaluate the quality of the bounds for the threshold, which we derive in this paper.

Extension of these results to a system with non-exponential service rate servers is difficult because the Markov structure of the problem is disrupted. However, when the service times are deterministic (as in the case of packet-switched systems), we take advantage of the simplicity and features of the constant service time to study the optimal control policy problem with reasonable success. Specifically,

(a) necessary conditions for the optimal control policy are derived that are identical to the conditions for the optimal threshold policy in the system with exponential servers;

(b) if we restrict our investigation to the class of threshold policies, lower and upper bounds for the value of the threshold are obtained, which have simple expression in terms of λ, T_1 and T_2;

(c) the calculation of the lower bound is applicable to the exponential server case as well, and, in that case it yields bounds that are remarkably close to the exact values of the threshold computed in [9].

III.1. Formulation and properties of the optimal policies

Let $x^0(t)$ denote the number of packets in the queue at time t and $r_i(t)$ the residual service time of server Si, $i = 1, 2$; clearly $0 \leq r_i(t) \leq T_i$. The vector $x(t) = [x^0(t), r_1(t), r_2(t)]$ is a Markov process and an appropriate state vector for the system. The number of the packets in the system is $|x(t)| = x^0(t) + x^1(t) + x^2(t)$ where $x^i(t) = \begin{cases} 0 & \text{if } r_i(t) = 0 \\ 1 & \text{if } r_1(t) > 0 \end{cases}$, $i = 1, 2$. Let π be a control policy, which at every $t \geq 0$ decides which idle server(s) to activate based upon $\{x(s), 0 \leq s \leq t\}$. A routing policy π is optimal if it minimizes $J_\pi(x)$, the long-run average cost associated with π, which is defined as: $J_\pi(x) = \limsup \frac{1}{T} E_x^\pi \left[\int_0^T |x(t)|\, dt\right] \forall x$, where E_x^π is the expectation under π and x is the starting state.

In the system under consideration (constant rate servers) the continuous-space description produces a continuum of transitions (i.e., in addition to the transitions at arrival and service completion times there are continuous transitions as the $r_1(t)$'s change value with time). The search for an optimal policy under these conditions is hopelessly difficult and of minimal practical value because of implementation issues. Nonetheless, we are able to find properties of the optimal policy (restricted within the class of Markov policies) and show that the optimal Markov policy shares crucial properties with the threshold-type policy which is optimal in the case of exponential servers. Thus, we provide strong indications that in this case, as well, the optimal Markov policy is of threshold type (modulo a variation of the threshold value as a function of the residual service time $r_1(t)$). Next we derive the properties of the optimal policy.

In our proofs we make pathwise comparisons between the evolution of the state process $x(t)$ under an arbitrary policy π, with that under an alternative policy $\tilde{\pi}$ that has the claimed properties. We show that there is improvement under $\tilde{\pi}$ and hence the optimal policy must have these properties. The details of the proofs involve careful accounting of gains and losses for each sample path under the two policies. The length and level of detail needed is considerable. Consequently, we provide the complete arguments selectively to illustrate our method.

We begin by proving the following lemma.

Lemma 1. *The optimal policy has the following properties:*
(1) *If both servers are idle, one of them must be activated without delay.*
(2) *If the slow server is busy, the fast server must be activated without delay.*
(3) *If both servers are idle, the fast one must activate first.*

Proof.

1. Assume that at the time $t = 0$ both servers are idle and the queue is not empty; assume that policy π activates one server, say S1, at time $t = \tau > 0$. (The same exact arguments will apply if the server S2 is activated.) We consider another policy $\tilde{\pi}$ which activates the server S1 without delay, at $t = 0$. Server S1 completes its service at time T_1. We consider two cases:
   i) $\tau < T_1$. In this case we let $\tilde{\pi}$ follow π's actions after τ. Observe that in the time interval $(T_1, \tau + T_1)$ there is one packet less in the tilde system, whereas after $\tau + T_1$ both systems have the same state space. Thus, the average waiting time in the tilde system is reduced by τ.
   ii) $\tau > T_1$. In this case we let, as before, $\tilde{\pi}$ to follow π's actions after τ, and introduce at time T_1 a dummy packet to replace the departed one and keep the two queues in tune. This dummy packet does not load the system since it stays simply in queue and does not get service. Its sole purpose is to keep $|\tilde{x}(t)| = |x(t)|$, for $t > T_1$. The duration of its presence corresponds to the reduction of the total waiting cost under $\tilde{\pi}$. It is easy to see that we must keep this dummy packet in $\tilde{\pi}$'s queue until there are no other packets to be served (i.e., to the end of the current busy period). Thus, the reduction of the waiting cost is the duration of the remaining busy period, which may be quite substantial.

Thus, a policy π which delays the activation of a server cannot be optimal.

2. The above arguments apply verbatim in this case as well.
3. Assume that at $t = 0$ both servers are idle and that the policy π activates the slow server S2, while policy $\tilde{\pi}$ activates the fast server S1. Let τ be the time of the next server activation in π. We consider three cases:
   i) $\tau > T_2$. In this case we let $\tilde{\pi}$ to follow π's actions after τ. The two state processes for π and $\tilde{\pi}$ coincide after τ. Observe that the

first packet leaves the original system at T_2 and the tilde system at $T_1 < T_2$; thus, there is a reduction in the waiting cost under $\tilde{\pi}$ equal to $T_2 - T_1$.

ii) $T_1 < \tau < T_2$. In this case both policies activate server S1 at time τ and we let gain $\tilde{\pi}$ follow π's actions at subsequent transitions; thus, again the two queues remain in identical states for $t > \tau$. The advantage of the tilde system is again the difference in the service time of the first packet.

iii) $\tau < T_1$. This is a more difficult case and requires a more careful and complete definition of the tilde policy. At time τ policy π will activate necessarily the fast server S1. We define $\tilde{\pi}$ to activate the available server at $t = \tau$. From then on the policy $\tilde{\pi}$ will keep the fast server continuously busy (if there are packets in the queue) and will follow π's actions concerning the slow server. This means that if π activates S2, $\tilde{\pi}$ will activate the same as soon as it becomes available to it (i.e., with a maximum delay of τ).

We first show that for any given time t, such that the system has not yet emptied, the number of packets served under policy $\tilde{\pi}$ is greater than or equal to that under π. Therefore, under $\tilde{\pi}$ the performance index given in Section II is improved. It also follows that server S1 will finish with a queue empty first under $\tilde{\pi}$ and then under π. Let t_0 be the time at which S1 empties under $\tilde{\pi}$, i.e., $t_0 = \min\{t : (x^0(t), r_1, r_2) = (0, 0, r_2)\}$. We must still trace what happens with respect to the finishing of the service by S2 after t_0. In all cases, a simple argument (essentially reiterating the argument that shows the gains of $\tilde{\pi}$ during a time of continuing service without idle periods) establishes that $\tilde{\pi}$ continues to dominate π.

So we now consider a time t such that S1 has been busy under both π and $\tilde{\pi}$ for all times in $(0, t]$. Recall that under $\tilde{\pi}$ at $t = 0$ the first packet is sent to S1 while under π that packet is sent to S2. Clearly, the interesting case is when the next arrival occurs at $\tau < T_1$, which we consider here. The number of packets served by S1 until t under $\tilde{\pi}$ (including fractional packets) is given by $\tilde{n}_1(t) = t/T_1$. In addition, the number of customers served by server S2 under π is equal to $n_2(t)$. Hence, the number served by S2 under $\tilde{\pi}$ is: $\tilde{n}_2(t) = n_2(t) - \frac{\tau}{T_2}$. Similarly, the number $n_1(t)$ served by S1 under π is $n_1(t) \leq \tilde{n}_1(t) - \frac{\tau}{T_1}$. Hence, the total number of packets served $N(t)$ and $\tilde{N}(t)$ by π and $\tilde{\pi}$ respectively satisfy

$$\tilde{N}(t) - N(t) = \tilde{n}_1(t) + \tilde{n}_2(t) - n_1(t) - n_2(t) \geq \frac{\tau}{T_1} - \frac{\tau}{T_2} > 0.$$

Lemma 1 implies that the optimal policy makes identical decisions concerning the activation of the fast server in both the constant-rate servers and exponential rate servers cases: it keeps that server busy as much as possible. Thus, we have to find out under what conditions the optimal policy activates the slow server.

Because we cannot use Markov decision theory to establish that the optimal policy is Markov and stationary here, we limit our search now within the reduced, but more interesting for implementation purposes, class of Markov policies.

To this end we prove the following Lemma 2:

Lemma 2. *The optimal Markov policy has the following property:*

If a policy activates the slow server when the system is in states $(y^0, r_1, 0)$ *and* $(z^0, r_1, 0)$, *for some* $y^0 < z^0$, *then it can be improved if it also activates that server from the state* $(x^0, r_1, 0)$ *for any* $y^0 < x^0 < z^0$.

Proof. Assume that the policy π complies with the assumptions of Lemma 2 but does not activate the slow server from the state $(x^0, r_1, 0)$ as defined above. Assume also that at $t = 0$ the system is in the state $(x^0, r_1, 0)$ and that the tilde system activates the slow server. Introduce a dummy packet in the tilde system and let $\tilde{\pi}$ follow π's actions for $t > 0$. After some time T the original system will reach state $(y^0, r_1, 0)$ or $(z^0, r_1, 0)$ and it will send a packet to the slow server. At this time the dummy packet is not needed and is removed. The time T that it stayed in the tilde system is the reduction of the waiting cost under $\tilde{\pi}$.

III.2. Bounds for the threshold value of a threshold policy

Here we concentrate on computing the value of the threshold. So far we have shown that within the class of Markov policies, the optimal policy is almost of threshold type (Lemma 2). Thus, we limit ourselves further to the class of threshold policies, i.e., policies that for a given residual time r_1 activate the slow server if the queue size exceeds a specific value m_0.

Lemma 3. *The optimal threshold value* m_0 *for a threshold policy satisfies:*

$$m_0 \geq \left[1 - \frac{r_1}{T_1} + \left[\frac{T_2}{T_1} - \frac{r_1}{T_1} - 1\right] \cdot (1 - \lambda \cdot T_1)\right] = m_{LB}(\lambda, T_1, T_2, r_1),$$

where λ *is the Poisson arrival rate, and* r_1 *the residual time of the fast server.*

Proof. Assume that policy π never activates the slow server. Consider $\tilde{\pi}$ to be a threshold policy with threshold m for a given r_1. Consider a busy period of policy π. Clearly, the fast server under $\tilde{\pi}$ will become idle no later than the end of π's busy period starting from the same state. Assume that at time $t = 0$, $(x^0, r_1, r_2) = (m, r_1, 0)$ under both policies. Let k be the number of times $\tilde{\pi}$ will activate the slow server within the busy period, and let $t_1 = 0 < t_2 < t_3 \cdots < t_k$ be the instants of these activations. Clearly $t_i - t_{i-1} > T_2$ since to activate the slow server for a subsequent time, the service time of the previous activation must have terminated. Let L_i, $i = 1, \ldots, k$ be the number of packets served by S1 under π after time t_i. Clearly, $L_1 > L_2 > \cdots > L_k$. The gain experienced by policy $\tilde{\pi}$ over π is equal to:

$$G = \sum_{i=1}^{k} [L_i \cdot T_1 - (T_2 - T_1 - r_1)],$$

since for every packet sent to serve S2 all L_i packets save T_1 seconds each while there is a loss of $(T_2 - T_1 - r_1)$ for the dispatched packet. Clearly, the expression can be bounded as $G \le \sum_{i=1}^{k} [L_1 \cdot T_1 - (T_2 - T_1 - r_1)]$ (we have just replaced the L_i's with the larger L_1). Now if the expected value of $[L_i \cdot T_1 - (T_2 - T_1 - r_1)]$ is negative then $E[G] < 0$ as well. But $E[L_i \cdot T_1 - (T_2 - T_1 - r_1)] = T_1 E[L_1] - (T_2 - T_1 - r_1)$ where the expected value is taken conditional on given r_1. But L_1 is no greater than the number served by π during its busy period starting from state $(m - 1, r_1)$.

Thus, from [5] we obtain

$$T_1 \cdot E[L_1] \le \frac{(m-1) \cdot T_1 + r_i}{1 - \lambda \cdot T_1}.$$

Therefore, if

$$T_1 \cdot E[L_1] - (T_2 - T_1 - r_1) \le \frac{(m-1) \cdot T_1 + r_1}{1 - \lambda \cdot T_1} - (T_2 - T_1 - r_1) \le 0,$$

then

$$E[G] < 0.$$

Consequently, if

$$m \le \left[1 - \frac{r_1}{T_1} + \left[\frac{T_2}{T_1} - \frac{r_1}{T_1} - 1\right] \cdot (1 - \lambda \cdot T_1)\right] = m_{LB},$$

policy $\tilde{\pi}$ is inferior to π. Therefore, the above expression must be a lower bound to the optimal threshold value.

We can easily see that these arguments can be applied to the exponential service case. The only difference in this case is that the residual time $r_1(t)$ of the fast server is a random variable with the same distribution as the service time of the fast server: i.e., exponential with mean $T_1 = 1/\mu_1$. Thus, we have the following result:

Corollary. *A lower bound to the threshold value m_0 for the exponential servers case is:*

$$m_0 \geq \left[\left[\frac{\mu_1}{\mu_2} - 2 \right] \cdot \left(1 - \frac{\lambda}{\mu_1} \right) \right] = m_{LB}(\lambda, \mu_1, \mu_2).$$

Remark. The dependence of the lower bound, m_{LB}, on the residual service time in S1 is small and we can easily see that the maximum difference between the values of the bounds for different residual times is less than 2. Indeed

$$m_{LB}[r_1 = 0] - m_{LB}[r_1 = 1] = 2 - \lambda T_1.$$

The lower bound is surprisingly tight considering the simplifying assumptions we have made. One possible explanation is that the threshold in the optimal policy is set in such a way so that for given λ and μ_i's there are very few activations of the slow server during the busy period of the fast server.

A somewhat different argument yields an upper bound for the threshold. That is given by

Lemma 4. *The optimal threshold value m_0 satisfies $m_0 \leq \frac{T_2}{T_1}(1 + \lambda T_1)$.*

Table 1 provides a comparison between the upper and lower bound values as a function of the arrival rate λ, parametrized by three different sets of values of T_1 and T_2.

It is seen that the bounds are reasonably close and thus yield the means for easily predicting a good threshold value for the slow server activation problem. Thus, a theoretical result that was based on a simple model (but very difficult to prove) becomes usable as a practical method for implementing bandwidth allocation. The "wisdom" offered by this analysis can now be applied in more realistic environments in which analysis is prohibitive and yet guarantee reasonable performance.

Table 1. Upper and Lower Bound Values

	$T_1 = 1, T_2 = 10$		$T_1 = 1, T_2 = 5$		$T_1 = 1, T_2 = 3$	
Arrival Rate λ	Lower Bound	Upper Bound	Lower Bound	Upper Bound	Lower Bound	Upper Bound
0.0	8	10	3	5	1	3
0.1	8	10	3	5	1	3
0.2	7	9	3	5	1	3
0.3	6	8	3	4	1	3
0.4	5	8	2	4	1	3
0.5	4	7	2	4	1	2
0.6	4	7	2	4	1	2
0.7	3	6	1	3	1	2
0.8	2	6	1	3	1	2
0.9	1	6	1	3	1	2
1.0	0	5	0	3	0	2

IV. Conclusions

In this paper an effort was made to highlight the use of techniques from probability theory in areas of application that are very current and topical. Thus, it can be viewed as an effort to pay tribute to S. Cambanis' work that, although mainly theoretical in nature, always kept a close connection to practice and engineering applications through a careful choice of topics with special attention to motivation by need for, and impact on practice.

References

[1] A. Ephremides, P. Varaiya, and J. Walrand, A simple dynamic routing problem, *IEEE Trans. on Automatic Control*, August 1990.

[2] W. Lin and P. R. Kumar, Optimal control of a queueing systems with two heterogeneous servers, *IEEE Trans. on Automatic Control* Vol. AC-29, No. 8, August 1984, 696–703.

[3] J. Walrand, A note on 'Optimal Control of a Queueing Systems with Two Heterogeneous Services', *Systems and Control Letters* **4**, May 1984, 131–134.

[4] I. Viniotis and A. Ephremides, On the optimal dynamic switching of voice and data in communication networks, *Proc. of Computer Networking Symposium*, April 1988, 8–16.

[5] P. Bhattacharya and A. Ephremides, Optimal scheduling with strict deadlines, *IEEE Trans. on Automatic Control* **34** (1989), 721–728.

[6] P. Bhattacharya, and A. Ephremides, Optimal allocation of a server

between two queues with due times, *IEEE Trans. on Automatic Control* **36**, No. 12, December 1991, 1417–1423.

[7] D. Stoyan, *Comparison Methods for Queues and Other Stochastic Models*, Wiley, New York, 1983.

[8] T. Kamae, V. Krengel, and G. L. O'Brien, Stochastic inequalities on partially covered spaces, *Annals of Probability* **6**, No. 6, (1978), 1044–1049.

[9] M. Rubinovich, The slow server problem: A queue with stalling, *Journal of Applied Probability* **22**, (1985), 879–892.

Department of Electrical Engineering
University of Maryland
College Park, MD 20742

On the Conditional Variance-Covariance of Stable Random Vectors, II

Stergios B. Fotopoulos

In Memory of Stamatis Cambanis

If he is indeed wise, he does not bid you enter the house of his own wisdom, but rather leads to the threshold of your own mind.

Kahlil Gibran

Abstract

Under the assumption that $\mathbf{X} = (\mathbf{X_1}, \mathbf{X_2})$ a $(n_1 + n_2)$-dimensional vector is strictly α-stable distributed, the conditional variance-covariance of $\mathbf{X_2}$ given $\mathbf{X_1}$ is expressed in terms of the spectral measure Γ. Moreover, if some additional assumptions on the vector $\mathbf{X_1}$ are imposed such that the coordinates are statistically independent, then an additive expression for the conditional variance-covariance is found. A trigonometric unified method is presented for establishing these expressions.

AMS 1991 subject classifications: 60E07, 60E10
Key words and phrases: Stable distributions, conditional moments, nonlinear regression.

1. Introduction

Let $\mathbf{X_1} \in \mathbb{R}^{n_1}$ and $\mathbf{X_2} \in \mathbb{R}^{n_2}$ be random vectors, such that $\mathbf{X} = (\mathbf{X}_1', \mathbf{X}_2')' \in \mathbb{R}^n$, $n = n_1 + n_2$, has a joint α-stable symmetric ($S\alpha S$), $1 < \alpha < 2$, distribution. That is, its characteristic function is of the form

$$\begin{aligned}\phi_{\mathbf{X_1},\mathbf{X_2}}(\mathbf{t_1}, \mathbf{t_2}) : &= E[\exp(i(\langle \mathbf{t_1}, \mathbf{X_1} \rangle + \langle \mathbf{t_2}, \mathbf{X_2} \rangle))] \\ &= \exp\left(-\int_{S_{n-1}} |\langle \mathbf{t_1}, \mathbf{s_1} \rangle + \langle \mathbf{t_2}, \mathbf{s_2} \rangle|^\alpha \Gamma(d\mathbf{s})\right),\end{aligned}$$

where Γ is the spectral measure of $\mathbf{X}$; it is a finite symmetric measure on the Borel sets of the unit circle $S_{n-1} = \{\mathbf{x} \in \mathbb{R}^n : ||\mathbf{x}|| = 1\}$ ($|| \cdot ||$) denotes the usual Euclidean distance). We assume that the components of $\mathbf{X}$ are non-degenerate and linearly independent. The objective is to obtain necessary and sufficient conditions for the existence of the conditional variance-covariance of $\mathbf{X_2}$ given $\mathbf{X_1}$ and derive an exact representation of the covariance with respect to spectral measure. In addition, we provide sufficient conditions in order to characterize conditional variance-covariance under independence for $S\alpha S$ vectors. We will show that the condition of independence reveals an additive expression. This work focuses on the case of $1 < \alpha < 2$.

Our overall approach is akin to that used in Cambanis and Wu (1992) and Cambanis and Fotopoulos (1995). However, various new ideas are integrated and a classical trigonometric approach is emphasized. The conditional variance when $n_1 = n_2 = 1$ was investigated by Wu and Cambanis (1992) and Cioczek–Georges and Taqqu (1995).

Throughout this work we shall use vector notation. For $\mathbf{a_1} = (a_{11}, \cdots, a_{1n})' \in \mathbb{R}^n$, $\mathbf{b_1} \in \mathbb{R}^n$, $\langle \mathbf{a_1}, \mathbf{b_1} \rangle$ denotes the usual inner product and $||\mathbf{a_1}||^2 = \langle \mathbf{a_1}, \mathbf{a_1} \rangle$ and $b^{\langle a \rangle} = \text{sign}(b)|b|^a$, for $b \in \mathbb{R}$. The layout of the remainder of the paper is as follows: Section 2 contains the main results of the investigation; Section 3 presents the proofs; and Section 4 states several useful technical lemmas which are repeatedly used in Section 3.

2. Results

It is easy to verify that when the conditional distribution of $\mathbf{X_2} \in \mathbb{R}^{n_2}$ given $\mathbf{X_1} \in \mathbb{R}^{n_1}$ is stable with index α, then $E\left[|X_{2j}|^p|\mathbf{X_1} = \mathbf{x_1}\right] < \infty$, $j = 1, \cdots, n_2$, but only for $p < \alpha$, and not necessarily for $p = \alpha$. On the other hand, Samorodnitsky and Taqqu (1991), Cioczek–Georges and Taqqu (1993) and Cambanis and Fotopoulos (1995) obtained necessary and sufficient conditions for which $E\left[|X_{2j}|^p|\mathbf{X_1} = \mathbf{x_1}\right]$, $j = 1, \cdots, n_2$, is finite almost everywhere, even for some $p \geq \alpha$. In the same framework, necessary and sufficient conditions for which $\text{Cov}(\mathbf{X_2}|\mathbf{X_1} = \mathbf{x_1})$ exist *a.e.* are obtained and, consequently, expressions of the conditional covariance with respect to the spectral measure $\Gamma(\cdot)$ are displayed.

These results supplement the work of Cambanis and Fotopoulos (1995). The methods for obtaining the conditional covariance deviate from the previous study. New ideas are added and various extensions are utilized to broaden the theory under conditioning.

The following theorem extends Theorem 1 in Cambanis and Fotopoulos, and shows that under the same conditions, we may extrapolate some of the previous results.

Theorem 1. *Let* $\mathbf{X_1} \in \mathbb{R}^{n_1}$, $n_1 \geq 1$, *and* $\mathbf{X_2} \in \mathbb{R}^{n_2}$, $n_2 \geq 1$, *be random vectors such that the components of* $\mathbf{X} = (\mathbf{X_1'}, \mathbf{X_2'})'$ *are linearly independent. Let* $\mathbf{X} \equiv (\mathbf{X_1}, \mathbf{X_2}) \in \mathbb{R}^n$, $n = n_1 + n_2$, *be* $S\alpha S$ *with* $1 < \alpha < 2$. *Then*

$E\left[\|\mathbf{X}_2\|^2|\mathbf{X}_1=\mathbf{x}_1\right]<\infty$ *for all* $\mathbf{x}_1\in\mathbb{R}^{n_1}$, *if and only if*

$$\int_{\mathbf{R}^{n_1}}\phi_{\mathbf{X}_1}(\mathbf{t}_1)\int_{S_{n-1}}\frac{\|\mathbf{s}_2\|^2\Gamma(ds)}{|\langle\mathbf{t}_1,\mathbf{s}_1\rangle|^{2-\alpha}}d\mathbf{t}_1<\infty\,.$$

Also for all $\mathbf{x}_1\in\mathbb{R}^{n_1}$,

I. $$E\left[X_{2j}^2|\mathbf{X}_1=\mathbf{x}_1\right]=\frac{1}{(2\pi)^{n_1}f_{\mathbf{X}_1}(\mathbf{x}_1)}\int_{\mathbf{R}^{n_1}}e^{-i\langle\mathbf{t}_1,\mathbf{x}_1\rangle}\phi_{\mathbf{X}_1}(\mathbf{t}_1)d\mathbf{t}_1$$

$$\left\{\alpha(\alpha-1)\int_{S_{n-1}}\frac{s_{2j}^2\Gamma(ds)}{|\langle\mathbf{t}_1,\mathbf{s}_1\rangle|^{2-\alpha}}-\left[\alpha^2\int_{S_{n-1}}s_{2j}\langle\mathbf{t}_1,\mathbf{s}_1\rangle^{\alpha-1}\Gamma(ds)\right]^2\right\},$$

for any $j=1,\cdots,n_2$, *and*

II. $E\left[X_{2j}X_{2k}|\mathbf{X}_1=\mathbf{x}_1\right]$

$$=\frac{4\alpha(\alpha-1)\cot\frac{\alpha\pi}{2}}{(2\pi)^{n_1}f_{\mathbf{X}_1}(\mathbf{x}_1)}\int_{\mathbf{R}^{n_1}}e^{-i\langle\mathbf{t}_1,\mathbf{x}_1\rangle}\phi_{\mathbf{X}_1}(\mathbf{t}_1)d\mathbf{t}_1\int_{S_{n-1}}\frac{s_{2j}s_{2k}\Gamma(ds)}{|\langle\mathbf{t}_1,\mathbf{s}_1\rangle|^{2-\alpha}}d\mathbf{t}_1\,,$$

for any $j,k=1,2,\cdots,n_2$, *and* $j\neq k$.

The motivation to derive expressions of the kind I and II in Theorem 1 is not only because they complete many earlier results in the theory of stable vectors, but also because they upgrade various related issues. For example, the behavior of the stable distributions would be better understood. This was also suggested by Hardin et al. (1991) who recommended that in order to comprehend the distribution of stable vectors, one needs to study the conditional behavior "of one stable variable given the observation of others." The complexity of the distribution of stable vectors is once again verified by viewing the explicit analytic formulae of I and II in Theorem 1, where one notices how different their functional forms are. This, of course, is not the case when Gaussian vectors are considered instead of stable ones, in which case degenerate expressions are obtained for both conditional variance and covariance. Another issue of concern is how to employ these formulae in miscellaneous applications of stochastic processes, especially in the stable motions.

Stronger sufficient conditions than the ones given in Theorem 1 are provided in Cambanis and Fotopoulos (1995). To avoid overlaps, we shall continue with the situation where the components of $\mathbf{X}_1=(X_{11},\cdots,X_{1n_1})$ are independent. Then, using similar techniques as in Cambanis and Fotopoulos (1995), additive expressions are derived.

To justify a second order differentiation of $\log\phi_{\mathbf{X}_1,X_{2j},X_{2k}}(\mathbf{t}_1,t_{2j},t_{2k})$ and $\log\phi_{\mathbf{X}_1,X_{2j}}(\mathbf{t}_1,t_{2j})$, with respect to t_{2j} and t_{2k} at $t_{2j}=t_{2k}=0$, for $j,k=1,2,\cdots,n_2,j\neq k$, the following proposition is offered.

Proposition 1. *Let* $\Lambda(\mathbf{t_1}, t_{2j}) = \int_{S_{n-1}} |\langle \mathbf{t_1}, \mathbf{s_1} \rangle + t_{2j}s_{2j}|^{\alpha}\Gamma(d\mathbf{s})$, $\mathbf{t_1} \in \mathbb{R}^{n_1}$, $t_{2j} \in \mathbb{R}$ *and* $1 < \alpha < 2$. *Then, for almost all* $t_{2j} \in \mathbb{R}$

$$I.\ \frac{\partial \Lambda}{\partial t_{2j}}(\mathbf{t_1}, t_{2j}) = \alpha \int_{S_{n-1}} s_{2j} \left\{\langle \mathbf{t_1}, \mathbf{s_1} \rangle + t_{2j}s_{2j}\right\}^{\langle \alpha - 1 \rangle} \Gamma(d\mathbf{s}), \ \ and$$

$$II.\ \frac{\partial^2 \Lambda}{\partial t_{2j}^2}(\mathbf{t_1}, t_{2j}) = \alpha(\alpha - 1) \int_{S_{n-1}} s_{2j} \frac{s_{2j}^2 \Gamma(d\mathbf{s})}{|\langle \mathbf{t_1}, \mathbf{s_1} \rangle + s_{2j}t_{2j}|^{2-\alpha}} .$$

Furthermore, for a fixed $\mathbf{t_1} \in \mathbb{R}^{n_1}$, *both* $\frac{\partial \Lambda}{\partial t_{2j}}(\mathbf{t_1}, t_{2j})$ *and* $\frac{\partial^2 \Lambda}{\partial t_{2j}^2}(\mathbf{t_1}, t_{2j})$ *are continuous at* $t_{2j} = 0$.

An additive property for the conditional covariance is revealed when the components of $\mathbf{X_1}$ are independent. We will show that both the conditional second moment, $E\left[X_{2j}^2|\mathbf{X_1} = \mathbf{x_1}\right]$, and cross products, $E\left[X_{2j}X_{2k}|\mathbf{X_1} = \mathbf{x_1}\right]$, $j, k = 1, \cdots, n_2, j \neq k$, will be expressed as a sum of $E\left[X_{2j}^2|X_{1i} = x_{1i}\right]$ and $E\left[X_{2j}X_{2k}|X_{1i} = x_{1i}\right]$, respectively. It is notable that the following two propositions enable us to study the conditional variance-covariance matrix for all multivariate vectors and not just for the stable case. Thus, we will use $\Lambda(\mathbf{t_1}, t_{2j}, t_{2k}) = -\log \phi_{\mathbf{X_1}, X_{2j}, X_{2k}}(\mathbf{t_1}, t_{2j}, t_{2k})$ instead of the $\Lambda(\mathbf{t_1}, t_{2j}, t_{2k}) = \int_{S_{n-1}} |\langle \mathbf{t_1}, \mathbf{s_1} \rangle + t_{2j}s_{2j} + t_{2k}s_{2k}|^{\alpha}\Gamma(ds)$. Similar methods were also used for Propositions 1 and 1′ in Cambanis and Wu (1992).

Proposition 2. *Assume that the components of* $\mathbf{X_1} \in \mathbb{R}^{n_1}$ *are independent and their corresponding characteristic functions* $\phi_{X_{11}}, \cdots, \phi_{X_{1n_1}}$ *are integrable and nonzero a.e. Further assume that* $\int_{\mathbb{R}^{n_1}} \left|E\left[||\mathbf{X_2}||^2 e^{i\langle \mathbf{X_1}, \mathbf{t_1} \rangle}\right]\right| d\mathbf{t_1} < \infty$. *Then*

I. $E\left[||\mathbf{X_2}||^2|\mathbf{X_1}\right] = \sum_{i=1}^{n_1} E\left[||\mathbf{X_2}||^2|X_{1i}\right]$, *a.s., if and only if for all* $\mathbf{t_1} \in \mathbb{R}^{n_1}$ *and* $\forall j = 1, \cdots, n_2$,

$$\frac{\partial^2 \Lambda}{\partial t_{2j}^2}(\mathbf{t_1}, t_{2j})|_{t_{2j}=0} + \left[\frac{\partial \Lambda}{\partial t_{2j}}(\mathbf{t_1}, t_{2j})|_{t_{2j}=0}\right]^2$$

$$= \sum_{i=1}^{n_1} \left\{ \frac{\partial^2 \Lambda}{\partial t_{2j}^2}(0, \cdots, t_{1i}, \cdots, 0, t_{2j})|_{t_{2j}=0} + \left[\frac{\partial \Lambda}{\partial t_{2j}}(0, \cdots, t_{1i}, \cdots, 0, t_{2j})|_{t_{2j}=0}\right]^2 \right\},$$

and

II. $E[X_{2j}X_{2k}|\mathbf{X_1}] = \sum_{i=1}^{n_1} E[X_{2j}X_{2k}|X_{1i}]$ *a.s., if and only if for all* $\mathbf{t_1} \in \mathbb{R}^{n_1}$

$$\frac{\partial^2\Lambda}{\partial t_{2j}t_{2k}}(\mathbf{t_1}, t_{2j}, t_{2k})\Big|_{\substack{t_{2j}=0\\ t_{2k}=0}} = \sum_{i=1}^{n_1} \frac{\partial^2\Lambda}{\partial t_{2j}t_{2k}}(0, \cdots, t_{1i}, \cdots, 0, t_{2j}, t_{2k})\Big|_{\substack{t_{2j}=0\\ t_{2k}=0}},$$

where $\Lambda(\mathbf{t_1}, t_{2j}, t_{2k}) = -\log \phi_{\mathbf{X_1}, X_{2j}, X_{2k}}(\mathbf{t_1}, t_{2j}, t_{2k})$ *and* $j, k = 1, \cdots, n_2,\ j \neq k$.

Under the same arguments used in Proposition 2, we may also establish the following more general result.

Proposition 3. *Let* $\mathbf{X_{k+1}} \in \mathbb{R}^{n_{k+1}}$, $n_{k+1} \geq 1$, *and let the vectors* $\mathbf{X_1} \in \mathbb{R}^{n_1}$, $n_1 \geq 1, \cdots, \mathbf{X_k} \in \mathbb{R}^{n_k}$, $n_k \geq 1$, *be independent with their characteristic functions being integrable and nonzero a.e. Let* $\int_{\mathbb{R}^n} \left|E\left[||\mathbf{X_{k+1}}||^2 \exp(i\langle \mathbf{t}, \mathbf{X}\rangle)\right]\right| d\mathbf{t} < \infty$, *with* $\mathbf{X} = (\mathbf{X}'_{\mathbf{n_1}}, \cdots, \mathbf{X}'_{\mathbf{n_k}})', \mathbf{X} \in \mathbb{R}^n, n = n_1 + \cdots + n_k$.

Then

I. $E\left[||\mathbf{X_{k+1}}||^2|\mathbf{X}\right] = \sum_{i=1}^{k} E\left[||\mathbf{X_{k+1}}||^2|\mathbf{X_i}\right]$, *a.s., if and only if for all* $\mathbf{t} \in \mathbb{R}^n$ *and* $\forall\ j = 1, \cdots, n_{k+1}$

$$\frac{\partial^2\Lambda}{\partial t_{k+1,j}^2}(\mathbf{t}, t_{k+1,j})|_{t_{k+1,j}=0} + \left[\frac{\partial\Lambda(\mathbf{t}, t_{k+1,j})}{\partial t_{k+1,j}}|_{t_{k+1,j}=0}\right]^2$$

$$= \sum_{i=1}^{k}\left\{\frac{\partial^2\Lambda}{\partial t_{k+1,j}^2}(0, \cdots, \mathbf{t_i}, \cdots, 0, t_{k+1,j})|_{t_{k+1,2j}=0} + \left[\frac{\partial\Lambda}{\partial t_{k+1,j}}(0, \cdots, \mathbf{t_i}, \cdots, 0, t_{k+1,j})|_{t_{k+2,j}=0}\right]^2\right\},$$

and

II. $E[X_{k+1,j}X_{k+1,\ell}|\mathbf{X}] = \sum_{i=1}^{k} E[X_{k+1,j}X_{k+1,\ell}|\mathbf{X_i}]$ *a.s., if and only if for all* $\mathbf{t} \in \mathbb{R}^n$

$$\frac{\partial^2\Lambda}{\partial t_{k+1,j}t_{k+1,\ell}}(\mathbf{t}, t_{k+1,j}, t_{k+1,\ell})\Big|_{\substack{t_{k+1,j}=0\\ t_{k+1,\ell}=0}} = \sum_{i=1}^{k} \frac{\partial^2\Lambda}{\partial t_{k+1,j}t_{k+1,\ell}}(0, \cdots, \mathbf{t_i}, \cdots, 0, t_{k+1,j}, t_{k+1,\ell})\Big|_{\substack{t_{k+1,j}=0\\ t_{k+1,\ell}=0}},$$

for all $j, \ell = 1, \cdots, n_{k+1}, j \neq \ell$.

The propositions developed can now be adapted for deriving necessary and sufficient conditions under which expressions for the conditional variance-covariance matrix with respect to spectral measures are characterized. At the same time, the propositions also enable us to provide a link to the results of Cambanis and Fotopoulos (1995).

Theorem 2. *If* $\mathbf{X_1} \in \mathbb{R}^{n_1}$, $n_1 \geq 1$, *and* $\mathbf{X_2} \in \mathbb{R}^{n_2}$, $n_2 \geq 1$ *are jointly SaS with* $1 < \alpha < 2$, *and the components of* $\mathbf{X_1}$ *are independent, and if* $\int_{\mathbb{R}^{n_1}} \phi_{\mathbf{X_1}}(\mathbf{t_1}) \int_{S_{n-1}} \frac{||\mathbf{s_2}||^2\Gamma(d\mathbf{s})}{|\langle \mathbf{t_1}, \mathbf{s_1} \rangle|^{2-\alpha}} d\mathbf{t_1} < \infty$, *then the following statements are always true:*

I. 1. *For all* $i = 1, \cdots, n_1$ $\quad E\left[||\mathbf{X_2}||^2 | X_{1i} = x_{1i}\right] < \infty$, *a.e.,*

2. *for all* $i = 1, \cdots, n_1$
$$\int_{S_{n-1}\cap\{||\mathbf{s_2}||^2+s_{1j}^2=1, s_{1j}\neq 0\}} \frac{||\mathbf{s_2}||^2\Gamma(d\mathbf{s})}{|s_{1j}|^{2-\alpha}} < \infty,$$

3. $E\left[||\mathbf{X_2}||^2 | \mathbf{X_1} = \mathbf{x_1}\right] = \sum_{i=1}^{n_1} E\left[||\mathbf{X_2}||^2 | X_{1i} = x_{1i}\right]$, *a.e.,*

4. *for all* $\mathbf{t_1} \in \mathbb{R}^{n_1}$
$$\int_{S_{n-1}} \frac{||\mathbf{s_2}||^2\Gamma(d\mathbf{s})}{|\langle \mathbf{t_1}, s_1 \rangle|^{2-\alpha}}$$
$$= \sum_{i=1}^{n_1} \frac{1}{|t_{1i}|^{2-\alpha}} \int_{S_{n-1}\cap\{||\mathbf{s_2}||^2+s_{1j}^2=1, s_{1j}\neq 0\}} \frac{||\mathbf{s_2}||^2\Gamma(d\mathbf{s})}{|s_{1j}|^{2-\alpha}},$$

and

II. 1. $E\left[X_{2j}X_{2k} | \mathbf{X_1} = \mathbf{x_1}\right] = \sum_{i=1}^{n_1} E\left[X_{2j}X_{2k} | X_{1i} = x_{1i}\right]$, *a.e.,* $j, k = 1, \cdots, n_2, j \neq k$, *and*

2. *for all* $\mathbf{t_1} \in \mathbb{R}^{n_1}$
$$\int_{S_{n-1}} \frac{s_{2j}s_{2k}\Gamma(d\mathbf{s})}{|\langle \mathbf{t_1}, s_1 \rangle|^{2-\alpha}}$$
$$= \sum_{i=1}^{n_1} \frac{1}{|t_{1i}|^{2-\alpha}} \int_{S_{n-1}\cap\{||\mathbf{s_2}||^2+s_{1j}^2=1, s_{1j}\neq 0\}} \frac{s_{2j}s_{2k}\Gamma(d\mathbf{s})}{|s_{1j}|^{2-\alpha}},$$

for all $j, k = 1, \cdots, n_2, j \neq k$.

Next, a generalized version of Theorem 2 is given.

Theorem 3. *If* $\mathbf{X_{k+1}} \in \mathbb{R}^{n_{k+1}}, n_{k+1} \geq 1$ *and* $\mathbf{X_1} \in \mathbb{R}^{n_1}, n_1 \geq 1, \cdots, \mathbf{X_k} \in \mathbb{R}^{n_k}, n_k \geq 1$, *are jointly SαS with* $1 < \alpha < 2$, *and the components of* $\mathbf{X}' = (\mathbf{X}_1', \cdots, \mathbf{X}_k')'$ *are independent, and if* $\int_{\mathbb{R}^n} \phi_{\mathbf{X}}(\mathbf{t}) \int_{S_{n_{k+1}+n-1}} \frac{||\mathbf{s_{k+1}}||^2 \Gamma(d(\mathbf{s}, \mathbf{s_{k+1}}))}{|\langle \mathbf{t}, \mathbf{s} \rangle|^{2-\alpha}} d\mathbf{t} < \infty$, *then the following statements are always true:*

I. 1. *For all* $i = 1, \cdots, k$ $\quad E\left[||\mathbf{X_{k+1}}||^2 | \mathbf{X_i} = \mathbf{x_i}\right] < \infty$, *a.e.,*

2. *for all* $\mathbf{t} \in \mathbb{R}^n$

$$\int_{S_{n-1} \cap \{||\mathbf{s}_2||^2 + ||\mathbf{s}_j||^2 = 1, \mathbf{s}_j \neq \mathbf{0}\}} \frac{||\mathbf{s_{k+1}}||^2 \Gamma(d(\mathbf{s}, \mathbf{s_{k+1}}))}{|\langle \mathbf{t_i}, \mathbf{s_i} \rangle|^{2-\alpha}} < \infty, \ i = 1, \cdots, k$$

3. $E\left[||\mathbf{X_{k+1}}||^2 | \mathbf{X}\right] = \sum_{i=1}^{k} E\left[||\mathbf{X_{k+1}}||^2 | \mathbf{X_i}\right]$, *a.e., and*

4. *for all* $\mathbf{t} \in \mathbb{R}^n$

$$\int_{S_{n_{k+1}+n-1}} \frac{||\mathbf{s_{k+1}}||^2 \Gamma(d(\mathbf{s}, \mathbf{s_{k+1}}))}{|\langle \mathbf{t}, \mathbf{s} \rangle|^{2-\alpha}} d\mathbf{t}$$

$$= \sum_{i=1}^{k} \frac{1}{|t_{1i}|^{2-\alpha}} \int_{S_{n-1} \cap \{||\mathbf{s}_2||^2 + ||\mathbf{s}_i||^2 = 1, \mathbf{s}_i \neq \mathbf{0}\}} \frac{||\mathbf{s_{k+1}}||^2 \Gamma(d(\mathbf{s}, \mathbf{s_{k+1}}))}{|\langle \mathbf{t_i}, \mathbf{s_i} \rangle|^{2-\alpha}},$$

and

II. 1. $E\left[X_{k+1,j} X_{k+1,\ell} | \mathbf{X} = \mathbf{x}\right] = \sum_{i=1}^{k} E\left[X_{k+1,j} X_{k+1,\ell} | \mathbf{X_i} = \mathbf{x_i}\right]$, *a.e., and*

2. *for all* $\mathbf{t} \in \mathbb{R}^n$

$$\int_{S_{n_{k+1}+n-1}} \frac{s_{k+1,j} s_{k+1,\ell} \Gamma(d(\mathbf{s}, \mathbf{s_{k+1}}))}{|\langle \mathbf{t}, \mathbf{s} \rangle|^{2-\alpha}}$$

$$= \sum_{i=1}^{k} \int_{S_{n-1} \cap \{||\mathbf{s}_2||^2 + ||\mathbf{s}_i||^2 = 1, \mathbf{s}_i \neq \mathbf{0}\}} \frac{s_{k+1,j} s_{k+1,\ell} \Gamma(d(\mathbf{s}, \mathbf{s_{k+1}}))}{|\langle \mathbf{t_i}, \mathbf{s_i} \rangle|^{2-\alpha}},$$

for all $j, k = 1, \cdots, n_{k+1}, j \neq k$.

Wu and Cambanis (1992) noticed that in the bivariate strictly stable case with $1 < \alpha < 2$, the conditional variance is affected only through a global multiplier which depends on the joint distribution. Cambanis and Fotopoulos (1995) showed that if the coordinates of $\mathbf{X_1}$ are statistically independent, then $\operatorname{Var}(X_{2j}|\mathbf{X_1} = \mathbf{x_1}) = \sum_{k=1}^{n_1} D^2_{n|k}(\alpha) S_1^2(\frac{x_{1k}}{\sigma_{1k}}; \alpha)$, where $D^2_{n|k}(\alpha)$, σ^{α}_{1k}, and $S_1^2(x;\alpha)$ are defined in the following corollary. The quantity $S_1^2(x;\alpha)$ was first investigated by Wu and Cambanis (1992) and Cioczek–Georges and Taqqu (1995) in the bivariate setup; Cambanis et al., (1997) studied a more complex version in the multivariate case. Here we adopt some of these ideas to produce the following corollary.

Corollary. *Let the vectors* $\mathbf{X_1} \in \mathbb{R}^{n_1}$, $n_1 \geq 1$, *and* $\mathbf{X_2} \in \mathbb{R}^{n_2}$, $n_2 \geq 1$ *have linearly independent components and be jointly* $S\alpha S$ *with* $1 < \alpha < 2$ *and let the components of* $\mathbf{X_1}$ *be stochastically independent.*
If $E\left[X^2_{2j}|\mathbf{X_1} = \mathbf{x_1}\right] < \infty$ *for all* $\mathbf{x_1} \in \mathbb{R}^{n_1}$, *then*

$$I. \qquad Var\,(X_{2j}|\mathbf{X_1} = \mathbf{x_1}) = \sum_{k=1}^{n_1} D^2_{n|k}(\alpha) S_1^2(\frac{x_{1k}}{\sigma_{1k}};\alpha), \text{ and}$$

$$II. \qquad Cov\,(X_{2j}X_{2\ell}|\mathbf{X_1} = \mathbf{x_1}) = \sum_{k=1}^{n_1} \left\{ M_{n|k}(\alpha)\frac{x^2_{1k}}{\sigma^2_{1k}} + F_{n|k}(\alpha) R_1(\frac{x_{1k}}{\sigma_{1k}};\alpha) \right\},$$

for $j, \ell = 1, \cdots, n_2$, $j \neq \ell$ *and for all* $\mathbf{x_1} \in \mathbb{R}^{n_1}$, *where the universal "standard deviation"* $S_1^2(x;\alpha)$ *and* $R_1(x;\alpha)$ *are*

$$\begin{aligned} S_1^2(x;\alpha) &= \alpha(\alpha-1)\frac{\int_0^\infty \cos(xt)e^{-t^\alpha}t^{\alpha-2}dt}{\int_0^\infty \cos(xt)e^{-t^\alpha}dt}, \\ R_1(x;\alpha) &= \frac{\alpha^2\int_0^\infty \cos(xt)e^{-t^\alpha}t^{2\alpha-2}dt}{\int_0^\infty \cos(xt)e^{-t^\alpha}dt} \end{aligned}$$

and the coefficients $D^2_{n|k}(\alpha)$, $M_{n|k}(\alpha)$ *and* $F_{n|k}(\alpha)$ *depend on the joint distribution of* X_{1k} *and* $\mathbf{X_2}$, $k = 1, \cdots, n_1$, *in the following ways:*

$$\begin{aligned} D^2_{n|k}(\alpha) &= \sigma_{1k}^{-2(\alpha-1)}\Bigg\{\sigma^\alpha_{1k} \int_{S_{n-1}\cap\{||\mathbf{s}_2||^2+s^2_{1k}=1, s_{1k}\neq 0\}} |s_{1k}|^{2-\alpha}s^2_{2j}\Gamma(d\mathbf{s}) \\ &\quad - \left(\int_{S_{n-1}} s_{1k}^{\langle 1-\alpha\rangle} s_{2j}\Gamma(d\mathbf{s})\right)^2\Bigg\}, \end{aligned}$$

$$F_{n|k}(\alpha) = 4\cot\frac{\alpha\pi}{2}\sigma_{1k}^{2-\alpha} \int_{S_{n-1}\cap\{||\mathbf{s}_2||^2+s^2_{1k}=1, s_{1k}\neq 0\}} \frac{s_{2j}s_{2\ell}\Gamma(d\mathbf{s})}{|s_{1k}|^{2-\alpha}}, \text{ and}$$

$$M_{n|k}(\alpha) = F_{n|k}(\alpha) - \int_{S_{n-1}} s_{2j}s_{1k}^{\langle\alpha-1\rangle}\Gamma(d\mathbf{s}) \int_{S_{n-1}} s_{2\ell}s_{1k}^{\langle\alpha-1\rangle}\Gamma(d\mathbf{s}),$$

and $\sigma_{1k}^{\alpha} = \int_{S_{n-1}} |s_{1k}|^{\alpha}\Gamma(d\mathbf{s})$ *is the scale of* X_{1k} *such that* $\phi_{X_{1k}}(t_{1k}) = e^{-\sigma_{1k}^{\alpha}|t_{1k}|^{\alpha}}$. *Moreover, for* $1 < \alpha < 2$

$$\begin{array}{ll} (i) & S_1^2(x;\alpha) = \dfrac{\alpha^2 \int_0^\infty \cos(xt)e^{-t^\alpha}t^{2\alpha-2}dt}{\int_0^\infty \cos(xt)e^{-t^\alpha}dt} + x^2, \\ (ii) & S_1^2(x;\alpha) - x^2 \to -\infty, \text{ as } x \to -\infty, \\ (iii) & S_1(x;\alpha) = x + o(x), \text{ as } x \to -\infty, \text{and} \\ (iv) & S_1(x;\alpha) = S_1(0;\alpha) + C(\alpha)x^2 + o(x^2), \text{ as } x \to 0, \end{array}$$

where

$$\begin{aligned} S_1(0;\alpha) &= \left[\frac{\alpha(\alpha-1)\Gamma(1-1/\alpha)}{\Gamma(1/\alpha)}\right]^{1/2}, \\ C(\alpha)S_1(0;\alpha) &= (\alpha - 1/4)\left[-1 + \frac{\alpha\Gamma(3/\alpha)\Gamma(1-1/\alpha)}{\Gamma^2(1/\alpha)}\right] \geq 0. \end{aligned}$$

Remark. Cioczek–Georges and Taqqu (1995) have extended these results to the case where $1/2 < \alpha \leq 1$.

3. Proofs

Some of the ingredients in the following proofs are straightforward adaptations of the original ideas from Cambanis and Fotopoulos (1995). For the worst part, we will skip over those, emphasizing the differences which arise in our case.

Proof of Theorem 1. Since the characteristic function $\phi_{\mathbf{X}_1,\mathbf{X}_2}(\mathbf{t}_1,\mathbf{t}_2)$ is integrable in $\mathbf{t}_1$ for each $\mathbf{t}_2$, then it is easy to show that the characteristic function of the probability density $f_{\mathbf{X}_2|\mathbf{X}_1}(\mathbf{x}_2|\mathbf{x}_1)$ is given by

$$\Psi(\mathbf{t}_2;\mathbf{x}_1) := E\left[e^{i\langle \mathbf{t}_2,\mathbf{x}_2\rangle}|\mathbf{X}_1 = \mathbf{x}_1\right] = \frac{\int_{\mathbb{R}^{n_1}} e^{-i\langle \mathbf{t}_1,\mathbf{x}_1\rangle}\phi_{\mathbf{X}_1,\mathbf{X}_2}(\mathbf{t}_1,\mathbf{t}_2)d\mathbf{t}_1}{\int_{\mathbb{R}^{n_1}} e^{-i\langle \mathbf{t}_1,\mathbf{x}_1\rangle}\phi_{\mathbf{X}_1}(\mathbf{t}_1)d\mathbf{t}_1}, \quad (3.1)$$

for any $\mathbf{t}_2 \in \mathbb{R}^{n_2}$ and $\mathbf{x}_1 \in \mathbb{R}^{n_1}$. Note that from the inversion theorem, the denominator of the expression in the right-hand side of (3.1) is just $(2\pi)^{n_1} f_{\mathbf{X}_1}(\mathbf{x}_1)$.

To show that $E[\mathbf{X}_2\mathbf{X}_2'|\mathbf{X}_1 = \mathbf{x}_1]$ exists, it suffices to show that $E\left[\text{trace}(\mathbf{X}_2\mathbf{X}_2')|\mathbf{X}_1 = \mathbf{x}_1\right] = \sum_{j=1}^{n_2} E\left[X_{2j}^2|\mathbf{X}_1 = \mathbf{x}_1\right] < \infty$. The last formula requires the existence of each $E\left[X_{2j}^2|\mathbf{X}_1 = \mathbf{x}_1\right]$. We proceed by first showing the existence of $E\left[X_{2j}^2|\mathbf{X}_1 = \mathbf{x}_1\right]$, $j = 1, \cdots, n_2$, and then finding expressions for both $E\left[X_{2j}^2|\mathbf{X_1}=\mathbf{x_1}\right]$ and $E\left[X_{2j}X_{2k}|\mathbf{X_1}=\mathbf{x_1}\right]$, $j,k=1,\cdots,n_2,\ j\neq k$ with respect to spectral measure $\Gamma(\cdot)$.

In view of Lemma 1, we need to show that for $j=1,\cdots,n_2$

$$\lim_{t_{2j}\to+0} I(t_{2j};\mathbf{x_1}) = \lim_{t_{2j}\to+0} \frac{\{2-\Psi(t_{2j};\mathbf{x_1})-\Psi(-t_{2j};\mathbf{x_1})\}}{t_{2j}^2} \quad \text{exists.} \tag{3.2}$$

As in Cambanis and Fotopoulos (1995),

$$\begin{aligned} I(t_{2j};\mathbf{x_1}) &= \frac{1}{(2\pi)^{n_1} f_{\mathbf{X_1}}(\mathbf{x_1})} \\ &\qquad \int_{\mathbb{R}^{n_1}} e^{-\langle \mathbf{t_1},\mathbf{x_1}\rangle}\phi_{\mathbf{X_1}}(\mathbf{t_1})\left\{J_1(\mathbf{t_1},t_{2j})+J_2(\mathbf{t_1},t_{2j})\right\}d\mathbf{t_1} \\ &:= I_1(t_{2j};\mathbf{x_1})+I_2(t_{2j};\mathbf{x_1}),\ \mathbf{x_1}\in\mathbb{R}^{n_1}, \end{aligned} \tag{3.3}$$

where

$$\begin{aligned} J_1(\mathbf{t_1},t_{2j}) &:= \frac{1}{t_{2j}^2}\Big\{[1-\Delta(\mathbf{t_1},t_{2j})-\exp(-\Delta(\mathbf{t_1},t_{2j}))] \\ &\qquad +[1-\Delta(\mathbf{t_1},-t_{2j})-\exp(-\Delta(\mathbf{t_1},-t_{2j}))]\Big\} \\ J_2(\mathbf{t_1},t_{2j}) &:= \frac{1}{t_{2j}^2}\left\{\Delta(\mathbf{t_1},t_{2j})+\Delta(\mathbf{t_1},-t_{2j})\right\} \end{aligned}$$

and

$$\Delta(\mathbf{t_1},t_{2j}) := \int_{S_{n-1}} [|\langle \mathbf{t_1},\mathbf{s_1}\rangle + t_{2j}s_{2j}|^\alpha - |\langle \mathbf{t_1},\mathbf{s_1}\rangle|^\alpha]\Gamma(d\mathbf{s})\,.$$

The existence of both $I_1(t_{2j};\mathbf{x_1})$ and $I_2(t_{2j};\mathbf{x_1})$ were answered in Cambanis and Fotopoulos (1995). In this article, we obtain their limits as $t_{2j}\to+0$ and consequently, provide the exact expression for both the variance and covariance.

We now evaluate $\Delta(\cdot,\cdot)$ with respect to $\lambda_i(\cdot,\cdot)$. Note that by Lemma 3 it follows that

$$\Delta(\mathbf{t_1},t_{2j}) = K_1(\mathbf{t_1},t_{2j})+K_2(\mathbf{t_1},t_{2j})\,,$$

and

$$\Delta(\mathbf{t_1},-t_{2j}) = K_1(\mathbf{t_1},t_{2j})-K_2(\mathbf{t_1},t_{2j})\,,$$

where

$$\begin{aligned} K_2(\mathbf{t_1},t_{2j}) &:= \int_{S_{n-1}} \lambda_1\left(\langle \mathbf{t_1},\mathbf{s_1}\rangle, t_{2j}s_{2j}\right)\Gamma(d\mathbf{s}) \\ &= \int_{S_{n-1}} \Gamma(d\mathbf{s})\mathrm{sgn}(\langle \mathbf{t_1},\mathbf{s_1}\rangle)\mathrm{sgn}(t_{1j},s_{1j})\int_0^\infty \frac{\sin(r|\langle \mathbf{t_1},\mathbf{s_1}\rangle|)}{r^{1+\alpha}}dr, \end{aligned}$$

and

$$\begin{aligned} K_1(\mathbf{t_1}, t_{2j}) &:= \int_{S_{n-1}} \lambda_2(\langle \mathbf{t_1}, \mathbf{s_1} \rangle, t_{2j}s_{2j}) \Gamma(d\mathbf{s}) \\ &= \int_{S_{n-1}} \Gamma(d\mathbf{s}) \int_0^\infty \frac{[1-\cos(rt_{2j}s_{2j})]\cos(r\langle \mathbf{t_1}, \mathbf{s_1} \rangle)}{r^{1+\alpha}} dr \,. \end{aligned}$$

Observe that

$$\begin{aligned} &\Delta(\mathbf{t_1}, t_{2j}) + \Delta(\mathbf{t_1}, -t_{2j}) = 2K_1(\mathbf{t_1}, t_{2j}), \text{ and} \\ &\Delta^2(\mathbf{t_1}, t_{2j}) + \Delta^2(\mathbf{t_1}, -t_{2j}) = 2\left(K_1^2(\mathbf{t_1}, t_{2j}) + K_2^2(\mathbf{t_1}, t_{2j})\right). \end{aligned} \tag{3.4}$$

Thus, it turns out that

$$\lim_{t_{2j}\to+0} J_1(\mathbf{t_1}, t_{2j}) = -\lim_{t_{2j}\to+0} \frac{2}{t_{2j}^2}\left(K_1^2(\mathbf{t_1}, t_{2j}) + K_2^2(\mathbf{t_1}, t_{2j})\right), \quad \text{and} \tag{3.5}$$

$$\lim_{t_{2j}\to+0} J_2(\mathbf{t_1}, t_{2j}) = -\lim_{t_{2j}\to+0} \frac{2K_1(\mathbf{t_1}, t_{2j})}{t_{2j}^2}.$$

In view of Lemma 2, it can be seen that

$$\begin{aligned} &\lim_{t_{2j}\to+0} \tfrac{K_2(\mathbf{t_1}, t_{2j})}{t_{2j}} \\ &= \lim_{t_{2j}\to+0} C_{2,\alpha} \textstyle\int_{S_{n-1}} \Gamma(d\mathbf{s}) \int_0^\infty \frac{\sin(rs_{2j}t_{2j})}{rs_{2j}t_{2j}} \frac{s_{2j}\sin(r\langle \mathbf{t_1}, \mathbf{s_1} \rangle)}{r^\alpha} dr \\ &= C_{3,\alpha}C_{2,\alpha} \int_{S_{n-1}} \operatorname{sgn}(\langle \mathbf{s_1}, \mathbf{t_1} \rangle) |\langle \mathbf{t_1}, \mathbf{s_1} \rangle|^{\alpha-1} s_{2j} \Gamma(d\mathbf{s}) \\ &= \alpha \int_{S_{n-1}} s_{2j} \langle \mathbf{s_1}, \mathbf{t_1} \rangle^{\langle \alpha-1 \rangle} \Gamma(d\mathbf{s}), \text{ since } C_{3,\alpha}C_{2,\alpha} = \alpha \,. \end{aligned} \tag{3.6}$$

Since $1-\cos x = 2\sin^2\dfrac{x}{2}$, and from Lemma 2

$$\begin{aligned} \lim_{t_{2j}\to+0} \tfrac{2K_1(\mathbf{t_1}, t_{2j})}{t_{2j}} &= \lim_{t_{2j}\to+0} C_{2,\alpha} \textstyle\int_{S_{n-1}} \Gamma(d\mathbf{s}) \int_0^\infty \frac{s_{2j}^2 \cos(r\langle \mathbf{t_1}, \mathbf{s_1} \rangle)}{r^{\alpha-1}} dr \\ = C_{2,\alpha}C_{4,\alpha} \int_{S_{n-1}} \frac{s_{2j}^2 \Gamma(d\mathbf{s})}{|\langle \mathbf{t_1}, \mathbf{s_1} \rangle|^{2-\alpha}} &= \alpha(\alpha-1) \int_{S_{n-1}} \frac{s_{2j}^2 \Gamma(d\mathbf{s})}{|\langle \mathbf{t_1}, \mathbf{s_1} \rangle|^{2-\alpha}} \,. \end{aligned} \tag{3.7}$$

The last statement follows because $C_{2,\alpha}C_{4,\alpha} = \alpha(\alpha-1)\dfrac{\sin(\frac{\pi}{2}-\frac{\alpha\pi}{2})}{\cos(\pi-\frac{\alpha\pi}{2})} = \alpha(\alpha-1)$. We also have that

$$\lim_{t_{2j}\to 0} \frac{2K_1^2(\mathbf{t_1}, \mathbf{t_{2j}})}{t_{2j}^2} = 0. \tag{3.8}$$

Combining (3.5)–(3.8), together with the arguments of existence in Theorem 1, as in Cambanis and Fotopoulos (1995), and the condition of Theorem 1, it can be shown that $I(t_{2j};\mathbf{x}_1)$, $\mathbf{X}_1 \in \mathbb{R}^{n_1}$ is bounded by an $L(\mathbb{R}^{n_1})$ function. Thus, by the Lebesgue dominated convergence theorem, we have

$$\begin{aligned}\lim_{t_{2j}\to+0} I(t_{2j};\mathbf{x}_1) &= E\left[X_{2j}|\mathbf{X}_1=\mathbf{x}_1\right]\\ &= \frac{1}{(2\pi)^{n_1} f_{\mathbf{X}_1}(\mathbf{x}_1)}\int_{\mathbb{R}^{n_1}} e^{-i\langle \mathbf{t}_1,\mathbf{x}_1\rangle}\phi_{\mathbf{X}_1}(\mathbf{t}_1)\\ &\quad\times\left\{\alpha(\alpha-1)\int_{S_{n-1}}\frac{s_{2j}^2\Gamma(ds)}{|\langle \mathbf{t}_1,\mathbf{s}_1\rangle|^{2-\alpha}}-\left[\alpha\int_{S_{n-1}} s_{2j}\langle \mathbf{t}_1,\mathbf{s}_1\rangle^{\langle\alpha-1\rangle}\Gamma(d\mathbf{s})\right]^2\right\}.\end{aligned} \tag{3.9}$$

This completes the proof of part I. ■

To obtain the $E\left[X_{2j}X_{2k}|\mathbf{X}_1=\mathbf{x}_1\right]$, $j,k=1,\cdots,n_2, j\neq k$, we need to show that for all $\mathbf{x}_1\in\mathbb{R}^{n_1}$,

$$\begin{aligned}&\lim_{\substack{t_{2j}\to+0\\ t_{2k}\to+0}} I(t_{2j},t_{2k},\mathbf{x}_1)\\ &=\lim_{\substack{t_{2j}\to+0\\ t_{2k}\to+0}}\Big\{\Psi(t_{2j},t_{2k};\mathbf{x}_1)\\ &\qquad-\Psi(-t_{2j},t_{2k};\mathbf{x}_1)-\Psi(t_{2j},-t_{2k};\mathbf{x}_1)+\Psi(-t_{2j},-t_{2k};\mathbf{x}_1)\Big\}/t_{2j}t_{2k}\end{aligned} \tag{3.10}$$

exists, where

$$\begin{aligned}\Psi(t_{2j},t_{2k};\mathbf{x}_1) &= E\left[e^{-i(t_{2j}X_{2j}+t_{2k}X_{2k})}|\mathbf{X}_1=\mathbf{x}_1\right]\\ &= \frac{1}{(2\pi)^{n_1} f_{\mathbf{X}_1}(\mathbf{x}_1)}\int_{\mathbb{R}^{n_1}} e^{-i\langle \mathbf{t}_1,\mathbf{x}_1\rangle}\phi_{\mathbf{X}_1,X_{2j},X_{2k}}(\mathbf{t}_1,t_{2j},t_{2k})d\mathbf{t}_1\end{aligned}$$

for $t_{2j},t_{2k}\in\mathbb{R}$. For convenience, we introduce the following:

$$I(t_{2j},t_{2k};\mathbf{x}_1) := \frac{1}{(2\pi)^{n_1} f_{\mathbf{X}_1}(\mathbf{x}_1)}\int_{\mathbb{R}^{n_1}} e^{-i\langle \mathbf{t}_1,\mathbf{x}_1\rangle}\phi_{\mathbf{X}_1}(\mathbf{t}_1)J(\mathbf{t}_1,t_{2j},t_{2k})d\mathbf{t}_1\,,$$

where

$$\begin{aligned}J(\mathbf{t}_1,t_{2j},t_{2k}) &:= \frac{1}{t_{2j}t_{2k}}\Big\{e^{-\Delta(\mathbf{t}_1,t_{2j},t_{2k})}-e^{-\Delta(\mathbf{t}_1,-t_{2j},t_{2k})}\\ &\qquad -e^{-\Delta(\mathbf{t}_1,t_{2j},-t_{2k})}+e^{-\Delta(\mathbf{t}_1,-t_{2j},-t_{2k})}\Big\}\,,\end{aligned}$$

and

$$\Delta(\mathbf{t}_1,t_{2j},t_{2k}) := \int_{S_{n-1}}\left[|\langle \mathbf{t}_1,\mathbf{s}_1\rangle+t_{2j},s_{2j}+t_{2k}s_{2k}|^{\alpha}-|\langle \mathbf{t}_1,\mathbf{s}_1\rangle|^{\alpha}\right]\Gamma(ds)\,.$$

Thus by Lemma 3 we proceed as

$$\lim_{\substack{t_{2j}\to+0\\ t_{2k}\to+0}} J(\mathbf{t}_1, t_{2j}, t_{2k})$$
$$= \lim_{\substack{t_{2j}\to+0\\ t_{2k}\to+0}} \frac{1}{t_{2j}t_{2k}}\{e^{-\Delta(\mathbf{t}_1,t_{2j},t_{2k})} - e^{-\Delta(\mathbf{t}_1,-t_{2j},t_{2k})}$$
$$- e^{-\Delta(\mathbf{t}_1,t_{2j},-t_{2k})} + e^{-\Delta(\mathbf{t}_1,-t_{2j},-t_{2k})}\}$$
$$= \lim_{\substack{t_{2j}\to+0\\ t_{2k}\to+0}} 4C_{2,\alpha}\int_{S_{n-1}}\Gamma(ds)\int_0^\infty \frac{\sin rt_{2j}s_{2j}}{rt_{2j}s_{2j}}\frac{\sin rt_{2k}s_{2k}}{rt_{2k}s_{2k}}\frac{s_{2j}s_{2k}\cos(r\langle \mathbf{t}_1, \mathbf{s}_1\rangle)}{r^{\alpha-1}}dr$$
$$= 4C_{3,2-\alpha}C_{2,\alpha}\int_{S_{n-1}}\frac{s_{2j}s_{2k}\Gamma(ds)}{|\langle \mathbf{t}_1, \mathbf{s}_1\rangle|^{2-\alpha}} = 4\alpha(\alpha-1)\cot\frac{\alpha\pi}{2}\int_{S_{n-1}}\frac{s_{2j}s_{2k}\Gamma(ds)}{|\langle \mathbf{t}_1, \mathbf{s}_1\rangle|^{2-\alpha}}, \tag{3.11}$$

where the last equation follows because

$$4C_{2,\alpha}C_{3,2-\alpha} = \frac{4\alpha(\alpha-1)}{\Gamma(2-\alpha)\sin\frac{(1-\alpha)\pi}{2}}\Gamma(2-\alpha)\sin\frac{(2-\alpha)\pi}{2} = 4\alpha(\alpha-1)\cot\frac{\alpha\pi}{2}.$$

Now, the rest of the proof of II easily follows.

Proof of Proposition 1. Define the following functions:

$$h(t_{2j}) = \alpha\int_{S_{n-1}} s_{2j}\{\langle \mathbf{t}_1, \mathbf{s}_1\rangle + t_{2j}s_{2j}\}^{\langle\alpha-1\rangle}\Gamma(ds), \tag{3.12}$$

and

$$s(t_{2j}) = \alpha(\alpha-1)\int_{S_{n-1}}\frac{s_{2j}^2\Gamma(ds)}{|\langle \mathbf{t}_1, \mathbf{s}_1\rangle + t_{2j}s_{2j}|^{2-\alpha}}. \tag{3.13}$$

From Holder's inequality $|h(t_{2j})| < \infty$, $\forall t_{2j} \in \mathbb{R}$. It suffices to show that $\frac{\partial}{\partial t_{2j}}\Lambda(\mathbf{t}_1, t_{2j}) = h(t_{2j})$ and $h'(t_{2j}) = s(t_{2j})$ for almost all $t_{2j} \in \mathbb{R}$. Simple calculations show that for arbitrary $t_{2j}, c, d \in \mathbb{R}$

$$\int_0^{t_{2j}} c(d+cb)^{\langle\alpha-1\rangle}db = \frac{1}{\alpha}\{|d+ct_{2j}|^\alpha - |d|^\alpha\}, \tag{3.14}$$

and

$$\int_0^{t_{2j}}\frac{c^2db}{|d+cb|^{2-\alpha}} = \frac{c}{\alpha-1}\left\{(d+ct_{2j})^{\langle\alpha-1\rangle} - d^{\langle\alpha-1\rangle}\right\}. \tag{3.15}$$

The integral part of (3.13) is nonnegative. By Fubini's theorem and (3.15), we have that $\int_0^{t_{2j}} s(b)db = h(t_{2j}) - h(0)$. Thus, $s(\cdot)$ is an absolutely integrable function on $[0, t_{2j})$ and h is its primitive on this segment. Then, by Theorem 8 in Kolmogorov and Fomin (1970), $h'(t_{2j}) = s(t_{2j})$. Similarly, by Fubini's theorem and (3.14), we obtain that $\int_0^{t_{2j}} h(b)db = \Lambda(\mathbf{t}_1, t_{2j}) - \Lambda(\mathbf{t}_1, 0)$, and by

Theorem 8 in Kolmogorov and Fomin (1970) $\frac{\partial \Lambda}{\partial t_{2j}}(\mathbf{t}_1, t_{2j}) = h(t_{2j})$ for almost all $t_{2j} \in \mathbb{R}$.

To show the continuity of $\frac{\partial \Lambda}{\partial t_{2j}}$ and $\frac{\partial^2 \Lambda}{\partial t_{2j}^2}$, we use the following arguments. From Cambanis and Wu (1992) we have that $\left||1+z|^b - 1\right| \leq |z|$, for $0 < b \leq 1$ and $z \in \mathbb{R}$, or equivalently $\frac{1}{|1+z|^{2-\alpha}} - 1 \leq \frac{|z|}{|1+z|}$, where the right-hand side tends to zero as $z \to 0$, for $\alpha \in (1,2)$. This implies that $\frac{\partial^2 \Lambda}{\partial t_{2j}^2}(\mathbf{t}_1, t_{2j})$ tends uniformly to $\alpha(\alpha-1)\int_{S_{n-1}} \frac{s_{2j}^2 \Gamma(ds)}{|\langle \mathbf{t}_1, \mathbf{s}_1 \rangle|^{2-\alpha}}$ as $t_{2j} \to 0$.

Similarly, the continuity of $\frac{\partial \Lambda}{\partial t_{2j}}(\mathbf{t}_1, t_{2j})$ is obvious. This completes the proof of Proposition 1. ■

Proof of Proposition 2. Since $E\left[X_{2j}^2 e^{i\langle \mathbf{t}_1, \mathbf{X}_1 \rangle}\right] \in L^1(\mathbb{R}^{n_1})$, and from (3.1), it is not hard to see that

$$E\left[X_{2j}^2 | \mathbf{X}_1 = \mathbf{x}_1\right] = \frac{1}{(2\pi)^{n_1} f_{\mathbf{X}_1}(\mathbf{x}_1)} \times \int_{\mathbb{R}^{n_1}} e^{-i\langle \mathbf{t}_1, \mathbf{X}_1 \rangle} \phi_{\mathbf{X}_1}(\mathbf{t}_1) \left\{ \frac{\partial^2 \Lambda}{\partial t_{2j}^2}(\mathbf{t}_1, t_{2j})|_{t_{2j}=0} + \left[\frac{\partial \Lambda}{\partial t_{2j}}(\mathbf{t}_1, t_{2j})|_{t_{2j}=0}\right]^2 \right\} d\mathbf{t}_1 . \tag{3.16}$$

Arguing in exactly the same way as in Cambanis and Fotopoulos (1992) for their Proposition 1, and using (3.16) but now only for the bivariate situation, we have (because of the independence of the components of $\mathbf{X}_1$) that for $j = 1, \cdots, n_2$,

$$E\left[X_{2j}^2 | X_{1j} = x_{1j}\right] = \frac{1}{2\pi f_{X_{1j}}(x_{1j})} \int_{\mathbb{R}} e^{-it_{1i}X_{1i}} \phi_{X_{1i}}(t_{1i}) \times \left\{ \frac{\partial^2 \Lambda}{\partial t_{2j}^2}(0, \cdots, t_{1i}, \cdots, 0, t_{2j})|_{t_{2j}=0} + \left[\frac{\partial \Lambda}{\partial t_{2j}}(0, \cdots, t_{1i}, \cdots, 0, t_{2j})|_{t_{2j}=0}\right]^2 \right\} dt_{1j}$$
$$= \frac{1}{(2\pi)^{n_1} f_{\mathbf{X}_1}(\mathbf{x}_1)} \int_{\mathbb{R}^{n_1}} e^{-i\langle \mathbf{t}_1, \mathbf{X}_1 \rangle} \phi_{\mathbf{X}_1}(\mathbf{t}_1) \times \left\{ \frac{\partial^2 \Lambda}{\partial t_{2j}^2}(0, \cdots, t_{1i}, \cdots, 0, t_{2j})|_{t_{2j}=0} + \left[\frac{\partial \Lambda}{\partial t_{2j}}(0, \cdots, t_{1i}, \cdots, 0, t_{2j})|_{t_{2j}=0}\right]^2 \right\} d\mathbf{t}_1. \tag{3.17}$$

In view of (3.16) and (3.14), we have that for $j = 1, \cdots, n_2$,

$$\begin{aligned}
E\Big[X_{2j}^2 | \mathbf{X}_1 = \mathbf{x}_1\Big] &= \sum_{i=1}^{n_1} E\Big[X_{2j}^2 | X_{1j} = x_{1j}\Big] \\
&= \frac{1}{(2\pi)^{n_1} f_{\mathbf{X}_1}(\mathbf{x}_1)} \int_{\mathbf{R}^{n_1}} e^{-t\langle \mathbf{t}_1, \mathbf{X}_1 \rangle} \phi_{\mathbf{X}_1}(\mathbf{t}_1) \\
&\quad \times \Bigg[\left\{ \frac{\partial^2 \Lambda}{\partial t_{2j}^2}(\mathbf{t}_1, t_{2j})|_{t_{2j}=0} + \left[\frac{\partial \Lambda}{\partial t_{2j}}(\mathbf{t}_1, t_{2j})|_{t_{2j}=0} \right]^2 \right\} \\
&\quad - \sum_{i=1}^{n_1} \left\{ \frac{\partial^2 \Lambda}{\partial t_{2j}^2}(0, \cdots, t_{1j}, \cdots, 0, t_{2j})|_{t_{2j}=0} \right. \\
&\quad \left. + \left[\frac{\partial \Lambda}{\partial t_{2j}}(0, \cdots, t_{1i}, \cdots, 0, t_{2j}) \Big|_{t_{2j}=0} \right]^2 \right\} \Bigg] d\mathbf{t}_1,
\end{aligned}$$

which follows from the uniqueness of the Fourier transform. ■

Proof of Theorem 2. Since $\mathbf{X} = (\mathbf{X}_1', \mathbf{X}_2')' \in \mathbf{R}^n$ is a jointly $S\alpha S$ vector, then I and II in Proposition 1 are satisfied. We assume that $\Gamma(\cdot)$ is nontrivial. If $\Gamma(S_{n-1} \cap \{\mathbf{s}_1 = \mathbf{0}\}) > 0$, then this implies that $\Gamma(S_{n-1} \cap \{\mathbf{s}_2 \neq \mathbf{0}\}) > 0$. However, this implies that for $j = 1, \cdots, n_2$, $\int_{\mathbf{R}^{n_1}} \left| E\left[X_{2j}^2 e^{i\langle \mathbf{t}_1, \mathbf{X}_1 \rangle}\right] \right| d\mathbf{t}_1 = \infty$, or equivalently $\int_{S_{n-1}} \frac{s_{2j}^2 \Gamma(ds)}{|\langle \mathbf{t}_1, \mathbf{s}_1 \rangle|^{2-\alpha}} d\mathbf{t}_1 = \infty$. Thus, $\mathbf{s}_1 \neq \mathbf{0}$. Now, since the coordinates of $\mathbf{X}_1$ are statistically independent, we may partition the event $S_{n-1} \cap \{\mathbf{s}_1 \neq \mathbf{0}\}$ into $\bigcup_{i=1}^{n_1} S_{n-1} \cap \left\{ ||\mathbf{s}_2||^2 + s_{1j}^2 = 1, s_{1j} \neq 0 \right\}$. This yields that for $i = 1, \cdots, n_1$, and $j = 1, \cdots, n_2$,

$$\frac{\partial^2 \Lambda}{\partial t_{2j}^2}(0, \cdots, t_{1i}, \cdots, 0, t_{2j})|_{t_{2j}=0} = \frac{1}{|t_{1j}|^{2-\alpha}} \int_{S_{n-1} \cap \left\{ ||\mathbf{s}_2||^2 + s_{1j}^2 = 1, s_{1j} \neq 0 \right\}} \frac{s_{2j}^2 \Gamma(ds)}{|s_{1j}|^{2-\alpha}}.$$

The last statement is in agreement with Theorem 2 in Cambanis and Fotopoulos (1995). The same arguments could apply for II in order to establish an expression for $E\left[X_{2j} X_{2k} | \mathbf{X}_1\right]$, $j, k = 1, \cdots, n_2, j \neq k$. This completes the proof of Theorem 2. ■

Proof of Corollary. The proof of part I was given by Wu and Cambanis (1992) and Cioczek–Georges and Taqqu (1995). Thus, we pay attention here only to the derivation of the covariance.

By a simple alteration of Kanter's (1972) result, it can be seen that

$$E\left[X_{2j}|X_{1\ell}=x_{1\ell}\right]=\frac{\int_{S_{n-1}} s_{2j}s_{1\ell}^{\langle\alpha-1\rangle}\Gamma(ds)}{\int_{S_{n-1}}|s_{1\ell}|^{\alpha}\Gamma(ds)}x_{1\ell}=\int_{S_{n-1}} s_{2j}s_{1\ell}^{\langle\alpha-1\rangle}\Gamma(ds)\frac{x_{1\ell}}{\sigma_{1\ell}^{\alpha}} \tag{3.18}$$

for $j=1,\cdots,n_2$, and $\ell=1,\cdots,n_1$. Furthermore, using Corollary (i), Theorem 1. II, and Theorem 2. II, it follows that

$$\begin{aligned} &E\left[X_{2j}X_{2k}|X_{1\ell}=x_{1\ell}\right] \\ &= 4\cot\tfrac{\alpha\pi}{2}\sigma_{1\ell}^{2-\alpha} \\ &\times \int\limits_{S_{n-1}\cap\left\{||\mathbf{s}_2||^2+s_{1\ell}^2=1,s_{1\ell}\neq 0\right\}} \frac{s_{2j}s_{2k}\Gamma(ds)}{|s_{1\ell}|^{2-\alpha}}S_1^2(\frac{x_{1\ell}}{\sigma_{1\ell}};\alpha) \\ &= 4\cot\tfrac{\alpha\pi}{2}\sigma_{1\ell}^{2-\alpha} \\ &\times \int\limits_{S_{n-1}\cap\left\{||\mathbf{s}_2||^2+s_{1\ell}^2=1,s_{1\ell}\ ne0\right\}} \frac{s_{2j}s_{2k}\Gamma(ds)}{|s_{1\ell}|^{2-\alpha}} \\ &\times\left\{\frac{\alpha^2\int_0^\infty \cos t_{1\ell}x_{1\ell}e^{-\sigma_{1\ell}^{\alpha}t_{1\ell}^{\alpha}}t_{1\ell}^{2\alpha-2}dt_{1\ell}}{\int_0^\infty \cos t_{1\ell}x_{1\ell}e^{-\sigma_{1\ell}^{\alpha}t_{1\ell}^{\alpha}}dt_{1\ell}}+\frac{x_{1\ell}^2}{\sigma_{1\ell}^{2\alpha}}\right\} \end{aligned} \tag{3.19}$$

Combining (3.18) and (3,19), the proof easily follows. ∎

4. Auxiliary Results

In this section, we collect all the lemmas which are used in the proofs of Theorem 1-3 and Propositions 1-3.

Lemma 1 (Wolf, 1973). *If f is the characteristic function of a random variable X, then the following statements are equivalent*

(i) $D_2f(0)=\lim_{t\to+0}\rho_2(t)/(2t)^2$ *exists, and*

(ii) $E\left[X^2\right]<\infty$,

where $\rho_2(t)=f(2t)+f(-2t)-2$.

Lemma 2 (Gradshteyn and Ryzhik, 1980).

(i) $$\int_0^\infty \frac{\sin bx \sin cx}{x^{1+\alpha}}dx=C_{1,\alpha}\left[|c-b|^{\alpha}-|c+b|^{\alpha}\right],$$

for $b,c>0$, $b\neq c$, $-1<\alpha<2$ *with* $C_{1,\alpha}=\frac{1}{2}\Gamma(-\alpha)\cos\frac{\alpha\pi}{2}$,

(ii) $$C_{2,\alpha}\int_0^\infty\frac{1-\cos bx}{x^{1+\alpha}}dx=|b|^{\alpha},$$

for $b > 0, 0 < \alpha < 2$, *with* $C_{2,\alpha} = \left\{\frac{\Gamma(2-\alpha)}{\alpha(1-\alpha)} \sin \frac{\pi(1-\alpha)}{2}\right\}^{-1}$,

(iii) $$\int_0^\infty \frac{\sin bx}{x^\alpha} dx = C_{3,\alpha} b^{\alpha-1}, \text{ eq. 3.761.4},$$

for $b > 0, 0 < \alpha < 1$, *with* $C_{3,1-\alpha} = \Gamma(1-\alpha) \sin \frac{(1-\alpha)\pi}{2}$, *and*

(iv) $$\int_0^\infty \frac{\cos bx}{x^{1+\alpha}} dx = C_{4,\alpha} |b|^{\alpha-2}, \text{ eq.3.761.9},$$

for $1 < \alpha < 2$, *with* $C_{4,\alpha} = \Gamma(2-\alpha) \cos \frac{(2-\alpha)\pi}{2}$.

The following result plays an important role in evaluating and obtaining expressions for the covariance matrix in Theorem 1.

Lemma 3 *For* $0 < \alpha < 2$

(i) $$|b \pm c|^\alpha - |b|^\alpha = \lambda_1(b, c) + \lambda_2(b, c)$$

where

$$\lambda_1(b, c) = C_{2,\alpha} \int_0^\infty \frac{(1 - \cos rc) \cos rb}{r^{1+\alpha}} dr ,$$

and

$$\lambda_2(b, c) = \text{sgn}(b)\text{sgn}(c) C_{2,\alpha} \int_0^\infty \frac{\sin r|b| \sin r|c|}{r^{1+\alpha}} dr ,$$

and

(ii) $$\begin{aligned} &|b + c + d|^\alpha - |b - c + d|^\alpha - |b + c - d|^\alpha + |b - c - d|^\alpha \\ &\quad = 4C_{2,\alpha} \text{sgn}(d)\text{sgn}(c) \int_0^\infty \frac{\cos rb \sin rc \sin rd}{r^{1+\alpha}} dr. \end{aligned}$$

Proof. From Lemma 2 it follows that

$$\begin{aligned} &|b \pm c|^\alpha - |b|^\alpha \\ &= C_{2,\alpha} \int_0^\infty \frac{\cos(rb) - \cos(r(b \pm c))}{r^{1+\alpha}} dr \\ &= \pm 2C_{2,\alpha} \int_0^\infty \frac{\sin(r(b \pm \frac{1}{2}c)) \sin(\frac{r}{2}c)}{r^{1+\alpha}} dr \\ &= \pm 2C_{2,\alpha} \int_0^\infty \frac{\left\{\sin(rb) \cos(\frac{1}{2}rc) \pm \cos(rb) \sin(\frac{1}{2}rc)\right\} \sin(\frac{1}{2}rc)}{r^{1+\alpha}} dr \\ &= C_{2,\alpha} \int_0^\infty \frac{\sin(rb) \sin(\pm rc) + \cos(rb)(1 - \cos(rc))}{r^{1+\alpha}} dr. \end{aligned}$$

This completes the proof of part (i). ■

Similarly,

$$|b+c+d|^\alpha - |b-c+d|^\alpha - |b+c-d|^\alpha + |b-c-d|^\alpha$$
$$= C_{2,\alpha}\int_0^\infty \left(\frac{\cos(r(b+c-d)) + \cos(r(b-c+d))}{r^{1+\alpha}}\right)dr$$
$$-\left(\frac{\cos(r(b+c+d)) + \cos(r(b-c-d))}{r^{1+\alpha}}\right)dr.$$

The rest of the proof follows by trigonometric manipulation of the cosines and sines.

Acknowledgements. The author is grateful to Professors B. Rajput and M. Taqqu for their invaluable assistance and to the referee for a careful reading of the original manuscript and his valuable suggestions.

References

Cambanis, S. and Fotopoulos, S. (1995). Conditional variance for stable random vectors. *Probab. Math. Statist.*, **15,** 195–214.

Cambanis, S., Fotopoulos, S., and He, L. (1997). On the conditional variance for scale mixtures of normal distributions. *Technical Report*, Washington State University.

Cambanis, S. and Wu, W. (1992). Multiple regression on stable vectors. *J. Multivariate Anal.* **41,** 243–272.

Cioczek–Georges R. and Taqqu M. (1993). Form of the Conditional Variance for Stable Random Variables. *Technical Report*, Boston University.

Hardin, C., Samorodnitsky, G. and Taqqu, M. (1991) Nonlinear progression of stable random variables. *Ann. Appl. Prob.* **1**, 582–612.

Kolmogorov, A.N., and Fomin, S.V.(1970). *Introductory Real Analysis.* Dover Publ., Inc.

Samorodnitsky, G and Taqqu, M.S. (1991) Conditional moments and linear regression for stable random variables. *Stochast. Proc. Applic.*, **39,** 183–199.

Wu, W. and Cambanis, S. (1991). Conditional Variance of Symmetric Stable Variables. In G. Samorodnitsky, S. Cambanis, and M. S. Taqqu, Editors, *Stable and Related Topics*, **25,** *Progress in Probability*, **85-99,** Birkhäuser Boston.

Department of Management and Systems and Program in Statistics
Washington State University
Pullman, WA 99164-4726
fotopo@wsu.edu / http://www.wsu.edu:8000/~fotopo

Interacting Particle Approximation for Fractal Burgers Equation[1]

T. Funaki and W.A. Woyczyński

Abstract

The paper reports on the existence of McKean's nonlinear processes and the related propagation of chaos results for a class of one-dimensional (1-D) generalized Burgers-type equations with a fractional power of the Laplacian in the principal part and a quadratic nonlinearity. Such equations naturally appear in continuum mechanics.

1. Introduction, physical motivation

In this paper we consider the 1-D fractal Burgers equation

$$u_t = \nu \Delta_\alpha u - u \nabla u, \qquad u(x,0) = u_0(x), \quad x \in \mathbf{R},\ t \geq 0, \tag{1.1}$$

where

$$\begin{aligned} u : \mathbf{R} \times \mathbf{R}^+ \to \mathbf{R},\ u_t &= \partial_t u \equiv \partial u/\partial t,\ \nu > 0,\ \Delta_\alpha \\ &= -(-\partial^2/\partial x^2)^{\alpha/2},\ 0 < \alpha \leq 2,\ \nabla = \partial/\partial x. \end{aligned}$$

In the 1-D case, the (one-sided) fractional derivative of order β can be defined by the singular integral

$$\partial^\beta v(x) = \frac{1}{\Gamma(-\beta)} \int_{-\infty}^{x} \frac{v(x')}{(x-x')^{\beta+1}}\, dx'. \tag{1.2}$$

A pedestrian introduction to fractal calculus can be found in Saichev and Woyczyński (1997b). However, for our purposes (and in multidimensional extensions), the fractional power $\Delta_\alpha \equiv -(-\Delta)^{\alpha/2} = -(-\partial^2/\partial x^2)^{\alpha/2}$ of the second spatial derivative (Laplacian) is more conveniently defined via the Fourier transform $\hat{}\equiv \mathcal{F}$:

$$\Delta_\alpha v(x) = -\mathcal{F}^{-1}(|\xi|^\alpha \hat{v}(\xi))(x). \tag{1.3}$$

[1] Supported, in part, by grants from ONR, JSPS and NSF.

In the case where $\alpha \in (0,2)$ it also has another representation

$$\Delta_\alpha v(x) = K \int_{\mathbf{R}} \left\{ v(x+y) - v(x) - \nabla v(x) \cdot \frac{y}{1+|y|^2} \right\} \frac{dy}{|y|^{1+\alpha}}, \tag{1.4}$$

for some positive constant $K = K_\alpha$, which identifies it as the infinitesimal generator for the symmetric α-stable Lévy process (see e.g., Stroock (1975), Komatsu (1984), Dawson and Gorostiza (1990)).

Equation (1.1) is a generalization of the classical 1-D Burgers equation where $\alpha = 2$ (see e.g., Burgers (1974), Smoller (1994)), and our interest in this extension is motivated by fractal (anomalous) diffusion related to the Lévy flights (see e.g., Stroock (1975), Bardos et al. (1979), Dawson and Gorostiza (1990), Sugimoto (1991, 1992), Shlesinger et al. (1995), Zaslavsky (1994), Zaslavsky and Abdullaev (1995), Woyczyński (1997), and the references quoted therein) that appear in models of relaxation phenomena, acoustics and anomalous interface growth with trapping.

The classical (1-D and d-D) Burgers equation (i.e. equation (1.1) with $\alpha = 2$) has been extensively used to model a variety of physical phenomena where shock creation is an important ingredient, from the growth of molecular interfaces, through traffic jams to the mass distribution for the large scale structure of the universe (see e.g., Kardar et al. (1986), Gurbatov et al. (1991), Vergassola et al. (1994) and Molchanov et al. (1997)). In the latter application, the Burgers equation is coupled with the continuity equation to consider the problem of passive tracer transport in Burgers velocity flows (E, Rykov and Sinai (1996), Saichev and Woyczyński (1997a)). Recent work on Burgers turbulence, i.e., the theory of statistical properties of solutions to the Burgers equation with random initial data and/or forcing is reviewed in Woyczyński (1997).

A great deal of analysis of the classical Burgers equation and its multidimensional counterparts is based on the global functional Hopf–Cole formula linearizing the Burgers equation to the heat equation. This crucial simplification is no longer available in the general fractal case of (1.1); and, of course, the major difference with the classical Burgers equation is the presence in (1.1) of the singular integro-differential operator Δ_α. The fractal Burgers equation is no longer local. The numerical handling of such singular nonlocal and nonlinear equations is quite tricky, hence the need for efficient and controllable approximation schemes.

The composition of the paper is as follows: A summary of the needed existence and uniqueness results for the weak solutions of the fractal Burgers equation (1.1) is provided in Section 2. Section 3 describes the main results of this paper: the existence of a McKean's nonlinear Markov process associated with the fractal Burgers equation (Theorem 3.1) which permits interpretation of (1.1) as a nonlinear Fokker–Planck–Kolmogorov equation, and the stochastic interacting particle system approximation scheme, including error estimates, for solutions of a regularization of the fractal Burgers equation (Theorem 3.2). The latter theorem is in the spirit of the "propagation of chaos" results, also pioneered by McKean (1967). Section 4 contains the proof of our main Theorem 3.2.

2. Existence and uniqueness of weak solutions

In this section we will explain what is meant by the (weak) solutions of the fractal Burgers equation (1.1), and summarize recent results on the existence and uniqueness of such solutions. Roughly speaking, the local in time existence (i.e., the existence in a time interval $[0,T]$, where $T = T(u_0) > 0$ depends on the initial datum u_0) is assured for any $\alpha \in (1/2, 2]$, but the global in time existence is guaranteed only for $\alpha \in (3/2, 2]$.

Throughout this paper we use the standard notation: $|u|_p$ for the Lebesgue $L^p(\mathbf{R})$-norms of functions, and $\|u\|_\beta \equiv \|u\|_{\beta,2}$ for the norms of Hilbert–Sobolev spaces $H^\beta(\mathbf{R}) \equiv W^{\beta,2}(\mathbf{R})$. The constants independent of solutions considered will be denoted by the same letter C, even if they may vary from line to line. For various interpolation inequalities we refer to Adams (1975), Ladyženskaja et al. (1968), Lions and Magenes (1972) and Henry (1982).

Definition 2.1. The *weak* solutions of the Cauchy problem (1.1) are functions

$$u \in V_2 \equiv L^\infty((0,T); L^2(\mathbf{R})) \cap L^2((0,T); H^1(\mathbf{R})) \tag{2.1}$$

satisfying the integral identity

$$\int_{\mathbf{R}} u(x,t)\phi(x,t)dx - \int_{\mathbf{R}} u_0(x)\phi(x,0)dx$$
$$= \int_0^t \int_{\mathbf{R}} u\,\phi_t\,dx\,ds + \int_0^t \int_{\mathbf{R}} \left(-\Delta_{\alpha/2} u\,\Delta_{\alpha/2}\phi + \frac{1}{2}u^2\phi_x\right) dx\,ds \tag{2.2}$$

for a.e. $t \in (0,T)$ and each test function $\phi \in H^1(\mathbf{R} \times (0,T))$. The viscosity ν is assumed to be 1 without loss of generality.

Observe that we assume $u(t) \in H^1(\mathbf{R})$ a.e. in $t \in (0,T)$, instead of just $u(t) \in H^{\alpha/2}(\mathbf{R})$ a.e. in t, which could be expected from a straightforward generalization of the definition of the weak solution of a parabolic second order equation (see e.g., Ladyženskaja et al. (1968)). We need this supplementary regularity to slightly simplify our construction; for the initial data $u_0 \in H^1(\mathbf{R})$ it is a consequence of the assumptions.

Theorem 2.1. (Biler, Funaki, and Woyczyński (1997)) Let $\alpha \in (3/2, 2]$, $T > 0$, and $u_0 \in H^1(\mathbf{R})$. Then the Cauchy problem (1.1) has a unique weak solution $u \in V_2$. Moreover, u enjoys the following regularity properties:

$$u \in L^\infty((0,T); H^1(\mathbf{R})) \cap L^2((0,T); H^{1+\alpha/2}(\mathbf{R})),$$

and

$$u_t \in L^\infty((0,T); L^2(\mathbf{R})) \cap L^2((0,T); H^{\alpha/2}(\mathbf{R}))$$

for each $T > 0$. For $t \to \infty$, this solution decays so that

$$\lim_{t\to\infty} |\Delta_{\alpha/2} u(t)|_2 = \lim_{t\to\infty} |u(t)|_\infty = 0.$$

If $\alpha \in (1/2, 2]$ then unique local in time solutions exist and depend continuously on the initial data (as a mapping $H^1 \to V_2$).

Remark 2.1. Recall that Δ_α, $0 < \alpha \leq 2$, is the infinitesimal generator of an analytic semigroup $(e^{t\Delta_\alpha})$, $t \geq 0$, called the *Lévy semigroup*, and the properties of its convolution kernel are well known (see e.g. Komatsu (1984), Biler, Funaki, and Woyczyński (1998)). For $u_0 \in H^{\alpha/2}$, the function $t \mapsto e^{t\Delta_\alpha} u_0$ can then be interpreted as a global solution of the Cauchy problem for the (linear) fractal heat equation $u_t = \Delta_\alpha u$, $u(x,0) = u_0(x)$.

Remark 2.2. The above mentioned paper by Biler, Funaki, and Woyczyński (1998) also contains a discussion of several other approaches to the solvability problem of the fractal Burgers equation including parabolic regularization, L^2-type estimates, travelling wave and self-similar solutions and, finally, the mild solutions, which involve Morrey and Besov spaces, and interpret the Cauchy problem (1.1) with $\nu = 1$ as the integral equation

$$u(t) = e^{t\Delta_\alpha} u_0 - \int_0^t \nabla e^{(t-s)\Delta_\alpha} (u^2(s)/2)\, ds,$$

which is a consequence of the variation of parameters formula.

Remark 2.3. Theorem 2.1 establishes the existence of solutions which are L^∞ in variable t, i.e., a.e. bounded functions of t with values in $L^2(\mathbf{R})$. Actually, this assertion can be slightly reformulated and one can prove that, $H^{\alpha/2}(\mathbf{R}) \hookrightarrow E$ is a continuous embedding and if $H^1(\mathbf{R}) \hookrightarrow E$ is a compact embedding, then these solutions $u \in C([0,T]; E)$, as long as $u_0 \in H^1(\mathbf{R})$

and $\alpha \in (1/2, 2]$. This follows from the fact that, for every (or, respectively, some) $T > 0$, the sequence of Galerkin approximations $\{u^n(t)\}$ is relatively compact in $C([0,T]; E)$ if $\alpha \in (3/2, 2]$ (or, respectively, if $\alpha \in (1/2, 3/2]$). Indeed, the sequence $\{u^n(t)\}$ is equicontinuous in $H^{\alpha/2}(\mathbf{R})$ in view of the inequality

$$\int_0^T \|\partial_t u^n(t)\|_{\alpha/2}^2 dt \le C = C_T, \qquad \forall t \in [0,T], \quad \forall n. \tag{2.3}$$

Now, the embedding $H^1(\mathbf{R}) \hookrightarrow E$ is compact, and

$$\|u^n(t)\|_1 \le C = C_T, \qquad \forall t \in [0,T], \quad \forall n. \tag{2.4}$$

This gives the compactness statement. Inequalities (2.3–4) are routine consequences of the standard Sobolev spaces embedding theorems (see the proof of Theorem 2.1 in Biler, Funaki, and Woyczyński (1997)). The above argument especially implies $u \in C([0,T];\ H^{\alpha/2}(-\ell, \ell))$ for every $\ell < \infty$, if $\alpha \in (1/2, 2)$, by Rellich's theorem.

Remark 2.4. For a much broader class of d-D equations with fractal diffusion and general algebraic and nonlocal nonlinearities, the issues of the existence of global and exploding solutions, self-similar solutions, and phase transitions have been studied in Biler, Funaki, and Woyczyński (1997) and Biler and Woyczyński (1997).

3. Nonlinear Markov processes and approximating interacting particle systems for the fractal Burgers equation

We begin with the construction of a nonlinear Markov process for which the fractal Burgers equation serves as the nonlinear Fokker–Planck–Kolmogorov equation. The standing assumption in what follows is that $\alpha \in (1, 2)$, which permits us to operate with expectations of the corresponding α-stable processes.

Let $u = u(x,t) \ge 0$ be a local in time weak solution of the fractal Burgers equation with viscosity $\nu = 1$, i.e.,

$$u_t = \Delta_\alpha u - u\nabla u, \tag{3.1}$$

where $x \in \mathbf{R}$, $t \in [0,T]$. We can assume that the solution is bounded, that is

$$\sup_{x\in\mathbf{R},t\in[0,T]} |u(x,t)| < \infty. \tag{3.2}$$

Indeed, (3.2) follows from Theorem 2.1 when $\alpha \in (1/2, 2]$ and, furthermore, $u(x,t)$ is continuous in (x,t) from Remark 2.3 when $\alpha \in (1, 2]$; note that $H^{\beta}(\mathbf{R})$ is continuously embedded in $L^{\infty}(\mathbf{R}) \cap C(\mathbf{R})$ if $\beta > 1/2$.

Next, consider a solution $X(t)$ of the stochastic differential equation

$$\begin{aligned} &dX(t) = dS(t) + \frac{1}{2}u(X(t),t)\,dt, \qquad t \in [0,T], \\ &X(0) \sim u(x,0)\,dx, \qquad \text{in law,} \end{aligned} \tag{3.3}$$

where $S(t)$ is a standard ($E\exp(i\xi S(t)) = \exp(-t|\xi|^{\alpha})$) symmetric $\alpha-$ stable process with independent increments. Then, since the coefficient $u(x,t)$ is bounded, the stochastic differential equation (3.3) has a unique weak solution (see e.g., Komatsu (1985)). The measure-valued function of t

$$v(dx,t) := P\{X(t) \in dx\}, \qquad t \in [0,T],$$

satisfies the weak forward equation

$$\begin{aligned} &\frac{d}{dt}\langle v(t),\varphi\rangle = \langle v(t),\mathcal{L}_{u(t)}\varphi\rangle, \qquad \forall\varphi \in \mathcal{S}(\mathbf{R}) \\ &v(0) = u(x,0)\,dx, \end{aligned} \tag{3.4}$$

where $\mathcal{S}(\mathbf{R})$ denotes the Schwartz class of functions on $\mathbf{R}$, and the operator

$$\mathcal{L}_u = \Delta_\alpha + \frac{1}{2}u(x)\nabla, \qquad u = u(x).$$

Theorem 3.1. *Let $1 < \alpha < 2$. Process $X(t)$ is the McKean process (nonlinear Markov process) corresponding to the fractal Burgers equation (3.1), i.e., it satisfies condition*

$$P\{X(t) \in dx\} = u(x,t)\,dx. \tag{3.5}$$

Proof. In view of results of Echeverria (see e.g., Funaki (1984)) the following two statements are equivalent:

(a) The martingale problem for the operator $\mathcal{L}_{u(t)}$ is well posed, and

(b) the existence and uniqueness theorem holds for the corresponding weak forward equation (3.4).

In our case, (a) holds for the martingale problem associated with (3.3). However, $u(dx,t) := u(x,t)\,dx$ is also a solution of (3.4) since, by (3.1) (or, (2.2)),

$$\begin{aligned}\frac{d}{dt}\langle u(t),\varphi\rangle &= \langle u(t),\Delta_\alpha\varphi\rangle + \langle u(t)^2/2,\nabla\varphi\rangle\\ &= \langle u(t),\mathcal{L}_{u(t)}\varphi\rangle.\end{aligned}$$

Now, the uniqueness for the (linear) equation (3.4) implies that

$$v(dx,t) = u(dx,t),$$

which, consequently, yields (3.5). ■

To formulate our main result, an analogue of the "random vortex" method for 2-D Navier–Stokes equations, let us introduce independent, symmetric, real-valued Lévy α-stable processes $\{S^i(t),\ i=1,2,\dots,n\}$ with the common infinitesimal generator $\Delta_\alpha = -(-\Delta)^{\alpha/2}$, and let

$$\delta_\epsilon(x) := \frac{1}{\sqrt{2\pi\epsilon}}\exp\left[-\frac{x^2}{2\epsilon}\right], \qquad \epsilon>0, \tag{3.6}$$

be a regularizing kernel. Consider a system of n interacting particles with positions

$$\{X^i(t)\}_{i=1,\dots,n} \equiv \{X^{i,n,\epsilon}(t)\}_{i=1,\dots,n}, \tag{3.7}$$

and the corresponding measure-valued process (empirical distribution)

$$\bar{X}^n(t) \equiv \bar{X}^{n,\epsilon}(t) := \frac{1}{n}\sum_{i=1}^{n}\delta(X^{i,n,\epsilon}(t)), \tag{3.8}$$

with the dynamics provided by the system of regularized singular stochastic differential equations

$$dX^i(t) = dS^i(t) + \frac{1}{2n}\sum_{j\neq i}\delta_\epsilon(X^i(t)-X^j(t))\,dt, \quad i=1,\dots,n. \tag{3.9}$$

Theorem 3.2. *Let $\alpha \in (1,2)$, $\epsilon > 0$, and assume that the initial particles' positions $\{X^i(0)\}_{i=1,\dots,n}$ satisfy the following condition:*

$$\sup_n \sup_{\lambda\in\mathbf{R}} \frac{n^{1-1/\alpha}}{1+|\lambda|^a} E\Big[\langle \bar{X}^{n,\epsilon}(0) - u^\epsilon(x,0), \chi_\lambda\rangle\Big] < \infty, \tag{3.10}$$

for some $a \geq 0$, where $\chi_\lambda(x) = e^{i\lambda x}$. Then:

(i) *For each $\epsilon > 0$, the empirical process*

$$\bar{X}^{n,\epsilon}(t) \Longrightarrow u^\epsilon(x,t)\,dx, \quad \text{in probability, as} \quad n \to \infty, \tag{3.11}$$

where $\Rightarrow$ denotes the weak convergence of measures, and the limit density $u^\epsilon \equiv u^\epsilon(x,t)$, $t > 0$, $x \in \mathbf{R}$, satisfies the regularized fractal Burgers equation

$$u_t^\epsilon = \Delta_\alpha u^\epsilon - \frac{1}{2}\nabla\Big((\delta_\epsilon * u^\epsilon)\cdot u^\epsilon\Big). \tag{3.12}$$

(ii) *For each $\epsilon > 0$, there exists a constant $C_\epsilon > 0$ such that, for any $\phi \in \mathcal{S}(\mathbf{R})$,*

$$E\Big|\langle \bar{X}^{n,\epsilon}(t) - u^\epsilon(t), \phi\rangle\Big| \leq C_\epsilon n^{-1+1/\alpha} \int_{\mathbf{R}} (1+|\lambda|^a)|\hat{\phi}|(d\lambda), \tag{3.13}$$

where $|\hat{\phi}|(d\lambda)$ denotes the total variation measure of $\hat{\phi}(d\lambda)$.

(iii) *For each $\alpha \in (3/2, 2)$ there exists a sequence $\epsilon(n) \to 0$ such that, for each ϕ satisfying the condition $\int_{\mathbf{R}}(1+|\lambda|^a)|\hat{\phi}|(d\lambda) < \infty$,*

$$E\Big|\langle \bar{X}^{n,\epsilon(n)}(t) - u(t), \phi\rangle\Big| \longrightarrow 0, \tag{3.14}$$

as $n \to \infty$, where $u(t) = u(x,t)$ is a solution of the nonregularized fractal Burgers equation (3.1).

Remark 3.1. The first propagation of chaos result for the classical Burgers equation $u_t = \Delta u - u\nabla u$ was proved by Gutkin and Kac (1983), following the pioneering work of McKean (1967), and their version can be formulated as follows: Consider n interacting diffusions with singular interactions described by the infinitesimal operator

$$L_n = \Delta_n + \frac{1}{n}\sum_{1\leq i<j\leq n} \delta(x_i - x_j)\left(\frac{\partial}{\partial x_i} + \frac{\partial}{\partial x_j}\right),$$

$n = 1, 2, \ldots$. Then the solutions of the evolution equations

$$\partial_t f_n = L_n^* f_n,$$

with the product-form initial conditions

$$f_n(x_1, x_2, \ldots, x_n; 0) = u_0(x_1)u_0(x_2)\cdots u_0(x_n), \qquad \int_{\mathbf{R}} u_0(x)dx = 1,$$

have the following property: the limit

$$\lim_{n\to\infty} \int_{\mathbf{R}^{n-1}} f_n(x, x_2, \ldots, x_n; t)\, dx_2 \cdots dx_n = u(x,t),$$

exists and satisfies the Burgers equation $u_t = \Delta u - u\nabla u$ with $u(x,0) = u_0(x)$. Moreover, for each $l = 2, 3, \ldots$,

$$\lim_{n\to\infty} \int_{\mathbf{R}^{n-l}} f_n(x_1, x_2, \ldots, x_n; t) dx_{l+1} \cdots dx_n$$
$$= u(x_1,t)u(x_2,t)\cdots u(x_l,t),$$

so that the independence in the initial data is propagated and preserved asymptotically. A detailed explanation of the connection between the Gutkin and Kac's result and the formulation of Theorem 3.2 can be found in Sznitman (1991) and Méléard (1996).

Gutkin and Kac's result was later reformulated and expanded by Calderoni and Pulvirenti (1983), Oelschläger (1985), Sznitman (1986), Osada (1986), Zheng (1995), and our approach follows, in a sense, that later tradition of propagation of chaos results. A number of authors have also considered propagation of chaos results for other nonlocal interacting particle systems including McKean–Vlasov and Boltzmann models (see Méléard (1996) and references therein). None of them however address the situation where the corresponding integral operator kernels have singularities corresponding to those of the infinitesimal generators of α-stable Lévy semigroups.

Remark 3.2. Our initial intention, following the example of the classical Burgers equation results of Kotani and Osada (1985), Sznitman (1986), Osada (1986), was to prove the propagation of chaos result for the nonregularized interacting particle system

$$dX^i(t) = dS^i(t) + \frac{1}{2n}\sum_{j\neq i} \delta(X^i(t) - X^j(t))\, dt, \quad i = 1, \ldots, n, \qquad (3.15)$$

instead of (3.9), and we thought about Theorem 3.2 as only a preliminary, although numerically satisfactory, step in that direction. However, our recent conversations with Professor H. Osada of Tokyo University, whom we thank for his interest in our project, seem to indicate that, after all, the formulation of Theorem 3.2 is the natural one in the fractal Laplacian context; the latter, for $\alpha < 2$, is too "weak" to control the quadratic nonlinearity.

Remark 3.3. Theorem 3.2 can be extended to a class of more general multidimensional nonlinear integro-differential equations of the form

$$u_t = -\nu\Delta_\alpha u - c \cdot \nabla(u^r)$$

where $x \in \mathbf{R}^d, d = 1, 2, \ldots, u : \mathbf{R}^d \times \mathbf{R}^+ \to \mathbf{R}, r \geq 1$, and $c \in \mathbf{R}^d$ is a fixed vector, which were studied in Biler, Funaki, and Woyczyński (1998). These results will be published elsewhere.

4. Proof of Theorem 3.2

Recall, that the α-stable Lévy process $S(t)$ has a representation

$$S(t) = \int_0^{t+} \int_{0<|y|<1} y\,\tilde{N}(ds\,dy) + \int_0^{t+} \int_{|y|\geq 1} y\,N(ds\,dy), \tag{4.1}$$

where $N(ds\,dy)$ is a Poisson point process with intensity $\hat{N}(ds\,dy) = ds\,\nu(dy)$, $\nu(dy) = K dy/|y|^{1+\alpha}$ is the Lévy measure and $\tilde{N}(ds\,dy) = N(ds\,dy) - \hat{N}(ds\,dy)$ (see e.g., Ikeda and Watanabe (1981)).

Proof of Theorem 3.2(i-ii). The assertion (3.11) essentially follows from (3.13); recall that $\alpha > 1$. Therefore, to prove Theorem 3.2(i), it suffices to demonstrate the speed of convergence result (3.13), the boundedness i.e.,

$$\eta^n(t) \equiv \eta^{n,\epsilon}(t) := n^\beta\Big(\bar{X}^{n,\epsilon}(t) - u^\epsilon(t)\Big) \tag{4.2}$$

with $\beta = 1 - 1/\alpha$, and we precede the proof of that error estimate, i.e., Theorem 3.2.(ii), by a proof of the following lemma which is a consequence of the Itô's formula for α-stable processes (see e.g., Ikeda and Watanabe (1981)).

Lemma 4.1. *For each $\phi \in C_b^\infty(\mathbf{R})$,*

$$\langle \eta^n(t), \phi\rangle = \langle \eta^n(0), \phi\rangle + m^n(\phi; t) + \int_0^t b^n(\phi; s)\,ds + \int_0^t \langle \eta^n(s), \Delta_\alpha\phi\rangle\,ds, \tag{4.3}$$

where

$$m^n(\phi;t) = n^{\beta-1}\sum_{i=1}^{n}\int_0^{t+}\int_{\mathbf{R}}\{\phi(X^i(s-)+y)-\phi(X^i(s-))\}\tilde{N}^i(dsdy) \quad (4.4)$$

with $\tilde{N}^i = N^i - \hat{N}$, *and* $N^i = N^i(dsdy)$ *being independent, Poisson random measures with identical intensity* $\hat{N}(dsdy) = Kdsdy/|y|^{1+\alpha}$, *and where*

$$\begin{aligned} b^n(\phi;t) =& \frac{1}{2}n^{-\beta}\Big\langle \eta^n(dx,t)\eta^n(dy,t), \delta_\epsilon(x-y)\nabla\phi(x)\Big\rangle \\ &+ \frac{1}{2}\Big\langle u^\epsilon(dx,t)\eta^n(dy,t) + u^\epsilon(dy,t)\eta^n(dx,t), \delta_\epsilon(x-y)\nabla\phi(x)\Big\rangle \\ &- \frac{1}{2}n^{\beta-1}\delta_\epsilon(0)\langle \bar{X}^n(t), \nabla\phi\rangle. \end{aligned} \quad (4.5)$$

Proof. Observe that

$$\langle \eta^n(t), \phi\rangle = n^\beta\left\{\frac{1}{n}\sum_{i=1}^{n}\phi(X^i(t)) - \langle u^\epsilon(t),\phi\rangle\right\}.$$

Then, by (4.1) and Itô's formula,

$$\begin{aligned} &\langle \eta^n(t),\phi\rangle - \langle \eta^n(0),\phi\rangle \\ &= n^{\beta-1}\sum_{i=1}^{n}\int_0^{t+}\int_{\mathbf{R}}\{\phi(X^i(s-)+y)-\phi(X^i(s-))\}\tilde{N}^i(dsdy) \\ &\quad + n^{\beta-1}\sum_{i=1}^{n}\int_0^t\Big\{\Delta_\alpha\phi(X^i(s)) \\ &\quad + \frac{1}{2n}\sum_{j\neq i}\delta_\epsilon(X^i(s)-X^j(s))\nabla\phi(X^i(s))\Big\}ds \\ &\quad - n^\beta\int_0^t\Big\langle \Delta_\alpha u^\epsilon(s) - \frac{1}{2}\nabla(\delta_\epsilon * u^\epsilon(s)\cdot u^\epsilon(s)),\phi\Big\rangle ds \end{aligned}$$

$$= m^n(\phi;t) + n^\beta \int_0^t \langle \bar{X}^n(s), \Delta_\alpha \phi\rangle ds$$

$$+ \frac{1}{2} n^{\beta-2} \sum_{i=1}^n \sum_{j\neq i} \int_0^t \delta_\epsilon(X^i(s) - X^j(s)) \nabla\phi(X^i(s)) ds$$

$$- n^\beta \int_0^t \langle u^\epsilon(s), \Delta_\alpha\phi + \frac{1}{2}(\delta_\epsilon * u^\epsilon(s)) \cdot \nabla\phi\rangle ds$$

$$= m^n(\phi;t) + \int_0^t \langle \eta^n(s), \Delta_\alpha\phi\rangle ds + \frac{1}{2}\int_0^t \tilde{b}^n(\phi;s) ds,$$

where

$$\tilde{b}^n(\phi;t) = n^{\beta-2} \sum_{i=1}^n \sum_{j\neq i} \delta_\epsilon(X^i(t) - X^j(t)) \nabla\phi(X^i(t))$$
$$- n^\beta \langle u^\epsilon(t), (\delta_\epsilon * u^\epsilon(t)) \cdot \nabla\phi\rangle.$$

Since

$$\frac{1}{n}\sum_{j\neq i} \delta_\epsilon(X^i(t) - X^j(t)) = \langle \bar{X}^n(t), \delta_\epsilon(X^i(t) - \cdot)\rangle - \frac{1}{n}\delta_\epsilon(0)$$
$$= \langle n^{-\beta}\eta^n(t), \delta_\epsilon(X^i(t) - \cdot)\rangle + (u^\epsilon(t) * \delta_\epsilon)(X^i(t)) - \frac{1}{n}\delta_\epsilon(0),$$

we obtain that

$$\tilde{b}^n(\phi;t)$$
$$= n^{\beta-1} \sum_{i=1}^n \Big\{ n^{-\beta}\langle \eta^n(t), \delta_\epsilon(X^i(t) - \cdot)\rangle$$
$$+ (u^\epsilon(t) * \delta_\epsilon)(X^i(t)) - \frac{1}{n}\delta_\epsilon(0) \Big\} \nabla\phi(X^i(t))$$
$$- n^\beta \langle u^\epsilon(t), (\delta_\epsilon * u^\epsilon(t)) \cdot \nabla\phi\rangle$$
$$= \Big\langle \bar{X}^n(dx,t)\eta^n(dy,t), \delta_\epsilon(x-y)\nabla\phi(x)\Big\rangle$$
$$+ n^\beta \Big\langle \bar{X}^n(t), (u^\epsilon(t) * \delta_\epsilon)\nabla\phi\Big\rangle$$
$$- n^{\beta-1}\delta_\epsilon(0)\langle \bar{X}^n(t), \nabla\phi\rangle - n^\beta\langle u^\epsilon(t), (u^\epsilon(t) * \delta_\epsilon)\nabla\phi\rangle.$$

Therefore,

$$
\begin{aligned}
\tilde{b}^n(\phi;t) =& n^{-\beta}\Big\langle \eta^n(dx,t)\eta^n(dy,t), \delta_\epsilon(x-y)\nabla\phi(x)\Big\rangle \\
&+ \Big\langle u^\epsilon(dx,t)\eta^n(dy,t), \delta_\epsilon(x-y)\nabla\phi(x)\Big\rangle \\
&+ \langle \eta^n(t), (u^\epsilon(t) * \delta_\epsilon)\nabla\phi\rangle - n^{\beta-1}\delta_\epsilon(0)\langle \bar{X}^n(t), \nabla\phi\rangle \\
=& 2b^n(\phi;t),
\end{aligned}
$$

which completes the proof of Lemma 4.1. ■

Proof of Theorem 3.2(i-ii) continued. We provide appropriate estimates for $\eta^n(t)$ defined in (4.2). Observe first that, by Lemma 4.1 $\phi(x) = \chi_\lambda(x) = e^{i\lambda x}$,

$$
\begin{aligned}
\langle \eta^n(t), \chi_\lambda\rangle =& \langle \eta^n(0), \chi_\lambda\rangle + m^n(\chi_\lambda;t) \\
&+ \int_0^t b^n(\chi_\lambda;s)\,ds - c_\alpha|\lambda|^\alpha \int_0^t \langle \eta^n(s), \chi_\lambda\rangle\,ds,
\end{aligned}
$$

since

$$
\Delta_\alpha \chi_\lambda = -c_\alpha|\lambda|^\alpha \chi_\lambda,
$$

where c_α is a positive constant given by

$$
c_\alpha = -K\int_{\mathbf{R}} \left(e^{iy} - 1 - \frac{iy}{1+y^2}\right)\frac{dy}{|y|^{1+\alpha}}.
$$

Hence, we have

$$
d\left[e^{c_\alpha|\lambda|^\alpha t}\langle \eta^n(t), \chi_\lambda\rangle\right] = e^{c_\alpha|\lambda|^\alpha t}\Big(dm^n(\chi_\lambda;t) + b^n(\chi_\lambda;t)\,dt\Big)
$$

and, as a result

$$
\langle \eta^n(t), \chi_\lambda\rangle = I_1^{n,\lambda}(t) + I_2^{n,\lambda}(t) + I_3^{n,\lambda}(t), \tag{4.6}
$$

where the three terms in the right-hand side are defined as

$$
\begin{aligned}
I_1^{n,\lambda}(t) &= e^{-c_\alpha|\lambda|^\alpha t}\langle \eta^n(0), \chi_\lambda\rangle, \\
I_2^{n,\lambda}(t) &= \int_0^{t+} e^{-c_\alpha|\lambda|^\alpha (t-s)}dm^n(\chi_\lambda;s), \\
I_3^{n,\lambda}(t) &= \int_0^t e^{-c_\alpha|\lambda|^\alpha (t-s)}b^n(\chi_\lambda;s)\,ds.
\end{aligned}
$$

Our goal is deriving a uniform estimate for

$$g^n(t) := \sup_{\lambda\in\mathbf{R}} \frac{1}{1+|\lambda|^a} E[|\langle \eta^n(t), \chi_\lambda\rangle|], \qquad t\in[0,T], \tag{4.7}$$

remembering that the basic assumption of Theorem 3.2 was that $\sup_n g^n(0) < \infty$ for some $a \geq 0$. To this end we give estimates for $I_i^{n,\lambda}(t), i = 2,3$, separately. First, we show the following lemma for $I_2^{n,\lambda}(t)$.

Lemma 4.2.

$$\sup_n \sup_{\lambda\in\mathbf{R}} \sup_{t\in[0,T]} E[|I_2^{n,\lambda}(t)|] < \infty. \tag{4.8}$$

Proof. Since $\alpha > 1$, the L^1-norm is dominated by the weak L^α-norm and we have

$$\begin{aligned} E[|\tilde{I}_2^{n,\lambda}(t)|] &\leq \sup_{z>0} z\left\{P\left(|\tilde{I}_2^{n,\lambda}(t)| > z\right)\right\}^{1/\alpha} \\ &\leq C\left[E\int_0^t \sum_{i=1}^n \left|e^{c_\alpha|\lambda|^\alpha s} F^i\right|^\alpha ds\right]^{1/\alpha}, \end{aligned} \tag{4.9}$$

where

$$F^i = n^{\beta-1} \sup_{\substack{s\leq t: S^i(s) \\ \neq S^i(s-)}} \left|\frac{\chi_\lambda(X^i(s-)+S^i(s)-S^i(s-)) - \chi_\lambda(X^i(s-))}{S^i(s)-S^i(s-)}\right|.$$

The second inequality in (4.9) is shown by the vector-valued (finite-dimensional) version

$$\sup_{z>0} z^\alpha P\left(\sup_{s\leq t}\int_0^s F(r)\cdot dS(r) > z\right) \leq c(\alpha) E\int_0^t \|F(s)\|^\alpha ds,$$

of Theorem 9.5.3 from Kwapien and Woyczyński (1992, p. 272). The proof of Lemma 4.2 is completed by noting that $|F^i|^\alpha \leq n^{-1}|\lambda|^\alpha$. ■

The next lemma will be helpful in the estimation of $I_3^{n,\lambda}(t)$.

Lemma 4.3.

$$\begin{aligned} &|b^n(\chi_\lambda;t)| \\ &\leq \frac{|\lambda|}{2}\left[\int_{\mathbf{R}}\left\{3|\langle\eta^n(t),\chi_{-\xi}\rangle| + |\langle\eta^n(t),\chi_{\lambda+\xi}\rangle|\right\}\hat{\delta}_\epsilon(\xi)d\xi + n^{\beta-1}\delta_\epsilon(0)\right]. \end{aligned} \tag{4.10}$$

Proof. We shall estimate the three terms on the right-hand side of (4.5) separately. For the first one we get

$$\begin{aligned}
&n^{-\beta}\Big|\Big\langle \eta^n(dx,t)\eta^n(dy,t), \delta_\epsilon(x-y)\nabla\chi_\lambda(x)\Big\rangle\Big| \\
=&n^{-\beta}\Big|\int_{\mathbf{R}} d\xi \hat{\delta}_\epsilon(\xi)\Big\langle \eta^n(dx,t)\eta^n(dy,t), e^{i\xi(x-y)}\cdot i\lambda e^{i\lambda x}\Big\rangle\Big| \\
\leq&|\lambda|\int_{\mathbf{R}} d\xi \hat{\delta}_\epsilon(\xi)|\langle \eta^n(t), \chi_{-\xi}\rangle|\, n^{-\beta}|\langle \eta^n(t), \chi_{\lambda+\xi}\rangle| \\
\leq&2|\lambda|\int_{\mathbf{R}} d\xi \hat{\delta}_\epsilon(\xi)|\langle \eta^n(t), \chi_{-\xi}\rangle|,
\end{aligned}$$

since $n^{-\beta}|\langle \eta^n(t), \chi_{\lambda+\xi}\rangle| \leq 2$. The second term in (4.5) has two summands and

$$\begin{aligned}
&\Big|\Big\langle u^\epsilon(dx,t)\eta^n(dy,t), \delta_\epsilon(x-y)\nabla\chi_\lambda(x)\Big\rangle\Big| \\
=&\Big|\int_{\mathbf{R}} d\xi \hat{\delta}_\epsilon(\xi)\Big\langle u^\epsilon(dx,t)\eta^n(dy,t), e^{i\xi(x-y)}\cdot i\lambda e^{i\lambda x}\Big\rangle\Big| \\
\leq&|\lambda|\int_{\mathbf{R}} d\xi \hat{\delta}_\epsilon(\xi)|\langle \eta^n(t), \chi_{-\xi}\rangle||\langle u^\epsilon(t), \chi_{\lambda+\xi}\rangle| \\
\leq&|\lambda|\int_{\mathbf{R}} d\xi \hat{\delta}_\epsilon(\xi)|\langle \eta^n(t), \chi_{-\xi}\rangle|,
\end{aligned}$$

since $|\langle u^\epsilon(t), \chi_{\lambda+\xi}\rangle| \leq 1$, and in a similar way

$$\begin{aligned}
&\Big|\Big\langle u^\epsilon(dy,t)\eta^n(dx,t), \delta_\epsilon(x-y)\nabla\chi_\lambda(x)\Big\rangle\Big| \\
&\leq |\lambda|\int_{\mathbf{R}} d\xi \hat{\delta}_\epsilon(\xi)|\langle \eta^n(t), \chi_{\lambda+\xi}\rangle|.
\end{aligned}$$

Finally the estimate of the third term in (4.5) follows from the fact that $\nabla\chi_\lambda = i\lambda\chi_\lambda$. Thus Lemma 4.3 has been proved. ■

Proof of Theorem 3.2(i-ii) continued. By Lemma 4.3,

$$\begin{aligned}
E[|I_3^{n,\lambda}(t)|] \leq&\frac{1}{2}\int_0^t |\lambda| e^{-c_\alpha|\lambda|^\alpha(t-s)}ds \\
&\times\int_{\mathbf{R}} E[|\langle \eta^n(s), \chi_\xi\rangle|]\{3\hat{\delta}_\epsilon(\xi) + \hat{\delta}_\epsilon(\xi-\lambda)\}d\xi \qquad (4.11) \\
&+\frac{1}{2}\int_0^t |\lambda| e^{-c_\alpha|\lambda|^\alpha(t-s)} n^{-1/\alpha}\delta_\epsilon(0)\, ds.
\end{aligned}$$

However, for each $a \geq 0$ and all $\lambda, \xi \in \mathbf{R}$,

$$\begin{aligned} &3\hat{\delta}_\epsilon(\xi) + \hat{\delta}_\epsilon(\xi - \lambda) \\ \leq &C_{\epsilon,a}\left(\frac{1}{1+|\xi|^a}\hat{\delta}_{\epsilon/2}(\xi) + \exp\Big(-\frac{\epsilon}{4}(\xi-\lambda)^2\Big)\hat{\delta}_{\epsilon/2}(\xi-\lambda)\right) \\ \leq &C_{\epsilon,a}\left(\frac{1}{1+|\xi|^a}\hat{\delta}_{\epsilon/2}(\xi) + C_{\epsilon,a}\frac{1+|\lambda|^a}{1+|\xi|^a}\hat{\delta}_{\epsilon/2}(\xi-\lambda)\right). \end{aligned} \tag{4.12}$$

Here, the last inequality can be justified as follows: if $|\xi - \lambda| \leq |\xi|/2$ then $|\xi| \leq 2|\lambda|$, so that $1 + |\lambda|^a \geq 1 + (|\xi|/2)^a \geq 2^{-a}(1 + |\xi|^a)$. On the other hand, if $|\xi - \lambda| \geq |\xi|/2$, then, for some $C_{\epsilon,a}$,

$$\exp\left(-\frac{\epsilon}{2}(\xi-\lambda)^2\right) \leq \exp\left(-\frac{\epsilon}{8}|\xi|^2\right) \leq C_{\epsilon,a}\frac{1+|\lambda|^a}{1+|\xi|^a}.$$

Therefore, noting that

$$|\lambda|e^{-c_\alpha|\lambda|^\alpha t} \leq C_{\alpha,T}t^{-1/\alpha}, \qquad \lambda \in \mathbf{R},\ t \in (0,T],$$

for some $C_{\alpha,T}$ and recalling the definition (4.7) of $g^n(t)$, we have, by (4.6), (4.8), (4.11) and (4.12),

$$g^n(t) \leq g^n(0) + C + c_\epsilon \int_0^t (t-s)^{-1/\alpha} g^n(s)\, ds + \bar{C}_\epsilon, \tag{4.13}$$

where (recall that $\alpha > 1$)

$$\bar{C}_\epsilon = \sup_n \sup_{\lambda \in \mathbf{R}} \frac{1}{2(1+|\lambda|^a)} \int_0^t |\lambda| e^{-c_\alpha|\lambda|^\alpha(t-s)} n^{-1/\alpha}\delta_\epsilon(0)\, ds < \infty.$$

Our basic assumption on the initial distribution is $\sup_n g^n(0) < \infty$, that is

$$\sup_n \sup_{\lambda \in \mathbf{R}} \frac{1}{1+|\lambda|^a} E[|\langle \eta^n(0), \chi_\lambda\rangle|] < \infty$$

for some $a \geq 0$. Then, together with Gronwall's lemma and Hölder's inequality, (4.13) gives the existence of a constant C_ϵ, independent of n and λ, such that

$$E[|\langle \eta^{n,\epsilon}(t), \chi_\lambda\rangle|] \leq C_\epsilon(1 + |\lambda|^a).$$

Since $\phi(x) = \int_R \chi_\lambda(x)\hat{\phi}(d\lambda)$, and thus

$$\langle \eta, \phi \rangle = \int_{\mathbf{R}} \langle \eta, \chi_\lambda \rangle \hat{\phi}(d\lambda),$$

we finally obtain that

$$E[|\langle \eta^{n,\epsilon}(t), \phi \rangle|] \leq \int_{\mathbf{R}} E[|\langle \eta^{n,\epsilon}(t), \chi_\lambda \rangle|] |\hat{\phi}|(d\lambda) \leq C_\epsilon \int_{\mathbf{R}} (1 + |\lambda|^a) |\hat{\phi}|(d\lambda),$$

which proves (3.13) and, therefore, Theorem 3.2(i-ii). ∎

Proof of Theorem 3.2(iii). Note that to prove (3.14) we need to show a purely analytic statement:

$$|\langle u^\epsilon(t) - u(t), \phi \rangle| \to 0 \tag{4.20}$$

as $\epsilon \to 0$. Indeed, once (4.20) is established, one can immediately conclude that for some $\epsilon(n) \to 0$,

$$E[|\langle \bar{X}^{n,\epsilon(n)}(t) - u(t), \phi \rangle|] \leq C_{\epsilon(n)} n^{-1+1/\alpha} + |\langle u^{\epsilon(n)}(t) - u(t), \phi \rangle| \to 0 \tag{4.21}$$

as $n \to \infty$ for ϕ satisfying $\int_{\mathbf{R}} (1 + |\lambda|^a) |\hat{\phi}|(d\lambda) < \infty$.

To prove (4.20) we will need a series of auxiliary propositions. Recall that $|\,.\,|_p$ denotes the Lebesgue $L^p(\mathbf{R})$-norm and $\|\,.\,\|_\beta$ denotes the usual Hilbert–Sobolev $H^\beta(\mathbf{R})$-norm.

Proposition 4.1. *Let $3/2 < \alpha < 2$ and let u^ϵ be a solution of the Cauchy problem for the regularized fractal Burgers equation (3.12) with the initial condition $u^\epsilon(0) = u_0$. Then, for any $T > 0$, there exists $\epsilon_0 = \epsilon_0(u_0) > 0$ such that*

$$\sup_{0<\epsilon<\epsilon_0} \sup_{t \in [0,T]} |u^\epsilon(t)|_2 < \infty.$$

Proof. We begin by an estimate of the time derivative of the L^2 norm of u^ϵ and note that

$$\begin{aligned} \frac{d}{dt} |u^\epsilon|_2^2 &= 2\langle u^\epsilon, \Delta_\alpha u^\epsilon - \frac{1}{2} \nabla (\delta_\epsilon * u^\epsilon \cdot u^\epsilon) \rangle \\ &= -2\|u^\epsilon\|_{\alpha/2}^2 + \frac{1}{2} \langle \nabla (u^\epsilon)^2, \delta_\epsilon * u^\epsilon \rangle, \end{aligned} \tag{4.22}$$

where for the last term, in view of the identity $\langle\nabla(u^\epsilon)^2, u^\epsilon\rangle = 0$, we have, for each $\beta > 0$, the estimate

$$\begin{aligned}\langle\nabla(u^\epsilon)^2, \delta_\epsilon * u^\epsilon\rangle &\le \|\nabla(u^\epsilon)^2\|_{-\beta}\|\delta_\epsilon * u^\epsilon - u^\epsilon\|_\beta \\ &\le \|(u^\epsilon)^2\|_\gamma\|\delta_\epsilon * u^\epsilon - u^\epsilon\|_\beta \end{aligned}\tag{4.23}$$

with $\beta + \gamma = 1$.

Introduce the "intrinsic" norm on $H^s(\mathbf{R}), 0 < s < 1$, via the formula

$$\|u\|_{(s)}^2 := |u|_2^2 + \int_{\mathbf{R}}\int_{\mathbf{R}} \frac{|u(x) - u(y)|^2}{|x-y|^{1+2s}}\, dx\, dy. \tag{4.24}$$

The norms $\| u \|_s$ and $\| u \|_{(s)}$ are equivalent. Also, recall the standard embeddings

$$H^s(\mathbf{R}) \hookrightarrow L^p(\mathbf{R}), \qquad \text{if} \qquad s \ge 1/2 - 1/p,\ p \ge 2, \tag{4.25}$$

$$H^s(\mathbf{R}) \hookrightarrow L^\infty(\mathbf{R}), \qquad \text{if} \qquad s > 1/2, \tag{4.26}$$

and the interpolation inequality

$$\|u\|_s \le C\, \|u\|_{\alpha/2}^{2s/\alpha}\, |u|_2^{1-(2s/\alpha)}, \qquad \text{if} \qquad 0 < s < \alpha/2. \tag{4.27}$$

Now, we can return to the needed estimate of $\|(u^\epsilon)^2\|_\gamma$ in (4.23) and prove that, for each $\delta > 1/2$, there exists a $C = C_\delta$ such that

$$\|(u^\epsilon)^2\|_\gamma^2 \le C\, \|u^\epsilon\|_{\alpha/2}^{(2\gamma+2\delta)/\alpha}\, |u^\epsilon|_2^{2-(2\gamma+2\delta)/\alpha}, \text{ if } \alpha > 1,\ 1/4 \le \gamma < \alpha/2. \tag{4.28}$$

Indeed, in view of (4.24–26), for $s \ge 1/4, \delta > 1/2$,

$$\begin{aligned}\|(u^\epsilon)^2\|_\gamma &\le C\left(|u^\epsilon|_4^4 + \int_{\mathbf{R}}\int_{\mathbf{R}} \frac{|u^\epsilon(x) + u^\epsilon(y)|^2|u^\epsilon(x) - u^\epsilon(y)|^2}{|x-y|^{1+2\gamma}}\, dx\, dy\right) \\ &\le C\left(|u^\epsilon|_4^4 + (2|u|_\infty)^2 \int_{\mathbf{R}}\int_{\mathbf{R}} \frac{|u^\epsilon(x) - u^\epsilon(y)|^2}{|x-y|^{1+2\gamma}}\, dx\, dy\right) \\ &\le C\left(\|u^\epsilon\|_s^4 + \|u^\epsilon\|_\delta^2\|u^\epsilon\|_\gamma^2\right) \le C\|u^\epsilon\|_\delta^2\|u^\epsilon\|_\gamma^2,\end{aligned}$$

as long as $\gamma \ge 1/4$. Hence, by (4.27),

$$\|(u^\epsilon)^2\|_\gamma \le C\|u^\epsilon\|_{\alpha/2}^{2\delta/\alpha}\, |u^\epsilon|_2^{1-(2\delta/\alpha)}\, \|u^\epsilon\|_{\alpha/2}^{2\gamma/\alpha}\, |u^\epsilon|_2^{1-(2\gamma/\alpha)},$$

if $1/2 < \delta < \alpha/2, 1/4 \le \gamma < \alpha/2$, which gives (4.28).

As far as the term $\|\delta_\epsilon * u^\epsilon - u^\epsilon\|_\beta$ is concerned, we have

$$\begin{aligned}\|\delta_\epsilon * u^\epsilon - u^\epsilon\|_\beta &\le \epsilon^{a/2}\|u^\epsilon\|_{\beta+a} \\ &\le C\epsilon^{a/2}\|u^\epsilon\|_{\alpha/2}^{2(\beta+a)/\alpha}|u^\epsilon|_2^{1-2(\beta+a)/\alpha}, \end{aligned} \tag{4.29}$$

for each, small enough $a > 0$ such that $0 < \beta + a < \alpha/2$. Indeed, the second inequality is a direct consequence of the interpolation formula (4.27), and the first inequality can be proved as follows:

$$\begin{aligned}\|\delta_\epsilon * u^\epsilon - u^\epsilon\|_\beta^2 &= \int (1+|\xi|^2)^\beta |\widehat{\delta_\epsilon * u^\epsilon} - \widehat{u^\epsilon}|^2(\xi)\, d\xi \\ &= \int (1-e^{-\epsilon|\xi|^2})^2(1+|\xi|^2)^\beta |\widehat{u^\epsilon}(\xi)|^2\, d\xi \\ &\le \epsilon^a \int (1+|\xi|^2)^{\beta+a}|\widehat{u^\epsilon}(\xi)|^2\, d\xi = \epsilon^a\|u^\epsilon\|_{\beta+a}^2,\end{aligned}$$

because, for $0 < a/2 < 1$, we have $(1-e^{-\epsilon|\xi|^2})^2 \le \epsilon^{a/2}|\xi|^a$.

Putting the estimates (4.28–29) together we get that, for any $\delta > 1/2$ and $a > 0$ (remember that $\beta + \gamma = 1, 1/4 \le \gamma < \alpha/2$),

$$\frac{d}{dt}|u^\epsilon|_2^2 \le -2\|u^\epsilon\|_{\alpha/2}^2 + C\epsilon^{a/2}\|u^\epsilon\|_{\alpha/2}^{(2+2\delta+2a)/\alpha}|u^\epsilon|_2^{3-(2+2\delta+2a)/\alpha}.$$

Since $\alpha > 3/2$, one can choose $\delta > 1/2$ and $a > 0$ such that $(2+2\delta+2a)/\alpha < 2$, so that, for sufficiently large N,

$$\begin{aligned}\frac{d}{dt}|u^\epsilon|_2^2 &\le -2\|u^\epsilon\|_{\alpha/2}^2 + C\epsilon^{a/2}\left(\|u^\epsilon\|_{\alpha/2}^2 + |u^\epsilon|_2^{2N}\right) \\ &\le -\|u^\epsilon\|_{\alpha/2}^2 + C\epsilon^{a/2}|u^\epsilon|_2^{2N}.\end{aligned}$$

In particular,

$$\frac{d}{dt}|u^\epsilon|_2^2 \le C\epsilon^{a/2}|u|_2^{2N},$$

so that

$$|u^\epsilon|_2^2 \le \left(|u_0|^{-2(N-1)} - (N-1)C\epsilon^{a/2}t\right)^{-1/(N-1)}.$$

This proves Proposition 4.1. ■

Proposition 4.2. *Let $3/2 < \alpha < 2$ and let u^ϵ be a solution of the Cauchy problem for the regularized fractal Burgers equation (3.12) with the initial condition $u^\epsilon(0) = u_0$. Then, for any $T > 0$, there exists $\epsilon_0 = \epsilon_0(u_0) > 0$ such that*

$$\sup_{0<\epsilon<\epsilon_0} \sup_{t\in[0,T]} \|u^\epsilon(t)\|_1 < \infty.$$

Proof. Note that

$$\begin{aligned}\frac{d}{dt}\|u^\epsilon\|_1^2 &= 2\left\langle -(u^\epsilon)'', \Delta_\alpha u^\epsilon - \frac{1}{2}\nabla(\delta_\epsilon * u^\epsilon \cdot u^\epsilon)\right\rangle \\ &= -2\|u^\epsilon\|_{1+\alpha/2}^2 + \langle (u^\epsilon)'', \nabla(\delta_\epsilon * u^\epsilon \cdot u^\epsilon)\rangle. \end{aligned} \tag{4.30}$$

Since $\langle (u^\epsilon)'', ((u^\epsilon)^2)'\rangle = -\int_{\mathbf{R}}((u^\epsilon)')^3 dx$,

$$\begin{aligned}&\langle (u^\epsilon)'', \nabla(\delta_\epsilon * u^\epsilon \cdot u^\epsilon)\rangle \\ &\quad \le \langle (u^\epsilon)'', \nabla((\delta_\epsilon * u^\epsilon - u^\epsilon)\cdot u^\epsilon)\rangle + |(u^\epsilon)'|_3^3 \\ &\quad \le \|(u^\epsilon)'''\|_{-2+\alpha/2}\|(\delta_\epsilon * u^\epsilon - u^\epsilon)\cdot u^\epsilon\|_{2-\alpha/2} + |(u^\epsilon)'|_3^3 \\ &\quad \le \|u^\epsilon\|_{1+\alpha/2}\|(\delta_\epsilon * u^\epsilon - u^\epsilon)\cdot u^\epsilon\|_{2-\alpha/2} + |(u^\epsilon)'|_3^3. \end{aligned} \tag{4.31}$$

The last term is estimated as follows (see the proof of Theorem 2.1 in Biler, Funaki, and Woyczyński (1998)): if $\alpha > 3/2$ then there exist m and $C > 0$ such that

$$|(u^\epsilon)'|_3^3 \le \|u^\epsilon\|_{1+\alpha/2}^2 + C|u^\epsilon|_2^m. \tag{4.32}$$

On the other hand, if $1 < s < 2$, for any $\delta > 1/2$,

$$\|uv\|_s \le C\big(\|u\|_s\|v\|_\delta + \|u\|_\delta\|v\|_s + \|u\|_{1+\delta}\|v\|_{s-1} + \|u\|_{s-1}\|v\|_{1+\delta}\big). \tag{4.33}$$

Indeed, for $0 < s < 1$,

$$\begin{aligned}\|uv\|_{(s)}^2 &= |uv|_2^2 + \int_{\mathbf{R}}\int_{\mathbf{R}} \frac{|u(x)v(x) - u(y)v(y)|^2}{|x-y|^{1+2s}}\, dx\, dy \\ &\le C\big(|u|_\infty\|v\|_s + |v|_\infty\|u\|_s\big)^2\end{aligned}$$

because $|uv|_2^2 \le |u|_\infty^2|v|_2^2$, and because the double integral $\le |v|_\infty^2\|u\|_s^2 + |u|_\infty^2\|v\|_s^2$. Hence, in view of the embedding (4.26), for any $0 < s \le 1$ and

$\delta > 1/2$,

$$\|uv\|_s \leq C(\|u\|_\delta \|v\|_s + \|v\|_\delta \|u\|_s),$$

so that, for any $1 < s < 2$,

$$\begin{aligned}\|uv\|_s = \|uv\|_1 + \|(uv)'\|_{s-1} &\leq \|uv\|_1 + \|u'v\|_{s-1} + \|uv'\|_{s-1} \\ &\leq C(\|u\|_\delta \|v\|_1 + \|v\|_\delta \|u\|_1) \\ &\quad + C(\|u\|_{1+\delta} \|v\|_{s-1} + \|v\|_\delta \|u\|_s) \\ &\quad + C(\|u\|_\delta \|v\|_s + \|v\|_{1+\delta} \|u\|_{s-1}),\end{aligned}$$

which gives (4.33).

Therefore, noting that $2 - \alpha/2 < 1 + \alpha/2$ and $1 + \delta < 1 + \alpha/2$,

$$\|(\delta_\epsilon * u^\epsilon - u^\epsilon) \cdot u^\epsilon\|_{2-\alpha/2} \leq C\epsilon^{a/4} \|u^\epsilon\|^\gamma_{1+\alpha/2} |u^\epsilon|_2^{2-\gamma}, \tag{4.34}$$

with $\gamma = (2\delta + 4 - \alpha + 2a)/(2 + \alpha)$. Indeed, after using (4.33), we obtain four terms and, for instance, the term $\|u^\epsilon\|_\delta \|\delta_\epsilon * u^\epsilon - u^\epsilon\|_{2-\alpha/2}$ can be further estimated by interpolation inequality (4.27),

$$\|u^\epsilon\|_\delta \leq |u^\epsilon|_2^{1-2\delta/(2+\alpha)} \|u^\epsilon\|_{1+\alpha/2}^{2\delta/(2+\alpha)}$$

and, by the estimate (4.29) and again inequality (4.27),

$$\begin{aligned}\|\delta_\epsilon * u^\epsilon - u^\epsilon\|_{2-\alpha/2} &\leq \epsilon^{a/4} \|u^\epsilon\|_{2-\alpha/2+a} \\ &\leq \epsilon^{a/4} |u^\epsilon|_2^{1-2(2-\alpha/2+a)/(2+\alpha)} \|u^\epsilon\|_{1+\alpha/2}^{2(2-\alpha/2+a)/(2+\alpha)}.\end{aligned}$$

The other three terms can be estimated similarly, and we have (4.34). Summarizing the above estimates we get

$$\begin{aligned}\frac{\partial}{\partial t} \|u^\epsilon\|_1^2 &\leq -\|u^\epsilon\|^2_{1+\alpha/2} + C|u^\epsilon|_2^m + C\epsilon^{a/4} \|u^\epsilon\|^{1+\gamma}_{1+\alpha/2} |u^\epsilon|_2^{2-\gamma} \\ &\leq C(|u^\epsilon|_2^m + |u^\epsilon|_2^n),\end{aligned} \tag{4.35}$$

for some $n > 0$, since $\alpha > 3/2$ implies that one can choose $\delta 1/2$ and $a > 0$ such that $1 + \gamma < 2$.

Finally, Proposition 4.1 implies that the quantity (4.35) remains bounded, which completes the proof of Proposition 4.2. ■

Proposition 4.3. *Let $3/2 < \alpha < 2$ and let u^ϵ be a solution of the Cauchy problem for the regularized fractal Burgers equation (3.12) with the initial* *condition* $u^\epsilon(0) = u_0 \in H^\alpha$. *Then, for any* $T > 0$ *there exists* $\epsilon_0 = \epsilon_0(u_0) > 0$ *such that*

$$\sup_{0<\epsilon<\epsilon_0} \left\{ \sup_{t\in[0,T]} |u_t^\epsilon(t)|_2 + \int_0^T \|u_t^\epsilon(t)\|_{\alpha/2}^2 \, dt \right\} < \infty.$$

Proof. Observe that

$$\begin{aligned} \frac{\partial}{\partial t}|u_t^\epsilon|_2^2 &= 2\langle u_t^\epsilon, u_{tt}^\epsilon\rangle = 2\langle u_t^\epsilon, \Delta_\alpha u_t^\epsilon - \frac{1}{2}\nabla((\delta_\epsilon * u^\epsilon)\cdot u^\epsilon)_t\rangle \\ &\le -2\|u_t^\epsilon\|_{\alpha/2}^2 + \|u_t^\epsilon\|_{\alpha/2}\cdot \|\delta_\epsilon * u_t^\epsilon \cdot u^\epsilon + \delta_\epsilon * u^\epsilon \cdot u_t^\epsilon\|_{1-\alpha/2}. \end{aligned} \tag{4.36}$$

Now, for each $\delta > 1/2$,

$$\begin{aligned} \|\delta_\epsilon * u_t^\epsilon \cdot u^\epsilon\|_{1-\alpha/2} &\le C\Big(\|\delta_\epsilon * u_t^\epsilon\|_\delta \|u^\epsilon\|_{1-\alpha/2} + \|\delta_\epsilon * u_t^\epsilon\|_{1-\alpha/2}\|u^\epsilon\|_\delta\Big) \\ &\le C\|u_t^\epsilon\|_{\delta\vee(1-\alpha/2)}\|u^\epsilon\|_1 \\ &\le C\|u_t^\epsilon\|_{\alpha/2}^a \, |u_t^\epsilon|_2^{1-a}\, \|u^\epsilon\|_1, \end{aligned} \tag{4.37}$$

for some $a \in (0,1)$. The first inequality follows from (4.33) since $0 < 1-\alpha/2 < 1$. The second is implied by (4.29) which, with $a = 0$, gives $\|\delta_\epsilon * u^\epsilon - u^\epsilon\|_\beta \le \|u^\epsilon\|_\beta$, so that $\|\delta_\epsilon * u_t^\epsilon\|_\delta \le 2\|u_t^\epsilon\|_\delta$ and $\|\delta_\epsilon * u_t^\epsilon\|_{1-\alpha/2} \le 2\|u_t^\epsilon\|_{1-\alpha/2}$. The third inequality follows from the interpolation inequality (4.27) since $1/2 < \delta < \alpha/2$ and $1-\alpha/2 < \alpha/2$.

In a similar fashion, using the interpolation inequality (4.27), and the estimates $\|\delta_\epsilon * u^\epsilon\|_\delta \le 2\|u^\epsilon\|_1$ and $\|\delta_\epsilon * u^\epsilon\|_{1-\alpha/2} \le 2\|u^\epsilon\|_1$, we get, for some $a \in (0,1)$,

$$\|\delta_\epsilon * u^\epsilon \cdot u_t^\epsilon\|_{1-\alpha/2} \le C\|u_t^\epsilon\|_{\alpha/2}^a |u_t^\epsilon|_2^{1-a}\|u^\epsilon\|_1. \tag{4.38}$$

Therefore,

$$\begin{aligned} \frac{\partial}{\partial t}|u_t^\epsilon|_2^2 &\le -2\|u_t^\epsilon\|_{\alpha/2}^2 + C\|u_t^\epsilon\|_{\alpha/2}^{1+a}\, |u_t^\epsilon|_2^{1-a}\, \|u^\epsilon\|_1 \\ &\le -\|u_t^\epsilon\|_{\alpha/2}^2 + C|u_t^\epsilon|_2^2\, \|u^\epsilon\|_1^{2/(1-a)}. \end{aligned} \tag{4.39}$$

The second inequality is a corollary to the elementary Young's inequality $xy \le x^p/p + y^q/q$, $1/p + 1/q = 1$, with $p = 2/(1+a), q = 2/(1-a)$ and $1 + a < 2$.

The norm $\|u^\epsilon\|_1$ remains bounded by Proposition 4.1, so that, neglecting the term $-\|u_t^\epsilon\|_{\alpha/2}^2$ in (4.39), and applying Gronwall's lemma, we have

$$\sup_{0<\epsilon<\epsilon_0} \sup_{t\in[0,T]} |u_t^\epsilon(t)|_2 < \infty.$$

Then, since

$$\int_0^T \|u_t^\epsilon\|_{\alpha/2}^2 \, dt \le |u_t^\epsilon(0)|_2^2 + C \int_0^T |u_t^\epsilon(t)|_2^2 \|u^\epsilon(t)\|_1^{2/(1-a)} \, dt, \tag{4.40}$$

we get the conclusion of Proposition 4.3. Note that the quantity $\sup_\epsilon |u_t^\epsilon(0)|_2$ appearing in (4.40) is finite if $u^\epsilon(0) = u(0) \in H^\alpha$. Indeed,

$$\begin{aligned} |u_t^\epsilon(0)|_2 &= |\Delta_\alpha u^\epsilon(0) + \frac{1}{2}\nabla(\delta_\epsilon * u^\epsilon \cdot u^\epsilon)(0)|_2 \\ &\le \|u(0)\|_\alpha + \frac{1}{2}\|\delta_\epsilon * u(0) \cdot u(0)\|_1, \end{aligned}$$

and the latter is bounded in ϵ. ■

Proof of Theorem 3.2(iii) continued. Now the conclusion of the proof relies on two observations:

(a) The family $\{u^\epsilon(t)\}_{0<\epsilon<\epsilon_0, t\in[0,T]}$ is equicontinuous as a family of $H^{\alpha/2}(\mathbf{R})$-valued functions. Indeed, for each $0 \le t_1 < t_2 \le T$,

$$\begin{aligned} \|u^\epsilon(t_2) - u^\epsilon(t_1)\|_{\alpha/2} &\le \int_{t_2}^{t_2} \|\partial_t u^\epsilon(t)\|_{\alpha/2} \, dt \\ &\le (t_2 - t_1)^{1/2} \left(\int_0^T \|\partial_t u^\epsilon(t)\|_{\alpha/2}^2 \, dt \right)^{1/2} \end{aligned}$$

and the latter integral remains bounded in ϵ.

(b) In view of Proposition 4.2, the quantity $\sup_\epsilon \sup_t \|u^\epsilon(t)\|_1 < \infty$, and the embedding $H^1(\mathbf{R}) \hookrightarrow H_{-\lambda}^{\alpha/2}(\mathbf{R})$, $\lambda > 0$, where $H_\lambda^\alpha = \{v : ve^{\lambda B} \in H^\alpha\}$, $\theta(x) \sim |x|$, $\theta \in C^\infty$, is compact by Rellich's theorem (see Funaki (1995), Lemma 9.21).

It follows from (a) and (b) that the family $\{u^\epsilon(t)\}_{0<\epsilon<\epsilon_0}$ is relatively compact in $C([0,T], H_\lambda^{\alpha/2}(\mathbf{R}))$, for each $\lambda > 0$. Its every limit is a weak solution of the fractal Burgers equation (1.1), and that solution is unique. Therefore we have, as $\epsilon \to 0$,

$$\langle u^\epsilon(t) - u(t), \phi \rangle \to 0,$$

for any $\phi \in H_\lambda^{-\alpha/2}(\mathbf{R})$ (see Funaki (1995)). This concludes the proof of Theorem 3.2(iii). ∎

References

[1] Adams R.A. (1975), *Sobolev Spaces*, New York: Academic Press.

[2] Bardos C., Penel P., Frisch U., Sulem P.L. (1979), Modified dissipativity for a nonlinear evolution equation arising in turbulence, *Arch. Rat. Mech. Anal.* **71**, 237–256.

[3] Biler P., Funaki T., Woyczyński, W.A. (1998), Fractal Burgers equations, *J. Diff. Equations*, to appear.

[4] Biler P., Woyczyński W.A. (1998), Global and exploding solutions for nonlocal quadratic evolution problems, *SIAM J. Appl. Math.*, to appear.

[5] Bossy M., Talay D. (1996), Convergence rate for the approximation of the limit law of weakly interacting particles: application to the Burgers equation, *Ann. Appl. Prob.* **6**, 818–861.

[6] Burgers J. (1974), *The Nonlinear Diffusion Equation*, Dordrecht.

[7] Calderoni P., Pulvirenti M. (1983), Propagation of chaos for Burgers' equation, *Ann. Inst. H. Poincaré* **A39**, 85–97.

[8] Dawson D., Gorostiza (1990), Generalized solutions of a class of nuclear space valued stochastic evolution equations, *Appl. Math. Optim.* **22**, 241–264.

[9] E W., Rykov Yu.G., Sinai Ya.G. (1996), General variational principles, global weak solutions and behavior with random initial data for systems of conservation laws arising in adhesion particle dynamics, *Comm. Math. Phys.* **177**, 349–380.

[10] Echeverria P. (1982), A criterion for invariant measures of Markov processes, *Z. Wahr. verw. Gebiete* **61**, 1–16.

[11] Funaki T. (1984), A certain class of diffusion processes associated with nonlinear parabolic equations, *Z. Wahr. verw. Gebiete* **67**, 331–348.

[12] Funaki T. (1995), The scaling limit for a stochastic PDE and the separation of phases, *Prob. Theory Rel. Fields* **102**, 221–288.

[13] Funaki T., Surgailis D., Woyczyński W.A. (1995), Gibbs–Cox random fields and Burgers' turbulence, *Ann. Applied Probability* **5**, 701–735.

[14] Gurbatov S., Malakhov A., Saichev A. (1991), *Nonlinear Random Waves and Turbulence in Nondispersive Media: Waves, Rays and Particles*, Manchester: University Press.

[15] Gutkin E., Kac M. (1983), Propagation of chaos and the Burgers' equation, *SIAM J. Appl. Math.* **43**, 971–980.

[16] Henry D.B. (1982), How to remember the Sobolev inequalities, Springer's *Lecture Notes in Math.* **957**, 97–109.

[17] Ikeda N., Watanabe, S. (1981), *Stochastic Differential Equations and Diffusion Processes*, Amsterdam-Tokyo: North-Holland/Kodansha.

[18] Kardar M., Parisi G., Zhang Y.-C. (1986), Dynamic scaling of growing interfaces, *Phys. Rev. Lett.* **56**, 889–892.

[19] Komatsu T. (1984), On the martingale problem for generators of stable processes with perturbations, *Osaka J. Math.* **21**, 113–132.

[20] Komatsu T. (1984), Pseudo-differential operators and Markov processes, *J. Math. Soc. Japan* **36**, 387–418.

[21] Kotani S., Osada H. (1985), Propagation of chaos for Burgers' equation, *J. Math. Soc. Japan*, **37**, 275–294.

[22] Kwapien S., Woyczyński W.A. (1992), *Random Series and Stochastic Integrals: Single and multiple*, Boston: Birkhäuser.

[23] Ladyženskaja O.A., Solonnikov V.A., Ural'ceva N.N. (1968), *Linear and Quasilinear Equations of Parabolic Type,* Providence: Amer. Math. Soc.

[24] Lions J.L., Magenes E. (1972), *Non-homogeneous boundary value problems and applications, Vol. I*, Berlin: Springer.

[25] McKean, H.P. (1967), Propagation of chaos for a class of nonlinear parabolic equations, in *Lecture Series in Differential Equations*, Session 7, pp. 177–194, Catholic University, Washington D.C.

[26] Méléard S. (1996), Asymptotic behavior of some interacting particle systems; McKean–Vlasov and Boltzmann models, Springer's *Lecture Notes in Math.* **1627**, 42–95.

[27] Molchanov S.A., Surgailis D., Woyczyński W.A. (1997), The large-scale structure of the Universe and quasi-Voronoi tessellation of shock fronts in forced Burgers turbulence in R^d, *Ann. Appl. Prob.* **7**, 200–228.

[28] Oelschläger K. (1985), A law of large numbers for moderately interacting diffusion processes, *Z. Wahr. verw. Gebiete* **69**, 279–322.

[29] Osada H. (1986), Propagation of chaos for the two dimensional Navier–Stokes equation, in *Probabilistic Methods in Mathematical Physics* (eds. Itô K., Ikeda N.), 303–334.

[30] Saichev A.S., Woyczyński W.A. (1997a), Advection of passive and reactive tracers in multidimensional Burgers' velocity field, *Physica D*, **100**, 119–142.

[31] Saichev A.S., Woyczyński W.A. (1997b), *Distributions in the Physical and Engineering Sciences, Vol.1, Distributional and Fractal Calculus, Integral Transforms and Wavelets,* Boston: Birkhäuser.

[32] Shlesinger M.F., Zaslavsky G.M., Frisch U., Eds. (1995), *Lévy Flights and Related Topics in Physics*, Lect. Notes in Phys. **450**, Springer.

[33] Smoller J. (1994), *Shock Waves and Reaction-Diffusion Equations,* Springer.

[34] Stroock D.W. (1975), Diffusion processes associated with Lévy generators, *Z. Wahr. verw. Gebiete* **32**, 209–244.

[35] Sugimoto N. (1991), Burgers equation with a fractional derivative; hereditary effects on nonlinear acoustic waves, *J. Fluid Mech.* **225**, 631–653.

[36] Sugimoto N. (1992), Propagation of nonlinear acoustic waves in a tunnel with an array of Helmholtz resonators, *J. Fluid Mech.* **244**, 55–78.

[37] Sznitman A. (1986), A propagation of chaos result for Burgers' equation, *Probab. Th. Rel. Fields* **71**, 581–613.

[38] Sznitman A. (1991), Topics in propagation of chaos, Springer's *Lecture Notes in Math.* **1464**.

[39] Vergassola M., Dubrulle B.D., Frisch U., Nullez A. (1994), Burgers equation, devil's staircases and mass distribution for the large-scale structure, *Astr. and Astrophys.* **289**, 325–356.

[40] Woyczyński W.A. (1993), Stochastic Burgers' Flows, in *Nonlinear Waves and Weak Turbulence*, Boston: Birkhäuser, pp. 279–311.

[41] Woyczyński W.A. (1997), *Göttingen Lectures on Burgers-KPZ Turbulence*, to appear.

[42] Zaslavsky G.M. (1994), Fractional kinetic equations for Hamiltonian chaos, *Physica D* **76**, 110–122.

[43] Zaslavsky, G.M., Abdullaev S.S. (1995), Scaling properties and anomalous transport of particles inside the stochastic layer, *Phys. Rev. E* **51**(5), 3901–3910.

[44] Zheng W. (1995), Conditional propagation of chaos and a class of quasilinear PDE's, *Ann. Probab.* **23**, 1389–1413.

Department of Mathematical Sciences
University of Tokyo
3-8-1 Komaba, Meguro, Tokyo 153
Japan
funaki@ms.u-tokyo.ac.jp

Department of Statistics and Center for Stochastic and Chaotic Processes
in Science and Technology
Case Western Reserve University
Cleveland, Ohio 44106
waw@po.cwru.edu

Optimal Transformations for Prediction in Continuous-Time Stochastic Processes

B. Gidas and A. Murua *

This paper is dedicated to the memory of Stamatis Cambanis–a good friend and a wonderful person. In the early stages of this paper (when we barely understood the problem), Stamatis provided a driving impetus through long and joyful conversations that would go on till 2:00 or 3:00 am (during a four day visit to Brown). He seems to have enjoyed these conversations himself: "I enjoyed my visit thoroughly, specially the long late night discussions! Perhaps something may come out" (e-mail to B. G.).

Abstract

In the classical Wiener–Kolmogorov prediction problem, one *fixes* a functional of the "future" and seeks its best predictor (in the L^2-sense). In this paper we treat a variant of this problem, whereby we seek the "most predictable" non-trivial functional of the future *and* its best predictor. In contrast to the Wiener-Kolmogorov problem, our problem may not have solutions, and if solutions exist, they might not be unique. We prove the existence of solutions for linear functionals under appropriate conditions on the spectral function of weakly stationary, continuous-time processes.

1. Introduction

Let $\{X_t : t \in \mathbb{R}\}$ be a real-valued stationary stochastic process on a probability space $(\Omega, \mathcal{F}, \mathcal{P})$. $L^2(\Omega, \mathcal{F}, \mathcal{P})$ will denote the usual Hilbert space of real or complex functions of $\{X_t\}$ which are measurable with respect to the σ-algebra $\mathcal{F}$, and square-integrable with respect to $\mathcal{P}$; its norm and inner product will be denoted by $||\cdot||_{\mathcal{P}}$ and $< \cdot, \cdot >_{\mathcal{P}}$, respectively. For any bounded or unbounded interval $I \subset \mathbb{R}$, $\mathcal{F}_I = \sigma(\{X_t : t \in I\})$ will denote the smallest σ-algebra generated by $\{X_t : t \in I\}$, and $L^2(\Omega, \mathcal{F}_I, \mathcal{P}) = \{\xi \in L^2(\Omega, \mathcal{F}, \mathcal{P}) : \xi \text{ is } \mathcal{F}_I\text{-measurable}\}$. Throughout this paper, the intervals $I_0 = [\Delta, \Delta + \tau]$ with $\Delta > 0$, $0 < \tau \leq +\infty$, and $I_1 = [-T, 0]$ with $0 < T \leq +\infty$, will play

*This research was partially supported by ARO Grant DAAH04-95-1-0101, ONR Grant NOOO14-91-J-1021, and ARPA via ARLMDA972-93-1-0012. The manuscript was prepared using computer facilities supported in part by The University of Chicago Block Fund.

the role of "future" and "past", respectively. Our work [8, 9] on a speech recognition problem lead to the following optimization problem

$$\inf\{||\xi_0 - \xi_1||^2_{\mathcal{P}} : \xi_i \in L^2(\Omega, \mathcal{F}_{I_i}, \mathcal{P}), E\{\xi_i\} = 0, i = 0, 1, ||\xi_0||_{\mathcal{P}} = 1\} \tag{1.1}$$

and its linear version to be described below. This problem may be viewed as a variant of the classical Wiener-Kolmogorov nonlinear prediction problem where one fixes a functional ξ_0 in the future and seeks its *best predictor* (in the L^2-sense) in terms of the values of the process in the past. Indeed, in problem (1.1), we seek the *most predictable, non-trivial* functional of the future *and* its best predictor. In contrast to the classical Wiener-Kolmogorov problem which always has a unique solution, problem (1.1) may not have solutions or it may have multiple solutions (see §2). We shall refer to the solutions of problem (1.1) (if they exist) as *fully nonlinear optimal transformations for prediction,* in analogy with the optimal transformations for regression [2].

The linear version of problem (1.1) corresponds to restricting ξ_i, $i = 0, 1$, to be linear functionals of $\{X_t\}$. More precisely, let $\{X_t : t \in \mathbb{R}\}$ be a weakly stationary, real-valued, mean-zero, finite-variance process, and for any interval $I \subset \mathbb{R}$, let $\mathcal{H}_I$ = closure in $|| \cdot ||_{\mathcal{P}}$ of $< \{X_t : t \in I\} >$, where $< \{X_t : t \in I\} >$ denotes the linear span of $\{X_t : t \in I\}$; for $I = \mathbb{R}$, we will write $\mathcal{H}$ instead of $\mathcal{H}_{\mathbb{R}}$. Then the linear analogue of (1.1) is

$$\inf\{||\xi_0 - \xi_1||_{\mathcal{P}} : \xi_i \in \mathcal{H}_{I_i}, i = 0, 1, ||\xi_0||_{\mathcal{P}} = 1\} \tag{1.2}$$

As in the case of (1.1), problem (1.2) may not have any solution, and if a solution exists, it might not be unique. Solutions of problem (1.2), if they exist, will be referred to as *linear optimal transformations for prediction.* For Gaussian stationary processes, Kolmogorov and Rozanov [18, p. 181] showed that problems (1.1) and (1.2) are equivalent (see also [4, p. 65]); in §3.1 we provide a new proof of this result, using the Ito-Wiener decomposition and the notion of *quantization* of operators.

Problems (1.1) and (1.2) are closely related to the *maximal correlation problem* defined by

$$\sup\{|E\{\xi_0\bar{\xi}_1\}| : ||\xi_i||_{\mathcal{P}} = 1, \;\; E\{\xi_i\} = 0, \;\; i = 0, 1\} \tag{1.3}$$

with $\xi_i \in L^2(\Omega, \mathcal{F}_{I_i}, \mathcal{P})$ for problem (1.1), and $\xi_i \in \mathcal{H}_{I_i}$ for problem (1.2). Problems (1.1)–(1.3) may be cast into an abstract Hilbert space framework to be presented in §2. Within this abstract framework we establish (see §2.2) that problems (1.1)–(1.3) have solutions iff a certain self-adjoint bounded operator has its norm as an eigenvalue.

The central objective of this paper is to study the existence of solutions for problem (1.2) with $T = \tau = +\infty$, under appropriate conditions on the spectral function F of the weakly stationary process $\{X_t : t \in \mathbb{R}\}$. The case $T < +\infty$, $\tau < +\infty$ appears to be more subtle and is treated in [7]. Our study of problem (1.1) uses the well-known equivalence [4, 11] (see also §3.2) between the spaces $\mathcal{H}_I$, $I \subset \mathbb{R}$, and certain function spaces $\mathcal{L}_I(F)$ (to be introduced in §3). We have not been able to find a satisfactory solution of the *fully nonlinear* problem (1.1).

For discrete-time processes, the analogue of problem (1.2) with $T < +\infty$, $\tau < +\infty$, is straightforward. For $T = \tau = +\infty$, the (discrete-time) problem was first considered by Helson and Szegö [10] in connection with a Functional Analysis problem in trigonometric series; their work stimulated a great deal of mathematical research in the theory of bounded analytic, BMO (Bounded Mean Oscillation) functions, and other problems (see [19], [13, pp. 249–287], [5, pp. 144, 254], and references cited therein); these studies are fundamentally different in spirit from the study in this paper. The existence of solutions for the discrete-time analogue of (1.2) with $T = \tau = +\infty$ leads to a study of Hankel (and Toeplitz) operators [1, 19] on Hardy spaces of the unit disc; similarly, the existence of solutions of problem (1.2) with $T = \tau = +\infty$ lead to Hankel type operators on Hardy spaces of the half-plane. In contrast, the existence of solutions of problem (1.2) with $T < +\infty$, $\tau < +\infty$, leads to the study of certain integral operators on spaces of entire functions of exponential type. If the spectral density is rational, then the solutions of problem (1.2) are obtained as solutions of differential equations on the space of generalized functions; the results for this case will appear in [7]. In [6], we study the statistical estimation of solutions (if they exist) of problem (1.2), from a finite set of samples, and establish the consistency of the estimators as the sampling rate goes to zero and the sample size goes to infinity.

The organization of this paper is as follows: Section 2 contains the abstract Hilbert space framework for problems (1.1) and (1.2). Our main results for problem (1.2) are contained in §3; §3.1 contains a new proof that for Gaussian stationary processes, problems (1.1) and (1.2) are equivalent; §3.2 summarizes the function spaces $\mathcal{L}_I(F)$; and §3.3 contains the results for $T = \tau = +\infty$.

2. Abstract Framework for Optimal Transformations

Let $\mathcal{L}$ be a Hilbert space and $\mathcal{L}_0$, $\mathcal{L}_1$ be two closed subspaces of $\mathcal{L}$; the norm and inner product on $\mathcal{L}$ will be denoted by $||\cdot||$ and $< \cdot, \cdot >$, respectively. The minimization problem

$$\sigma^2 \ = \ \sigma^2(\mathcal{L}_0, \mathcal{L}_1) \ = \ \inf\left\{||\phi_0 - \phi_1||^2 : \phi_i \in \mathcal{L}_i, i = 0, 1, ||\phi_0|| = 1\right\} \quad (2.1)$$

unifies problems (1.1) and (1.2). We will see that this problem is equivalent to the *maximization* problem

$$\rho \;=\; \rho(\mathcal{L}_0, \mathcal{L}_1) \;=\; \sup\{|<\phi_0, \phi_1>| : \phi_i \in \mathcal{L}_i, ||\phi_i|| = 1, i = 0, 1\} \quad (2.2)$$

which is the abstract version of problem (1.3). Note that ρ may be interpreted as the cosine of the angle $0 \leq \theta \leq \pi/2$ between the subspaces $\mathcal{L}_0$ and $\mathcal{L}_1$. Subsection 2.1 establishes some basic properties of problems (2.1) and (2.2), while §2.2 establishes that σ^2 (or ρ) is attained iff a certain self-adjoint bounded operator has its norm as an eigenvalue; at the end of §2.2 we give an example in which σ^2 (equivalently ρ) is not attained.

2.1 Basic Properties of Problems (2.1) and (2.2). Let P_i be the orthogonal projection from $\mathcal{L}$ onto $\mathcal{L}_i$, $i = 0, 1$, and define

$$P_{01} = P_1|_{\mathcal{L}_0} : \mathcal{L}_0 \longrightarrow \mathcal{L}_1, \qquad P_{10} = P_0|_{\mathcal{L}_1} : \mathcal{L}_1 \longrightarrow \mathcal{L}_0 \quad (2.3)$$
$$A_0 = P_{10}P_{01} = P_0P_1|_{\mathcal{L}_0}, \qquad A_1 = P_{01}P_{10} = P_1P_0|_{\mathcal{L}_1}, \quad (2.4)$$
$$B_0 = P_0P_1P_0 : \mathcal{L} \longrightarrow \mathcal{L}_0, \qquad B_1 = P_1P_0P_1 : \mathcal{L} \longrightarrow \mathcal{L}_1. \quad (2.5)$$

It is easily seen that $P_{10} = P_{01}^*$, and consequently A_i, $i = 0, 1$, are self-adjoint. Note that B_i, $i = 0, 1$, are also self-adjoint.

Lemma 2.1 *(a)* $\rho = \sup\{||P_1\phi_0|| : \phi_0 \in \mathcal{L}_0, ||\phi_0|| = 1\} = \sup\{||P_0\phi_1|| : \phi_1 \in \mathcal{L}_1, ||\phi_1|| = 1\}$. *(b)* $\sigma^2 = 1 - \rho^2$. *(c)* $\rho = ||P_{01}|| = ||P_{10}|| = \sqrt{||A_i||} = ||P_0P_1|| = ||P_1P_0|| = \sqrt{||B_i||}$, $i = 0, 1$.

Proof. Let $\phi_i \in \mathcal{L}_i$, $||\phi_i|| = 1$, $i = 0, 1$. The inequalities $|<\phi_0, \phi_1>| = |<P_1\phi_0, \phi_1>| \leq ||P_1\phi_0||$ and $||P_1\phi_0|| = <P_1\phi_0/||P_1\phi_0||, \phi_0> \;\leq \rho$, yield the first equality in part (a); the second equality is obtained similarly. Part (b) is a consequence of part (a) and of the following inequalities

$$||P_1\phi_0||^2 \;=\; 1 - ||\phi_0 - P_1\phi_0||^2 \leq 1 - \sigma^2$$
$$||\phi_0 - \phi_1||^2 \;\geq\; ||\phi_0 - P_1\phi_0||^2 = 1 - ||P_1\phi_0||^2 \geq 1 - \sigma^2.$$

Part (a) is equivalent (by the definition of $||P_{01}||$, $||P_{10}||$) to $\rho = ||P_{01}|| = ||P_{10}||$; since $A_0 = P_{10}P_{10}^* = P_{01}^*P_{01}$ and $A_1 = P_{01}P_{01}^* = P_{10}^*P_{10}$, we have $||A_0|| = ||A_1|| = ||P_{01}||^2 = ||P_{10}||^2$. To obtain the last three equalities in part (c), note that $\rho = ||P_{01}|| \leq ||P_0P_1||$ and $||P_1P_0|| = \sup\{||P_1P_0\psi|| : \psi \in \mathcal{L}, ||\psi|| = 1\} = \sup\{||P_1(P_0\psi/||P_0\psi||)|| \times ||P_0\psi|| : \psi \in \mathcal{L}, ||\psi|| = 1\} \leq \rho$. Hence $\rho = ||P_1P_0||$; similarly $\rho = ||P_0P_1||$; these together with $B_0 = (P_0P_1)(P_0P_1)^*$ and $B_1 = (P_1P_0)(P_1P_0)^*$ yield the last equality in part (c) □

Remarks. 1) Part (b) of Lemma 2.1 implies $\sigma^2(\mathcal{L}_0, \mathcal{L}_1) = \sigma^2(\mathcal{L}_1, \mathcal{L}_0)$, and $0 \le \sigma^2 \le 1$, since obviously $0 \le \rho \le 1$. 2) Clearly $\sigma^2 = 1$ (equivalently $\rho = 0$) iff $\mathcal{L}_0$ and $\mathcal{L}_1$ are orthogonal to each other (i.e. $\mathcal{L}_0 \perp \mathcal{L}_1$). If $\mathcal{L}_0 \perp \mathcal{L}_1$, then σ^2 is attained by any pair (ϕ_0, $\phi_1 = 0$) with $||\phi_0|| = 1$. On the other hand, the interpretation of $\sigma^2 = 0$ is more subtle; if $\mathcal{L}_0 \cap \mathcal{L}_1 \neq \{0\}$, then $\sigma^2 = 0$, but one may have $\sigma^2 = 0$ even if $\mathcal{L}_0 \cap \mathcal{L}_1 = \{0\}$ (see the example at the end of §2.2). The question of when is $\sigma^2 > 0$ (equivalently $\rho < 1$) is at the heart of the problem studied by Helson and Szegö [10]. Below we give some geometric and operator theoretic conditions for having $\sigma^2 > 0$. The question of when is $\sigma^2 > 0$ is distinct from the question of when is σ^2 attained, which is treated in this paper.

The following proposition provides a necessary and sufficient condition for having $\sigma^2 > 0$.

Proposition 2.1 *$\sigma^2 > 0$ iff $\mathcal{L}_0 \cap \mathcal{L}_1 = \{0\}$ and $\mathcal{L}_0 + \mathcal{L}_1$ is closed. In particular: (a) If $\mathcal{L}_0 \cap \mathcal{L}_1 = \{0\}$, then $\sigma^2 = 0$ iff $\mathcal{L}_0 + \mathcal{L}_1$ is* not *closed; (b) If $\sigma^2 = 0$, then either $\mathcal{L}_0 \cap \mathcal{L}_1 \neq \{0\}$, or else $\mathcal{L}_0 \cap \mathcal{L}_1 = \{0\}$* and *$\mathcal{L}_0 + \mathcal{L}_1$ is* not *closed.*

Proof. An easy calculation shows that for $0 \le \rho \le 1$, we have

$$(1-\rho)(||\phi_0||^2 + ||\phi_1||^2) \le ||\phi_0 + \phi_1||^2, \quad \text{for all } \phi_i \in \mathcal{L}_i, \quad i = 0,1 \quad (2.6)$$

Now suppose that $\rho < 1$, then necessarily $\mathcal{L}_0 \cap \mathcal{L}_1 = \{0\}$ (since $\mathcal{L}_0 \cap \mathcal{L}_1 \neq \{0\}$ implies $\sigma^2 = 0$). The closeness of $\mathcal{L}_0 + \mathcal{L}_1$ is an easy consequence of (2.6). Next suppose that $\mathcal{L}_0 \cap \mathcal{L}_1 = \{0\}$ *and* $\mathcal{L}_0 + \mathcal{L}_1$ is closed. Notice that

$$\sigma^2 = 1 - \rho^2 = \inf\{||(I - P_0)\phi_1||^2 : \phi_1 \in \mathcal{L}_1, \quad ||\phi_1|| = 1\} \quad (2.7)$$

Let $Q = (I - P_0)|_{\mathcal{L}_1} : \mathcal{L}_1 \to \mathcal{L}_0^{\perp}$, where $\mathcal{L}_0^{\perp}$ denotes the orthogonal complement of $\mathcal{L}_0$ in $\mathcal{L}$. Since $\mathcal{L}_0 + \mathcal{L}_1 = \mathcal{L}_0 + (I - P_0)\mathcal{L}_1$ is closed and $\mathcal{L}_0 \perp (I - P_0)\mathcal{L}_1$, we conclude that the range $\operatorname{Ran} Q = (I - P_0)\mathcal{L}_1$ is closed. On the other hand, $\mathcal{L}_0 \cap \mathcal{L}_1 = \{0\}$ easily implies that $Q : \mathcal{L}_1 \to \operatorname{Ran} Q$ is one-to-one. Since $\operatorname{Ran} Q$ is closed and $Q : \mathcal{L}_1 \to \operatorname{Ran} Q$ is one-to-one and onto, there exists a *bounded* operator $R : \operatorname{Ran} Q \to \mathcal{L}_1$ so that $RQ =$ identity on $\mathcal{L}_1$ and $QR =$ identity on $\operatorname{Ran} Q$. Consequently

$+\infty > ||R|| = \sup_{\phi_1 \in \mathcal{L}_1}(||RQ\phi_1||/||Q\phi_1||) = \sup_{\phi_1 \in \mathcal{L}_1}(||\phi_1||/||Q\phi_1||) = \sigma^{-2}$

where we have used (2.7) to obtain the last equality; hence $\sigma^2 > 0$. □

The next proposition gives a sufficient condition for $\mathcal{L}_0 + \mathcal{L}_1$ to be closed, and it is an easy consequence of Proposition 2.1 and Theorem 2.1 below.

Proposition 2.2 *If $\mathcal{L}_0 \cap \mathcal{L}_1 = \{0\}$ and $||A_0||$ is an eigenvalue of A_0, then $\mathcal{L}_0 + \mathcal{L}_1$ is closed.*

Remark. It can be shown (see Lemma 2.3 below) that if P_{01} is compact, then so is A_0, and so $\|A_0\|$ is an eigenvalue of A_0.

2.2 Existence of Optimal Solutions. Since $\sigma^2 = 1$ iff $\mathcal{L}_0 \perp \mathcal{L}_1$, the existence problem in this case is trivial; so throughout this section we will assume that $\sigma^2 < 1$ ($\rho > 0$). The proof of the next lemma is straightforward (see [7]).

Lemma 2.2 *(a) If $\phi_i \in \mathcal{L}_i$, $i = 0, 1$ with $||\phi_0|| = 1$ attain σ^2 ($\neq 1$), then $\rho^2 = ||\phi_1||^2 = ||P_1\phi_0||^2 = ||P_0\phi_1||$, and $\phi_1 = P_1\phi_0$, $\phi_0 = P_0\phi_1/||P_0\phi_1|| = \rho^{-2}P_0\phi_1$. (b) If $\psi_i \in \mathcal{L}_i$, $||\psi_i|| = 1$, $i = 0, 1$, attain ρ ($\neq 0$), then $\rho = ||P_1\psi_0|| = ||P_0\psi_1||$, and $\psi_1 = \gamma_1 P_1\psi_0/||P_1\psi_0||$, $\psi_0 = \gamma_0 P_0\psi_1/||P_0\psi_1||$, with some complex numbers γ_0, γ_1 such that $|\gamma_0| = |\gamma_1| = 1$. (c) If $\phi_i \in \mathcal{L}_i$, $i = 0, 1$, with $||\phi_0|| = 1$ attain σ^2 ($\neq 1$), then $\psi_0 = \phi_0$, $\psi_1 = \rho^{-1}\phi_1$. attain ρ. Conversely, if $\psi_i \in \mathcal{L}_i$, $||\psi_i|| = 1$, $i = 0, 1$, attain ρ ($\neq 0$), then $\phi_0 = \psi_0$, $\phi_1 = \bar{\gamma}_1\rho\psi_1$, with γ_1 as in part (b), attain σ^2.*

Remark. If $\psi_i \in \mathcal{L}_i$, $||\psi_i|| = 1$, $i = 0, 1$, attain ρ, then for any complex numbers μ_0 and μ_1 with $|\mu_0| = |\mu_1| = 1$, the pair $\tilde{\psi}_0 = \mu_0\psi_0$, $\tilde{\psi}_1 = \mu_1\psi_1$ also attains ρ. Moreover, if one chooses μ_0 and μ_1 so that $\bar{\mu}_0\mu_1\gamma_1 = 1$, with γ_1 as in part (b) of the lemma, then $\tilde{\psi}_1 = P_1\tilde{\psi}_0/||P_1\tilde{\psi}_0||$, which implies that $\rho =< \tilde{\psi}_0, \tilde{\psi}_1 >$ (no absolute value!).

The next theorem is the central result in this section.

Theorem 2.1 *(a) If $\phi_i \in \mathcal{L}_i$, $i = 0, 1$, $||\phi_0|| = 1$ attain σ^2 ($\neq 1$), then ϕ_i is an eigenvector of A_i with eigenvalue $\rho^2 = 1 - \sigma^2 = ||A_i||$, i.e. $A_i\phi_i = \rho^2\phi_i$, $i = 0, 1$. (b) Conversely, suppose that $\rho^2 = ||A_i||$, $i = 0, 1$ is strictly positive (equivalently, $\sigma^2 < 1$). If ρ^2 is an eigenvalue of A_0 with a normalized eigenvector ϕ_0, then the pair ϕ_0 and $\phi_1 = P_1\phi_0$ attains σ^2, and ϕ_1 is an eigenvector of A_1 with eigenvalue ρ^2. Similarly, if ρ^2 is an eigenvalue of A_1 with an eigenvector ϕ_1 normalized so that $||\phi_1|| = \rho$, then the pair $\phi_0 = P_0\phi_1/||P_0\phi_1||$ and ϕ_1 attain σ^2, and ϕ_0 is an eigenvector of A_0 with eigenvalue ρ^2.*

Proof. Part (a) is a direct consequence of Lemma 2.2 (part (a)). Part (b) is simple: if $A_0\phi_0 = \rho^2\phi_0$ with $||\phi_0|| = 1$, then $\rho^2 =< \phi_0, A_0\phi_0 >= ||P_1\phi_0||^2$ and $A_1P_1\phi_0 = \rho^2P_1\phi_0$, i.e. $\phi_1 = P_1\phi_0$ is an eigenvector of A_1 with eigenvalue ρ^2. Next notice that $||\phi_0 - \phi_1||^2 = ||\phi_0 - P_1\phi_0||^2 = 1 - ||P_1\phi_0||^2 = 1 - \rho^2 = \sigma^2$, which implies that ϕ_0, ϕ_1 attain σ^2. The proof for A_1 is the same. □

Remarks. 1) Theorem 2.1 is an extension of Theorem 5.3 of [2, p. 591]; note that we do not assume A_0, A_1 to be compact, or that $\rho = ||A_i||$ is an isolated eigenvalue. 2) The operators A_0, A_1 are (obviously) self-adjoint, bounded and non-negative; hence $||A_i||$ belongs to the spectrum of A_i, but it might not be an eigenvalue of A_i, $i = 0, 1$. The question then arises: under what conditions is $||A_i||$ an eigenvalue of A_i? This is definitely the case if A_0 (or A_1) is compact. In this latter case the multiplicity of solutions of problem (2.1) is finite. 3) It can be easily shown that A_0 and A_1 have the same strictly positive eigenvalues (if such eigenvalues exist!); moreover, there exists a one-to-one correspondence between the eigenspaces of A_0 and A_1 corresponding to strictly positive eigenvalues; the correspondence is given by $\psi_0 = P_0\psi_1/||P_0\psi_1||$, $\psi_1 = P_1\psi_0/||P_1\psi_0||$, for normalized eigenvectors ψ_0, ψ_1 of A_0, A_1, respectively. The set of positive eigenvalues (if they exist) of A_0 (equivalently, A_1) and the corresponding eigenvectors are very useful in applications; these eigenvalues may be used [3, 9] to design a *nonlinear* discriminant analysis that generalizes the classical *linear* discriminant analysis. 4) If $\rho^2 = ||A_0|| = ||A_1|| > 0$ is an eigenvalue of A_0 (respectively, A_1), then the question arises: is it an isolated eigenvalue? We have not been able to construct an example where $\rho^2 = ||A_i||$ is *not* an isolated eigenvalue, but we suspect that this occurs.

The proof of the next lemma is a consequence of well-known properties of compact operators (see [17]).

Lemma 2.3 *(a) If one of the operators P_{01}, P_{10}, A_0, A_1, B_0, B_1, P_0P_1, P_1P_0 is compact, then so are all the others. (b) If one of the operators A_0, A_1, B_0, B_1 is trace class, then so are all the others. Moreover $tr\,A_0 = tr\,A_1 = tr\,B_0 = tr\,B_1 = \sum_{n,m=1}^{+\infty} |<\phi_0^{(n)}, \phi_1^{(m)}>|^2$, where tr denotes the trace of an operator, and $\{\phi_0^{(n)}\}$, $\{\phi_1^{(n)}\}$, are arbitrary orthonormal bases of $\mathcal{L}_0$ and $\mathcal{L}_1$, respectively.*

Remark. If A_0 (equivalently, A_1) is trace class, then by definition (see for example [22, p. 175]) P_{01} and P_{10} are Hilbert-Schmidt operators. In §3 we will see that P_{10} (or P_{01}) is an integral kernel operator, and we shall give a necessary and sufficient condition for it to be a Hilbert-Schmidt operator.

We end this section with an example in which σ^2 is not attained: let $\mathcal{L}$ be a Hilbert space with an orthonormal (o.n.) basis $\{e_n : n \geq 0\}$, and $\mathcal{L}_0$ be the subspace with o.n. basis $\{e_{2n} : n \geq 0\}$. Fix a real number α, and a sequence of real numbers $\{\alpha_n : n \geq 0\}$ so that $\alpha_n^2 > \alpha^2$, $n \geq 0$ and $\alpha_n \to \alpha$ as $n \to +\infty$. Let $\mathcal{L}_1$ be the subspace with o.n. basis $\left\{(1+\alpha_n^2)^{-1/2}(e_{2n} + \alpha_n e_{2n+1}) : n \geq 0\right\}$. We will show that $\rho = \rho(\mathcal{L}_0, \mathcal{L}_1) = (1+\alpha^2)^{-1/2}$, and that ρ is not attained. Indeed, for any $\phi_i \in \mathcal{L}_i$, $||\phi_i|| = 1$, one can easily show, using $\alpha_n^2 > \alpha^2$, that $|<\phi_0, \phi_1>| < (1+\alpha^2)^{-1/2}$; on the other hand, $< e_{2n}, (1+\alpha_n^2)^{-1/2}(e_{2n} +$

$\alpha_n e_{2n+1}) > = (1+\alpha_n^2)^{-1/2} \longrightarrow (1+\alpha^2)^{-1/2}$ as $n \to +\infty$. Hence $\rho = (1+\alpha^2)^{-1/2}$ and is not attained. We note that for $\alpha = 0$, we have $\rho = 1$; on the other hand $\mathcal{L}_0 \cap \mathcal{L}_1 = \{0\}$.

3. Existence of Optimal Linear Transformations for Weakly Stationary Continuous-Time Processes

Throughout this section (except §3.1) we assume that $\{X_t : t \in \mathbb{R}\}$ is a weakly stationary, mean-zero, finite-variance stochastic process with spectral function $F(\omega)$, $\omega \in \mathbb{R}$. We establish the existence of optimal linear transformations for prediction (i.e. solutions of problem (1.2)) under appropriate conditions on F. Our study uses the well-known equivalence between $\mathcal{H}_I$ (defined in §1) and certain function spaces to be defined in §3.2. Subsection §3.1 contains a new proof of the Kolmogorov-Rozanov result that problems (1.1) and (1.2) are equivalent for Gaussian stationary processes.

3.1 Equivalence of Problems (1.1) and (1.2) for Gaussian Stationary Processes. Let $\{X_t : t \in \mathbb{R}\}$ be a Gaussian stationary process. The intervals I_i, $i = 0, 1$, the spaces $\mathcal{H}_{I_i}$ and $\mathcal{H}$ and the σ-algebras $\mathcal{F}_{I_i}$ are as in §1. Let $\mathcal{M} = \{\xi \in L^2(\Omega, \mathcal{F}, \mathcal{P}) : E\{\xi\} = 0\}$ and $\mathcal{M}_{I_i} = \{\xi \in L^2(\Omega, \mathcal{F}_{I_i}, \mathcal{P}) : E\{\xi\} = 0\}$, $i = 0, 1$. Using the notation of (2.2), the Kolmogorov-Rozanov theorem [18, p. 181], [4, p.67], reads:

Theorem 3.1 *For Gaussian processes* $\rho(\mathcal{H}_{I_0}, \mathcal{H}_{I_1}) = \rho(\mathcal{M}_{I_0}, \mathcal{M}_{I_1})$

Proof. Let Q_i (respectively, P_i) be the orthogonal projection of $\mathcal{M}$ (respectively, $\mathcal{H}$) onto $\mathcal{M}_{I_i}$ (respectively, $\mathcal{H}_{I_i}$); as in §2.1, we define $Q_{01} = Q_1|_{\mathcal{M}_{I_0}}$ and $P_{01} = P_1|_{\mathcal{H}_{I_0}}$. By part (c) of Lemma 2.1, it suffices to prove that $||Q_{01}|| = ||P_{01}||$. We shall prove this by using the Wiener-Ito decomposition, and the notion of *quantization* of operators (see [20, p. 25]). The former states that

$$\mathcal{M} = \oplus_{n=1}^{+\infty} \mathcal{M}^{(n)}, \quad \mathcal{M}_{I_i} = \oplus_{n=1}^{+\infty} \mathcal{M}_{I_i}^{(n)}$$

where $\mathcal{M}^{(n)}$ (respectively, $\mathcal{M}_{I_i}^{(n)}$), $n \geq 1$, is the closed subspace generated by the Wick polynomials (see [20, p. 12]) $: X_{t_1} \cdots X_{t_n} :$, $n \geq 1$, with $t_1, \ldots, t_n \in \mathbb{R}$ (respectively, $t_1, \ldots, t_n \in I_i$); the t_k's are not necessarily distinct. Note that $\mathcal{M}^{(1)} = \mathcal{H}$, $\mathcal{M}_{I_i}^{(1)} = \mathcal{H}_{I_i}$, $i = 0, 1$.

The (second) quantization $\Gamma(P_{01})$ of the operator P_{01} is an operator from $\mathcal{M}_{I_0}$ into $\mathcal{M}_{I_1}$, and is uniquely defined by $\Gamma(P_{01}) : X_{t_1} \cdots X_{t_n} :=: P_{01}X_{t_1} \cdots P_{01}X_{t_n} : =: P_1X_{t_1} \cdots P_1X_{t_n} :$, $n \geq 1$, with $t_1, \ldots, t_n \in I_0$ (see [20, p. 25]). To prove the theorem, it suffices to show

$$Q_{01} = \Gamma(P_{01}) \qquad ||\Gamma(P_{01})|| = ||P_{01}|| \tag{3.1}$$

Proof of (3.1). As in the case of $\Gamma(P_{01})$, the (second) quantization $\Gamma(P_1)$ of P_1 is uniquely defined by

$$\Gamma(P_1) : X_{t_1} \cdots X_{t_n} :=: P_1 X_{t_1} \cdots P_1 X_{t_n} : \quad n \geq 1, \ t_1, \dots, t_n \in \mathbb{R}$$

From which we easily conclude that

$$(\Gamma(P_1)|_{\mathcal{M}^{(n)}})^2 = \Gamma(P_1)|_{\mathcal{M}^{(n)}}, \qquad \Gamma(P_1)|_{\mathcal{M}_{I_1}^{(n)}} = \mathbf{I}_{\mathcal{M}_{I_1}^{(n)}}, \quad n \geq 1 \quad (3.2)$$

On the other hand, if Π_n denotes the set of permutations of $(1, 2, \dots, n)$, then, with $s_1, \dots, s_n \in \mathbb{R}$, $t_1, \dots, t_n \in \mathbb{R}$

$$\begin{aligned} &< \Gamma(P_1) : X_{t_1} \cdots X_{t_n} :, : X_{s_1} \cdots X_{s_n} :>_{\mathcal{P}} \\ &= \sum_{\pi \in \Pi_n} < P_1 X_{t_{\pi(1)}}, X_{s_1} >_{\mathcal{P}} \cdots < P_1 X_{t_{\pi(n)}}, X_{s_n} >_{\mathcal{P}} \\ &=<: X_{t_1} \cdots X_{t_n} :, \Gamma(P_1) : X_{s_1} \cdots X_{s_n} :>_{\mathcal{P}} \quad (3.3) \end{aligned}$$

which implies that $(\Gamma(P_1)|_{\mathcal{M}^{(n)}})^* = \Gamma(P_1)|_{\mathcal{M}^{(n)}}$. This together with (3.2) yield $Q_1|_{\mathcal{M}^{(n)}} = \Gamma(P_1)|_{\mathcal{M}^{(n)}}$, and hence $Q_1 = \Gamma(P_1)$. It is also obvious from the definition of $\Gamma(P_1)$ and $\Gamma(P_{01})$ that $\Gamma(P_1)|_{\mathcal{M}_{I_0}^{(n)}} = \Gamma(P_{01})|_{\mathcal{M}_{I_0}^{(n)}}$. Therefore $Q_{01} = \Gamma(P_{01})$. Next we prove the second equality in (3.1). Since $\mathcal{H}_{I_0} = \mathcal{M}_{I_0}^{(1)} \subset \mathcal{M}_{I_0}$, we have that $||P_{01}|| \leq ||\Gamma(P_{01})||$. On the other hand, a computation as in (3.3) shows that $||\Gamma(P_{01})|_{\mathcal{M}_{I_0}^{(n)}}|| \leq ||P_{01}||^n \leq ||P_{01}||$; thus for any $\xi = (\xi_1, \xi_2, \dots) \in \mathcal{M}_{I_0}$, $\xi_n \in \mathcal{M}_{I_0}^{(n)}$, we have

$$||\Gamma(P_{01})\xi||^2 = \textstyle\sum_{n=1}^{+\infty} ||\Gamma(P_{01})\xi_n||^2 \leq ||P_{01}||^2 ||\xi||^2$$

which implies the reverse inequality $||\Gamma(P_{01})|| \leq ||P_{01}||$. □

Remark. Property (3.1) may also be deduced from Theorem 3.9 of [21, p. 27]; while the inequality $||\Gamma(P_{01})|| \leq ||P_{01}||$ may be deduced from Theorem 3.10 of [21, p. 28], and the fact (see Lemma 2.1, (b)) that $||P_{01}|| = ||P_0 P_1|| = ||P_1 P_0||$.

3.2 The function spaces. Let $F(\omega)$, $\omega \in \mathbb{R}$, be the spectral function of the weakly stationary process $\{X_t : t \in \mathbb{R}\}$. Let $\mathcal{L}(F) = L^2(\mathbb{R}, F)$ be the usual weighted L^2 space on $\mathbb{R}$ with weight F; its norm and inner product will be denoted by $|| \cdot ||_F$ and $< \cdot, \cdot >_F$, respectively. This space is also characterized by $\mathcal{L}(F)$ = closure in $|| \cdot ||_F$ of $<\{e^{i\omega t} : t \in \mathbb{R}\}>$, $\omega \in \mathbb{R}$, where $< \{e^{i\omega t} : t \in A\} >$ for any $A \subseteq \mathbb{R}$ denotes the linear span of $\{e^{i\omega t} : t \in A\}$. For any bounded or unbounded interval $I \subseteq \mathbb{R}$, the space $\mathcal{L}_I(F)$ is defined by $\mathcal{L}_I(F)$ = closure in $|| \cdot ||_F$ of $<\{e^{i\omega t} : t \in I\}>$, $\omega \in \mathbb{R}$. For $I = \mathbb{R}$, we have $\mathcal{L}(F) = \mathcal{L}_{\mathbb{R}}(F)$. It is well-known [11, 4] that $\mathcal{L}_I$ is isometrically isomorphic to the space $\mathcal{H}_I$ defined in §1. The one-to-one correspondence is given [11, p. 17] by $\xi = \int_{\mathbb{R}} \phi(\omega)\, dZ(\omega)$, $\xi \in \mathcal{H}_I$, $\phi \in \mathcal{L}_I(F)$, where $dZ(\omega)$ is the random

spectral measure of $\{X_t : t \in \mathbb{R}\}$. One may also define the spaces
$\mathcal{L}^0(F) =$ closure in $||\cdot||_F$ of $< \{(i\omega)^n : \int_{\mathbb{R}} \omega^{2n}\, dF(\omega) < +\infty, n \geq 0\} >$
$\mathcal{L}^{0+}(F) = \bigcap_{\varepsilon>0} \mathcal{L}_{[-\varepsilon,\varepsilon]}(F)$, $\mathcal{L}^a(F) = e^{i\omega a}\mathcal{L}^0(F)$, and $\mathcal{L}^{a+}(F) = e^{i\omega a}\mathcal{L}^{0+}(F)$, $a \in \mathbb{R}$. Clearly $\mathcal{L}^a(F) \subseteq \mathcal{L}^{a+}(F)$. These subspaces are also isometrically isomorphic to properly defined subspaces of $\mathcal{H}$ associated with the point $a \in \mathbb{R}$; they may be finite or infinite dimensional; if $\{X_t : t \in \mathbb{R}\}$ is infinitely differentiable (stochastically) at, say, $t = 0$, then $\mathcal{L}^0(F)$ is infinite dimensional.

If $\xi_i \in \mathcal{H}_{I_i}$, $\phi_i \in \mathcal{L}_{I_i}(F)$, are such that $\xi_i = \int_{\mathbb{R}} \phi_i(\omega)\, dZ(\omega)$, $i = 0, 1$, then problems (1.2) and (1.3) are equivalent to

$$\sigma^2 = \inf\{||\phi_0 - \phi_1||_F^2 : \phi_i \in \mathcal{L}_{I_i}(F), i = 0, 1, ||\phi_0||_F = 1\} \tag{3.4}$$

$$\rho = \sup\{| < \phi_0, \phi_1 >_F | : \phi_i \in \mathcal{L}_{I_i}(F), ||\phi_i||_F = 1, i = 0, 1\} \tag{3.5}$$

Existence of solutions for problem (3.4) (or equivalently, problem (3.5)) is treated in §3.3 with $T = \tau = +\infty$. The case $T < +\infty$, $\tau < +\infty$, as well as the cases when one of the intervals is replaced by a single point (e.g. for the pair of subspaces $\mathcal{L}_{[-T,0]}(F)$ and $\mathcal{L}^{\Delta}(F)$ or $\mathcal{L}^{\Delta+}(F)$, $\Delta > 0$), or when both intervals are replaced by single points (e.g. for the subspaces $\mathcal{L}^0(F)$ or $\mathcal{L}^{0+}(F)$ and $\mathcal{L}^{\Delta}(F)$ or $\mathcal{L}^{\Delta+}(F)$), are treated in [7] together with the case when $dF(\omega) = f(\omega)\, d\omega$, and the spectral density f is rational.

3.3 The case $T = +\infty$, $\tau = +\infty$. In this subsection we study problem (2.1) for $\mathcal{L} = \mathcal{L}(F)$ and subspaces

$$\mathcal{L}_0 = \mathcal{L}_{[\Delta,+\infty)}(F) = e^{i\omega\Delta}\mathcal{L}_{[0,+\infty)}(F), \qquad \mathcal{L}_1 = \mathcal{L}_{(-\infty,0]}(F). \tag{3.6}$$

Let $dF = dF^{(a)} + dF^{(s)}$ be the decomposition of dF into its absolutely continuous $dF^{(a)}(\omega) = f(\omega)\, d\omega$ and singular $dF^{(s)}(\omega)$ parts. Theorem 2 (parts 1 and 2) of [11, p. 36], easily yields the following lemma.

Lemma 3.1 *(a) If $dF^{(s)}$ is non-trivial (i.e. if $dF(\omega)$ is not absolutely continuous w.r.t. $d\omega$), then $\mathcal{L}_0 \cap \mathcal{L}_1 \neq \{0\}$. In particular, $\sigma^2 = 0$ and is attained.* *(b) If $dF(\omega) = f(\omega)\, d\omega + dF^{(s)}(\omega)$ and $\int_{\mathbb{R}} (1 + \omega^2)^{-1} \log f(\omega)\, d\omega = -\infty$, then $\mathcal{L}_0 = \mathcal{L}_1$. In particular, $\sigma^2 = 0$ and is attained.*

Hence we focus only in the case $dF(\omega) = f(\omega)\, d\omega$ with

$$\int_{\mathbb{R}} (1 + \omega^2)^{-1} \log f(\omega)\, d\omega > -\infty\,. \tag{3.7}$$

This will be assumed in the remaining of this subsection; F in (3.6) as well as in the inner product $< \cdot, \cdot >_F$ and norm $||\cdot||_F$ will be replaced by f. We also recall that $f \geq 0$, and $f \in L^1(\mathbb{R})$.

It is well-known (see [5, p. 66]) that (3.7) holds iff $f(\omega) = |h(\omega)|^2$ with some outer function $h \in \mathcal{H}^{2+}$, where $\mathcal{H}^{2+}$ is the Hardy space on the (open) upper-half plane. For any complex number z with $\Im z < 0$, $\bar{h}(z) = \overline{h(\bar{z})}$ belongs to the Hardy space $\mathcal{H}^{2-}$ in the (open) lower-half plane. For various properties (to be used below) of the spaces $\mathcal{H}^{2\pm}$ as well as of the Hardy spaces $\mathcal{H}^{1\pm}$ and $\mathcal{H}^{\infty\pm}$, we refer the reader to [13, pp. 226–248].

Since $f(\omega) = f(-\omega)$ (by the weak stationarity of $\{X_t\}$), we may choose $h(\omega)$ so that $\bar{h}(\omega) = \overline{h(\omega)} = h(-\omega)$, $\omega \in \mathbb{R}$. The function

$$g(\omega) = e^{i\omega\Delta}\bar{h}(\omega)(h(\omega))^{-1} = e^{i\omega\Delta}h(-\omega)(h(\omega))^{-1} \tag{3.8}$$

will play a key role in this subsection. For $\varepsilon > 0$ we also define

$$g_\varepsilon(\omega) = g(\omega)(1 - i\varepsilon\omega)^{-1}, \quad \omega \in \mathbb{R}. \tag{3.9}$$

Note that $g_\varepsilon(\omega) \in L^2(\mathbb{R})$ and its inverse Fourier transform

$$\check{g}_\varepsilon(t) = (2\pi)^{-1} \textstyle\int_{\mathbb{R}} e^{-i\omega t} g_\varepsilon(\omega)\, d\omega$$

is well-defined. Since $\overline{g_\varepsilon(\omega)} = g_\varepsilon(-\omega)$, $\check{g}_\varepsilon(t)$ is real. Let P_{01} be the operator associated with the spaces (3.6) (see (2.3)). Our first main result in this section is

Theorem 3.2 P_{01} *is Hilbert-Schmidt iff*

$$\lim_{\varepsilon \downarrow 0} \int_0^{+\infty} t|\check{g}_\varepsilon(-t)|^2\, dt < +\infty. \tag{3.10}$$

In particular, if this holds, then σ^2 is attained.

Proof. Using a well-known [11, p. 38] representation of $\mathcal{L}_{[0,+\infty)}(f)$ and $\mathcal{L}_{(-\infty,0]}(f)$ in terms of the Hardy spaces $\mathcal{H}^{2\pm}$, we have $\mathcal{L}_0 = e^{i\omega\Delta}h^{-1}\mathcal{H}^{2+}$, $\mathcal{L}_1 = \bar{h}^{-1}\mathcal{H}^{2-}$. Let $\pi^\pm$ be the orthogonal projection of $L^2(\mathbb{R})$ onto $\mathcal{H}^{2\pm}$. Then the projections P_i, $i = 0, 1$ from $\mathcal{L}(f)$ onto $\mathcal{L}_i$, $i = 0, 1$, are given by $P_0 = e^{i\omega\Delta}h^{-1}\pi^+e^{-i\omega\Delta}h$, $P_1 = \bar{h}^{-1}\pi^-\bar{h}$. If

$$\phi_0 = e^{i\omega\Delta}h^{-1}\phi_+ \in \mathcal{L}_0,\ \phi_+ \in \mathcal{H}^{2+}, \qquad \phi_1 = \bar{h}^{-1}\phi_- \in \mathcal{L}_1,\ \phi_- \in \mathcal{H}^{2-} \tag{3.11}$$

then $||\phi_0||_f = ||\phi_+||_2$, $||\phi_1||_f = ||\phi_-||_2$ and (see notation in §2.1)

$$P_{01}\phi_0 = P_1\phi_0 = \bar{h}^{-1}\pi^- g\phi_+ = \bar{h}^{-1}H_g\phi_+ \tag{3.12}$$

$$P_{10}\phi_1 = P_0\phi_1 = e^{i\omega\Delta}h^{-1}\pi^+\bar{g}\phi_- = e^{i\omega\Delta}h^{-1}H_g^*\phi_- \tag{3.13}$$

where g is the function in (3.8), $\bar{g}(\omega) = \overline{g(\omega)}$, and the operator H_g and its adjoint H_g^* are defined by $H_g : \mathcal{H}^{2+} \longrightarrow \mathcal{H}^{2-}$, $H_g\phi_+ = \pi^- g\phi_+$, and

$H_g^* : \mathcal{H}^{2-} \longrightarrow \mathcal{H}^{2+}$, $H_g^*\phi_- = \pi^+\bar{g}\phi_-$. The operator H_g is a *Hankel* type operator [19]. The operator A_0 (see (2.4)) is then given by

$$A_0\phi_0 = e^{i\omega\Delta}h^{-1}H_g^*H_g\phi_+ \tag{3.14}$$

which implies that $< A_0\phi_0, \phi_0 >_f = < H_g^*H_g\phi_+, \phi_+ >_2$; hence, A_0 is trace class (equivalently, P_{01} is Hilbert-Schmidt) iff $H_g^*H_g$ is trace class as an operator on $\mathcal{H}^{2+}$. Now let A_ε be defined by replacing g by g_ε in (3.14). Theorem 3.1 is implied by the next two lemmas.

Lemma 3.2 *$tr\, A_0 = \lim_{\varepsilon\downarrow 0} tr\, A_\varepsilon$, where tr denotes the trace of the operators.*

Lemma 3.3 *$tr\, A_\varepsilon = \int_0^{+\infty} t|\breve{g}_\varepsilon(-t)|^2\, dt$*

Proof of Lemma 3.2. We shall prove shortly that for all $\phi_+ \in \mathcal{H}^{2+}$

$$< H_{g_\varepsilon}^*H_{g_\varepsilon}\phi_+, \phi_+ >_2 \quad \leq \quad < H_g^*H_g\phi_+, \phi_+ >_2 \tag{3.15}$$

$$\lim_{\varepsilon\downarrow 0} < H_{g_\varepsilon}^*H_{g_\varepsilon}\phi_+, \phi_+ >_2 \quad = \quad < H_g^*H_g\phi_+, \phi_+ >_2 . \tag{3.16}$$

If $\{\phi_{+,n}\}_{n=1}^{+\infty}$ is an orthonormal basis for $\mathcal{H}^{2+}$, then (3.15) implies $\mathrm{tr}\, H_{g_\varepsilon}^*H_{g_\varepsilon} \leq \mathrm{tr}\, H_g^*H_g$; this together with (use (3.16))

$$\begin{aligned}\mathrm{tr}\, H_g^*H_g &= \lim_{N\to+\infty}\sum_{n=1}^{N} < H_g^*H_g\phi_{+,n}, \phi_{+,n} >_2 \\ &= \lim_{N\to+\infty}\lim_{\varepsilon\downarrow 0}\sum_{n=1}^{N} < H_{g_\varepsilon}^*H_{g_\varepsilon}\phi_{+,n}, \phi_{+,n} >_2 \leq \lim_{\varepsilon\downarrow 0} \mathrm{tr}\, H_{g_\varepsilon}^*H_{g_\varepsilon}\end{aligned}$$

implies the lemma.

Proof of (3.15). Let $\alpha(\omega) = g(\omega)\phi_+(\omega)$, $\alpha_\varepsilon(\omega) = g_\varepsilon(\omega)\phi_+(\omega) = \alpha(\omega)(1 - i\varepsilon\omega)^{-1}$, and $\breve{\alpha}(t)$, $\breve{\alpha}_\varepsilon(t)$ be their inverse Fourier transforms. An easy calculation gives that $\breve{\alpha}_\varepsilon(t) = (2\pi)^{-1}\int_{\mathbb{R}} e^{-i\omega t}\alpha_\varepsilon(\omega)\, d\omega = \int_0^\infty \breve{\alpha}(t - \varepsilon s)e^{-s}\, ds$. Since $(H_{g_\varepsilon}\phi_+)(\omega) = (\pi^-\alpha_\varepsilon)(\omega) = \int_{-\infty}^0 e^{i\omega t}\breve{\alpha}_\varepsilon(t)\, dt$ the inequality in (3.15) is obtained from

$$\begin{aligned}||H_{g_\varepsilon}\phi_+||_2^2 = 2\pi\int_{-\infty}^0 |\breve{\alpha}_\varepsilon(t)|^2\, dt &= \int_0^{+\infty} e^{-s}\, 2\pi\int_{-\infty}^{-\varepsilon s} |\breve{\alpha}(t)|^2\, dtds \\ &\leq 2\pi\int_{-\infty}^0 |\breve{\alpha}(t)|^2\, dt = ||H_g\phi_+||_2^2.\end{aligned}$$

Proof of (3.16). By Parseval and Schwartz inequalities, we have

$$\begin{aligned}||H_{g_\varepsilon}\phi_+ - H_g\phi_+||_2^2 &= 2\pi\int_{-\infty}^0 \left|\int_0^{+\infty} \{\breve{\alpha}(t - \varepsilon s) - \breve{\alpha}(t)\}e^{-s}\, ds\right|^2 dt \\ &\leq \int_0^{+\infty} e^{-s}\, 2\pi\int_{-\infty}^0 |\breve{\alpha}(t - \varepsilon s) - \breve{\alpha}(t)|^2\, dtds \\ &\leq \int_{\mathbb{R}} |\phi_+(\omega)|^2\int_0^{+\infty} e^{-s}|1 - e^{i\varepsilon\omega s}|^2\, dsd\omega\end{aligned}$$

which, by an application of the Lebesgue dominated convergence theorem, yields that $||H_{g_\varepsilon}\phi_+ - H_g\phi_+||_2 \to 0$, as $\varepsilon \downarrow 0$. This easily implies (3.16). □

Proof of Lemma 3.3 Let $\check{\phi}_+(t)$ be the inverse Fourier transform of ϕ_+, i.e. $\phi_+(\omega) = \int_0^{+\infty} e^{i\omega t}\check{\phi}_+(t)\,dt$. Then $(H_{g_\varepsilon}\phi_+)(\omega) = \int_{-\infty}^{0} e^{i\omega t}(\check{g}_\varepsilon * \check{\phi}_+)(t)\,dt = \int_0^{+\infty} e^{-i\omega t}(\check{g}_\varepsilon * \check{\phi}_+)(-t)\,dt$. Note that for $t \geq 0$ $(K_\varepsilon\check{\phi}_+)(t) = (\check{g}_\varepsilon * \check{\phi}_+)(-t) = \int_0^{+\infty} \check{g}_\varepsilon(-t-s)\check{\phi}_+(s)\,ds$ defines an integral operator K_ε on $L^2(0,+\infty)$; clearly

$$\operatorname{tr} K_\varepsilon^* K_\varepsilon = \int_0^{+\infty}\int_0^{+\infty} |\check{g}_\varepsilon(-t-s)|^2\,dtds = \int_0^{+\infty} t|\check{g}_\varepsilon(-t)|^2\,dt.$$

But by Parseval's equality $||H_{g_\varepsilon}\phi_+||^2 = 2\pi||K_\varepsilon\check{\phi}_+||_{L^2(0,+\infty)}$, and $||\phi_+||_2^2 = 2\pi||\check{\phi}_+||_{L^2(0,+\infty)}$. These imply that $\operatorname{tr} H_{g_\varepsilon}^* H_{g_\varepsilon} = \operatorname{tr} K_\varepsilon^* K_\varepsilon$ and hence the lemma. □

Remarks. 1) Note that condition (3.10) involves $\check{g}_\varepsilon(t)$ for $t \leq 0$ only; the reason for this will become evident below. 2) Since $g \in L^\infty(\mathbb{R})$ (in fact, $|g(\omega)| = 1$, $\omega \in \mathbb{R}$), g has a well-defined inverse Fourier transform $\check{g}$ in the sense of distributions [17]. If $\check{g}(-t)$ for $t > 0$ is an ordinary function, then the limit in (3.10) may be taken under the integral. This can be shown by following the proof above of Lemma 3.3 This fact will be used in the example below.

Example. Let n be an integer larger than 1, $\nu = n/2$, and $f(\omega) = (1+\omega^2)^{-\nu}$. Then $h(\omega) = (1 - i\omega)^{-\nu}$, and

$$g(\omega) = e^{i\omega\Delta}(1 - i\omega)^{\nu}(1 + i\omega)^{-\nu} = e^{i\omega\Delta}(1 - i\omega)^n f(\omega)\,.$$

Hence, for $t > 0$ $\check{g}(-t) = (2\pi)^{-1}\int e^{i\omega t}g(\omega)\,d\omega = (2\pi)^{-1}(1 - \frac{d}{dt})^n R(t + \Delta)$, where $R(\cdot)$ is the covariance function corresponding to $f(\omega)$. $R(t)$ may be expressed in terms of the modified Bessel function $K_{\nu-\frac{1}{2}}(|t|)$ (see [15]); it is well-known that $R(t)$ is positive, belongs to $L^1(\mathbb{R})$, is continuous for all t, analytic for all t except $t = 0$, and satisfies, with some constant $C > 0$, $\left|\left(1 - \frac{d}{dt}\right)^n R(t)\right| \leq Ce^{-(|t|/2)}$, for all $|t| \geq 1$. Hence $\int_0^{+\infty} t|\check{g}(-t)|^2\,dt < +\infty$, and so, P_{01} is Hilbert-Schmidt.

Remarks. 1) If $u \in \mathcal{H}^{\infty+}$, then $u\mathcal{H}^{2+} \subseteq \mathcal{H}^{2+}$. Thus, if $g = g_0 + u$ with $g_0 \in L^\infty(\mathbb{R})$ and $u \in \mathcal{H}^{\infty+}$, then $P_{01}\phi_0 = \bar{h}^{-1}\pi^- g\phi_+ = \bar{h}^{-1}\pi^- g_0\phi_+$, i.e. P_{01} (and hence σ^2 and ρ) is independent of u; this is related to the fact that condition (3.10) depends only on $\check{g}(t)$ with $t < 0$. 2), Using (3.11) one easily obtains that $\sigma^2(\mathcal{L}_0, \mathcal{L}_1) = \sigma^2(g\mathcal{H}^{2+}, \mathcal{H}^{2-})$, and $\rho(\mathcal{L}_0, \mathcal{L}_1) = \rho(g\mathcal{H}^{2+}, \mathcal{H}^{2-})$. Moreover, by the definition of $||H_g||$,

$$\rho = \sup\{| < H_g\phi_+, \phi_- >_2 | : \phi_\pm \in \mathcal{H}^{2\pm}, ||\phi_\pm||_2 = 1\} = ||H_g||\,. \tag{3.17}$$

Also, by a well-known property of Hardy spaces, if $\phi_\pm \in \mathcal{H}^{2\pm}$, then $\theta = \phi_+\bar{\phi}_- \in \mathcal{H}^{1+}$ and $||\theta||_1 \leq ||\phi_+||_2\,||\phi_-||_2$; hence

$$\rho = \sup\{|\int_{\mathbb{R}} g(\omega)\theta(\omega)\,d\omega| : \theta \in \mathcal{H}^{1+},\ ||\theta||_1 \leq 1\}\,. \tag{3.18}$$

The integral $\Lambda_g(\theta) = \int_{\mathbb{R}} g(\omega)\theta(\omega)\,d\omega$, $\theta \in \mathcal{H}^{1+}$, defines a bounded linear functional Λ_g on $\mathcal{H}^{1+}$. But it is well-known [13, p. 264] that the dual $(\mathcal{H}^{1+})^*$ of $\mathcal{H}^{1+}$ is precisely the quotient space $L^\infty/\mathcal{H}^{\infty+}$. Hence (3.18) is equivalent to

$$\rho = ||g + \mathcal{H}^{\infty+}||_\infty = \inf\{||g+u||_\infty : u \in \mathcal{H}^{\infty+}\}. \tag{3.19}$$

The infimum in (3.19) is always attained (see [13, p. 266]), that is $\rho = ||g - u_0||_\infty$ for some $u_0 \in \mathcal{H}^{\infty+}$ (u_0 might not be unique); by contrast, the supremum in (3.17) might not be attained. 3) From (3.12) we see that $P_{01}\phi_0 = \bar{h}^{-1}(g\phi_+ - \pi^+ g\phi_+)$ But (see [13]) $(\pi^+ g\phi_+)(\omega) = (1/2)\{(g\phi_+)(\omega) + i(Hg\phi_+)(\omega)\}$, where H is the Hilbert transform, defined by $(Hg\phi_+)(\omega) = \pi^{-1}\,\mathrm{P.V.}\int_{\mathbb{R}} g(\omega')\,\phi(\omega')\,(\omega-\omega')^{-1}\,d\omega'$ and P.V. denotes the Principal Value of the integral. A straightforward algebraic manipulation gives $(P_{01}\phi_0)(\omega) = (i/2\pi)\,\bar{h}^{-1}\,\mathrm{P.V.}\int_{\mathbb{R}}(g(\omega)-g(\omega'))\,(\omega-\omega')^{-1}\,\phi_+(\omega')\,d\omega'$. Thus, P_{01} is a *singular integral operator;* we suspect that this representation could be used to provide an alternative proof, or derive an alternative but equivalent form, of Theorem 3.2; in a sense, the "regularization" of $g(\omega)$ via $g_\varepsilon(\omega)$, $\varepsilon > 0$, serves to control the singular integral above.

The next theorem may be deduced from Theorem 3.4 below, but we provide a direct proof of it using the Hahn–Banach theorem. Let $C(\mathbb{R})$ be the space of continuous functions on $\mathbb{R}$ and $C_0(\mathbb{R}) = \{u \in C(\mathbb{R}) : \lim_{|\omega|\to\infty} u(\omega) = 0\}$.

Theorem 3.3 *If $g \in C_0(\mathbb{R}) + \mathcal{H}^{\infty+}$ then σ^2 (equivalently, ρ) is attained.*

Proof. Let $\mathcal{M}_0 = C_0(\mathbb{R}) \cap \mathcal{H}^{\infty+}$ (this is a strict subspace of $C_0(\mathbb{R})$). Equip $C_0(\mathbb{R})$ and $\mathcal{M}_0$ with the sup norm. By a duality theorem (see [12, p. 193]), the dual $(C_0(\mathbb{R})/\mathcal{M}_0)^*$ of $C_0(\mathbb{R})/\mathcal{M}_0$ is precisely $\mathcal{H}^{1+}$. Then an application of the Hahn-Banach theorem shows (see [17, Lemma 1.1, p. 133]) that $||g+\mathcal{M}_0||_\infty = \sup\{|\int g\theta| : \theta \in \mathcal{H}^{1+},\ \ ||\theta||_1 \le 1\}$. Thus by (3.19), $\rho = ||g+\mathcal{H}^{\infty+}||_\infty = ||g+\mathcal{M}_0||_\infty$. By another application of the Hahn-Banach theorem, there exists (see [17, Corollary 2, p. 77]) a linear functional Λ on $C_0(\mathbb{R})/\mathcal{M}_0$ of norm 1, so that $\Lambda(g+\mathcal{M}_0) = ||g+\mathcal{M}_0||_\infty$. Since Λ is a linear function and $||\Lambda|| = 1$, there exists a $\theta_0 \in \mathcal{H}^{1+}$ such that $||\theta_0||_1 = 1$ and $\Lambda(g+\mathcal{M}_0) = \int (g+\mathcal{M}_0)\theta_0\,d\omega = \int g(\omega)\theta_0(\omega)\,d\omega$, i.e. $\rho = \int_{\mathbb{R}} g(\omega)\theta_0(\omega)\,d\omega$. Since $\theta_0 \in \mathcal{H}^{1+}$, there exist (see [13, p. 240]) $\phi_\pm \in \mathcal{H}^{2\pm}$ so that $\theta_0 = \phi_+\bar{\phi}_-$ and $||\phi_+||_2 = ||\phi_-||_2 = \sqrt{||\theta_0||_1} = 1$. Then $\rho = \int g\phi_+\bar{\phi}_- = \langle g\phi_+, \phi_- \rangle$, proving the theorem. □

The next theorem implies Theorem 3.3; it is known in the literature as Hartman's theorem (see [16, Corollary 4.10, p. 46]).

Theorem 3.4 *P_{01} is compact iff $g \in C_0(\mathbb{R}) + \mathcal{H}^{\infty+}$.*

Remark. Since $\rho = \rho(\mathcal{L}_0, \mathcal{L}_1) = \rho(g\mathcal{H}^{2+}, \mathcal{H}^{2-})$, $\rho = 0$ iff $g\mathcal{H}^{2+}$ is orthogonal to $\mathcal{H}^{2-}$; a natural question then is: when do we have $g\mathcal{H}^{2+} \perp \mathcal{H}^{2-}$? By (3.19) $\rho = 0$ iff $g = u$ for some $u \in \mathcal{H}^{\infty+}$; since $|g(\omega)| = 1$, this means that u is an *inner* function. But this happens iff h is an entire function of exponential type less or equal to Δ (see [4, Exercise 4, p. 100]).

References

[1] A. Böttcher and B. Silbermann. *Analysis of Toeplitz Operators.* Springer-Verlag, 1990.

[2] L. Breiman and J. H. Friedman. Estimating optimal transformations for multiple regression and correlation (with discussion). *Journal of the American Statistical Association*, **80**:580–619, 1985.

[3] L. Breiman and R. Ihaka. Nonlinear discriminant analysis via scaling and ACE. Technical report, University of California, Berkeley, Dept. of Statistics, 1988.

[4] H. Dym and H. P. McKean. *Gaussian Processes, Function Theory, and the Inverse Spectral Theorem.* Academic Press, Inc., New York, 1976.

[5] J. Garnett. *Bounded Analytic Functions.* Academic Press, New York, 1981.

[6] B. Gidas and A. Murua. Estimation and consistency for Linear functionals of continuous-time processes from a finite data set, II: Optimal transformations for prediction. Preprint.

[7] B. Gidas and A. Murua. Optimal transformations for prediction in continuous-time stochastic processes with rational spectral densities. In preparation.

[8] B. Gidas and A. Murua. Classification and clustering of stop consonants via nonparametric transformations and wavelets. In *ICASSP-95*, volume 1, pages 872–875, 1995.

[9] B. Gidas and A. Murua. Stop consonants discrimination and clustering using nonlinear transformations and wavelets. In S. E. Levinson and L. Shepp, editors, *Image Models (And Their Speech Model Cousins)*, volume 80 of *IMA*, pages 13–62. Springer-Verlag, 1996.

[10] H. Helson and G. Szegö. A problem in prediction theory. *Ann. Mat. Pura Appl.*, **51**:107–138, 1960.

[11] I. A. Ibragimov and Y. A. Rozanov. *Gaussian Random Processes.* Springer-Verlag, New York, 1978.

[12] P. Koosis. *Introduction to Hp Spaces.* Cambridge University Press, 1980.

[13] P. Koosis. *The Logarithmic Integral*, volume 2. Cambridge University Press, 1988.

[14] N. Levinson and H. P. McKean. Weighted trigonometrical approximation on the line with application to the germ field of a stationary gaussian noise. *Acta. Math.*, 112:99–143, 1964.

[15] W. Magnus and F. Oberthettinger. *Formulas and Theorems for the Special Functions of Mathematical Physics.* Springer-Verlag, 1966.

[16] J. R. Partington. *An Introduction to Hankel Operators.* Cambridge University Press, 1988.

[17] M. Reed and B. Simon. *Methods of Modern Mathematical Physics*, volume 1. Academic Press, New York, 1972.

[18] Y. A. Rozanov. *Stationary Random Processes.* Holden-Day, San Francisco, 1967.

[19] D. Sarason. *Function theory on the unit disc.* Virginia Polytechnique Institute and State University, Blacksburg, Virginia, 1978.

[20] B. Simon. *The* $P(\phi)_2$ *Euclidean (Quantum) Field Theory.* Princeton University Press, Princeton, New Jersey, 1974.

[21] B. Simon. *Functional Integration and Quantum Physics.* Academic Press, New York, 1979.

[22] J. Weidman. *Linear Operators in Hilbert Spaces.* Springer-Verlag, New York, 1980.

Basilis Gidas
Division of Applied Mathematics
Brown University
Providence, Rhode Island 02912

Alejandro Murua
Department of Statistics
The University of Chicago
Chicago, Illinois 60637

Algebraic Methods Toward Higher-Order Probability Inequalities

Kenneth I. Gross and Donald St. P. Richards

1. Introduction

In work motivated by problems in the analysis of variance, Kimball [15] proved that if $V_i = Q_i/Q_0$, $i = 1, \ldots, n$, where the Q_i are mutually independent positive random variables, then $V_1, \ldots, V_n$ are *positively upper orthant dependent* (PUOD),

$$P(\bigcap_{i=1}^{n}\{V_i \geq v_i\}) \geq \prod_{i=1}^{n} P(V_i \geq v_i), \quad v_1, \ldots, v_n \geq 0, \tag{1.1}$$

and also are *positively lower orthant dependent* (PLOD),

$$P(\bigcap_{i=1}^{n}\{V_i \leq v_i\}) \geq \prod_{i=1}^{n} P(V_i \leq v_i), \quad v_1, \ldots, v_n \geq 0. \tag{1.2}$$

These results motivated much research on other inequalities; cf. [4], [10], [20], [24].

In Section 2 we use the notion of *similarly ordered* functions ([1], [6], [24]), and simple algebraic identities, to obtain higher-order extensions of (1.1) and (1.2) valid for *positively related* random variables. As special cases we deduce new proofs of (1.1) and (1.2) for some multivariate normal, t, chi-square and F-distributions.

In Section 3 we utilize the theory of Tchebycheff systems to derive inequalities for random vectors (X, Y) whose probability density functions are totally positive of order r (TP_r). The more basic of these inequalities are of the form: $\det(A) \geq 0$ where $A = (a_{i,j})$ is an $r \times r$ matrix of probabilities. As an example of these results, we prove that if $a_1 < a_2$, $b_1 < b_2$, and (X, Y) is TP_2 then

$$\begin{aligned} &P(X \geq a_1, Y \geq b_1)P(X \geq a_2, Y \geq b_2) \\ &\quad - P(X \geq a_1, Y \geq b_2)P(X \geq a_2, Y \geq b_1) \geq 0. \end{aligned} \tag{1.3}$$

As a limiting case of (1.3), we will show that (X, Y) is PUOD.

Using the theory of the characters of the symmetric group, and a remarkable theorem of Stembridge [23], we obtain new inequalities starting with those derived from the theory of total positivity. We prove that the distribution function of a vector of exchangeable random variables satisfies a higher-order generalized total positivity (GTP) property, extending a result in [13]; moreover, this higher-order GTP property holds with respect to any irreducible character of the symmetric group.

In Section 4 we utilize some generalizations of the concept of total positivity, formulated by Gross and Richards [8], to develop new inequalities. These generalizations were motivated by the theory of finite reflection groups (cf. [2], [5]). We derive inequalities within the new total positivity framework, and give examples for the two most basic formulations. As a special case we obtain a new proof, and generalizations, of the bivariate case of Šidák's inequality (cf. [10], p. 41).

2. Similarly ordered functions

Throughout, measure-theoretic or integrability issues are not germane; we assume that all functions considered are measurable on some measure space $\mathcal{X}$, and all integrals converge absolutely; in all cases, the conditions required for absolute convergence will be clear from the context. We denote by $E_{X_1,\ldots,X_n}$ expectation with respect to the joint distribution of a set of random variables $X_1, \ldots, X_n$; and we denote by $E_{X_1} \cdots E_{X_n}$ expectation with respect to the product of marginal distributions of $X_1, \ldots, X_n$.

2.1. Definition. Two functions $f_1, f_2 : \mathcal{X} \to \mathbb{R}$ are *similarly ordered* if, for all $x_1, x_2 \in \mathcal{X}$, $[f_1(x_1) - f_1(x_2)][f_2(x_1) - f_2(x_2)] \geq 0$. A family of functions $f_1, \ldots, f_n$ is *similarly ordered* if, for any $i, j \in \{1, \ldots, n\}$, f_i, f_j are similarly ordered.

The inequalities (1.1) and (1.2) can be proved using an inequality of Chebyshev. This result is as follows (cf. [6], [15], [24]).

2.2. Proposition. (P. L. Chebyshev) *If the random variable* $X \in \mathcal{X}$, *and* $f_1, f_2 : \mathcal{X} \to \mathbb{R}$ *are similarly ordered, then* $\operatorname{Cov}(f_1(X), f_2(X)) \geq 0$.

2.3. Lemma. *Let* $f_1, f_2, f_3 : \mathcal{X} \to \mathbb{R}_+$ *be a set of nonnegative similarly ordered functions. Then* f_1 *and* $f_2 f_3$ *are similarly ordered.*

Proof. Let $x_1, x_2 \in \mathcal{X}$. In the algebraic identity $ac - bd \equiv \frac{1}{2}[(a-b)(c+d) + (a+b)(c-d)]$, valid for all $a, b, c, d \in \mathbb{R}$, set $a = f_2(x_1)$, $b = f_2(x_2)$,

$c = f_3(x_1)$ and $d = f_3(x_2)$. Then

$$\begin{aligned}[f_1(x_1)-f_1(x_2)][f_2(x_1)f_3(x_1) - f_2(x_2)f_3(x_2)] \\ = \frac{1}{2}[f_3(x_1) + f_3(x_2)][f_1(x_1) - f_1(x_2)][f_2(x_1) - f_2(x_2)] \\ + \frac{1}{2}[f_2(x_1) + f_2(x_2)][f_1(x_1) - f_1(x_2)][f_3(x_1) - f_3(x_2)],\end{aligned}$$

which is a sum of nonnegative terms; hence f_1 and $f_2 f_3$ are similarly ordered. ■

Now we extend Proposition 2.2 to n nonnegative similarly ordered functions.

2.4. Proposition. (Kimball [15]) *If* $f_i : \mathcal{X} \to \mathbb{R}_+$, $i = 1, \ldots, n$, *are similarly ordered, then*

$$E\prod_{i=1}^{n} f_i(X) \geq E\prod_{i=1}^{k} f_i(X) \cdot E\prod_{i=k+1}^{n} f_i(X), \qquad 1 \leq k \leq n. \tag{2.4.1}$$

In particular,

$$E\prod_{i=1}^{n} f_i(X) \geq \prod_{i=1}^{n} Ef_i(X). \tag{2.4.2}$$

Proof. By repeated application of Lemma 2.3, we see that $f_1 f_2 \cdots f_k$ and $f_{k+1} f_{k+2} \cdots f_n$ are similarly ordered, $1 \leq k \leq n$. Then (2.4.1) follows from Proposition 2.2, and repeated application of (2.4.1) leads to (2.4.2). ■

2.5. Definition. The random variables $V_1, \ldots, V_n$ are *positively related* if there holds the stochastic representation

$$(V_1, \ldots, V_n) \stackrel{\text{st}}{=} (\phi_1(W_1, X), \ldots, \phi_n(W_n, X)),$$

where $X, W_1, \ldots, W_n$ are mutually independent; $X \in \mathbb{R}$ and $W_1, \ldots, W_n \in \mathcal{X}$; and $\phi_i : \mathcal{X} \times \mathbb{R} \to \mathbb{R}$ are such that $\phi_i(w, x)$ are all strictly decreasing (or all strictly increasing) in $x \in \mathbb{R}$ for any fixed $w \in \mathcal{X}$.

2.6. Theorem. *Suppose* $V_1, \ldots, V_n$ *are positively related random variables. Then*

$$P(\bigcap_{i=1}^{n}\{V_i \leq v_i\}) \geq P(V_1 \leq v_1, \ldots, V_k \leq v_k)\, P(V_{k+1} \leq v_{k+1}, \ldots, V_n \leq v_n), \quad 1 \leq k \leq n. \tag{2.6.1}$$

In particular, $V_1, \ldots, V_n$ are PLOD.

Proof. Suppose $x \mapsto \phi_i(w, x)$, $x \in \mathbb{R}$, is strictly decreasing for any fixed $w \in \mathcal{X}$; then this function is invertible and we denote by $\psi_i(w, \cdot)$ the corresponding inverse function. Therefore $\phi_i(w, x) \leq v$ if and only if $\psi_i(w, v) \leq x$. Hence, for $v_1, \ldots, v_n \in \mathbb{R}$,

$$\begin{aligned} P(\bigcap_{i=1}^{n} \{V_i \leq v_i\}) &= P(\bigcap_{i=1}^{n} \{\phi_i(W_i, X) \leq v_i\}) = P(\bigcap_{i=1}^{n} \{\psi_i(W_i, v_i) \leq X\}) \\ &= E_{W_1, \ldots, W_n} E_{X | \{W_1 = w_1, \ldots, W_n = w_n\}} \prod_{i=1}^{n} f_i(X) \end{aligned} \tag{2.6.2}$$

where, for $i = 1, \ldots, n$ and fixed $w_1, \ldots, w_n \in \mathcal{X}$, $f_i(x) = 1$ or 0 according as $x \geq \psi_i(w_i, v_i)$ or $x < \psi_i(w_i, v_i)$, respectively. Since $f_1, \ldots, f_n$ are non-negative and similarly ordered then, by Lemma 2.3, $f_1 \cdots f_k$ and $f_{k+1} \cdots f_n$ also are similarly ordered. By Proposition 2.4

$$\begin{aligned} & E_{X | \{W_i = w_i, i=1, \ldots, n\}} \prod_{i=1}^{n} f_i(X) \\ & \geq E_{X | \{W_i = w_i, i=1, \ldots, k\}} \prod_{i=1}^{k} f_i(X) \cdot E_{X | \{W_i = w_i, i=k+1, \ldots, n\}} \prod_{i=k+1}^{n} f_i(X) \\ & = P(\bigcap_{i=1}^{k} \{\psi_i(w_i, v_i) \leq X\}) \, P(\bigcap_{i=k+1}^{n} \{\psi_i(w_i, v_i) \leq X\}), \end{aligned} \tag{2.6.3}$$

almost surely. Taking expectations on both sides of (2.6.3) with respect to $W_1, \ldots, W_n$, it follows from (2.6.2) that

$$\begin{aligned} P(\bigcap_{i=1}^{n} \{V_i \leq v_i\}) &\geq P(\bigcap_{i=1}^{k} \{\psi_i(W_i, v_i) \leq X\}) \, P(\bigcap_{i=k+1}^{n} \{\psi_i(W_i, v_i) \leq X\}) \\ &= P(\bigcap_{i=1}^{k} \{V_i \leq v_i\}) \, P(\bigcap_{i=k+1}^{n} \{V_i \leq v_i\}), \end{aligned}$$

where the last equality follows from the definition of ψ_i. In turn, the PLOD property follows by iterating (2.6.1).

Finally, the proof for the case in which all the $\phi_i(w, x)$ are strictly increasing in x for each fixed w follows along the same lines. ■

2.7. Remark. (i) In the case in which $\phi_i(w, x) = w/x$, $w, x \in \mathbb{R}_+$, $i = 1, \ldots, n$, we recover (1.1) and (1.2). In this case, the conclusion that $V_1, \ldots, V_n$ are PLOD remains valid under the assumption that $W_1, \ldots, W_n$

are PLOD and independent of X; this can be proved using an argument similar to the proof of Theorem 2.6.

If $W_1, \ldots, W_n$ also are i.i.d. then $V_1, \ldots, V_n$ are said to be *positively dependent by mixture* (PDM); cf. Shaked and Tong [22] and Tong [24] for inequalities for PDM random variables.

(ii) Positively related random variables $V_1, \ldots, V_n$ also satisfy the PUOD property; this result is proved using an argument similar to the proof of Theorem 2.5.

2.8. Examples.

(i) Suppose $(V_1, \ldots, V_n)$ is multivariate normal with mean $(0, \ldots, 0)$ and correlation matrix $R = (\rho_{ij})$, where $\rho_{ij} = \lambda_i \lambda_j$ for all $i \neq j$ ([10], p. 47). (Without loss of generality, we assume $\lambda_i \neq 0$, $i = 1, \ldots, n$.) If all λ_i are of the same sign then $(V_1, \ldots, V_n)$ are positively related with $X, W_1, \ldots, W_n$ being independent standard normal variables and $\phi_i(w, x) = w(1 - \lambda_i^2)^{1/2} - x\lambda_i$, $w, x \in \mathbb{R}$, $i = 1, \ldots, n$. Since the $\phi_i(w, x)$ all are strictly increasing (or all decreasing) in x for each fixed w, then $V_1, \ldots, V_n$ are positively related; therefore (2.6.1) holds.

If all λ_i have the same sign then $V_1, \ldots, V_n$ are positively correlated normal variables, hence are *associated* (cf. [4], Section 5.6); this leads to inequalities stronger than (2.6.1).

(ii) In Definition 2.5 suppose $X \sim \chi^2_\nu$, a (central or noncentral) chi-square distribution; $W_1, \ldots, W_n$ are normal; and, for $i = 1, \ldots, n$, $\phi_i(w, x) = (w + \delta_i)/(x/\nu)^{1/2}$, $w, \delta_i \in \mathbb{R}$, $x \in \mathbb{R}_+$. Then $(V_1, \ldots, V_n)$ follows a multivariate t-distribution ([10], p. 132 ff). Since each $\phi_i(w, x)$ is strictly decreasing in x for each fixed w, then the hypotheses of Theorem 2.6 are satisfied; hence (2.6.1) holds. Note also that, by Remark 2.7(i), (2.6.1) remains valid if $W_1, \ldots, W_n$ satisfy (2.6.1); this is the case if $(W_1, \ldots, W_n)$ is multivariate normal with a distribution of the type given in Example 2.8 (i). In this case the resulting multivariate t-distributions are given in [10], pp. 140–141, 144.

(iii) In Definition 2.5 suppose $X, W_1, \ldots, W_n$ are gamma distributed and $\phi_i(w, x) = \lambda_i(w + x)$, $i = 1, \ldots, n$. Then $(V_1, \ldots, V_n)$ has a multivariate gamma distribution of Type I ([10], pp. 216–220). Each $\phi_i(w, x)$ is strictly increasing in $x > 0$ for each fixed $w > 0$; therefore we obtain the inequality (2.6.1).

2.9. Proposition. *Suppose $(U_1, \ldots, U_n)$ and $(Y_1, \ldots, Y_n)$ are independent positively related, positive random vectors. Define $V_i = U_i/Y_i$, $i = 1, \ldots, n$. Then $(V_1, \ldots, V_n)$ is both PLOD and PUOD.*

Proof. We establish the PLOD property only, because the proof of the PUOD property is similar. Denoting by I_A the indicator function of a set

A, then, for $v_1, \ldots, v_n \in \mathbb{R}_+$,

$$P(V_1 \leq v_1, \ldots, V_n \leq v_n) = P(U_1 \leq v_1 Y_1, \ldots, U_n \leq v_n Y_n)$$
$$= E_{Y_1,\ldots,Y_n} E_{U_1,\ldots,U_n} \prod_{i=1}^{n} I_{(-\infty, v_i Y_i]}(U_i).$$

Clearly, $I_{(-\infty, v_i Y_i]}(U_i) \equiv I_{[U_i/v_i, \infty)}(Y_i)$, $i = 1, \ldots, n$. Since $\{I_{[a_i,\infty)}(x), x \in \mathbb{R}, i = 1, \ldots, n\}$, are similarly ordered, and $(U_1, \ldots, U_n)$ and $(Y_1, \ldots, Y_n)$ both are positively related, then

$$P(V_1 \leq v_1, \ldots, V_n \leq v_n) \geq E_{Y_1,\ldots,Y_n} \prod_{i=1}^{n} E_{U_i} I_{(-\infty, v_i Y_i]}(U_i)$$
$$= E_{U_1} \cdots E_{U_n} E_{Y_1,\ldots,Y_n} \prod_{i=1}^{n} I_{(-\infty, v_i Y_i]}(U_i)$$
$$\geq E_{U_1} \cdots E_{U_n} \prod_{i=1}^{n} E_{Y_i} I_{(-\infty, v_i Y_i]}(U_i)$$
$$= \prod_{i=1}^{n} E_{U_i} E_{Y_i} I_{[U_i/v_i, \infty)}(Y_i).$$

Noting that $E_{U_i} E_{Y_i} I_{[U_i/v_i,\infty)}(Y_i) = P(U_i/v_i \leq Y_i) = P(V_i \leq v_i)$, then the proof is complete. ∎

2.10. Example. Suppose $(U_1, \ldots, U_n)$ and $(Y_1, \ldots, Y_n)$ are independent multivariate gamma distributed random vectors, each being positively related. Set $V_i = U_i/Y_i$, $i = 1, \ldots, n$; then $(V_1, \ldots, V_n)$ follows a multivariate F-distribution (cf. [10], p. 242). By Proposition 2.9, $(V_1, \ldots, V_n)$ is PLOD and PUOD.

Another approach to (1.1) and (1.2) utilizes the theory of total positivity [12]. Hence we now look for a deeper connection between the approach based on similarly ordered functions and that based on the theory of totally positive distributions (cf. Definition 3.1).

In the sequel we denote by $\mathcal{X}^n$ the Cartesian product $\mathcal{X} \times \cdots \times \mathcal{X}$ (n copies).

2.11. Proposition. (Karlin [11]) *Let $f_i, g_i : \mathcal{X} \to \mathbb{R}$, $i = 1, \ldots, n$, and μ be a positive measure on $\mathcal{X}$. Then*

$$n!\, det\Big(\int_{\mathcal{X}} f_i(t) g_j(t) d\mu(t)\Big) = \int_{\mathcal{X}^n} det\,(f_i(t_j))\, det\,(g_i(t_j)) \prod_{i=1}^{n} d\mu(t_i). \tag{2.11.1}$$

If the space $\mathcal{X}$ also is totally ordered then

$$det\Big(\int_{\mathcal{X}} f_i(t)g_j(t)d\mu(t)\Big) = \int_{t_1<\cdots<t_n}\cdots\int det\big(f_i(t_j)\big)\, det\big(g_i(t_j)\big)\prod_{i=1}^{n} d\mu(t_i). \tag{2.11.2}$$

The *basic composition formula*, (2.11.2), is obtained from (2.11.1) by the standard procedure of decomposing $\mathcal{X}^n$ into $n!$ subsets, each of the form $t_{\sigma(1)} < \cdots < t_{\sigma(n)}$ where $\sigma \in S_n$, the symmetric group on n symbols, and then applying an invariance argument.

2.12. Corollary. *Suppose $f, g : \mathcal{X} \to \mathbb{R}$ are similarly ordered and X is a random variable on $\mathcal{X}$. Then, for $n = 1, 2, \ldots$, the $n \times n$ determinant $det(Ef(X)^{n-i}g(X)^{n-j})$ is nonnegative. If $f, g : \mathcal{X} \to \mathbb{R}_+$ then all minors of the matrix $\big(Ef(X)^{n-i}g(X)^{n-j}\big)$ are nonnegative.*

Proof. Let μ denote the probability distribution of X. By Proposition 2.11,

$$\begin{aligned} n!\det\big(Ef(X)^{n-i}g(X)^{n-j}\big) &= n!\ \det\Big(\int_{\mathcal{X}} f(t)^{n-i}g(t)^{n-j}d\mu(t)\Big) \\ &= \int_{\mathcal{X}^n} \det\big(f(t_i)^{n-j}\big)\det\big(g(t_i)^{n-j}\big)\prod_{i=1}^{n} d\mu(t_i) \\ &= \int_{\mathcal{X}^n} \prod_{1\le i<j\le m} \big(f(t_i)-f(t_j)\big)\big(g(t_i)-g(t_j)\big)\prod_{i=1}^{n} d\mu(t_i), \end{aligned}$$

where the last equality is due to the Vandermonde formula, $\det\big(z_i^{n-j}\big) = \prod_{i<j}(z_i - z_j)$. Since f and g are similarly ordered, the integrand is nonnegative; hence the result is proved.

Next, suppose f and g are nonnegative. For $1 \le p \le n$, choose p-tuples of integers $k_1, \ldots, k_p$ and $l_1, \ldots, l_p$ such that $n-1 \ge k_1 > \cdots > k_p \ge 0$ and $n-1 \ge l_1 > \cdots > l_p \ge 0$. By Proposition 2.11

$$p!\det\big(E(f(X)^{k_i}g(X)^{l_j}\big) = \int_{\mathcal{X}^p} \det\big(f(t_i)^{k_j}\big)\det\big(g(t_i)^{l_j}\big)\prod_{i=1}^{p} d\mu(t_i). \tag{2.12.2}$$

By the theory of symmetric functions (cf. Macdonald [19]),

$$\det\big(z_i^{k_j}\big) = P_{k_1,\ldots,k_p}(z_1,\ldots,z_p)\prod_{i<j}(z_i - z_j) \tag{2.12.3}$$

where the *Schur function* $P_{k_1,\ldots,k_p}(z_1,\ldots,z_p)$ is a polynomial symmetric in $z_1,\ldots,z_p$ and nonnegative if $z_i \geq 0$ for all $i = 1,\ldots,p$. Applying (2.12.3) to both determinants in (2.12.2), and using the fact that f and g are similarly ordered, we find that the integrand is nonnegative. This completes the proof. ∎

2.13. Remark. Suppose $\mathcal{X} = \mathbb{R}$ and $f(x) = g(x) = x$, $x \in \mathbb{R}$. Then it follows from Corollary 2.12 that the determinant of the *moment matrix* $\big(E(X^{2n-i-j})\big)$ is nonnegative. Moment matrices similar to these have arisen in several important contexts, cf. [17].

3. Total positivity

In this section we study higher-order inequalities in which we establish the nonnegativity of alternating sums of probabilities. In some instances we will show that these alternating sums reduce, in limiting cases, to inequalities studied in Section 2.

3.1. Definition. (Karlin [11]) A rectangular matrix $A = (a_{ij})$ is *totally positive of order r* (TP_r) if all $n \times n$ minors of A are nonnegative for all $n = 1,\ldots,r$. Further, A is *strictly totally positive of order r* (STP_r) if all $n \times n$ minors of A are positive for all $n = 1,\ldots,r$.

A function $K : \mathbb{R}^2 \to \mathbb{R}$ is said to be TP_r (resp. STP_r) if, for all $x_1 < \cdots < x_r$ and $y_1 < \cdots < y_r$, the $r \times r$ matrix $(K(x_i, y_j))$ is TP_r (resp. STP_r). If K is TP_r (resp. STP_r) for all $r = 1, 2, \ldots$, then K is said to be TP_∞ (resp. STP_∞).

The following examples are provided by Karlin [11], Chapter 1.

3.2. Example. (i) The function $K(x,y) = e^{xy}$, $x, y \in \mathbb{R}$, is STP_∞. In turn, it is simple to verify that, for $\rho > 0$, $K(x,y) = n_2(x, y; \mu_1, \mu_2, \sigma_1^2, \sigma_2^2, \rho)$, the bivariate normal density with correlation ρ, is also STP_∞.

(ii) Suppose $K(x,y) = 1$ or 0 according as $x \geq y$ or $x < y$, respectively; then K is TP_∞.

3.3. Definition. ([14],[11]) A set of functions $\phi_i : \mathbb{R} \to \mathbb{R}$, $i = 1,\ldots,r$, is a *weak Tchebycheff system* (WT-system) if, for all $x_1 < \cdots < x_r$, the $r \times r$ matrix $(\phi_i(x_j))$ is TP_r.

The following are two well-known WT-systems (cf. [14], Chapter 1).

3.4. Example. (i) The set of functions $\phi_i(x) = x^{i-1}$, $x \in \mathbb{R}$, $i = 1,\ldots,n$, is a WT-system. Indeed, by the Vandermonde formula, $\det(\phi_i(x_j)) = \prod_{i<j}(x_j - x_i)$.

(ii) For $a_1 < \cdots < a_r$, the set of indicator functions $\phi_i(x) = I_{[a_i,\infty)}(x)$, $x \in \mathbb{R}$, $i = 1, \ldots, r$, is a WT-system. This result follows from Example 3.2(ii); in fact,

$$\det(\phi_i(x_j)) = \begin{cases} 1, \text{ if } a_1 \le x_1 < a_2 \le x_2 < \cdots < a_r \le x_r \\ 0, \text{ otherwise.} \end{cases} \tag{3.4.1}$$

We now present our main result on higher-order probability inequalities for totally positive distributions.

3.5. Theorem. *Let $(X,Y) \in \mathbb{R}^2$ be a random vector with probability density function (p.d.f.) K, where K is TP_r, and let $\{\phi_1, \ldots, \phi_r\}$ and $\{\psi_1, \ldots, \psi_r\}$ be two WT-systems. Then the $r \times r$ matrix $(E\phi_i(X)\psi_j(Y))$ is TP_r.*

Proof. Suppose $1 \le k_1 < k_2 < \cdots < k_n \le r$ and $1 \le l_1 < l_2 < \cdots < l_n \le r$; then

$$\det(K(x_i, y_j)) \cdot \det(\phi_{k_i}(x_j)) \cdot \det(\psi_{l_i}(y_j)) \ge 0 \tag{3.5.1}$$

for $x_1 < \cdots < x_n$ and $y_1 < \cdots < y_n$, since all three determinants are nonnegative. Notice that the left-hand side of (3.5.1) is symmetric in $x_1, \ldots, x_n$ and in $y_1, \ldots, y_n$, and is identically zero if any two x_i or y_i are coincident. Therefore (3.5.1) is valid for all $x_1, \ldots, x_n$ and $y_1, \ldots, y_n$. Integrating (3.5.1) over all $x_1, \ldots, x_n$ and $y_1, \ldots, y_n$ in $\mathbb{R}$, we obtain

$$\begin{aligned} 0 &\le \int_{\mathbb{R}^n} \int_{\mathbb{R}^n} \det(K(x_i, y_j)) \cdot \det(\phi_{k_i}(x_j)) \cdot \det(\psi_{l_i}(y_j)) \prod_{i=1}^n dx_i\, dy_i \\ &= (n!)^2 \det\Big(\int_{\mathbb{R}^2} K(x,y)\phi_{k_i}(x)\psi_{l_j}(y)\, dx\, dy \Big), \end{aligned}$$

where the equality is obtained by two applications of (2.11.1), once to integrate over the x variables and again to integrate over the y variables. Since the sequences $k_1, \ldots, k_n$ and $l_1, \ldots, l_n$ were chosen arbitrarily, this proves that the matrix $(E\phi_i(X)\psi_j(Y))$ is TP_r. ■

3.6. Remark. (i) The idea of symmetrizing an antisymmetric function, which in our context is the determinant $\det(K(x_i, y_j))$, dates back to Frobenius [7], who applied this technique with fundamental effect in his pioneering work on the representation theory of the symmetric group. For remarks on the history of this technique, we refer to Ledermann [16], p. 104, who described Frobenius' brilliant application of this technique as "a stroke of genius that is surely unsurpassed even among the great masters of algebraic formalism ...." In the sequel we refer to this technique as the *Frobenius symmetrization device.*

(ii) We refer to Samuels and Studden [22] for another approach to the development of Bonferroni-type probability inequalities by means of the theory of Tchebycheff systems.

3.7. Example. (i) Consider the case in which $\phi_i(x) = \psi_i(x) = x^{i-1}$, $i = 1, \ldots, r$. Then for any TP_r random vector (X, Y), all minors of the matrix $(EX^{i-1}Y^{j-1})$ are nonnegative. This result is valid, in particular, if (X, Y) follows a bivariate normal distribution with positive correlation, a result which we have been unable to locate in the literature.

(ii) Suppose $a_1 < \cdots < a_r$ and $b_1 < \cdots < b_r$. Let $\phi_i(x) = I_{[a_i,\infty)}(x)$ and $\psi_i(x) = I_{[b_i,\infty)}(x)$, $i = 1, \ldots, r$. By Theorem 3.3, if (X, Y) is a TP_r random vector then the $r \times r$ matrix $(E\phi_i(X)\psi_j(Y)) = (P(X \geq a_i, Y \geq b_j))$ is TP_r. For $r = 2$, this result reduces to the inequality

$$\begin{vmatrix} P(X \geq a_1, Y \geq b_1) & P(X \geq a_1, Y \geq b_2) \\ P(X \geq a_2, Y \geq b_1) & P(X \geq a_2, Y \geq b_2) \end{vmatrix} \geq 0 \tag{3.7.1}$$

for $a_1 < a_2$ and $b_1 < b_2$. Denoting a_2 and b_2 by x and y, respectively, and letting $a_1 \to -\infty$ and $b_1 \to -\infty$, then (3.7.1) reduces to the inequality

$$P(X \geq x, Y \geq y) \geq P(X \geq x)P(Y \geq y),$$

a result which is well-known for TP_2 random variables (cf. [12]).

3.8. Remark. In the preceding results, we have established that for certain matrices $A = (a_{ij})$, where each a_{ij} is a probability, or an expectation, related to a TP_r bivariate random vector, the alternating sum

$$\det{}_\chi(a_{ij}) := \sum_{\sigma \in S_n} \chi(\sigma) \prod_{i=1}^{n} a_{i,\sigma(i)} \tag{3.8.1}$$

is nonnegative, where S_n denotes the group of permutations on n symbols, and $\chi(\sigma) \equiv \operatorname{sgn}(\sigma)$, the sign of the permutation σ.

Recall that the sign function is an *irreducible character* of S_n (cf. [3], [16]). That is, $\operatorname{sgn}(\sigma)$, $\sigma \in S_n$, is a multiplicative function; and if we write $\operatorname{sgn} \equiv \chi_1\chi_2$ where both χ_1 and χ_2 are characters then χ_1, say, is the trivial character, $\chi_1(\sigma) \equiv 1$. Therefore it is natural to search for inequalities in the case in which χ is any irreducible character. In this case, the function $\det_\chi$ is called an *immanant* [18], [23].

For general irreducible characters χ, we apply a fundamental result of Stembridge [23] to obtain the following inequalities.

3.9. Theorem. *Let (X, Y) be a TP_r random vector, and $\{\phi_1, \ldots, \phi_r\}$ and $\{\psi_1, \ldots, \psi_r\}$ be WT-systems. Suppose $1 \le n \le r$, $1 \le k_1 < \cdots < k_n \le r$, $1 \le l_1 < \cdots < l_n \le r$, and χ is an irreducible character of S_n. Then*

$$det_\chi(E\phi_{k_i}(X)\psi_{l_j}(Y)) := \sum_{\sigma \in S_n} \chi(\sigma) \prod_{i=1}^{n} E\phi_{k_i}(X)\psi_{\sigma(l_i)}(Y) \ge 0. \quad (3.9.1)$$

Proof. To establish (3.9.1) we must show that any immanant of the matrix with (i,j)th entry $E\phi_{k_i}(X)\psi_{l_j}(Y)$ is nonnegative. By Theorem 3.5 this matrix is TP_r and, by Stembridge [23], all immanants of TP matrices are nonnegative. Hence (3.9.1) is proved. ■

Therefore, once we establish the nonnegativity of the alternating sum (3.9.1) for all the minors of a matrix when χ is the sign character then, by Stembridge's theorem, we obtain nonnegativity again when the sign character is replaced by *any* irreducible character.

3.10. Remark. Let us describe the inequalities (3.9.1) in greater detail, assuming, without loss of generality, that $n = r$. We begin with some background material on the irreducible characters of S_n.

A *partition* of n is a vector $(m_1, \ldots, m_p)$ of nonnegative integers such that $m_1 \ge m_2 \ge \cdots \ge m_p$ and $m_1 + \cdots + m_p = n$. We denote by $p(n)$ the number of partitions of n. For example, $p(4) = 5$ and the corresponding partitions of $n = 4$ are (4), $(3,1)$, $(2,2)$, $(2,1,1)$ and $(1,1,1,1)$; in the literature, the standard notations for these (4), (31), (2^2), (21^2) and (1^4), respectively.

In the general case, it is well-known (cf. [3], [16], [18]) that there is a bijective correspondence between the set of irreducible characters of S_n and the set of partitions of n. Therefore Theorem 3.7 constitutes $p(n)$ different inequalities. One of these inequalities corresponds to the partition (n), in which case $\chi(\sigma) \equiv \operatorname{sgn}(\sigma)$, and the resulting inequality is given in Theorem 3.5. Another inequality corresponds to the partition (1^n), in which case $\chi(\sigma) \equiv 1$, the trivial character; and then the resulting inequality is trivial.

The remaining $p(n) - 2$ inequalities are new to the area of inequalities. We will describe them explicitly in the case in which $n = 3$.

3.11. Example. Suppose $n = 3$; by a table of characters for S_3 (cf. [16], p. 50; [18], p. 265) we find that if $A = (a_{ij})$ is a 3×3 matrix then $2a_{11}a_{22}a_{33} -$

$a_{12}a_{23}a_{31} - a_{21}a_{13}a_{32}$ is the immanant of A corresponding to (the character indexed by) the partition (21). Substituting $a_{ij} = E\phi_i(X)\psi_j(Y)$ then, if (X, Y) is TP_3,

$$\begin{aligned} 2\prod_{i=1}^{3} E\phi_i(X)\psi_i(Y) & \\ \geq (E\phi_1(X)\psi_2(Y))&(E\phi_2(X)\psi_3(Y))(E\phi_3(X)\psi_1(Y)) \\ + (E\phi_2(X)\psi_1(Y))&(E\phi_1(X)\psi_3(Y))(E\phi_3(X)\psi_2(Y)). \end{aligned} \tag{3.11.1}$$

In the case in which $\phi_i(x) = I_{[a_i,\infty)}(x)$ and $\psi_i(x) = I_{[b_i,\infty)}(x)$, $i = 1, 2, 3$, where $a_1 < a_2 < a_3$ and $b_1 < b_2 < b_3$, (3.11.1) reduces to the probability inequality

$$\begin{aligned} 2\prod_{i=1}^{3} P(X \geq a_i, Y \geq b_i) & \\ \geq P(X \geq a_1, Y \geq b_2)&P(X \geq a_2, Y \geq b_3)P(X \geq a_3, Y \geq b_1) \\ + P(X \geq a_2, Y \geq b_1)&P(X \geq a_1, Y \geq b_3)P(X \geq a_3, Y \geq b_2). \end{aligned}$$

For $n \geq 4$ we obtain new inequalities corresponding to the higher partitions.

3.12. Generalized total positivity. The theory of generalized total positivity (GTP), initiated by Karlin and Rinott [13], may be described as follows. Recall that the natural action of $\sigma \in S_n$ on $(x_1, \ldots, x_n) \in \mathbb{R}^n$ is given by

$$\sigma \cdot (x_1, \ldots, x_n) = (x_{\sigma(1)}, \ldots, x_{\sigma(n)}). \tag{3.12.1}$$

A function $K : \mathbb{R}^n \times \mathbb{R}^n \to \mathbb{R}$ is called *S_n-invariant* if, for all $(x_1, \ldots, x_n)$, $(y_1, \ldots, y_n)$ in $\mathbb{R}^n$ and $\sigma \in S_n$, $K(\sigma \cdot (x_1, \ldots, x_n); \sigma \cdot (y_1, \ldots, y_n)) = K((x_1, \ldots, x_n); (y_1, \ldots, y_n))$.

Now fix an integer p where $1 \leq p \leq n$. For indices $1 \leq k_1 < \cdots < k_p \leq n$, let $S_{\{k_1,\ldots,k_p\}}$ denote the subgroup of S_n containing all $\sigma \in S_n$ which permute only the indices $\{k_1, \ldots, k_p\}$ among themselves; i.e., $\sigma(k) = k$ for all $k \in \{1, \ldots, n\} \setminus \{k_1, \ldots, k_p\}$. Let χ be an irreducible character of $S_{\{k_1\ldots,k_p\}}$. An S_n-invariant function $K : \mathbb{R}^n \times \mathbb{R}^n \to \mathbb{R}$ is *χ-generalized sign consistent of order p* (χ-GSC_p) if, for all monotone p-tuples of indices $k_1 < \cdots < k_p$ and for all $(x_1, \ldots, x_n)$ and $(y_1, \ldots, y_n)$ in $\mathbb{R}^n$ such that

$x_{k_1} < \cdots < x_{k_p}$ and $y_{k_1} < \cdots < y_{k_p}$, there holds the inequality

$$\sum_{\sigma \in S_{\{k_1,\ldots,k_p\}}} \chi(\sigma) K((x_1,\ldots,x_n);\sigma\cdot(y_1,\ldots,y_n)) \geq 0.$$

The function K is *generalized totally positive of order* r (GTP_r) if K is χ-GSC_p for all irreducible characters χ of $S_{\{k_1\ldots,k_p\}}$, and for all $p = 1,\ldots,r$.

As an example, if $K((x_1,\ldots,x_n);(y_1,\ldots,y_n)) = \prod_{i=1}^n K_0(x_i,y_i)$ for some $K_0 : \mathbb{R}^2 \to \mathbb{R}$, then K is GTP_r iff K_0 is TP_r. Further, GTP_2 is also known as *decreasing-in-transposition* or *arrangement-increasing* [20], [21].

We now extend a result of Karlin and Rinott [13], p. 290, on a GTP_2 property of the distribution function of a set of exchangeable random variables.

3.13. Theorem. *Let $(V_1,\ldots,V_n)$ be a vector of exchangeable random variables. For $(x_1,\ldots,x_n)$ and $(y_1,\ldots,y_n)$ in $\mathbb{R}^n$, define*

$$K((x_1,\ldots,x_n);(y_1,\ldots,y_n)) = P\big(x_1 \leq V_1 \leq y_1,\ldots,x_n \leq V_n \leq y_n\big).$$

Then K is GTP_n.

Proof. Since $V_1,\ldots,V_n$ are exchangeable then, clearly, K is S_n-invariant. Notice also that

$$\begin{aligned} K((x_1,\ldots,x_n);(y_1,\ldots,y_n)) &= E\prod_{i=1}^n I_{[x_i,y_i]}(V_i) \\ &= E\prod_{i=1}^n I_{[x_i,\infty)}(V_i) I_{(-\infty,y_i]}(V_i). \end{aligned} \tag{3.13.1}$$

For $1 \leq p \leq n$ choose a monotone p-tuple of indices $1 \leq k_1 < \cdots < k_p \leq n$, and suppose $(x_1,\ldots,x_n)$ and $(y_1,\ldots,y_n)$ in $\mathbb{R}^n$ are such that $x_{k_1} < \cdots < x_{k_p}$ and $y_{k_1} < \cdots < y_{k_p}$. Since K is S_n-invariant, and hence also $S_{\{k_1,\ldots,k_p\}}$-invariant, then for any irreducible character χ of $S_{\{k_1,\ldots,k_p\}}$,

$$\begin{aligned} p! &\sum_{\sigma \in S_{\{k_1,\ldots,k_p\}}} \chi(\sigma) K((x_1,\ldots,x_n);\sigma\cdot(y_1,\ldots,y_n)) \\ &= \sum\sum_{\sigma_1,\sigma_2 \in S_{\{k_1,\ldots,k_p\}}} \chi(\sigma_1)\,\chi(\sigma_2) K(\sigma_1\cdot(x_1,\ldots,x_n);\sigma_2\cdot(y_1,\ldots,y_n)). \end{aligned} \tag{3.13.2}$$

Substituting (3.13.1) into (3.13.2), we see that (3.13.2) becomes

$$\sum\sum_{\sigma_1,\sigma_2\in S_{\{k_1,\ldots,k_p\}}} \chi(\sigma_1)\,\chi(\sigma_2) E\prod_{i=1}^{n} I_{[x_{\sigma_1(i)},\infty)}(V_i) I_{(-\infty,y_{\sigma_2(i)}]}(V_i)$$
$$= EK_1(x_1,\ldots,x_n;V_1,\ldots,V_n)K_2(y_1,\ldots,y_n;V,\ldots,V_n),$$

where

$$K_1(x_1,\ldots,x_n;v_1,\ldots,v_n) = \sum_{\sigma\in S_{\{k_1,\ldots,k_p\}}} \chi(\sigma)\prod_{i=1}^{n} I_{[x_{\sigma(i)},\infty)}(v_i)$$

and

$$K_2(y_1,\ldots,y_n;v_1,\ldots,v_n) = \sum_{\sigma\in S_{\{k_1,\ldots,k_p\}}} \chi(\sigma)\prod_{i=1}^{n} I_{(-\infty,y_{\sigma(i)}]}(v_i).$$

Denote by $l_{p+1} < \cdots < l_n$ the monotone $(n-p)$-tuple of indices complementary to $k_1,\ldots,k_p$; i.e., $\{l_{p+1},\ldots,l_n\}\cup\{k_1,\ldots,k_p\} = \{1,\ldots,n\}$. From the definition of $S_{\{k_1,\ldots,k_p\}}$ it follows that

$$K_1(x_1,\ldots,x_n;v_1,\ldots,v_n)$$
$$= \left(\sum_{\sigma\in S_p} \chi(\sigma)\prod_{i=1}^{p} I_{[x_{\sigma(k_i)},\infty)}(v_{k_i})\right)\prod_{j=p+1}^{n} I_{[x_{l_j},\infty)}(v_{l_j})$$
$$\equiv \det{}_\chi\left(I_{[x_{k_i},\infty)}(v_{k_j})\right)\prod_{j=p+1}^{n} I_{[x_{l_j},\infty)}(v_{l_j}).$$

Similarly,

$$K_2(y_1,\ldots,y_n;v_1,\ldots,v_n) = \det{}_\chi\left(I_{(-\infty,y_{k_i}]}(v_{k_j})\right)\prod_{j=p+1}^{n} I_{(-\infty,y_{l_j}]}(v_{l_j}).$$

Therefore (3.13.2) equals $E\Psi(V_1,\ldots,V_n)$ where

$$\Psi(v_1,\ldots,v_n)$$
$$= \det{}_\chi\left(I_{[x_{k_i},\infty)}(v_{k_j})\right)\det{}_\chi\left(I_{(-\infty,y_{k_i}]}(v_{k_j})\right)$$
$$\times\prod_{j=p+1}^{n} I_{[x_{l_j},\infty)}(v_{l_j}) I_{(-\infty,y_{l_j}]}(v_{l_j}). \tag{3.13.3}$$

Notice also that, by (3.13.3), $\Psi(\sigma \cdot (v_1, \ldots, v_n)) = \Psi(v_1, \ldots, v_n)$ for all $\sigma \in S_{\{k_1,\ldots,k_p\}}$.

Denote by τ the probability distribution of $V_1, \ldots, V_n$. Since τ is exchangeable then

$$\begin{aligned} E\Psi(V_1, \ldots, V_n) &= \int_{\mathbb{R}^n} \Psi(v_1, \ldots, v_n) d\tau(v_1, \ldots, v_n) \\ &= \sum_{\sigma \in S_{\{k_1,\ldots,k_p\}}} \int_{v_{k_1} < \cdots < v_{k_p}} \cdots \int \Psi(\sigma \cdot (v_1, \ldots, v_n)) d\tau(v_1, \ldots, v_n) \\ &= p! \int_{v_{k_1} < \cdots < v_{k_p}} \cdots \int \Psi(v_1, \ldots, v_n) d\tau(v_1, \ldots, v_n), \end{aligned} \tag{3.13.4}$$

where the second equality holds since $\Psi(v_1, \ldots, v_n) = 0$ if $v_{k_i} = v_{k_j}$ for any $i \neq j$, and the last equality follows from the fact that Ψ is $S_{\{k_1,\ldots,k_p\}}$-invariant. However, $\Psi(v_1, \ldots, v_n) \geq 0$ on the orthant $\{v_{k_1} < \cdots < v_{k_p}\}$; indeed, both χ-determinants in (3.13.3) are constructed from WT-systems so, by [23], they are nonnegative. Therefore (3.13.2) is nonnegative. Since p and χ were chosen arbitrarily, the proof is complete. ■

3.14. Remark. In Theorem 3.13 if $\chi \equiv \text{sgn}$ then, by (3.13.2)–(3.13.4),

$$\begin{aligned} \sum_{\sigma \in S_{\{k_1,\ldots,k_p\}}} & \chi(\sigma) K((x_1, \ldots, x_n); \sigma \cdot (y_1, \ldots, y_n)) \\ &= \int_{v_{k_1} < \cdots < v_{k_p}} \cdots \int \Psi(v_1, \ldots, v_n) d\tau(v_1, \ldots, v_n). \end{aligned} \tag{3.14.1}$$

Moreover, it follows from (3.13.3) and (3.4.1) that Ψ is the indicator function of the set

$$\begin{aligned} &\{(v_1, \ldots, v_n) : x_{k_1} \leq v_{k_1} < x_{k_2} \leq v_{k_2} < \cdots < x_{k_p} \leq v_{k_p}\} \\ &\cap \{(v_1, \ldots, v_n) : v_{k_1} \leq y_{k_1} < v_{k_2} \leq y_{k_2} < \cdots < v_{k_p} \leq y_{k_p}\} \\ &\cap \{(v_1, \ldots, v_n) : x_{l_j} \leq v_{l_j} \leq y_{l_j}, j = p+1, \ldots, n\}. \end{aligned} \tag{3.14.2}$$

Then (3.14.1) is the probability that $(V_1, \ldots, V_n)$ lies in the set (3.14.2). This realization of (3.14.1) as a probability implies also that it is bounded above by 1.

A close scrutiny of the proof of Theorem 3.13 reveals the following generalization.

3.15. Theorem. *Let $V_1, \ldots, V_n$ be exchangeable random variables, and $L_1, L_2 : \mathbb{R}^2 \to \mathbb{R}$ be TP_n functions. For $(x_1, \ldots, x_n)$ and $(y_1, \ldots, y_n)$ in $\mathbb{R}^n$ define*

$$K((x_1, \ldots, x_n); (y_1, \ldots, y_n)) = E \prod_{i=1}^{n} L_1(x_i, V_i) L_2(y_i, V_i).$$

Then K is GTP_n.

Proof. Suppose $1 \le k_1 < \cdots < k_p \le n$, χ is an irreducible character of $S_{\{k_1, \ldots, k_p\}}$, and $(x_1, \ldots, x_n)$ and $(y_1, \ldots, y_n)$ in $\mathbb{R}^n$ are such that $x_{k_1} < \cdots < x_{k_p}$ and $y_{k_1} < \cdots < y_{k_p}$, where $1 \le p \le n$. By modifying the proof of Theorem 3.13 we obtain

$$\begin{aligned} &\sum_{\sigma \in S_{\{k_1, \ldots, k_p\}}} \chi(\sigma) K((x_1, \ldots, x_n); \sigma \cdot (y_1, \ldots, y_n)) \\ &\qquad = \int \cdots \int_{v_{k_1} < \cdots < v_{k_p}} \Psi(v_1, \ldots, v_n) d\tau(v_1, \ldots, v_n) \end{aligned} \tag{3.15.1}$$

where, if $l_{p+1} < \cdots < l_n$ are the indices complementary to $k_1, \ldots, k_p$ then,

$$\begin{aligned} &\Psi(v_1, \ldots, v_n) \\ &\qquad = \det{}_\chi \Big(L_1(x_{k_i}, v_{k_j}) \Big) \det{}_\chi \Big(L_2(y_{k_i}, v_{k_j}) \Big) \\ &\qquad \times \prod_{j=p+1}^{n} L_1(x_{l_j}, v_{l_j}) L_2(y_{l_j}, v_{l_j}). \end{aligned} \tag{3.15.2}$$

By the definition of $\det_\chi$ and Stembridge's theorem, (3.15.2) is nonnegative; hence so is (3.15.1). Since p and χ were chosen arbitrarily, the proof is complete. ∎

If L_1 and L_2 are chosen as indicator functions, Theorem 3.15 reduces to Theorem 3.13.

4. Finite reflection groups

Observe, from Definition 3.1, that $K : \mathbb{R}^2 \to \mathbb{R}_+$ is TP_r if, for all $n = 1, \ldots, r$, and for all $x_1 < \cdots < x_r$ and $y_1 < \cdots < y_r$,

$$\sum_{\sigma \in S_n} \operatorname{sgn}(\sigma) \prod_{j=1}^{n} K(x_j, y_{\sigma(j)}) \ge 0.$$

The group S_n is an example of a *finite reflection* (or *Coxeter) group* acting on $\mathbb{R}^n$ (cf. [5], [9]). That is, endowed with its natural action, as given by (3.12.1), S_n may be viewed as a finite subgroup of $O(n)$, the group of $n \times n$ orthogonal linear transformations. Moreover, $\det(\sigma)$, the determinant of the linear transformation σ, is identical to $\operatorname{sgn}(\sigma)$.

A finite reflection group W is *irreducible* if it cannot be written as a direct product of two nontrivial subgroups; in the sequel, we consider only irreducible reflection groups. Given a finite reflection group W acting on $\mathbb{R}^n$, the action of W partitions $\mathbb{R}^n$ into a finite number of mutually congruent open subsets called *Weyl chambers*; and W acts simply transitively on the set of all Weyl chambers. It is customary to delineate a fixed chamber $\mathcal{C}$ and refer to it as the *fundamental Weyl chamber.*

4.1. Definition. (Gross and Richards [8]) A function $K : \mathbb{R}^2 \to \mathbb{R}$ is *totally positive with respect to* W if for all $(x_1, \dots, x_n)$ and $(y_1, \dots, y_n)$ in $\mathcal{C}$,

$$\sum_{w \in W} (\det w) \prod_{j=1}^{n} K(x_j, (w \cdot y)_j) \geq 0 \tag{4.1.1}$$

where $(w \cdot y)_j$ denotes the jth component of the vector $w \cdot (y_1, \dots, y_n)$.

4.2. Example. Suppose $W = S_n$, the symmetric group. The action of S_n partitions $\mathbb{R}^n$ into $n!$ Weyl chambers, each of the form $\{(x_1, \dots, x_n) \in \mathbb{R}^n : x_{\sigma(1)} < \cdots < x_{\sigma(n)}\}$, $\sigma \in S_n$. We take $\mathcal{C} = \{(x_1, \dots, x_n) \in \mathbb{R}^n : x_1 < \cdots < x_n\}$ to be a *fundamental Weyl chamber.*

Recall that, for $\sigma \in S_n$, $\det(\sigma)$ and $\operatorname{sgn}(\sigma)$ are identical. Then the sum in (4.1.1) reduces to the $n \times n$ determinant $\det(K(x_i, y_j))$. Hence, $K : \mathbb{R}^2 \to \mathbb{R}$ is TP_r in the classical sense of Definition 3.1 if and only if K is totally positive with respect to $S_1, S_2, \dots, S_r$.

A general theory of total positivity with respect to finite reflection groups W was developed in [8]. Of special importance are those finite reflection groups, called *Weyl groups*, which play a central role in the classification of the compact Lie groups (cf. [2], [25]). By means of harmonic analysis on the compact Lie groups, it was proved in [8] that the function $K(x, y) = e^{xy}$, $(x, y) \in \mathbb{R}^2$, is totally positive with respect to all Weyl groups, a result well-known in the classical case in which $W = S_n$.

To complete the paper we examine in detail these ideas for two important finite reflection groups, both of which are Weyl groups ([8], Section 4). We apply a generalization of the Frobenius symmetrization device to deduce higher-order inequalities for random vectors whose density functions are totally positive with respect to these two groups.

4.3. The reflection group $G(n)$. Let $W = G(n)$, consisting of all signed permutations of the set $\{1, \ldots, n\}$. Each $w \in W$ is of the form $w = \epsilon\sigma$ where $\epsilon = (\epsilon_1, \ldots, \epsilon_n)$ with $\epsilon_i = \pm 1$ for all $i = 1, \ldots, n$; $\sigma \in S_n$ is a permutation on n symbols; and $w = \epsilon\sigma \in W$ acts on $(x_1, \ldots, x_n) \in \mathbb{R}^n$ by $w \cdot (x_1, \ldots, x_n) = (\epsilon_1 x_{\sigma(1)}, \ldots, \epsilon_n x_{\sigma(n)})$; and $\det(w) = \epsilon_1 \cdots \epsilon_n \operatorname{sgn}(\sigma)$. The fundamental Weyl chamber is taken to be $C = \{(x_1, \ldots, x_n) \in \mathbb{R}^n : 0 < x_1 < \cdots < x_n\}$. Further (cf. [8], p. 78) the sum in (4.1.1) becomes

$$\begin{aligned}
&\sum_{\sigma \in S_n} \operatorname{sgn}(\sigma) \sum_{\epsilon_1=\pm 1, \ldots, \epsilon_n = \pm 1} \prod_{i=1}^{n} \epsilon_i K(x_i, \epsilon_{\sigma(i)} y_{\sigma(i)}) \\
&\quad = \sum_{\sigma \in S_n} (\det \sigma) \prod_{i=1}^{n} \Big(K(x_i, y_{\sigma(i)}) - K(x_i, -y_{\sigma(i)}) \Big) \\
&\quad = \det\left(K(x_i, y_j) - K(x_i, -y_j)\right).
\end{aligned}$$

Thus, $K : \mathbb{R}^2 \to \mathbb{R}$ is totally positive with respect to the reflection group $G(n)$ if, for all $0 < x_1 < \cdots < x_n$ and $0 < y_1 < \cdots < y_n$, $\det(K(x_i, y_j) - K(x_i, -y_j)) \geq 0$.

The following result is an analog of Theorem 3.5 for the reflection group $G(n)$.

4.4. Theorem. *Let $(X, Y) \in \mathbb{R}^2$ be a random vector with p.d.f. K, where K is totally positive with respect to $G(n)$. Let $\{\phi_1, \ldots, \phi_n\}$ and $\{\psi_1, \ldots, \psi_n\}$ be two WT-systems on $\mathbb{R}$. Then*

$$\begin{aligned}
\det\Big(&E(\phi_i(X)\psi_j(Y) | X > 0, Y > 0) P(X > 0, Y > 0) \\
&- E(\phi_i(X)\psi_j(-Y) | X > 0, Y < 0) P(X > 0, Y < 0) \Big) \geq 0.
\end{aligned} \tag{4.4.1}$$

Proof. By the Frobenius symmetrization device,

$$\det\left(K(x_i, y_j) - K(x_i, -y_j)\right) \det\left(\phi_i(x_j)\right) \det\left(\psi_i(y_j)\right) \geq 0 \tag{4.4.2}$$

at first, for all $0 < x_1 < \cdots < x_n$ and $0 < y_1 < \cdots < y_n$, and in turn for all $x_i > 0$, $y_i > 0$, $i = 1, \ldots, n$. Integrating (4.4.2) over all $x_i, y_i > 0$, we deduce from Proposition 2.11 that

$$\det\left(\int_{\mathbb{R}_+^2} (K(x, y) - K(x, -y)) \phi_i(x) \psi_j(y) \, dx \, dy \right) \geq 0;$$

this proves (4.4.1). ∎

4.5. Example. Let (X,Y) be bivariate normally distributed with mean vector $(0,0)$. Without loss of generality, assume $\mathrm{Var}(X) = \mathrm{Var}(Y) = 1$, and denote $\mathrm{Cov}(X,Y)$ by ρ. For $0 < \rho < 1$, $K(x,y)$, the p.d.f. of (X,Y), is totally positive with respect to $G(n)$ for all n ([8], Theorem 6.1).

For the WT-systems we choose $\phi_i(x) = I_{[a_i,\infty)}(x)$ and $\psi_i(y) = I_{[b_i,\infty)}(y)$, $i = 1,\dots,n$, where $a_1 < \cdots < a_n$ and $b_1 < \cdots < b_n$. By Theorem 4.4 and the well-known formulas

$$P(X>0,Y>0) = \frac{1}{2}\big(1 - \frac{1}{\pi}\cos^{-1}\rho\big)$$

and

$$P(X>0,Y<0) = \frac{1}{2\pi}\cos^{-1}\rho,$$

we obtain the probability inequality

$$\begin{aligned}\det\Big(&c_\rho P\big(X \ge a_i, Y \ge b_j | X>0, Y>0\big)\\ &- (\pi - c_\rho) P\big(X \ge a_i, Y \le -b_j | X>0, Y<0\big)\Big) \ge 0.\end{aligned} \tag{4.5.1}$$

where $c_\rho \equiv \pi - \cos^{-1}\rho$. If we also assume $a_1 > 0$ and $b_1 > 0$ then (4.5.1) reduces to

$$\det\Big(P\big(X \ge a_i, Y \ge b_j\big) - P\big(X \ge a_i, Y \le -b_j\big)\Big) \ge 0.$$

4.6. The reflection group $SG(n)$. Let $W = SG(n)$, the subgroup of $G(n)$ containing all signed permutations of the set $\{1,\dots,n\}$ but with an even number of sign changes. Each $w \in SG(n)$ is of the form $w = \epsilon\sigma$ where $\sigma \in S_n$; $\epsilon = (\epsilon_1,\dots,\epsilon_n)$ with $\epsilon_i = \pm 1$ for all $i = 1,\dots,n$ and, since the number of sign changes is even, $\epsilon_1 \cdots \epsilon_n = 1$. Note that for $w = \epsilon\sigma$, $\det(w) = \mathrm{sgn}(\sigma)$. A fundamental Weyl chamber is $C = \{(x_1,\dots,x_n) \in \mathbb{R}^n : 0 < |x_1| < x_2 < \cdots < x_n\}$.

From (4.1.1) we now deduce that $K : \mathbb{R}^2 \to \mathbb{R}$ is totally positive with respect to the reflection group $SG(n)$ if, for all $(x_1,\dots,x_n)$ and $(y_1,\dots,y_n)$ in C,

$$\sum_{\substack{\epsilon_1=\pm1,\dots,\epsilon_n=\pm1\\ \epsilon_1\cdots\epsilon_n=1}} \det\big(K(x_i,\epsilon_j y_j)\big) \;\ge\; 0. \tag{4.6.1}$$

The following result is an analog of Theorem 4.4 for $SG(n)$.

4.7. Theorem. *Let $(X,Y) \in \mathbb{R}^2$ be a random vector with p.d.f. K, where K is totally positive with respect to $SG(n)$; and let $\{\phi_1,\ldots,\phi_n\}$ and $\{\psi_1,\ldots,\psi_n\}$ be two WT-systems on $\mathbb{R}_+$. Then* $\det\big(E\phi_i(X^2)\psi_j(Y^2)\big) \geq 0$.

Proof. Denote the sum in (4.6.1) by $D(x_1,\ldots,x_n;y_1,\ldots,y_n)$. Here, the Frobenius symmetrization device takes the form

$$D(x_1,\ldots,x_n;y_1,\ldots,y_n)\det\left(\phi_i(x_j^2)\right)\det\left(\psi_i(y_j^2)\right) \geq 0 \qquad (4.7.2)$$

firstly, for all $0 < |x_1| < x_2 < \cdots < x_n$ and $0 < |y_1| < y_2 < \cdots < y_n$, and then for all $x_i, y_i \in \mathbb{R}$, $i = 1,\ldots,n$. Applying the basic composition formula to integrate (4.7.2) over all $x_i, y_i \in \mathbb{R}$, and performing some simple algebraic manipulations, we obtain

$$\det\Big(\int_{\mathbb{R}^2} K(x,y)\phi_i(x^2)\psi_j(y^2)\,dx\,dy\Big) \geq 0.$$

This completes the proof. ∎

4.8. Example. The bivariate normal p.d.f. in Example 4.5 is totally positive with respect to $SG(n)$ ([8], *loc. cit*). Choosing $\phi_i(x) = I_{[a_i^2,\infty)}(x)$ and $\psi_i(y) = I_{[b_i^2,\infty)}(y)$, $i = 1,\ldots,n$, where $0 < a_1 < \cdots < a_n$ and $0 < b_1 < \cdots < b_n$, we obtain from Theorem 4.7 the inequality

$$\det\left(P(|X| \geq a_i, |Y| \geq b_j)\right) \geq 0. \qquad (4.8.1)$$

Since the distribution of $(|X|,|Y|)$ remains the same if ρ is replaced by $-\rho$ then (4.8.1) is valid for all ρ. In the case in which $n = 2$ (4.8.1) reduces to

$$\begin{aligned} &P(|X| \geq a_1, |Y| \geq b_1)P(|X| \geq a_2, |Y| \geq b_2) \\ &\qquad \geq P(|X| \geq a_1, |Y| \geq b_2)P(|X| \geq a_2, |Y| \geq b_1), \end{aligned} \qquad (4.8.2)$$

where $0 < a_1 < a_2$, $0 < b_1 < b_2$. By the same argument as in Example 3.7(ii), it follows from (4.8.2) that $(|X|,|Y|)$ is PUOD; similarly, we deduce also that $(|X|,|Y|)$ is PLOD. This provides a new proof of the bivariate case of Šidák's inequality (cf. [10], p. 41).

Acknowledgements. Portions of this work were performed during a highly stimulating workshop on *Algebraic Methods in Multivariate Statistical Analysis* held in July, 1995, at the Mathematisches Forschungsingstitut, Oberwolfach, Germany. Richards is very grateful to the organizers, Michael Perlman and Friedrich Pukelsheim, for having given him the opportunity to participate in the workshop.

References

[1] T. Armstrong (1994). Chebyshev inequalities and comonotonicity. *Real Analysis Exchange* **19**, 266–268.

[2] T. Bröcker and T. tom Dieck (1985). *Representations of Compact Lie Groups.* Springer, New York.

[3] P. Diaconis (1988). *Group Representations in Probability and Statistics.* IMS Lecture Notes **11**. Institute of Mathematical Statistics, Hayward, CA.

[4] M. L. Eaton (1987). *Lectures on Topics in Probability Inequalities.* Centrum voor Wiskunde Informatica, Amsterdam.

[5] M. L. Eaton and M. D. Perlman (1977). Reflection groups, generalized Schur functions and the geometry of majorization. *Ann. Probab.* **5**, 829–860.

[6] A. M. Fink and M. Jodeit, Jr. (1984). On Chebyshev's other inequality. *IMS Lecture Notes* **5**, 115–120 Institite of Mathematical Statistics, Hayward, CA.

[7] G. Frobenius (1900). Über die Charaktere der symmetrischen Gruppe, *Sitz. Ber. Preuss. Akad. Berlin*, 516–534.

[8] K. I. Gross and D. St. P. Richards (1995). Total positivity, finite reflection groups and a formula of Harish-Chandra. *J. Approximation Theory* **82**, 60–87.

[9] L. C. Grove and C. T. Benson (1985). *Finite Reflection Groups.* 2nd. edition. Springer, New York.

[10] N. L. Johnson and S. Kotz (1972). *Distributions in Statistics: Continuous Multivariate Distributions.* Wiley, New York.

[11] S. Karlin (1968). *Total Positivity*, Stanford University Press, Stanford, CA.

[12] S. Karlin and Y. Rinott (1980). Classes of orderings of measures and related correlation inequalities. I. Multivariate totally positive distributions. *J. Multivariate Analysis* **10**, 467–498.

[13] S. Karlin and Y. Rinott (1988). A generalized Cauchy-Binet formula and applications to total positivity and majorization. *J. Multivariate Analysis* **27**, 284–299.

[14] S. Karlin and W. J. Studden (1966). *Tchebycheff Systems: With Applications in Analysis and Statistics.* Wiley-Interscience, New York.

[15] A. W. Kimball (1951). On dependent tests of significance in the analysis of variance. *Ann. Math. Statist.* **22**, 599–602.

[16] W. Ledermann (1977). *Introduction to Group Characters.* Cambridge University Press, Cambridge.

[17] B. G. Lindsay (1989). Moment matrices: Applications in mixtures. *Ann. Statist.* **17**, 722–740.

[18] D. E. Littlewood (1958). *The Theory of Group Characters and Matrix Representations of Groups*, second edition. Oxford University Press, London.

[19] I. G. Macdonald (1995). *Symmetric Functions and Hall Polynomials.* Oxford University Press, Oxford.

[20] A. W. Marshall and I. Olkin (1979). *Inequalities: Theory of Majorization and its Applications.* Academic Press, New York.

[21] P. R. Rosenbaum (1995). *Observational Studies.* Springer, New York.

[22] S. M. Samuels and W. J. Studden (1989). Bonferroni-type probability bounds as an application of the theory of Tchebycheff systems. In: *Probability, Statistics and Mathematics: Essays in Honor of Samuel Karlin*, pp. 271–289, (T. W. Anderson, et al., eds.), Academic Press, San Diego, CA.

[22] M. Shaked and Y. L. Tong (1985). Some partial orderings of exchangeable random variables by positive dependence. *J. Multivariate Analysis* **17**, 333–349.

[23] J. R. Stembridge (1991). Immanants of totally positive matrices are nonnegative. *Bull. London Math. Soc.* **23**, 422–428.

[24] Y. L. Tong (1980). *Probability Inequalities in Multivariate Distributions.* Academic Press, New York.

[25] H. Weyl (1939). *The Classical Groups, Their Invariants and Representations.* Princeton University Press, Princeton, NJ.

K.I. Gross
Department of Mathematics & Statistics
University of Vermont
Burlington, VT 05405

D. St. P. Richards
Division of Statistics
University of Virginia
Charlottesville, VA 22903

Comparison and Deviation from a Representation Formula

Christian Houdré

1. Introduction

Let $X \sim ID(b, \Sigma, \nu)$, i.e., let X be a d-dimensional infinitely divisible random vector with characteristic function

$$\varphi(t) = \exp\left\{ i\langle t, b\rangle - \frac{1}{2}\langle \Sigma t, t\rangle + \int_{\mathbb{R}^d} (e^{i\langle t,u\rangle} - 1 - i\langle t, u\rangle \mathbf{1}(|u| < 1))\nu(du) \right\}, \tag{1.1}$$

where $t, b \in \mathbb{R}^d$, Σ is a positive semidefinite $d \times d$ matrix and ν (the Lévy measure) is a positive measure on $\mathcal{B}(\mathbb{R}^d)$, the Borel σ-algebra of $\mathbb{R}^d$, without atom at the origin and such that $\int_{\mathbb{R}^d}(|u|^2 \wedge 1)\nu(du) < +\infty$ ($\langle \cdot, \cdot \rangle$ and $|\cdot|$ are respectively the Euclidean inner product and norm in $\mathbb{R}^d$). Let $A = A_X$ be the generator of the corresponding Lévy process defined by

$$\begin{aligned} Af(x) = \langle b, \nabla f(x)\rangle &+ \frac{1}{2} tr\Sigma\nabla^2 f(x) \\ &+ \int_{\mathbb{R}^d} (f(x+u) - f(x) - \langle u, \nabla f(x)\rangle \mathbf{1}(|u| < 1))\nu(du), \end{aligned} \tag{1.2}$$

where for any $f \in C_b^2$ (C_b^k is the space of k-times continuously differentiable functions on $\mathbb{R}^d$ with k bounded derivatives),

$$\nabla f(x) = \left(\frac{\partial f(x)}{\partial x_i}\right) \quad \text{and} \quad \nabla^2 f(x) = \left(\frac{\partial^2 f(x)}{\partial x_i \partial x_j}\right).$$

Note that

$$Ae^{i\langle t,x\rangle} = e^{i\langle t,x\rangle} \operatorname{Log} \varphi(t). \tag{1.3}$$

Given two vectors $X_i \sim ID(b_i, \Sigma_i, \nu_i)$, $i = 0, 1$, with corresponding characteristic function φ_i, $i = 0, 1$, and a real $0 \leq \alpha \leq 1$, let X_α be an i.d. vector with characteristic function

$$\varphi_\alpha(t) = \varphi_0^{1-\alpha}(t)\varphi_1^\alpha(t), \qquad t \in \mathbb{R}^d. \tag{1.4}$$

Clearly, $X_\alpha \sim ID(b_\alpha, \Sigma_\alpha, \nu_\alpha)$, where $b_\alpha = (1-\alpha)b_0+\alpha b_1$, $\Sigma_\alpha = (1-\alpha)\Sigma_0 + \alpha\Sigma_1$, $\nu_\alpha = (1-\alpha)\nu_0+\alpha\nu_1$. Write $\boldsymbol{E}_\alpha$ for the corresponding expectation, i.e., $\boldsymbol{E}\exp(i\langle t, X_\alpha\rangle) = \boldsymbol{E}_\alpha \exp(i\langle t, X\rangle) = \varphi_\alpha(t)$ and let A_α be the corresponding generator. Then,

$$\boldsymbol{E}_1 f(X) - \boldsymbol{E}_0 f(X) = \int_0^1 \boldsymbol{E}_\alpha[(A_1 - A_0)f(X)]d\alpha. \tag{1.5}$$

In view of (1.2) and of the definition of X_α, the integrand in the right-hand side of (1.5) can be rewritten as

$$\begin{aligned}\boldsymbol{E}\Big[&\langle b_1 - b_0, \nabla f(X_\alpha)\rangle + \frac{1}{2}tr(\Sigma_1 - \Sigma_0)\nabla^2 f(X_\alpha) \\ &+ \int_{\mathbb{R}^d} (f(X_\alpha + u) - f(X_\alpha) - \langle u, \nabla f(X_\alpha)\rangle \mathbf{1}(|u| < 1))(\nu_1 - \nu_0)(du)\Big].\end{aligned} \tag{1.6}$$

To prove (1.5), by linearity and a standard approximation argument, it is enough to show its validity for $f(x) = e^{i\langle t,x\rangle}$. But, this easily follows from (1.3) and (1.4) and from the fundamental theorem of calculus (see [HPAS]). With $X \sim ID(b, \Sigma, \nu)$, we associate the $2d$-dimensional i.d. vectors $(X_i, Y_i)'$, $i = 0, 1$, with characteristic functions $\varphi_1(t, s) = \varphi(t + s)$ and $\varphi_0(t, s) = \varphi(t)\varphi(s)$, $t, s \in \mathbb{R}^d$, where $\varphi(t) = \varphi_X(t)$ is given by (1.1). Let $(X_\alpha, Y_\alpha)'$ have characteristic function

$$\begin{aligned}\varphi_\alpha(t, s) &= [\varphi_0(t, s)]^{1-\alpha}[\varphi_1(t, s)]^\alpha \\ &= [\varphi(t)\varphi(s)]^{1-\alpha}[\varphi(t + s)]^\alpha,\end{aligned} \tag{1.7}$$

$0 \leq \alpha \leq 1$, and write $\boldsymbol{E}_\alpha$ for the corresponding expectation. Now introducing the notation $\Delta_u f(x) = f(x + u) - f(x)$, $x, u \in \mathbb{R}^d$, for any $f_1, f_2 \in C_b^1$, it is easily seen that (1.5) becomes

$$\begin{aligned}\mathrm{Cov}(f_1(X), f_2(X)) = \int_0^1 &\boldsymbol{E}_\alpha[\langle \Sigma\nabla f_1(X), \nabla f_2(X)\rangle \\ &+ \int_{\mathbb{R}^d} \Delta_u f_1(X)\Delta_u f_2(X)\nu(du)]d\alpha.\end{aligned} \tag{1.8}$$

The main purpose of [HPAS] was to easily derive from (1.8) some well known and some new correlation inequalities and identities. In particular, the (positive) association result of [Pit] and [LRS] follow from (1.8). Furthermore, easy consequences are not listed in [HPAS] the Gaussian correlation inequalities in [JDPP, Corollaries 2, 3] as well as the characterization of negative Gaussian (resp. stable) association obtained in [JDP] (resp. [LRS]). Also, as pointed out by the referee, a paper closely related to [HPAS] and the present one is [HP]. [HP] is concerned with a different set of questions but the techniques there are similar in spirit and the setup is considerably more general.

Here we will not pursue this type of study in correlation inequalities but first investigate how (1.5) can be used to derive some well known and some new comparisons theorems. To clarify what we mean by a comparison theorem, consider the following situation. Let X_0 and X_1 be two d-dimensional Gaussian vectors with mean vector μ_0 and μ_1 and respective covariance matrix Σ_0 and Σ_1, and let $f \in C_b^2$, then (1.5) becomes

$$\begin{aligned} \boldsymbol{E}f(X_1) - \boldsymbol{E}f(X_0) = &\int_0^1 \boldsymbol{E}\langle \mu_1 - \mu_0, \nabla f(X_\alpha)\rangle d\alpha \\ &+ \frac{1}{2}\int_0^1 \boldsymbol{E}tr(\Sigma_1 - \Sigma_0)\nabla^2 f(X_\alpha) d\alpha. \end{aligned} \tag{1.9}$$

Thus, if $\Sigma_1 - \Sigma_0$ is positive semi-definite and if f is convex

$$\boldsymbol{E}f(X_1) - \boldsymbol{E}f(X_0) \geq \int_0^1 \boldsymbol{E}\langle \mu_1 - \mu_0, \nabla f(X_\alpha)\rangle d\alpha.$$

This last inequality allows one to compare the expectation of classes of functions of Gaussian vectors and it is in this sense that we use the terminology comparison theorems. In particular, if $\mu_1 = \mu_0$, then $\boldsymbol{E}f(X_1) \geq \boldsymbol{E}f(X_0)$. In fact, under a symmetry assumption, a stronger result (see [T, p.170]) holds. It is also clear that if $\mu_1 \geq \mu_0$ (componentwise) and if f is non-decreasing (componentwise), then

$$\boldsymbol{E}f(X_1) - \boldsymbol{E}f(X_0) \geq \frac{1}{2}\int_0^1 \boldsymbol{E}tr(\Sigma_1 - \Sigma_0)\nabla^2 f(X_\alpha) d\alpha,$$

and again $\boldsymbol{E}f(X_1) \geq \boldsymbol{E}f(X_0)$, if $\Sigma_1 = \Sigma_0$. Again under a symmetry assumption a stronger conclusion holds (see [T, p. 170]).

Comparison theorems are part of Cambanis' early work ([CSS], [CS]) and the author dedicates this paper to his memory.

Besides comparisons theorems, (1.9) also easily provides some deviation inequality. These deviation inequalities are part of a joint work in progress (on the discrete cube) with S. Bobkov and F. Götze. Namely, it is shown in the third section of the present notes that for any Lipschitz function f on $\mathbb{R}^d$ with $\|f\|_{\mathrm{Lip}} \le 1$, and any $\lambda > 0$,

$$\boldsymbol{P}(|f(X) - \boldsymbol{E}f(X)| \ge \lambda) \le \boldsymbol{E}|f(X) - \boldsymbol{E}f(X)|\frac{e^{-\lambda^2/2}}{\lambda}, \tag{1.10}$$

where $X \sim N(0, I)$ is a standard normal vector. Since $\boldsymbol{E}|f(X) - \boldsymbol{E}f(X)| \le \sqrt{2}\,\boldsymbol{E}|\nabla f(X)|/\sqrt{\pi} \le \sqrt{2/\pi}$ ([Pis, Chap. 2]), (1.10) recovers a deviation inequality of isoperimetric nature. However the main feature of (1.10) besides the simplicity of its proof, is the fact that $\boldsymbol{E}|f(X) - \boldsymbol{E}f(X)|$ can be small and that it is valid for all $\lambda > 0$.

2. Comparison

We start by providing an heuristic argument (which can be made rigorous by standard approximation) which illustrates our use of the Gaussian formula (1.9) to derive a well known result, namely, Slepian's lemma.

Let X_0 and X_1 be two centered Gaussian vectors with respective covariance matrix Σ_0 and Σ_1 such that $\sigma_0^{i,i} = \sigma_1^{i,i}$ and let $f(x_1, \dots, x_d) = \prod_{i=1}^d \mathbf{1}_{(-\infty,\lambda_i]}(x_i)$, $\lambda_i \in \mathbb{R}$. Then the derivatives (in the sense of distributions) of f are given by $\frac{\partial f}{\partial x_i}(x) = -\delta_{\lambda_i}(x_i) \prod_{\substack{k=1\\k\neq i}}^n \mathbf{1}_{(-\infty,\lambda_k]}(x_k)$ and for $i \neq j$; $\frac{\partial^2 f}{\partial x_i \partial x_j}(x) = \delta_{\lambda_i}(x_i)\delta_{\lambda_j}(x_j) \prod_{\substack{k=1\\k\neq i,j}}^n \mathbf{1}_{(-\infty,\lambda_k]}(x_k)$. Hence applying (1.9) to this function f gives (since $\sigma_1^{i,i} = \sigma_0^{i,i}$)

$$\begin{aligned}
&\boldsymbol{P}(X_1^1 \le \lambda_1, \dots, X_1^d \le \lambda_d) - \boldsymbol{P}(X_0^1 \le \lambda_1, \dots, X_0^d \le \lambda_d) \\
&\quad = \frac{1}{2}\int_0^1 \boldsymbol{E}_\alpha \sum_{\substack{i=1\\i\neq j}}^n \sum_{j=1}^n (\sigma_1^{i,j} - \sigma_0^{i,j})\delta_{\lambda_i}(x_i)\delta_{\lambda_j}(x_j) \prod_{\substack{k=1\\k\neq i,j}}^n \mathbf{1}_{(-\infty,\lambda_k]}(x_k)\,d\alpha.
\end{aligned}$$

Now, $\delta_{\lambda_i}(x_i)\delta_{\lambda_j}(x_j) \prod_{\substack{k=1 \\ k\neq i,j}}^{n} \mathbf{1}_{(-\infty,\lambda_k]}(x_k)$ is a positive generalized function (measure), and so if $\sigma_1^{i,j} \geq \sigma_0^{i,j}$, $i \neq j$, it follows that

$$\sum_{\substack{i=1 \\ i\neq j}}^{n} \sum_{j=1}^{n} (\sigma_1^{i,j} - \sigma_0^{i,j})\delta_{\lambda_i}(x_i)\delta_{\lambda_j}(x_j) \prod_{\substack{k=1 \\ k\neq i,j}}^{n} \mathbf{1}_{(-\infty,\lambda_k]}(x_k) \geq 0.$$

Thus,

$$\boldsymbol{P}(X_1^1 \leq \lambda_1, \ldots, X_1^d \leq \lambda_d) \geq \boldsymbol{P}(X_0^1 \leq \lambda_1, \ldots, X_0^d \leq \lambda_d),$$

which is the conclusion of Slepian's Lemma. We also note here that the results in [K] are immediate using (1.9). This leads us to the following proposition which is inspired by results in [K] and [G1–3].

Proposition 2.1. *Let X_0 and X_1 be two centered d-dimensional Gaussian vectors such that*

$$\begin{aligned} &\boldsymbol{E}(X_1^i - X_1^j)^2 \geq \boldsymbol{E}(X_0^i - X_0^j)^2 && \textit{if } (i,j) \in A \\ &\boldsymbol{E}(X_1^i - X_1^j)^2 \leq \boldsymbol{E}(X_0^i - X_0^j)^2 && \textit{if } (i,j) \in B \\ &\boldsymbol{E}(X_1^i - X_1^j)^2 = \boldsymbol{E}(X_0^i - X_0^j)^2 && \textit{if } (i,j) \notin A \cup B \end{aligned}$$

$$\begin{aligned} &\boldsymbol{E}(X_1^i)^2 \geq \boldsymbol{E}(X_0^i)^2 && \textit{if } i \in C \\ &\boldsymbol{E}(X_1^i)^2 \leq \boldsymbol{E}(X_0^i)^2 && \textit{if } i \in D \\ &\boldsymbol{E}(X_1^i)^2 = \boldsymbol{E}(X_0^i)^2 && \textit{if } i \notin C \cup D. \end{aligned}$$

Let $f \in C_b^2$ such that $\frac{\partial^2 f}{\partial x_i \partial x_j} \leq 0$ for $(i,j) \in A$, $\frac{\partial^2 f}{\partial x_i \partial x_j} \geq 0$ for $(i,j) \in B$; $\sum_{j=1}^d \frac{\partial^2 f}{\partial x_i \partial x_j} \geq 0$ for $i \in C$, $\sum_{i=1}^d \frac{\partial^2 f}{\partial x_i \partial x_j} \leq 0$ for $i \in D$. Then,

$$\boldsymbol{E}f(X_1) \geq \boldsymbol{E}f(X_0). \tag{2.1}$$

Proof. Using (1.9) and setting $\sigma_1^{i,j} = \boldsymbol{E}X_1^iX_1^j$, $\sigma_0^{i,j} = \boldsymbol{E}X_0^iX_0^j$; the proof of this result is rather simple. Indeed, it is enough to note that $\boldsymbol{E}(X_1^i - X_1^j)^2 = \sigma_1^{i,i} + \sigma_1^{j,j} - 2\sigma_1^{i,j}$, and that $\boldsymbol{E}(X_0^i - X_0^j)^2 = \sigma_0^{i,i} + \sigma_0^{j,j} - 2\sigma_0^{i,j}$, to rewrite $\sigma_1^{i,j} - \sigma_0^{i,j} = \sigma_1^{i,j} - \frac{1}{2}\sigma_1^{i,i} - \frac{1}{2}\sigma_1^{j,j} - \sigma_0^{i,j} + \frac{1}{2}\sigma_0^{i,i} + \frac{1}{2}\sigma_0^{j,j} + \frac{1}{2}(\sigma_1^{i,i} - \sigma_0^{i,i} + \sigma_1^{j,j} - \sigma_0^{j,j})$ and to use the hypothesis.

Remark 2.2. Using a standard regularization argument, the conclusion persists when the second derivatives exist only in the sense of distributions and satisfy the corresponding positivity or negativity assumption also in the sense of distributions. In turn the above result also implies Slepian's lemma which corresponds to $A = \{1,\dots,d\}^2, B = \emptyset,\ C = D = \emptyset;\ f = \prod_{i=1}^{d} \mathbf{1}_{]-\infty,\lambda_i]},\ \lambda_i \in \mathbb{R}$. Our conclusions also imply results in [G1–3]. In particular, taking $A = \{1,\dots,d\}^2, B = \emptyset$, and $f(x_1,\dots,x_d) = \max(x_1,\dots,x_d)$ recovers the Sudakov-Fernique inequality since then (see [G2]) $\frac{\partial^2 \max}{\partial x_i \partial x_j} \le 0$, $i \ne j$ and since if e is the unit vector in $\mathbb{R}^d$, $\max(x + te) = (\max x) + t$, $t \in \mathbb{R}$, which implies that $\sum_{i=1}^{d} \frac{\partial \max}{\partial x_i} = 1$. Proposition 2.1 also extends more recent results in [Kh].

Except for its use of (1.9), Proposition 2.1 presents little novelty since indeed it has already essentially been presented elsewhere. We now present another result which seems to have some interesting consequences. The statement below involving positive definite functions is in turn inspired by [HPAS] and [Ko2].

Proposition 2.3. *Let X_0 and X_1 be two centered d-dimensional Gaussian vectors with respective covariance matrix $\Sigma_1 = \begin{pmatrix} \Sigma_{11} & \Sigma_{12} \\ \Sigma_{21} & \Sigma_{22} \end{pmatrix}$; $\Sigma_0 = \begin{pmatrix} \Sigma_{11} & 0 \\ 0 & \Sigma_{22} \end{pmatrix}$, where Σ_{11} is a $k \times k$; Σ_{22} is $(d-k) \times (d-k)$, for some $k \in \{1,\dots,d-1\}$. Let $f \in C_b^2$ be positive definite and such that $f(x,y) = f(x,-y)$, for all $x \in \mathbb{R}^k$, $y \in \mathbb{R}^{d-k}$. Then,*

$$\boldsymbol{E} f(X_1) \ge \boldsymbol{E} f(X_0) \tag{2.2}$$

Proof. Since f is positive definite and real, $f(x) = \int_{\mathbb{R}^d} e^{i\langle t,x\rangle}\mu(dt)$, for some finite positive Borel symmetric measure μ on $\mathbb{R}^d$. Hence, (1.9) becomes

$$\begin{aligned} \boldsymbol{E} f(X_1) - \boldsymbol{E} f(X_0) &= -\frac{1}{2}\int_0^1 \int_{\mathbb{R}^d} \sum_{\ell=1}^{d}\sum_{n=1}^{d} (\sigma_1^{\ell,n} - \sigma_0^{\ell,n}) t_\ell t_n \boldsymbol{E} e^{i\langle t, X_\alpha\rangle}\mu(dt)d\alpha \\ &= -\frac{1}{2}\int_0^1 \int_{\mathbb{R}^d} \langle(\Sigma_1 - \Sigma_0)t, t\rangle \varphi_\alpha(t)\mu(dt)d\alpha. \end{aligned} \tag{2.3}$$

Now set $t = (t_1, t_2)$ where $t_1 \in \mathbb{R}^k$, $t_2 \in \mathbb{R}^{d-k}$. Then (2.3) becomes

$$-\int_0^1 \int_{\mathbb{R}^d} \varphi_\alpha(t_1,t_2)\langle \Sigma_{12} t_1, t_2\rangle \mu(dt_1 dt_2) d\alpha$$

$$= -\int_0^1 \int_{\langle \Sigma_{12} t_1, t_2 \rangle \geq 0} \varphi_\alpha(t_1, t_2) \langle \Sigma_{12} t_1, t_2 \rangle \mu(dt_1 dt_2)$$

$$- \int_0^1 \int_{\langle \Sigma_{12} t_1, t_2 \rangle \leq 0} \varphi_\alpha(t_1, t_2) \langle \Sigma_{12} t_1, t_2 \rangle \mu(dt_1 dt_2)$$

$$= -\int_0^1 \int_{\langle \Sigma_{12} t_1, t_2 \rangle \geq 0} (\varphi_\alpha(t_1, t_2) - \varphi_\alpha(t_1, -t_2)) \langle \Sigma_{12} t_1, t_2 \rangle \mu(dt_1 dt_2),$$

since $\mu(dt_1 dt_2) = \mu(dt_1(-dt_2))$. Now $\varphi_\alpha(t_1, t_2) = e^{-\frac{1}{2}\langle \Sigma_0 t, t \rangle} e^{-\alpha \langle \Sigma_{12} t_1, t_2 \rangle}$ and this last expression becomes

$$\int_0^1 \int_{\langle \Sigma_{12} t_1, t_2 \rangle \geq 0} e^{-\frac{1}{2}\langle \Sigma_0 t, t \rangle} \langle \Sigma_{12} t_1, t_2 \rangle \times \left(e^{\alpha \langle \Sigma_{12} t_1, t_2 \rangle} - e^{-\alpha \langle \Sigma_{12} t_1, t_2 \rangle} \right) \mu(dt_1 dt_2) d\alpha ,$$

which is clearly non-negative.

Remark 2.4. The above proposition can be extended as in the previous remark to positive definite functions which are Fourier transform of tempered positive measures. So the crucial point becomes to find such functions of interest. Typical examples are norms on $\mathbb{R}^d$. For example (see [Ko2]) $\|x\|_\infty^p = (\max_{1 \leq i \leq d} |x_i|)^p$ is p.d. if and only if $p \in (-d, -d+3]$. Combining this result with Proposition 2.3 recovers (when extended to stable vectors) an inequality in [Ko2], and more generally for $\|x\|_q^p$, $\|x\|_q = \left(\sum_{k=1}^d |x_k|^q\right)^{1/q}$. Classes of function g such that $f = g(\|\cdot\|)$ is p.d. have also been characterized, e.g., see [Ko1]. Note finally that the condition $f(x,y) = f(x,-y)$ is always satisfied for such functions. It is clear from (2.3) that whenever f is merely continuous positive definite and that $\Sigma_1 - \Sigma_0$ is positive semi-definite, then again $\boldsymbol{E} f(X_1) \geq \boldsymbol{E} f(X_0)$.

In the above applications, we have limited ourselves to the Gaussian framework, but the infinitely divisible representation is also useful. Indeed, it can be used to easily recover most of the results in [ST1-2].

3. Deviation

If X is a standard d-dimensional Gaussian vector, the representation (1.8) can be rewritten as

$$\mathrm{Cov}(f(X), g(X)) = \int_0^1 \boldsymbol{E} \left\langle \nabla f(X), \nabla g(\alpha X + \sqrt{1-\alpha^2} Y) \right\rangle d\alpha, \tag{3.1}$$

where Y is a standard normal vector independent of X. Now, if μ_α denotes the Gaussian measure in $\mathbb{R}^{2d}$ which is the distribution of $(X, \alpha X + \sqrt{1-\alpha^2}Y)$, and if μ denotes the probability measure $\int_0^1 \mu_\alpha d\alpha$, then (3.1) can also be written as

$$\mathrm{Cov}(f(X), g(X)) = \boldsymbol{E}\langle \nabla f(X), \nabla g(Z)\rangle = \int_{\mathbb{R}^d}\int_{\mathbb{R}^d} \langle \nabla f(x), \nabla g(z)\rangle d\mu(x,z) \tag{3.2}$$

where (X, Z) is a random vector in $\mathbb{R}^{2d}$ distributed according to μ. Note that X and Z are standard normal vectors but that (X, Z) is not normal (i.e., the measure μ is not Gaussian).

Applying (3.2) to f and $g = e^{tf}$ where $\|f\|_{\mathrm{Lip}} \leq 1$, $\boldsymbol{E}f(X) = 0$ and $t \geq 0$ gives

$$\begin{aligned}\boldsymbol{E}f(X)e^{tf(X)} &= \mathrm{Cov}(f(X)), e^{tf(X)}) \\ &= t\boldsymbol{E}\langle \nabla f(X), \nabla f(Z)\rangle e^{tf(Z)} \leq t\boldsymbol{E}e^{tf(Z)}\ .\end{aligned}$$

Now, let the function u be defined via $\boldsymbol{E}e^{tf(X)} = e^{u(t)}$. Then $\boldsymbol{E}f(X)e^{tf(X)} = u'(t)e^{u(t)}$, so that the above inequality reads as $u'(t) \leq t$. Since $u(0) = 0$, we conclude that, $u(t) \leq t^2/2$, that is

$$\boldsymbol{E}e^{tf(X)} \leq e^{t^2/2}. \tag{3.3}$$

By symmetry, one can also apply (3.3) to $-f$, to obtain that (3.3) holds for all $t \in \mathbb{R}$ and for every Lipschitz function f on $\mathbb{R}^d$ with $\|f\|_{\mathrm{Lip}} \leq 1$ and $\boldsymbol{E}f(X) = 0$. It thus follows from (3.3) and Chebyshev's inequality, that for all f on $\mathbb{R}^d$ such that $\|f\|_{\mathrm{Lip}} \leq 1$ and all $\lambda \geq 0$,

$$\boldsymbol{P}\{f(X) - \boldsymbol{E}f(X) \geq \lambda\} \leq e^{-\lambda^2/2}. \tag{3.4}$$

The inequality (3.3) is proved in ([Pis, Chap. 2]) using stochastic integrals. This type of idea goes back to [CIS] where the functional $f(X)$ is expressed as $f(X) = \boldsymbol{E}f(X) + W(\tau)$ where W is the standard Wiener process and τ is a stopping time with $0 \leq \tau \leq 1$. Hence, $f(X) - \boldsymbol{E}f(X) \leq \sup_{0\leq t\leq 1} W(t)$ and since this last supremum is distributed as $|W(1)|$, it follows that:

$$\boldsymbol{P}\{f(X) - \boldsymbol{E}f(X) \geq \lambda\} \leq 2(1 - \Phi(\lambda)) = \frac{2}{\sqrt{2\pi}}\int_\lambda^\infty e^{-x^2/2}dx. \tag{3.5}$$

Applying (3.5) to $-f$, also gives

$$\boldsymbol{P}\{|f(X) - \boldsymbol{E}f(X)| \geq \lambda\} \leq 4(1 - \Phi(\lambda)). \tag{3.6}$$

At the level of deviation inequalities, (3.5) is better than (3.4) since $2(1 - \Phi(\lambda)) \leq e^{-\lambda^2/2}$ but (3.5) implies (3.3) with a worse multiplicative constant, namely, it implies $\boldsymbol{E}e^{tf(X)} \leq 2e^{t^2/2}$. Different proofs of (3.3) based on semigroup techniques and on Gross' logarithm Sobolev inequality have been given in ([L1], [L2]).

Inequalities, related to (3.5), for the deviation of $f(X)$ from its quantiles can also be obtained from the Gaussian isoperimetric inequality ([B], [ST]). In particular, letting $m(\cdot)$ denote a median of a random variable, the isoperimetric inequality gives for all $\lambda \geq 0$,

$$\boldsymbol{P}\{f(X) - m(f(X)) \geq \lambda\} \leq 1 - \Phi(\lambda), \tag{3.7}$$

with equality for linear Lipschitz functions f (note that such functions also provide equality in (3.3)). The relation between (3.7) and (3.5) is however not clear. It is known (see [CIS]) that $|\boldsymbol{E}f(X) - m(f(X))| \leq 1/\sqrt{2\pi}$ for any f on $\mathbb{R}^d$ such that $\|f\|_{\text{Lip}} \leq 1$, but this is not enough to recover (3.5) from (3.7).

We now precise the argument used in getting (3.3) and (3.4) from (3.2) to prove:

Proposition 3.1. *For every Lipschitz function f on $\mathbb{R}^d$ with $\|f\|_{\text{Lip}} \leq 1$, the function*

$$T_f(\lambda) = e^{\lambda^2/2}\boldsymbol{E}(f(X) - \boldsymbol{E}f(X))\mathbf{1}_{\{f(X) - \boldsymbol{E}f(X) \geq \lambda\}}$$

is non-increasing in $\lambda \geq 0$. In particular, for all $\lambda > 0$,

$$\boldsymbol{P}\{f(X) - \boldsymbol{E}f(X) \geq \lambda\} \leq \boldsymbol{E}(f(X) - \boldsymbol{E}f(X))^+ \frac{e^{-\lambda^2/2}}{\lambda}, \tag{3.8}$$

$$\boldsymbol{P}\{|f(X) - \boldsymbol{E}f(X)| \geq \lambda\} \leq \boldsymbol{E}|f(X) - \boldsymbol{E}f(X)| \frac{e^{-\lambda^2/2}}{\lambda}. \tag{3.9}$$

Proof. Without loss of generality let $\boldsymbol{E}f(X) = 0$. Moreover, one can assume that $f(X)$ has a continuous positive density p on the whole real line since otherwise the statement of the theorem can be applied to the functions $f_\delta(x_1, \ldots, x, x_{n+1}) = (1-\delta)f(x_1, \ldots, x_n) + \delta x_{n+1}$, $0 < \delta < 1$, and let δ tend to 0. As in getting (3.3), applying (3.2) to f and $g = U(f)$, where U is a non–decreasing (piecewise) differentiable function on $\mathbb{R}$, gives

$$\boldsymbol{E}f(X)U(f(X)) \leq \boldsymbol{E}U'(f(X)). \tag{3.10}$$

Let F be the distribution function of $f(X)$. Given $\lambda > 0$ and $\epsilon > 0$, applying (3.10) to the function $U(x) = \min((x-\lambda)^+, \epsilon)$ gives

$$\int_\lambda^{\lambda+\epsilon} x(x-\lambda)dF(x) + \epsilon \int_{\lambda+\epsilon}^\infty x dF(x) \le F(\lambda+\epsilon) - F(\lambda)$$

Dividing by ϵ and letting ϵ tend to 0, we get, for all $\lambda > 0$,

$$\int_\lambda^\infty x dF(x) \le p(\lambda).$$

Thus, the function $V(\lambda) = \int_\lambda^\infty x dF(x) = \int_\lambda^\infty x p(x) dx$ satisfies the differential inequality $V(\lambda) \le -V'(\lambda)/\lambda$, i.e.,

$$(\mathrm{Log} V(\lambda))' \le (-\lambda^2/2)'.$$

This last inequality is equivalent to saying that the function $\mathrm{Log} V(\lambda) + \lambda^2/2$ is non–increasing. Therefore, so is the function $T_f(\lambda) = V(\lambda)\exp(\lambda^2/2)$ and Proposition 3.1 is proved.

Remark 3.2. As in the case of the Gaussian isoperimetric inequality, the assertion "the function T_f is non–increasing in $\lambda \ge 0$" expresses an extremal property of the linear functionals f since T_f is constant when f is linear. We also note that, for λ large, (3.9) improves the constant 4 in (3.6). Indeed, by the Gaussian Poincaré inequality, we have $\boldsymbol{E}|f(X) - \boldsymbol{E}f(X)|^2 \le 1$, therefore, recalling the asymptotic

$$1 - \Phi(\lambda) = \frac{e^{-\lambda^2/2}}{\lambda\sqrt{2\pi}}(1 + o(1)), \qquad \text{as } \lambda \to +\infty,$$

one obtains from (3.9)

$$\boldsymbol{P}\{|f(X) - \boldsymbol{E}f(X)| \ge \lambda\} \le \frac{e^{-\lambda^2/2}}{\lambda} = \sqrt{2\pi}(1 - \Phi(\lambda))(1 + o(1)),$$

as $\lambda \to +\infty$; but $\sqrt{2\pi} \le 4$. A main feature of the Gaussian Chebyshev's inequality (3.9) is however the fact that the value $\boldsymbol{E}|f(X) - \boldsymbol{E}f(X)|$ can be small, and then (3.9) becomes essentially better than (3.9) or even than (3.7). The first natural example of such a situation is $f(x) = \max_{i \le n} x_i$ in which case $\|f\|_{\mathrm{Lip}} = 1$ while $\boldsymbol{E}|f(X) - \boldsymbol{E}f(X)|$ is majorized by $1/\sqrt{2\log n}$.

References

[B] Borel, C. (1975) The Brunn-Minkowski inequality in Gauss space. *Invent. Math.* **30**, 207–211.

[CSS] Cambanis, S. Simons, G., Stout. W. (1976) Inequalities for $Ek(X,Y)$ when the marginals are fixed. *Z. Wahrsch. Verw. Gebiete* **36**, 285–294.

[CS] Cambanis, S., Simons, G. (1982) Probability and expectation inequalities. *Z. Wahrsch. Verw. Gebiete* **59**, 1–25.

[CIS] Cirel'son, B. S., Ibragimov, I. A., Sudakov, V. N. (1976) Norms of Gaussian sample functions. *Proceedings of the Third Japan-USSR Symposium on Probability Theory. Lecture Notes in Math.* **550**, Springer-Verlag, 20–41.

[G1] Gordon, Y. (1985) Some inequalities for Gaussian processes and applications, *Israel J. Math.* **50**, 265–289.

[G2] Gordon, Y. (1987) Elliptically contoured distributions, *Prob. Theory Rel. Fields* **76**, 429–438.

[G3] Gordon, Y. (1992) Majorization of Gaussian processes and geometric applications. *Prob. Theory Rel. Fields* **91**, 251–267.

[HP] Herbst, I. and Pitt, L. D. (1991) Diffusion equation techniques in stochastic monotonicity and positive correlations. *Prob. Theory Rel. Fields* **87**, 275–312.

[HPAS] Houdré, C., Peréz-Abreu, V., Surgailis, D. (1996) Interpolation, Correlation Identities and Inequalities for Infinitely Divisible Variables. *Journal of Fourier Analysis and Applications*, to appear.

[JDPP] Joag-Dev, K., Perlman, M. D. and Pitt, L. D. (1983) Association of normal random variables and Slepian's inequality. *Ann. Probab.* **11**, 451–455.

[JDP] Joag-Dev, K. and Proschan, Y. Z. (1983) Negative association of random variable with applications. *Ann. Stat.* **11**, 286–295.

[K] Kahane, J. P. (1986) Une inégalité du type de Slepian et Gordon sur les processus Gaussiens. *Israel J. Math.* **55**, 109–110.

[Kh] Khaoulani, B. (1993) A vectorial Slepian type inequality. Applications. *Proc. Amer. Math. Soc.* **118**, 95–102.

[Ko1] Koldobsky, A. (1992) Schoenberg's problem on positive definite functions. *English translation in St. Petersburg Math. J.* **3**, 563–570. *Algebra and Analysis* **3** *(1991)*, 78–85.

[Ko2] Koldobsky, A. (1996) Positive definite distributions and subspaces of L_{-p} with applications to stable processes. Preprint.

[L1] Ledoux, M. (1994) Semigroup proof of the isoperimetric inequality in Euclidean and Gauss space. *Bull. Sci. Math.* **118**, 485–510.

[L2] Ledoux, M. (1996) Isoperimetry and Gaussian Analysis. *Ecole d'été de Probabilités de Saint-Flour. Lect. Notes in Math.* **1648**, 165–294.

[LRS] Lee, M. L. T., Rachev, S. T., and Samorodnitsky, G. (1990) Association of stable random variables. *Ann. Probab.* **18**, 1759–1764.

[Pis] Pisier, G. (1986) Probabilistic methods in the geometry of Banach spaces. Probability and Analysis, Varenna (Italy) 1985. *Lecture Notes Math.* **1206**, 167–241.

[Pit] Pitt, L. D. (1982) Positively correlated normal variables are associated. *Ann. Probab.* **10**, 496–499.

[ST1] Samorodnitsky, G. and Taqqu, M. (1993) Stochastic monotonicity and Slepian-type inequalities for infinitely divisible and stable random vectors. *Ann. Probab.* **21**, 143–160.

[ST2] Samorodnitsky, G. and Taqqu, M. (1994) Lévy measures of infinitely divisible random vectors and Slepian inequalities. *Ann. Probab.* **22**, 1930–1956.

[ST] Sudakov, V. N., Tsirel'son, B. S. (1978) Extremal properties of half-spaces for spherically invariant measures. *J. Soviet Math.* **9**, 9–18. Translated from: *Zap. Nauch. Sem. L.O.M.I.* **41** (1974), 14–24 (Russian).

[T] Tong, Y.L. (1990) The multivariate normal distribution. Springer Series in Statistics, *Springer-Verlag, New York.*

Southeast Applied Analysis Center
School of Mathematics
Georgia Institute of Technology
Atlanta, GA 30332 USA

Components of the Strong Markov Property

Olav Kallenberg

Summary

The strong Markov property of a process X at an optional time $\tau < \infty$ may be thought of as a combination of the conditional independence $X_{\tau+h} \perp\!\!\!\perp_{X_\tau} \mathcal{F}_\tau$ with the homogeneity $P[X_{\tau+h} \in \cdot|X_\tau] = \mu_h(X_\tau, \cdot)$ for a suitable set of probability kernels μ_h. In an earlier paper, a stronger version of the latter condition was shown to imply the former property. Our present aim is to examine to what extent the two properties are in fact equivalent.

1. Introduction

Recall that a process X adapted to some filtration $\mathcal{F} = (\mathcal{F}_t)$ is said to satisfy the *strong Markov property* at some optional time $\tau < \infty$, if

$$P[\theta_\tau X \in \cdot|\mathcal{F}_\tau] = P_{X_\tau} \quad \text{a.s.} \tag{1}$$

for a given probability kernel (P_x). Here the shifts θ_t are defined by $(\theta_t w)_s = w_{s+t}$, and for general τ we may require X to be $\mathcal{F}$-progressive to ensure the appropriate measurability in (1). The stated equation may be written in two steps as

$$P[\theta_\tau X \in \cdot|\mathcal{F}_\tau] = P[\theta_\tau X \in \cdot|X_\tau] = P_{X_\tau} \quad \text{a.s.,} \tag{2}$$

where the first relation is equivalent to the *conditional independence* $\theta_\tau X \perp\!\!\!\perp_{X_\tau} \mathcal{F}_\tau$, whereas the second relation expresses the associated *time homogeneity*, the fact that the same kernel (P_x) can be used for all τ.

Under suitable conditions on X and $\mathcal{F}$, it was shown in Kallenberg (1987) (cf. Kallenberg (1997), Theorem 7.23) that the homogeneity property alone, for all finite optional times, implies the strong Markov property (1). A strengthened version of this result will be established in Section 3 below, where we shall also derive an improved version of the associated zero–one law for absorption, originally appearing as Theorem 1 in Kallenberg (1987) (cf. Kallenberg (1997), Lemma 7.24).

Somewhat surprisingly, it is possible to obtain results even in the opposite direction. In Section 2 we shall see how the time-homogeneous Markov property can be deduced, under suitable hypotheses, from the conditional independence alone. Thus the homogeneity and conditional independence are in this case essentially equivalent.

Rather than discussing the strong Markov property in the form (1), it is convenient to consider the simpler but equivalent version

$$P[X_{\tau+h} \in \cdot|\mathcal{F}_\tau] = \mu_h(X_\tau, \cdot) \text{ a.s.}, \quad h > 0, \tag{3}$$

where $\mu_h(x, \cdot) = P_x\{w; w_h \in \cdot\}$. Just as in (2) we get for each $h > 0$ a decomposition

$$P[X_{\tau+h} \in \cdot|\mathcal{F}_\tau] = P[X_{\tau+h} \in \cdot|X_\tau] = \mu_h(X_\tau, \cdot) \text{ a.s.} \tag{4}$$

where the first equality is equivalent to the conditional independence $X_{\tau+h} \perp\!\!\!\perp_{X_\tau} \mathcal{F}_\tau$, and the second equality expresses the associated time homogeneity. Note that the former relation for arbitrary τ and h implies the previous version $\theta_\tau X \perp\!\!\!\perp_{X_\tau} \mathcal{F}_\tau$, by the chain rule for conditional independence (cf. Proposition 5.8 in Kallenberg (1997)). However, the present homogeneity condition seems to be strictly weaker than the functional version $P[\theta_\tau X \in \cdot|X_\tau] = P_{X_\tau}$, and for the full equivalence it may have to be supplemented by a condition for absorption probabilities of the form

$$P[\theta_\tau X \in I_h|X_\tau] = a_h(X_\tau), \quad h > 0, \tag{5}$$

where I_h denotes the set of paths w with $w_0 = w_h = w_{2h} = \cdots$.

It seems preferable to state even the homogeneity property in intrinsic form without reference to any given set of kernels μ_h. For this purpose, we may introduce a pairwise version of the homogeneity condition. Then put $\nu_\tau = P \circ X_\tau^{-1}$, and introduce for each optional time $\tau < \infty$ an associated set of probability kernels μ_h^τ satisfying

$$\mu_h^\tau(X_\tau, \cdot) = P[X_{\tau+h} \in \cdot|X_\tau], \quad h > 0. \tag{6}$$

Recall that such regular conditional distributions exist when the state space S is Borel (cf. Kallenberg (1997), Theorem 5.3).

The *pairwise homogeneity* of X may now be defined by the condition that, for any two optional times $\sigma, \tau < \infty$,

$$\mu_h^\sigma = \mu_h^\tau \text{ a.e. } \nu_\sigma \wedge \nu_\tau, \quad h > 0, \tag{7}$$

or more explicitly

$$(\nu_\sigma \wedge \nu_\tau)\{s \in S;\ \mu_h^\sigma(s, \cdot) \neq \mu_h^\tau(s, \cdot)\} = 0, \quad h > 0.$$

Here the minimum $\nu_\sigma \wedge \nu_\tau$ is defined as the largest measure ν with $\nu \le \nu_\sigma$ and $\nu \le \nu_\tau$. To prove the existence of such a ν, put $\mu = \nu_\sigma + \nu_\tau$, introduce the densities $f_\sigma = d\nu_\sigma/d\mu$ and $f_\tau = d\nu_\tau/d\mu$, and let $f = f_\sigma \wedge f_\tau$. Then $\nu = f \cdot \mu$, the measure with μ-density f, has clearly the stated property. Note that the condition in (7) is independent of any specific choice of kernels μ_h^τ.

In the next section we shall identify two sets of conditions, each of which ensures the conditional independence $X_{\tau+h} \perp\!\!\!\perp_{X_\tau} \mathcal{F}_\tau$ to imply the time-homogeneous Markov property. Conversely, we shall see in Section 3 how the pairwise homogeneity in (7) together with a pairwise version of condition (5) implies conditional independence and hence the time-homogeneous Markov property. For those results it is enough to consider simple optional times $\tau < \infty$, so it suffices to assume that X be adapted. Under the stronger hypothesis of progressive measurability, we may consider the corresponding conditions for more general optional times $\tau < \infty$, and deduce appropriate forms of the strong Markov property.

We conclude this section with some bibliographical notes. The history of the subject goes back to the elementary Proposition 8.2 of Blumenthal and Getoor (1968), where a homogeneity condition for extended valued optional times τ was shown to imply the strong Markov property. Such a statement obviously requires special conventions about the interpretation of X_τ, θ_τ, and P_{X_τ} when $\tau = \infty$. In Kallenberg (1982), a primitive version of the statement for finite optional times τ emerged from some general results for exchangeable sequences, applied to the set of excursions from an arbitrary fixed point. A more direct approach was introduced in Kallenberg (1987), where the results were further generalized to the context of homogeneity in a set $B \subset S$, an extension suggested by both exchangeability and excursion theory.

In the modern axiomatic development of Markov process theory (see, e.g., Kallenberg (1997), Chapters 7 and 17), one normally starts from a set of transition kernels μ_h satisfying the Chapman–Kolmogorov equation (or equivalent semigroup property) identically. In the present paper, we are not requiring *a priori* that the kernels (P_x) or μ_h be associated with any Markov semigroup, and the asserted Markov property will only hold in the primitive sense of equation (1) or (3). In order to tie our results to the usual axiomatic treatment, one needs to find versions of the kernels μ_h that form a semigroup. This leads to a difficult construction problem, for which we refer to Walsh (1972).

2. Homogeneity from independence

In this section we shall exhibit conditions ensuring that the conditional independence $X_{\tau+h} \perp\!\!\!\perp_{X_\tau} \mathcal{F}_\tau$, for suitable optional times $\tau < \infty$ and constants

$h > 0$, will imply the time-homogeneous Markov property. Our arguments are based on the simple device of randomization between two optional times σ and τ to form a new random time $\rho = 1_A\sigma + 1_{A^c}\tau$. Here ρ is again optional provided that $A \in \mathcal{F}_{\sigma\wedge\tau}$, so to guarantee a successful randomization we need to assume that $\mathcal{F}_0$ rich enough. This is not as restrictive as it may seem, since if $\mathcal{F}_\varepsilon$ satisfies the stated conditions for some $\varepsilon > 0$, the results will apply to the shifted process $\theta_\varepsilon X$ to yield the time-homogeneous Markov property at times $t > \varepsilon$. Though our results allow some extension to general optional times, we shall focus our attention on the simple Markov property, which requires a minimum of technicalities.

Throughout the paper, we consider a fixed process X defined on some probability space $(\Omega, \mathcal{A}, P)$ endowed with a right-continuous and complete filtration $\mathcal{F} = (\mathcal{F}_t)$. We shall assume that X is adapted to $\mathcal{F}$, which ensures X_τ to be $\mathcal{F}_\tau$-measurable for any simple optional time τ. For results involving more general optional times, stronger measurability conditions need to be imposed on X. A sub-σ-field $\mathcal{G} \subset \mathcal{A}$ is said to be *non-atomic*, if for any $A \in \mathcal{G}$ there exists some $B \in \mathcal{G}$ with $0 < P(A \cap B) < PA$. In that case there exists for every $\varepsilon > 0$ some finite partition $A_1, \ldots, A_n \in \mathcal{G}$ of Ω, such that $PA_k < \varepsilon$ for all k. Recall that $\nu_\tau = P \circ X_\tau^{-1}$ and $\mu_h(X_\tau, \cdot) = P[X_{\tau+h} \in \cdot | X_\tau]$ a.s. for appropriate optional times $\tau < \infty$.

Theorem 2.1. *Let X be an $\mathcal{F}$-adapted process in some Borel space S, such that $X_{\tau+h} \perp\!\!\!\perp_{X_\tau} \mathcal{F}_\tau$ for all simple optional times $\tau < \infty$ and constants $h > 0$. Then X is homogeneous $\mathcal{F}$-Markov under each of these conditions:*

(i) *S is countable and $\mathcal{F}_0$ is non-atomic;*

(ii) *there exist some $A \in \mathcal{F}_0$ with $0 < P[A|X] < 1$ a.s. and some σ-finite measure ν on S with $\nu_t \ll \nu$ for all $t \geq 0$.*

In particular it is enough to choose the set $A \in \mathcal{F}_0$ in (ii) such that $A \perp\!\!\!\perp X$ and $0 < PA < 1$.

To prepare for the proof of Theorem 2.1, we shall first consider a general result for families of σ-finite measures. Given any measures $\nu, \nu_1, \nu_2, \ldots$ on some measurable space $(S, \mathcal{S})$, we define the relation $\nu \ll (\nu_n)$ by the requirement that $\nu A = 0$ whenever $A \in \mathcal{S}$ with $\nu_n A = 0$ for all n.

Lemma 2.2. *For any σ-finite measures ν_t, $t \in T$, on some measurable space S, these conditions are equivalent:*

(i) *there exists some σ-finite measure ν on S with $\nu_t \ll \nu$ for all $t \in T$;*

(ii) *there exist some $t_1, t_2, \ldots \in T$ with $\nu_t \ll (\nu_{t_n})$ for all $t \in T$.*

Proof. (ii) ⇒ (i): Since the ν_t are σ-finite, we may choose some measurable functions $g_t : S \to (0, \infty)$ with $\nu_t g_t \le 1$. Assuming $\nu_t \ll (\nu_{t_n})$ for all t, we may define $\nu = \sum_n 2^{-n}(g_{t_n} \cdot \nu_{t_n})$, and we note that ν is bounded with $\nu_t \ll \nu$ for all t.

(i) ⇒ (ii): We may clearly assume $\nu S = 1$, and further that $\nu_t \le \nu$ for all t. Introduce some densities $f_t = d\nu_t/d\nu$, and define $c = \sup_{(t_n)} \nu \sup_n f_{t_n}$, where the outer supremum extends over all sequences $t_1, t_2, \ldots \in T$. Combining sequences (t_{mn}) with $\nu \sup_n f_{t_{mn}} \to c$ as $m \to \infty$, we note that the outer supremum is attained, so $\nu \sup_n f_{t_n} = c$ for some $t_1, t_2, \ldots \in T$.

To see that (t_n) has the stated property, let $A \in \mathcal{S}$ with $\nu_{t_n} A = 0$ for all n, and fix any $t \in T$. Then

$$\nu_t A + c = \nu(f_t 1_A) + \nu \sup_n f_{t_n} \le \nu(f_t \vee \sup_n f_{t_n}) \le c,$$

so $\nu_t A = 0$. Thus $\nu_t \ll (\nu_{t_n})$. ■

We shall now consider families of measurable functions f_t with associated measures ν_t, such that $f_s = f_t$ a.e. $\nu_s \wedge \nu_t$, in the sense described in the introduction. Under suitable conditions, we may use the last lemma to construct a common version f, such that $f_t = f$ a.e. ν_t for all t. The result will be useful both here and in the next section.

Lemma 2.3. *Fix a set T and two measurable spaces $(S, \mathcal{S})$ and $(S', \mathcal{S}')$, and consider some σ-finite measures ν and $\nu_t \ll \nu$, $t \in T$, and measurable functions $f_t \colon S \to S'$, $t \in T$. Then these conditions are equivalent:*

(i) *$f_s = f_t$ a.e. $\nu_s \wedge \nu_t$ for all $s, t \in T$;*

(ii) *there exists some measurable function $f \colon S \to S'$ with $f_t = f$ a.e. ν_t for all t.*

Proof. The implication (ii) ⇒ (i) is obvious, so it is enough to prove the converse. Thus assume (i).

First let $T = \mathbb{N}$. We shall first prove the existence of some measurable functions g_n on S, $n \in \mathbb{N}$, such that $f_k = g_n$ a.e. ν_k for all $k \le n$. For $n = 1$ we may clearly take $g_1 = f_1$. Proceeding by induction, assume for some n that g_n has the stated property. To extend the construction to $n + 1$, note that $\nu_{n+1} \perp \nu_1, \ldots, \nu_n$ on the set $D = \{g_n \ne f_{n+1}\}$. We may then choose some set $A \in \mathcal{S} \cap D$ with $(\nu_1 + \cdots + \nu_n)A = \nu_{n+1}(A^c \cap D) = 0$, and we note that $g_{n+1} = f_{n+1} 1_A + g_n 1_{A^c}$ has the desired property. This completes the induction. Now define $f = g_n$ whenever $g_n = g_{n+1} = \cdots$ for some n, and otherwise put $f = a$ for some fixed a. Then f is clearly measurable with $f_n = g_n = g_{n+1} = \cdots = f$ a.e. ν_n for each n.

In the general case, Lemma 2.2 shows that we may choose $t_1, t_2, \ldots \in T$ with $\nu_t \ll (\nu_{t_n})$ for all t. By the result in the countable case there exists some measurable function f on S, such that $f_{t_n} = f$ a.e. ν_{t_n} for all n. For any $t \in T$ we further have $f_t = f_{t_n}$ a.e. $\nu_t \wedge \nu_{t_n}$, so by combination $f_t = f$ a.e. $\nu_t \wedge \nu_{t_n}$. Since n is arbitrary, the stated absolute continuity yields $f_t = f$ a.e. ν_t, as desired. ■

As a first step towards a proof of the homogeneity condition, we record the following elementary observation.

Lemma 2.4. *Let X be $\mathcal{F}$-adapted with values in a Borel space S, and fix two optional times $\sigma, \tau < \infty$ and a constant $h > 0$ such that $X_{\sigma+h} \perp\!\!\!\perp_{X_\sigma} \mathcal{F}_\sigma$ and $X_{\tau+h} \perp\!\!\!\perp_{X_\tau} \mathcal{F}_\tau$. Then*

$$\mu_h^\sigma(X_\sigma, \cdot) = \mu_h^\tau(X_\tau, \cdot) \quad a.s. \text{ on } \{\sigma = \tau\}. \tag{8}$$

Proof. By Lemma 6.1 in Kallenberg (1997) we have $\{\sigma = \tau\} \in \mathcal{F}_\sigma \cap \mathcal{F}_\tau$, and furthermore $X_{\sigma+h} = X_{\tau+h}$ on $\{\sigma = \tau\}$. Using Lemma 5.2 in the same reference together with the postulated conditional independencies, we obtain for any $B \in \mathcal{S}$, a.s. on $\{\sigma = \tau\}$,

$$\mu_h^\sigma(X_\sigma, B) = P[X_{\sigma+h} \in B | \mathcal{F}_\sigma] = P[X_{\tau+h} \in B | \mathcal{F}_\tau] = \mu_h^\tau(X_\tau, B).$$

Since S is Borel, the exceptional null set can be chosen to be independent of B. ■

The randomization required for part (i) of Theorem 2.1 will be based on the following result.

Lemma 2.5. *Let $\mathcal{F}_0$ be non-atomic. Then for any [simple] optional times $\sigma, \tau < \infty$ and state $k \in S$ with*

$$P\{X_\sigma = k\} \wedge P\{X_\tau = k\} > 0, \tag{9}$$

there exists some [simple] optional time $\rho < \infty$ with

$$P\{X_\sigma = k,\ \sigma = \rho\} \wedge P\{X_\tau = k,\ \tau = \rho\} > 0.$$

Proof. Since $\mathcal{F}_0$ is non-atomic and

$$E\, P[X_\sigma = k | \mathcal{F}_0] = P\{X_\sigma = k\} > 0,$$

we may choose some set $A \in \mathcal{F}_0$ with

$$A \subset \{P[X_\sigma = k | \mathcal{F}_0] > 0\}, \qquad 0 < PA < P\{X_\tau = k\}.$$

Then $\rho = \sigma 1_A + \tau 1_{A^c}$ is optional since $A \in \mathcal{F}_0$, and we note that

$$\begin{aligned} P\{X_\sigma = k,\ \sigma = \rho\} &\geq P[X_\sigma = k; A] = E[P[X_\sigma = k|\mathcal{F}_0]; A] > 0, \\ P\{X_\tau = k,\ \tau = \rho\} &\geq P[X_\tau = k; A^c] \geq P\{X_\tau = k\} - PA > 0. \end{aligned}$$ ■

Combining the last two lemmas, we obtain the appropriate pairwise homogeneity.

Lemma 2.6. *Assume the hypotheses of Theorem* 2.1 (i), *and let* $\sigma, \tau < \infty$ *be simple optional times satisfying* (9) *for some* $k \in S$. *Then* $\mu_h^\sigma(k, \cdot) = \mu_h^\tau(k, \cdot)$.

Proof. Choosing ρ as in Lemma 2.5, we get by Lemma 2.4

$$\mu_h^\sigma(k, \cdot) = \mu_h^\rho(k, \cdot) = \mu_h^\tau(k, \cdot).$$ ■

An elementary further lemma will be needed for the proof of Theorem 2.1. (ii).

Lemma 2.7. *If* $P[A|\mathcal{F}] > 0$ *a.s. and* $B \in \mathcal{F}$ *with* $P(A \setminus B) = 0$, *then* $PB = 1$.

Proof. If $PB < 1$, we get the contradiction

$$0 = P(A \setminus B) = E[P[A|\mathcal{F}]; B^c] > 0.$$ ■

It is now easy to complete the proof of the main result.

Proof of Theorem 2.1. (i) Note that $\nu_t \ll \nu$ for all t, where ν is the counting measure on S. Fix any $h > 0$, and conclude from Lemma 2.6. that $\mu_h^s = \mu_h^t$ a.e. $\nu_s \wedge \nu_t$ for any $s, t \geq 0$. By Lemma 2.3. there exists some probability kernel μ_h on S with $\mu_h^t = \mu_h$ a.e. ν_t for all t, and we obtain

$$P[X_{t+h} \in \cdot|\mathcal{F}_t] = P[X_{t+h} \in \cdot|X_t] = \mu_h(X_t, \cdot) \text{ a.s.}$$

(ii) For any times s and t we note that the time $\tau = 1_A s + 1_{A^c} t$ is optional. By the conditional independence and Lemma 2.4. we obtain

$$\begin{aligned} \mu_h^s(X_s, \cdot) &= \mu_h^\tau(X_s, \cdot) \text{ a.s. on } A, \\ \mu_h^t(X_t, \cdot) &= \mu_h^\tau(X_t, \cdot) \text{ a.s. on } A^c. \end{aligned}$$

Since $0 < P[A|X] < 1$ a.s., Lemma 2.7. shows that the two relations remain true a.s. on Ω, so $\mu_h^s = \mu_h^t$ a.e. $\nu_s \wedge \nu_t$, where $\nu_t = P \circ X_t^{-1}$. Since $\nu_t \ll \nu$, an appeal to Lemma 2.3. completes the proof. ∎

3. Independence from homogeneity

In this section we shall consider the converse problem of deducing the conditional independence $X_{\tau+h} \perp\!\!\!\perp_{X_\tau} \mathcal{F}_\tau$ from suitable homogeneity conditions. This requires a preliminary zero–one law for absorption, which is stated as a separate theorem because of its possible independent interest. Though the present results are stronger than the corresponding statements in Kallenberg (1987), some of the earlier ideas will be useful for their proofs.

For any space S and number $h > 0$, write I_h for the set of paths $w : \mathbb{R}_+ \to S$ with $w_0 = w_h = w_{2h} = \cdots$, and note that $I = \bigcap_h I_h$ consists of all constant paths. Given a process X in S, we define as before $\nu_\tau = P \circ X_\tau^{-1}$ for any optional time τ. Motivated by (5), we may further introduce some measurable functions a^τ and a_h^τ, such that a.s.

$$a^\tau(X_\tau) = P[\theta_\tau X \in I | X_\tau], \qquad a_h^\tau(X_\tau) = P[\theta_\tau X \in I_h | X_\tau], \tag{10}$$

and form the associated sets

$$A^\tau = \{s \in S;\ a^\tau(s) = 1\}, \qquad A_h^\tau = \{s \in S;\ a_h^\tau(s) = 1\}.$$

Theorem 3.1. *Consider a process X in some Borel space S.*

(i) *If X is optional and such that $a^\sigma = a^\tau$ a.e. $\nu_\sigma \wedge \nu_\tau$ for any optional times $\sigma, \tau < \infty$, then for any such time τ*

$$a^\tau(X_\tau) = 1_{A^\tau}(X_\tau) = 1_I(\theta_\tau X) \quad a.s. \tag{11}$$

(ii) *If X is progressive [adapted] and such that $a_h^\sigma = a_h^\tau$ a.e. $\nu_\sigma \wedge \nu_\tau$ for any [any simple] optional times $\sigma, \tau < \infty$, then for any such time τ*

$$a_h^\tau(X_\tau) = 1_{A_h^\tau}(X_\tau) = 1_{I_h}(\theta_\tau X) \quad a.s. \tag{12}$$

Here part (i) is a partial improvement of Theorem 1 in Kallenberg (1987) (cf. Kallenberg (1997), Lemma 7.24), where X was assumed to be right-continuous with left-hand limits and to take values in a separable metric space, and where the full homogeneity $P[\theta_\tau X \in \cdot|X_\tau] = P_{X_\tau}$ was postulated for arbitrary optional times $\tau < \infty$. Part (ii) is new.

Proof of Theorem 3.1. (i) Since S is Borel, we may clearly take $S = \mathbb{R}$. First assume that $a^\tau \equiv a$ for some measurable function a. Fix any $n \in \mathbb{N}$, and introduce the optional set

$$D = \{(\omega, t);\ t \geq \tau,\ [nX_t] \neq [nX_\tau]\}.$$

By the optional section theorem (cf. Dellacherie & Meyer (1975), IV.84) applied to the sets $D \cap (\Omega \times [0, m])$, there exists for every $\varepsilon > 0$ some optional time $\sigma \in [\tau, \infty)$ with

$$P(\pi D \setminus \{[nX_\sigma] \neq [nX_\tau]\}) \leq \varepsilon, \tag{13}$$

where πD denotes the projection of D onto Ω. Next introduce the optional times

$$\tau_k = \tau + (\sigma - \tau) 1\{[nX_\tau] = k\}, \quad k \in \mathbb{Z}, \tag{14}$$

and note that

$$\{[nX_{\tau_k}] = k\} \subset \{[nX] = k \text{ on } [\tau, \infty)\} \cup (\pi D \setminus \{[nX_\sigma] \neq [nX_\tau]\}). \tag{15}$$

Writing $b = 1 - a$, we get as $\varepsilon \to 0$ and $n \to \infty$

$$\begin{aligned} E[b(X_\tau);\ \theta_\tau X \in I] &= \textstyle\sum_k E[b(X_\tau);\ \theta_\tau X \in I,\ [nX_\tau] = k] \\ &= \textstyle\sum_k E[b(X_{\tau_k});\ [nX_{\tau_k}] = k] \\ &= \textstyle\sum_k P\{\theta_{\tau_k} X \notin I,\ [nX_{\tau_k}] = k\} \\ &\leq \textstyle\sum_k P\{\theta_\tau X \notin I,\ [nX] = k \text{ on } [\tau, \infty)\} + \varepsilon \\ &\leq P\left\{\theta_\tau X \notin I,\ \sup_{t \geq \tau} |X_t - X_\tau| \leq n^{-1}\right\} + \varepsilon \\ &\to P\{\theta_\tau X \notin I,\ \theta_\tau X \in I\} = 0, \end{aligned}$$

where step two holds since $X_{\tau_k} = X_\tau$ on $\{\theta_\tau X \in I\}$, step three holds by (10), and step four holds by (13) and (15). Hence $a(X_\tau) = 1$ a.s. on $\{\theta_\tau X \in I\}$. Since also $Ea(X_\tau) = P\{\theta_\tau X \in I\}$ by (10), we obtain $a(X_\tau) = 1_I(\theta_\tau X)$ a.s., and (11) follows by the definition of A.

Now turn to the case of general functions a^τ. Fixing any τ, we note that the set of times τ_k for arbitrary $n, \varepsilon^{-1} \in \mathbb{N}$ and $k \in \mathbb{Z}$ is countable. Hence by Lemma 2.3. there exists some measurable function a, such that $a^\tau = a$ a.s. ν_τ and $a^{\tau_k} = a$ a.s. ν_{τ_k} for all n, ε, and k as above. Using that version we may proceed as before. By the choice of a it is clear that (11) remains true for the original function a^τ with associated set A^τ.

(ii) The proof in this case is similar to that of (i), so we shall only indicate the major changes. For any $m, n \in \mathbb{N}$ we may now define

$$\sigma = \inf\{t = \tau + rh,\ r \leq m;\ [nX_t] \neq [nX_\tau]\},$$

where r is restricted to $\mathbb{Z}_+$. For τ_k as in (14), we get in place of (15)

$$\{[nX_{\tau_k}] = k\} \subset \{[nX_{\tau+rh}] = k,\, r \le m\}.$$

Assuming first that $a_h^\tau = a_h$ for all τ and writing $b_h = 1 - a_h$, we get as before

$$E[b_h(X_\tau);\, \theta_\tau X \in I_h] \le P\left\{\theta_\tau X \notin I_h,\, \sup_r |X_{\tau+rh} - X_\tau| \le n^{-1}\right\},$$

which tends to zero as $n \to \infty$. Hence $a_h(X_\tau) = 1$ a.s. on $\{\theta_\tau X \in I_h\}$, and (12) follows by the same argument as for (i). Since only countably many optional times are involved, the result extends as before to the case of general functions a_h^τ. ■

We turn to the main result of this section, where the pairwise homogeneity from (7) together with the corresponding condition for the functions a_h in (10) will be shown to imply the conditional independence required for the strong Markov property. By combining the present statement with Theorem 2.1., we obtain conditions for the two relations in decompositions (2) and (4) to be equivalent.

Recall that $\nu_\tau = P \circ X_\tau^{-1}$, and let μ_h^τ and a_h^τ be given by (6) and (10).

Theorem 3.2. *Let X be a progressive [adapted] process in some Borel space S, and let $h > 0$ be such that $a_h^\sigma = a_h^\tau$ and $\mu_h^\sigma = \mu_h^\tau$ a.e. $\nu_\sigma \wedge \nu_\tau$ for any [any simple] optional times $\sigma, \tau < \infty$. Then $X_{\tau+h} \perp\!\!\!\perp_{X_\tau} \mathcal{F}_\tau$ for every such time τ.*

This is a partial improvement of Theorem 2 in Kallenberg (1987) (cf. Kallenberg (1997), Theorem 7.23), where X was again assumed to be right-continuous with left-hand limits and to take values in a separable metric space, and where homogeneity was required in the complete functional form $P[\theta_\tau X \in \cdot | X_\tau] = P_{X_\tau}$, for arbitrary optional times $\tau < \infty$.

Proof of Theorem 3.2. As before we may take $S = \mathbb{R}$. First assume that $a_h^\tau = a_h$ and $\mu_h^\tau = \mu_h$ for all τ. For any $B \in \mathcal{B}(\mathbb{R})$ we get by Theorem 3.1. (ii)

$$P[X_{\tau+h} \in B; F] = P[X_\tau \in B; F], \quad F \in \mathcal{F}_\tau \cap \{X_\tau \in A_h\},$$

so $P[X_{\tau+h} \in B | \mathcal{F}_\tau] = 1_B(X_\tau)$ a.s. on $\{X_\tau \in A_h\}$. Hence on the same set

$$P[X_{\tau+h} \in B | \mathcal{F}_\tau] = P[X_{\tau+h} \in B | X_\tau] = \mu_h(X_\tau, B) \text{ a.s.}$$

Now fix any $F \in \mathcal{F}_\tau \cap \{X_\tau \notin A_h\}$. Letting $m, n \in \mathbb{N}$ and writing

$$\begin{aligned} \sigma &= \inf\{t = \tau + rh,\, r \le m;\, [nX_t] \ne [nX_\tau]\}, \\ \tau_k &= \tau + (\sigma - \tau) 1_{F^c} 1\{[nX_\tau] = k\}, \quad k \in \mathbb{Z}, \end{aligned}$$

we note that

$$\{[nX_{\tau_k}] = k\} \setminus F \subset \{[nX_{\tau+rh}] = k,\, r \le m\}. \tag{16}$$

For any $B \in \mathcal{B}(\mathbb{R})$ we get as $m \to \infty$ and then $n \to \infty$

$$\begin{aligned}
&|E[1_B(X_{\tau+h}) - \mu_h(X_\tau, B); F]| \\
&\quad = \left|\textstyle\sum_k E[1_B(X_{\tau+h}) - \mu_h(X_\tau, B);\ [nX_\tau] = k, F]\right| \\
&\quad = \left|\textstyle\sum_k E[1_B(X_{\tau_k+h}) - \mu_h(X_{\tau_k}, B);\ [nX_{\tau_k}] = k,\ X_{\tau_k} \notin A_h, F]\right| \\
&\quad = \left|\textstyle\sum_k E[1_B(X_{\tau_k+h}) - \mu_h(X_{\tau_k}, B);\ [nX_{\tau_k}] = k,\ X_{\tau_k} \notin A_h, F^c]\right| \\
&\quad \le \textstyle\sum_k P(\{[nX_{\tau_k}] = k,\ X_{\tau_k} \notin A_h\} \setminus F) \\
&\quad \le \textstyle\sum_k P\{[nX_{\tau+rh}] = k,\ r \le m;\ \theta_\tau X \notin I_h\} \\
&\quad \le P\left\{\sup_{r \le m}|X_{\tau+rh} - X_\tau| \le n^{-1},\ \theta_\tau X \notin I_h\right\} \\
&\quad \to P\{\theta_\tau X \in I_h,\ \theta_\tau X \notin I_h\} = 0,
\end{aligned}$$

where step two holds since $\tau_k = \tau$ on F, step three holds by (6), step four holds since $0 \le \mu_h \le 1$, and step five holds by (16) together with Theorem 3.1. (ii) and the definition of I_h. Thus

$$P[X_{\tau+h} \in B|\mathcal{F}_\tau] = \mu_h(X_\tau, B) \ \text{ a.s. on } \{X_\tau \in A_h\}.$$

For general functions a_h^τ and kernels μ_h^τ, we note that $\nu_{\tau_k} \ll (\nu_{\tau+rh};\, r \in \mathbb{Z}_+)$. Hence by Lemma 2.3 there exist for any τ and h some measurable function a_h and probability kernel μ_h, such that $a_h^\tau = a_h$ and $\mu_h^\tau = \mu_h$ a.s. ν_τ, as well as $a_h^{\tau_k} = a_h$ and $\mu_h^{\tau_k} = \mu_h$ a.s. ν_{τ_k} for all $m, n \in \mathbb{N}$, $k \in \mathbb{Z}$, and $F \in \mathcal{F}_\tau$. Applying the previous argument to those versions yields $P[X_{\tau+h} \in \cdot|\mathcal{F}_\tau] = \mu_h(X_\tau, \cdot)$ a.s., which implies $X_{\tau+h} \perp\!\!\!\perp_{X_\tau} \mathcal{F}_\tau$. ■

References

Blumenthal, R.M., Getoor, R.K. *Markov Processes and Potential Theory.* Academic Press, New York (1968).

Dellacherie, C., Meyer, P.A. *Probabilités et Potentiel, Chap. I–IV.* Hermann, Paris (1975).

Kallenberg, O. Characterizations and embedding properties in exchangeability. *Z. Wahrscheinlichkeitstheorie verw. Gebiete* **60**, 249–281 (1982).

Kallenberg, O. Homogeneity and the strong Markov property. *Ann. Probab.* **15**, 213–240 (1987).

Kallenberg, O. *Foundations of Modern Probability.* Springer, New York (1997).

Walsh, J.B. Transition functions of Markov processes. In: *Séminaire de Probabilités VI, Lecture Notes in Mathematics* **258**, 215–232. Springer, Berlin (1972).

Department of Mathematics
218 Parker Hall
Auburn University
Auburn, AL 36849-5310, USA
olavk@mail.auburn.edu

The Russian Options*

G. Kallianpur

1. Introduction

In the European and American options of option pricing theory, the time period between the time the option is purchased and the time at or before which the option has to be exercised is fixed and known. If the purchase time is taken to be $t = 0$ and the exercise time $t = T$, then the European option pricing theory requires the option to be exercised at $t = T$ (the date of maturity) while under the American option, you can exercise it at any time up to T and moreover, the exercise time can be random.

A third approach to option pricing has recently been proposed by L.A. Shepp and A.N. Shiryaev and called, the Russian option. In fact, they study the put option and, in a later work, the call option from this point of view [1],[2]. In both options, the period before the option is exercised can be indefinitely long and cannot be predicted in advance. Also, the option can be exercised at a random time. Another feature of both the put and call options is that an explicit expression for the fair price of the option is obtained. By "fair price" is meant the optimal expected present value based on the option. The optimal strategy is obtained and shown to be unique.

As in the European and American options, the Russian option assumes that the asset fluctuation follows the geometric Brownian motion model.

The Russian options differ from the European and American options in that they are basically optimal stopping problems.

2. The Russian Put Option

In this paper we generally follow Shepp and Shiryaev's work and complement it by showing that their solution is unique. We do this by proving that the corresponding free boundary problem (FBP) has a unique solution. In [1] and [2], the FBP is indeed alluded to but is not exploited to obtain uniqueness. As the authors themselves say in [1], the solution is arrived at by "guessing."

We first consider the so-called put option. Here you, the buyer of the option can exercise your option at any time; in other words, the *exercise* time is up to you, can be random and the time period between buying the option and exercising it can be indefinitely long. We assume the Black and Scholes

*Research supported by the Army Research Office Grant No. DAAH 04 95 10042.

GBM model for the stock price:

$$dX_t = \mu X_t dt + \sigma X_t dW_t, \qquad t > 0, X_0 = x \tag{2.1}$$

or

$$X_t = xe^{(\mu - \frac{1}{2}\sigma^2)t + \sigma W_t} \tag{2.2}$$

where $\mu \in \mathbf{R}$ and $\sigma > 0$ are known constants. Let

$$S_t = \max\left\{s, \sup_{0 \le u \le t} X_u\right\}, t \ge 0, s \ge x. \tag{2.3}$$

If you stop at τ (τ being a finite stopping time) you receive the payoff S_τ discounted by $e^{-r\tau}$, i.e., $e^{-r\tau}S_\tau$, where $r > 0$ is the discount rate. You, as the owner of the option, want to seek a strategy that will maximize $E_{x,s}e^{-r\tau}S_\tau$. This quantity, maximized over all finite stopping times, may be regarded as a "fair" price for buying the option. Thus we have to find

$$V^*(x, s) := \sup_\tau E_{x,s}(e^{-r\tau}S_\tau)\ . \tag{2.4}$$

We shall assume that $r > \mu$ (for otherwise as will be seen below, $V^*(x, s)$ will be infinite).

3. The Free Boundary Problem for the Put Option

At this stage, we introduce the following free boundary problem associated with the Russian option.

$$\frac{1}{2}\sigma^2 x^2 \frac{\partial^2 V}{\partial x^2} + \mu x \frac{\partial V}{\partial x} - rV = 0 \tag{3.1}$$

if $g(s) < x \le s$;
(Conditions on the free boundary) g is continuous and differentiable $g(s) > 0$ if $s > 0$, and $g(0) = 0$; (3.2)

(i) $V(g(s), s) = s$,

(ii) $\frac{\partial V}{\partial x}(g(s), s) = 0$,

(Conditions on the known boundary) (3.3)
(i) $V(x, s) \ge s$; $\frac{\partial V}{\partial s}|_{x=s} = s$.
(ii) $\frac{\partial V}{\partial s}|_{x=s} = 0$.
Conditions (3.2) (i) and (ii) are conditions of smooth fit since A. N. Kolmogorov in the 1950's.

One way to arrive at the differential equation (3.1) is to follow the principle due, presumably, to Mikhailov [1]. The generator of the Markov process (t, X_t) is given by

$$L_t = \frac{\partial}{\partial t} + \mu x \frac{\partial}{\partial x} + \frac{1}{2}\sigma^2 x^2 \frac{\partial^2}{\partial x^2}.$$

$V(x,s)$ is then chosen to satisfy $L_t\{e^{-rt}V\} = 0$ in the continuation region $g(s) \leq x \leq s$. This immediately yields (3.1). Another way to derive (3.1) is given in [1]. To solve the FBP (3.1)–(3.3), we have to find V and the free boundary $g(s)$. Our aim is to show that there is a unique solution (V, g) and then use it to obtain the optimal stopping strategy and the fair price.

Shepp and Shiryaev [1],[2] guess at a solution to the above free boundary problem (FBP) and obtain an expression for $V^*(x,s)$ for $0 < x \leq s$. We will show that their solution is, in fact, the *unique* solution of the single phase Stefan-like problem (3.1)–(3.3).

Observe that (3.1) is really an ordinary differential equation. The so-called indicial equation (for $V = x^m$) is given by

$$\frac{1}{2}\sigma^2 m^2 + (\mu - \frac{1}{2}\sigma^2)m - r = 0.$$

The roots of this quadratic, denoted by γ_1 and γ_2, $\gamma_1 < 0 < 1 < \gamma_2$ are given by

$$\frac{(\frac{1}{2}\sigma^2 - \mu) \pm \sqrt{(\mu - \frac{1}{2}\sigma^2)^2 + 2\sigma^2 r}}{\sigma^2}.$$

The general solution

$$V(x,s) = A(s)x^{\gamma_1} + B(s)x^{\gamma_2}.$$

From condition (3.2) we have

$$\begin{aligned} Ag^{\gamma_1} + Bg^{\gamma_2} &= s, \\ A\gamma_1 g^{\gamma_1} + B\gamma_2 g^{\gamma_2} &= 0. \end{aligned}$$

Hence

$$\begin{aligned} A &= \frac{s\gamma_2}{(\gamma_2 - \gamma_1)g^{\gamma_1}}, \\ B &= -\frac{s\gamma_1}{(\gamma_2 - \gamma_1)g^{\gamma_2}}, \end{aligned}$$

and we obtain

$$V(x,s) = \frac{s}{\gamma_2 - \gamma_1}\left\{\gamma_2\left(\frac{x}{g(s)}\right)^{\gamma_1} - \gamma_1\left(\frac{x}{g(s)}\right)^{\gamma_2}\right\} \qquad (3.4)$$

if $g(s) < x \leq s$. For purposes of later comparison we write $V(x, s; g)$ for $V(x, s)$. We need a number of lemmas whose proofs will be given in Section 4.

Lemma 1. *If $g(S_t) < X_t \leq S_t$, then*

$$V(X_t, S_t; g) \leq V(S_t, S_t, g). \tag{3.5}$$

Lemma 2. *Assume that s is in a bounded closed interval $[0, \bar{s}]$ and $\bar{s}$ is arbitrary. Then*

$$g(s) \geq \frac{s}{H}, \tag{3.6}$$

where H is a constant greater than 1.

Lemma 3. $V(S_t, S_t; g) \leq A.s$

$$\text{where} \quad A = \frac{\gamma_2 H^{\gamma_1} - \gamma_1 H^{\gamma_2}}{\gamma_2 - \gamma_1} > 0.$$

Lemma 4. *If $X_t < S_t$ then $dS_t = 0$*

Apply Itó's formula to the process

$$Y_t := e^{-rt} V(X_t, S_t; g). \tag{3.7}$$

$$dY_t = e^{-rt}\left\{-rV(X_t, S_t)dt + \frac{\partial V}{\partial x}(X_t, S_t)dX_t \right. \tag{3.8}$$
$$\left. + \frac{1}{2}\frac{\partial^2 V}{\partial x^2}(X_t, S_t)d < X >_t + \frac{\partial V}{\partial s}(X_t, S_t)dS_t \right\}.$$

Then, recalling that we always work in the region $g(S_t) < X_t < S_t$, the last term in the curly brackets is zero since if $X_t = S_t$, $\frac{\partial V}{\partial s}(S_t, S_t, g) = 0$ by (3.3) (ii) and by Lemma 4. Since $dS_t = 0$ a.s. on $\{X_t < S_t\}$ by Lemma 4. $d < X >_t = \sigma^2 X_t^2 dt$, (3.8) becomes

$$dY_t = e^{-rt}\left\{-rV + \mu X_t \frac{\partial V}{\partial x} + \frac{1}{2}\sigma^2 X_t^2 \frac{\partial^2 V}{\partial x^2}\right\} dt + e^{-rt}\sigma X_t \frac{\partial V}{\partial x} dW_t. \tag{3.9}$$
$$= \sigma e^{-rt}\frac{\partial V}{\partial x} X_t dW_t,$$

and

$$Y_{t\wedge\tau} = Y_0 + \int_0^{t\wedge\tau} \sigma e^{-ru}\frac{\partial V}{\partial x} X_u dW_u$$

where τ is any finite stopping time. Letting $\sigma_n \uparrow \infty$ be a sequence of stopping times such that

$$\int_0^{t\wedge\tau\wedge\sigma_n} \sigma e^{-ru} \frac{\partial V}{\partial x} X_u dW_u$$

is a martingale, we see that $Y_{t\wedge\tau\wedge\sigma_n}$ is a positive martingale and

$$E_{x,s} Y_{t\wedge\tau\wedge\sigma_n} = E_{x,s} Y_0.$$

Making $\sigma_n \uparrow \infty$ and then $t \to \infty$ and applying Fatou's Lemma we have

$$E_{x,s} Y_\tau \leq E_{x,s} Y_0.$$

From the first condition in (3.3) and the above inequality we obtain

$$E_{x,s} e^{-r\tau} S_\tau \leq V(x, s), \tag{3.10}$$

and taking the sup over τ this yields the inequality

$$V^*(x, s) \leq V(x, s). \tag{3.11}$$

To show the opposite inequality we first define the stopping time

$$\tau^* = \inf \{t > 0; X_t = g(S_t)\} \tag{3.12}$$

as the time of the first exit from the continuation set. Our aim is to show that

$$E_{x,s} Y_{t\wedge\tau^*} = V(x, s; g). \tag{3.13}$$

It should be noted that, although $Y_{t\wedge\tau}$ is a local martingale for any τ (hence for τ^*) it does not automatically follow that (3.13) holds. Here again we need the necessary lemmas.

Lemma 5. $P[\tau^* < \infty] = 1$

Lemma 6. *For* $a > 0, b > 0,$

$$P\{W_t \leq at + b, \quad 0 \leq t < \infty\} \geq 1 - e^{-2ab}. \tag{3.14}$$

From Lemmas 3.1 and 3.3, if $g(S_t) < X_t \leq S_t$,

$$Y_t = e^{-rt} V(X_t, S_t; g) \leq A e^{-rt} S_t$$

where A is the constant in Lemma 3.3.

Hence to show that $\{Y_t\}_{0<t<\tau^*}$ is uniformly integrable it suffices to show that

$$E_{x,s} \left(\sup_{0<t<\infty} Y_t \right) < \infty$$

which follows if

$$E_{x,s} \sup_{0<t<\infty} e^{-rt} S_t < \infty. \tag{3.15}$$

This will be shown in Lemma 3.7 below.

From $E_{x,s} \sup_{0<t<\infty} Y_t < \infty$ it follows that for $T > t$,

$$E_{x,s} Y_{t\wedge\tau^*} = E_{x,s} Y_{t\wedge\tau^*} 1_{[\tau^* \leq T]} + E_{x,s} Y_t 1_{[\tau^* > T]} < \infty.$$

Thus, the local martingale $Y_{t\wedge\tau^*}$ is actually a martingale so that

$$E_{x,s} Y_{t\wedge\tau^*} = E_{x,s} Y_0. \tag{3.16}$$

Since $Y_{t\wedge\tau^*} \to Y_{\tau^*}$ a.s. as $t \to \infty$, by the uniform integrability established above, (3.16) yields

$$E_{x,s} Y_{\tau^*} = E_{x,s} Y_0. \tag{3.17}$$

Now

$$\begin{aligned} Y_{\tau^*} &= e^{-r\tau^*} V(X_{\tau^*}, S_{\tau^*}) \\ &= e^{-r\tau^*} V(g(S_{\tau^*}), S_{\tau^*}) \\ &= e^{-r\tau^*} S_{\tau^*}, \end{aligned}$$

the last equality following from (3.2) (i). Hence from (3.17) we have

$$E_{x,s} e^{-r\tau^*} S_{\tau^*} = V(x, s; g). \tag{3.18}$$

(3.18) immediately gives

$$V^*(x, s) \geq V(x, s; g). \tag{3.19}$$

Combining with (3.11) we finally obtain the desired equality

$$V(x, s; g) = V^*(x, s). \tag{3.20}$$

(3.18) and (3.20) further show that τ^* is the optimal stopping time.

It will be seen from the proof of Lemma 3.2 that $g'(0) = \frac{1}{c}$ where $c = \left(\frac{1-\frac{1}{\gamma_1}}{1-\frac{1}{\gamma_2}}\right)^{\frac{1}{\gamma_2-\gamma_1}}$, $c > 1$ and that $c \leq H$ since $\frac{g(s)}{s} \geq \frac{1}{H}$ for all $s > 0$. Letting $g_c(s) = \frac{s}{c}$ it is easily verified that g_c satisfies conditions (3.2) (i) and (ii) required of the free boundary.

We may now repeat the above procedure by replacing g by g_c and obtain, exactly as above the relation

$$V^*(x, s) = V(x, s; g_c) \quad \text{for} \quad \frac{s}{c} < x \leq s. \tag{3.21}$$

Comparing (3.20) and (3.21) we have

$$V(x,s;g_c) = V(x,s;g) \quad \text{for} \quad \max\{\frac{s}{c}, g(s)\} < x \le s, \tag{3.22}$$

i.e.

$$\frac{s}{\gamma_2-\gamma_1}\left\{\gamma_2(\frac{cx}{s})^{\gamma_1} - \gamma_1(\frac{cx}{s})^{\gamma_2}\right\} = \frac{s}{\gamma_2-\gamma_1}\left\{\gamma_2(\frac{x}{g})^{\gamma_1} - \gamma_1(\frac{x}{g})^{\gamma_2}\right\}.$$

Setting $x = s$ we have

$$\gamma_2 c^{\gamma_1} - \gamma_1 c^{\gamma_2} = \gamma_1(\frac{s}{g})^{\gamma_1} - \gamma_1(\frac{s}{g})^{\gamma_2}. \tag{3.23}$$

Differentiating (3.23) with respect to s at $s > 0$ we get

$$\left\{(\frac{s}{g})^{\gamma_1-1} - (\frac{s}{g})^{\gamma_2-1}\right\}\frac{d}{ds}(\frac{s}{g}) = 0.$$

As $g(s) > s$, we cannot have $(\frac{s}{g(s)})^{\gamma_1} = (\frac{s}{g(s)})^{\gamma_2}$. Hence $\frac{d}{ds}(\frac{s}{g}) = 0$ or $\frac{g'(s)}{g(s)} = \frac{1}{s}$ which gives $g(s) = \frac{s}{k}$ for $s > 0$, k being a constant of integration.

Since $g(0) = 0$, we have $g(s) = \frac{s}{k}$ for all s. To find k, substituting in the expression for $g'(s)$,

$$g'(s) = \frac{\frac{1}{\gamma_1}(\frac{s}{g})^{\gamma_1} - \frac{1}{\gamma_2}(\frac{s}{g})^{\gamma_2}}{(\frac{s}{g})^{\gamma_1+1} - (\frac{s}{g})^{\gamma_2+1}},$$

we find that

$$\frac{1}{k} = (\frac{1}{\gamma_1}k^{\gamma_1} - \frac{1}{\gamma_2}k^{\gamma_2})/(k^{\gamma_1+1} - k^{\gamma_2+1})$$

which gives the value

$$k = \left(\frac{1-\frac{1}{\gamma_1}}{1-\frac{1}{\gamma_2}}\right)^{\frac{1}{\gamma_2-\gamma_1}}. \tag{3.24}$$

The constant k is the same as α, which is Shepp and Shiryaev's value (and also $= c$). We have thus proved that the free boundary problem (3.1)–(3.3) has the unique solution $(V(x,s), g(s))$ where $g(s) = \frac{s}{\alpha}$.

Remark . It is interesting to note that the unique solution of our free boundary problem is obtained without having to use $V^*(x,s)$ in the region $0 < x \le \frac{s}{\alpha}$.

We can complete the option problem as follows. From

$$V^*(x,s) = \sup_\tau E_{x,s}e^{-r\tau}S_\tau,$$

since $S_t \ge s$ for all t,

$$E_{x,s}e^{-r\tau}S_\tau \ge sE_{x,s}e^{-r\tau}.$$

Now choose the stopping time $\tau_\epsilon = \epsilon\tau$ where τ is a finite stopping time and $\epsilon > 0$. Then $V^*(x,s) \geq sE_{x,s}e^{-r\tau_\epsilon} = sE_{x,s}e^{-\epsilon r\tau}$ for every $\epsilon > 0$.

Making $\epsilon \to 0$, since $E_{x,s}e^{-\epsilon r\tau} \to 1$ we have

$$V^*(x,s) \geq s. \tag{3.25}$$

(Alternately, we get (3.25) by taking $\tau \equiv 0$ if the latter is allowed to be a stopping time for the problem.)

Next, writing

$$Z_t = e^{-rt}S_t \quad \text{in} \quad 0 < X_t \leq \frac{S_t}{\alpha},$$

by Itó's formula,

$$\begin{aligned} dZ_t &= -re^{-rt}S_t dt + e^{-rt}dS_t \\ &= -rZ_t dt \end{aligned}$$

since $dS_t = 0$ by Lemma 3.4.

$$Z_t = Z_0 - \int_0^t rZ_u du \leq Z_0 \quad \text{since} \quad Z_u \geq 0.$$

Therefore,

$$E_{x,s}Z_{t\wedge\tau} \leq E_{x,s}Z_0 = s$$

where τ is any finite stopping time. By Fatou's Lemma as $t \to \infty$,

$$E_{x,s}e^{-r\tau}S_\tau = E_{x,s}Z_\tau \leq s. \tag{3.26}$$

Taking sup over τ of the left hand side, we get

$$V^*(x,s) \leq s. \tag{3.27}$$

This proves $V^*(x,s) = s$ if $0 < x \leq s/\alpha$. ■

We have proved the following result:

Theorem 1. *The free boundary problem (3.1)–(3.3) has the unique solution* $(V(x,s), g(s))$ *with* $g(s) = \frac{s}{\alpha}$. *Furthermore,*

$$V(x,s) = V^*(x,s) \quad in \quad \frac{s}{\alpha} < x \leq s.$$

It has also been shown that

$$V^*(x,s) = s \quad in \quad 0 < x \leq \frac{s}{\alpha}.$$

Note that $V(x,s) \to s$ *on* $x \searrow \frac{s}{\alpha}$.

Remark . Shepp and Shiryaev claim that the Russian option has "reduced risk" in the sense that you don't have "to worry about missing a good price in the recent past" because you get the "best price up to the settlement time." The *reduced* risk formulation is needed for, if one considers not S_τ but the current value X_τ, the problem has no interest, as is shown by Shepp and Shiryaev in their second paper. Here is their result.

Let $V(x) = \sup_\tau E_x e^{-r\tau} X_\tau$.

Theorem 2. *(i) If $r \geq \mu$ then $V(x) = x$;*
(ii) If $r < \mu$ then $V(x) = \infty$.

Proof. Write $Z_t = e^{-rt} X_t$. Then

$$dZ_t = e^{-rt} X_t \{(\mu - r)dt + \sigma dW_t\}.$$

(i) Proceeding as in the proof of Theorem 3.1 we have $E_x dZ_t \leq 0$ and $E_x Z_\tau \leq E_x Z_0 = x$ for any τ, so that $V(x) \leq x$. Taking $\tau = 0$ or using the stopping time τ_ϵ as in (3.25) we get $V(x) \geq x$.
(ii) Now $r < \mu$. Taking the fixed stopping time $\tau \equiv t$ we have

$$V(x) \geq E_x e^{-rt} X_t = e^{(\mu - r)t} \to \infty$$

as $t \to \infty$ giving $V(x) = \infty$. ■

4. Proofs of the Lemmas

Lemma 1. *Differentiating $V(x, s; g)$ with respect to x in $g(s) < x \leq s$ we get*

$$\frac{\partial V}{\partial x} = \frac{s\gamma_1\gamma_2}{\gamma_2 - \gamma_1} x^{-1} \left\{ (\frac{x}{g})^{\gamma_1} - (\frac{x}{g})^{\gamma_2} \right\} > 0$$

since $\gamma_1\gamma_2 < 0$ and $(\frac{x}{g})^{\gamma_1} - (\frac{x}{g})^{\gamma_2} < 0$.

Lemma 2. *For $s > 0$ we have from condition (3.3) (ii)*

$$g'(s) = \frac{\frac{1}{\gamma_1}(\frac{s}{g})^{\gamma_1} - \frac{1}{\gamma_2}(\frac{s}{g})^{\gamma_2}}{(\frac{s}{g})^{\gamma_1+1} - (\frac{s}{g})^{\gamma_2+1}} \equiv h(s), \qquad \textit{say.}$$

$$\textit{so that } \frac{1}{h(s)} = \frac{(\frac{s}{g})^{\gamma_2+1} - (\frac{s}{g})^{\gamma_1+1}}{\frac{1}{\gamma_2}(\frac{s}{g})^{\gamma_2} - \frac{1}{\gamma_1}(\frac{s}{g})^{\gamma_1}}.$$

The denominator is positive for $s > 0$ (and of course, continuous). Since $g(0) = 0$ by assumption,

$$\lim_{s\downarrow 0} \frac{s}{g(s)} = \lim_{s\downarrow 0} \frac{1}{g'(s)} = \frac{1}{g'(0)} = c,$$

where c is a positive finite number. In fact, from the expression for $g'(s)$ it is easily seen that

$$c = \left(\frac{1 - \frac{1}{\gamma_1}}{1 - \frac{1}{\gamma_2}} \right)^{\frac{1}{\gamma_2 - \gamma_1}}.$$

Hence, restricting s to an arbitrary compact interval $[0, \overline{s}]$ (it will turn out that this is no restriction at all), we get $\frac{1}{h(s)} \leq H$ for all s in $[0, \overline{s}]$. We then get

$$g(s) = \int_0^s h(u)du \geq \frac{s}{H}$$

where $H > 1$ since $g(s) < s$. This proves the lemma.

Lemma 3. *Let $0 < y < s$. Then*

$$\begin{aligned} V(s, s; y) &= \frac{s}{\gamma_2 - \gamma_1} \left\{ \gamma_2 (\frac{s}{y})^{\gamma_1} - \gamma_1 (\frac{s}{y})^{\gamma_2} \right\}. \\ \frac{dV(s, s; y)}{dy} &= \frac{s}{\gamma_2 - \gamma_1} \left\{ -\gamma_2 \gamma_1 s^{\gamma_1} y^{-\gamma_1 - 1} + \gamma_2 \gamma_1 s^{\gamma_2} y^{-\gamma_2 - 1} \right\} \\ &= \frac{s}{\gamma_2 - \gamma_1} \frac{\gamma_1 \gamma_2}{y} \left\{ (\frac{s}{y})^{\gamma_2} - (\frac{s}{y})^{\gamma_1} \right\} < 0. \end{aligned}$$

since $\gamma_1\gamma_2 < 0$ and $(\frac{s}{y})^{\gamma_2} > (\frac{s}{y})^{\gamma_1}$. The latter holds because $(\frac{s}{y})^{\gamma_2 - \gamma_1} > 1$ since $0 < y < s$..

Now since from Lemma 2, $g(s) \geq \frac{s}{H}$ we have $V(s, s; g) \leq \frac{s}{\gamma_2 - \gamma_1} (\gamma_2 H^{\gamma_1} - \gamma_1 H^{\gamma_2})$.

Note that we cannot write the right-hand expression as $V(s, s; g_H)$ where $g_H = \frac{s}{H}$ because g_H has not been shown to satisfy condition (3.2) (i).

Lemma 4. From

$$\begin{aligned} S_t &= \max\{s, \sup_{0 \leq u \leq t} X_u\} \\ \text{we have for} \quad h > 0, S_{t+h} &= \max\{S_t, \sup_{t \leq u \leq t+h} X_u\}. \\ \text{For} \quad u \geq t \quad \text{since } X_u &= X_t e^{(\mu - 1/2\sigma^2)(u-t) + \sigma(W_u - W_t)}. \end{aligned}$$

From the lim sup part of Lévy's version of the law of the iterated logarithm we have

$$\limsup_{0<t_2-t_1\to 0} \frac{W(t_2)-W(t_1)}{\left(2(t_2-t_1)\log\frac{1}{t_2-t_1}\right)^{1/2}} = 1 \quad \text{a.s.}$$

Hence, for $\delta > 0$,

$$W_u(\omega) - W_t(\omega) \le (1+\delta)(2(u-t)\log\left(\frac{1}{u-t}\right)^{1/2}$$

for $0 < u - t \le h_0(\delta, \omega)$ a.s.

Define Ω_1 to be the set of ω's for which the above inequality holds. Then $P(\Omega_1) = 1$ and for $\omega \in \Omega_1 \cap \{X_t < S_t\}$ and $0 < h \le h_0$,

$$\begin{aligned} S_{t+h} &= \max\left\{S_t, X_t \sup_{t<u\le t+h} e^{(\mu-\frac{1}{2}\sigma^2)(u-t)+\sigma(W_u-W_t)}\right\} \\ &\le \max\left\{S_t, X_t \sup_{0\le u-t\le h} e^{(\mu-\frac{1}{2}\sigma^2)(u-t)+\sigma(1+\delta)(2(u-t)\log\frac{1}{u-t})^{1/2}}\right\}. \end{aligned}$$

By choosing δ and hence h_0 sufficiently small, we obtain for $0 < h \le h_0$ and noting that $X_t(\omega) < S_t(\omega)$,

$$\begin{aligned} S_{t+h} &\le S_t. \\ \text{But} \quad S_{t+h} &\ge S_t \quad \text{a.s. Hence} \\ S_{t+h} - S_t &= 0 \quad \text{a.s. on the set} \quad \Omega_1 \cap \{X_t < S_t\}. \\ \text{That is,} \quad dS_t &= 0 \quad \text{a.s. on} \quad \Omega_1 \cap \{X_t < S_t\}. \end{aligned}$$

(In some of the above steps, ω has been suppressed for convenience).

Lemma 5. For $T > 0$, by the definition of τ^*,

$$\begin{aligned} \{\tau^* > T\} &= \{X_t > g(S_t), \quad 0 \le t \le T\} \\ &\subset \{X_t > \frac{S_t}{H}, \quad 0 \le t \le T\}, \end{aligned}$$

the last inclusion, a consequence of Lemma 4.2. Hence

$$P_{x,s}(\tau^* > T) \le P_{x,s}(X_t > \frac{S_t}{H}, \quad 0 \le t \le T).$$

Now for $0 \le t \le T, X_t > \frac{S_t}{H}$ implies $\log X_t > \log S_t + \log\frac{1}{H}$, i.e.

$$(\mu - \frac{1}{2}\sigma^2)t + \sigma W_t > (\mu - \frac{1}{2}\sigma^2)u + \sigma W_u + \log\frac{1}{H}$$

for $0 \le u \le t \le T$. Hence

$$P_{x,s}(\tau^* > T) \le P_{x,s}\left\{\sigma(W_t - W_u) + (\mu - \frac{1}{2}\sigma^2)(t-u) \ge \log\frac{1}{H}, \quad 0 \le u \le t \le T\right\}.$$

The right-hand side is

$$\begin{aligned}
&\le P_{x,s}\left\{W_t - W_u \ge \frac{-\log H}{\sigma} - \frac{|\mu - \frac{1}{2}\sigma^2|}{\sigma}(t-u), 0 \le u \le t \le T\right\} \\
&= P_{x,s}\left\{\tilde{W}_t - \tilde{W}_u \le \frac{\log H}{\sigma} + \frac{|\mu - \frac{1}{2}\sigma^2|}{\sigma}(t-u), 0 \le u \le t \le T\right\} \\
&\le P_{x,s}\left\{\tilde{W}_d \le b, \tilde{W}_{2d} - \tilde{W}_d \le b, \ldots, \tilde{W}_{nd} - \tilde{W}_{(n-1)d} \le b\right\}.
\end{aligned}$$

Here $\tilde{W}$ is written for $-W, T = nd$, $d > 0$ being fixed and $b = \frac{\log H}{\sigma} + \frac{|\mu - \frac{1}{2}\sigma^2|}{\sigma}d$. From the above chain of inequalities we get

$$P_{x,s}(\tau^* > T) \le \left(\frac{1}{\sqrt{2\pi d}}\int_{-\infty}^{b} e^{-\frac{x^2}{2d}}dx\right)^n \to 0$$

since $n \to \infty$ as $T \to \infty$. Hence $\tau^* < \infty$ a.s.

Lemma 6. *If $a > 0, b > 0$, then*

$$P_{x,s}\{W_t \le at + b, 0 \le t < \infty\} \ge 1 - e^{-2ab}.$$

This is a well known inequality and the proof is given below.

Let $Z_t = e^{2aW_t - 2a^2 t}$. Then Z_t is a martingale with $EZ_t = 1$.

$$\begin{aligned}
&P_{x,s}\{W_s \le as + b, 0 \le s \le t\} \\
&= P_{x,s}\left\{\sup_{0 \le s \le t}(2aW_s - 2a^2 s) \le 2ab\right\} \\
&= 1 - P_{x,s}\left\{\sup_{0 \le s \le t} Z_s > e^{2ab}\right\} \ge 1 - e^{-2ab}.
\end{aligned}$$

Lemma 7.

$$E\left(\sup_{t>0}, e^{-rt}S_t\right) < \infty$$

or

$$\int_0^\infty P\left[\sup_t e^{-rt}S_t > y\right]dy < \infty.$$

Proof. Let $y > s, y > x$.

$$\sup_t e^{-rt} S_t = \sup_t e^{-rt} \max\left\{x, \sup_{0\le u\le t} X_u\right\}.$$

Therefore

$$\begin{aligned}
\sup_t e^{-rt} S_t > y \quad &<=> \quad \sup_t e^{-rt} \sup_{0\le u\le t} X_u > y \\
&<=> \quad \sup_t \left[-rt + \log x + \sup_{0\le u\le t}\left\{\left(\mu - \frac{\sigma^2}{2}\right)u + \sigma W_u\right\}\right] > \log y \\
&<=> \quad \sup_t \left[\sup_{0\le u\le t}\left\{(\mu - \frac{\sigma^2}{2})u + \sigma W_u\right\} > \log\frac{y}{x} + rt\right].
\end{aligned}$$

Hence

$$P\left\{\sup_t e^{-rt} S_t > y\right\} = P\left\{\sup_t \left(\sup_{0\le u\le t}\left[(\mu - \frac{\sigma^2}{2})u + \sigma W_u\right] > \log\frac{y}{x} + rt\right)\right\}. \tag{4.1}$$

Writing

$$\begin{aligned}
a &= (r - \mu + \frac{1}{2}\sigma^2)/\sigma \quad (r > \mu), \\
b &= \frac{1}{\sigma}\log\frac{y}{x}
\end{aligned}$$

we have that if $W_t \le at + b \quad \forall t$, then

$$\begin{aligned}
&\sup_{0\le u\le t}\left\{(\mu - \frac{\sigma^2}{2})u + \sigma W_u\right\} \\
\le\ &\sup_{0\le u\le t}\left\{(\mu - \frac{\sigma^2}{2})u + \sigma(au + b)\right\} \\
=\ &\sup_{0\le u\le t}\left\{\log(\frac{y}{x}) + ru\right\} = \log(\frac{y}{x}) + rt, \quad \forall t. \qquad (4.2)
\end{aligned}$$

Noting that $ab = \frac{1}{\sigma^2}(r - \mu + \frac{1}{2}\sigma^2)\log\frac{y}{x} = \left(\frac{1}{2} + \frac{r-\mu}{\sigma^2}\right)\log\frac{y}{x}$ it follows from (4.1) and (4.2) that

$$\begin{aligned}
&P\left\{\sup_t e^{-rt} S_t > y\right\} \\
\le\ &1 - P\{W_t \le at + b \quad \forall t\} \\
\le\ &e^{-2ab} = e^{-(1+\frac{2(r-\mu)}{\sigma^2})\log\frac{y}{x}} \\
=\ &\left(\frac{y}{x}\right)^{-(1+\frac{2(r-\mu)}{\sigma^2})}.
\end{aligned}$$

Hence

$$\int_1^\infty P\left[\sup_t e^{-rt} S_t > y\right] dy \leq \int_1^\infty \left(\frac{y}{x}\right)^{-\{1+\frac{2(r-\mu)}{\sigma^2}\}} dy < \infty$$

if $r > \mu$. ■

5. The Russian Call Option (or Option for Selling Short)

In this section and the next, we consider a dual to the Russian put option. This is a *call* option for "selling short" studied by Shepp and Shiryaev. The idea is that the seller pays the *minimum* price (in inflated as opposed to discounted dollars) of the share or asset during the time period between the *selling* time ("now") and the *delivery* time, the latter to be determined by the seller using a stopping rule. The seller gets the best (i.e. minimum) price up to the settlement time. The problem is to find the optimal settlement time and to obtain an exact formula for the optimal expected fair price.

The stock price process is, as before, the geometric Brownian motion model

$$X_t = xe^{(\mu - 1/2\sigma^2)t + \sigma W_t} \tag{5.1}$$

as in the previous sections. Let $r > 0$ and $y \leq x$ be given and let τ be a finite stopping time which the seller chooses as the time of delivery. Let

$$Y_t := \min\{y, \inf_{0 \leq u \leq t} X_u\}, t \geq 0. \tag{5.2}$$

That is, Y_t is the minimum value, starting at y, for X. The mathematical problem of our "short selling" option is, starting with the initial values x and y, to minimize $E_{x,y} e^{r\tau} Y_\tau$ over all finite stopping times τ, i.e., to calculate

$$V^*(x, y) := \inf_\tau \mathbf{E}_{x,y} e^{r\tau} Y_\tau \tag{5.3}$$

and, further, to find the optimal stopping time $\hat{\tau}$ to achieve this

$$\mathbf{E}_{x,y} e^{r\hat{\tau}} Y_{\hat{\tau}} = V^*(x, y). \tag{5.4}$$

6. The F.B.P. for the Call Option

We shall first dispose of a trivial case when $V^* = 0$. Here we give Shepp and Shiryaev's proof [1], [2].

Proposition 1. $V^* = 0$ *if any of the following conditions is satisfied:*

$$\begin{aligned} r &< 0, & (6.1)\\ r &= 0, \quad \text{and} \quad \mu \le \frac{\sigma^2}{2}, & (6.2)\\ 0 &< r \le \frac{1}{2\sigma^2}\left(\mu - \frac{\sigma^2}{2}\right)^2, \mu < \frac{\sigma^2}{2}. & (6.3) \end{aligned}$$

Proof. If either (6.1) or (6.2) holds, the result follows since e^{rt} is bounded and $Y_T \to 0$ as $T \to \infty$. By taking $\tau \equiv T, V^*$ can be made arbitrarily small by T taking large enough.

Suppose (6.3) holds. Now use the formula

$$P(W_t \le at + b, 0 \le t \le T) = \Phi(a\sqrt{T} + \frac{b}{\sqrt{T}}) - e^{-2ab}\Phi(a\sqrt{T} - \frac{b}{\sqrt{T}}) \quad (6.4)$$

if $b > 0$, where Φ is the standard normal distribution function. Then it can be verified that

$$\begin{aligned} &\lim_{T\to\infty} e^{rT} E \inf_{0\le t\le T} \left\{e^{\sigma W_t + (\mu - \sigma^2/2)t}\right\} \\ = &\lim_{T\to\infty} e^{rT} \int_0^\infty e^{-x} d_x P\left\{\sigma W_t + (\mu - \sigma^2/2)t \ge -x, \quad 0 \le t \le T\right\} \quad (6.5)\\ = &\lim_{T\to\infty} e^{rT} \left[e^{\mu T}(1 + \frac{\sigma^2}{2\mu})\Phi(-(\frac{\sigma}{2} + \frac{\mu}{\sigma})\sqrt{T}) + (1 - \frac{\sigma^2}{2\mu})\Phi(-(\frac{\sigma}{2} - \frac{\mu}{\sigma})\sqrt{T})\right] \\ = &\ 0 \quad \text{since} \quad \mu < \frac{\sigma^2}{2} \quad \text{and} \\ &e^{rT}\left\{e^{\mu T} e^{-(\sigma/2 + \mu/\sigma)^2 T/2} + e^{-(\sigma/2 - \mu/\sigma)^2 \frac{T}{2}}\right\} \to 0. \end{aligned}$$

The latter holds since $r - \mu(\frac{\sigma}{2} + \frac{\mu}{\sigma})^2 \cdot \frac{1}{2} \le 0$ and $r - (\frac{\sigma}{2} - \frac{\mu}{\sigma})^2 \cdot \frac{1}{2} \le 0$. ■

The above proof is taken from Shepp and Shiryaev [2].

For the remaining values of r, μ and σ we consider the FBP:

$$\frac{1}{2}\sigma^2 x^2 \frac{\partial^2 V}{\partial x^2} + \mu x \frac{\partial V}{\partial x} + rV = 0 \quad \text{in} \quad 0 < y < x \le g(y) \quad (6.6)$$

with the following conditions

$$V(x, y) \le y, \quad \frac{\partial V}{\partial y}(x, y)|_{x=y} = 0. \quad (6.7)$$

The conditions on the free boundary are

$$\text{(a)} \qquad V(g(y), y) = y$$

$$\text{(b)} \qquad \frac{\partial V}{\partial x}(g(y), y) = 0. \tag{6.8}$$

Furthermore g is assumed continuous and differentiable with $g(s) > 0$ for $s > 0$ and $g(0) = 0$. We also assume that g has a right-hand derivative at $s = 0$. These conditions do not make sense when $g(y) \equiv \infty$. As we shall see, this case arises when $r = 0$. Conditions (6.8)(a) and (b) have then to be replaced by

$$\text{(a}'\text{)} \qquad \lim_{x\to\infty} V(x, y) = y$$

$$\text{(b}'\text{)} \qquad \lim_{x\to\infty} \frac{\partial V}{\partial x}(x, y) = 0. \tag{6.8$'$}$$

The indicial equation corresponding to (6.6) is the quadratic

$$\frac{1}{2}\sigma^2\gamma^2 + (\mu - \frac{\sigma^2}{2})\gamma + r = 0. \tag{6.9}$$

Several cases arise which have to be treated separately.

Case 1. $0 < r < \frac{1}{2\sigma^2}(\mu - \frac{\sigma^2}{2})^2, \mu > \frac{1}{2}\sigma^2$. Equation (6.6) has distinct real roots $\gamma_1 < \gamma_2 < 0$. Writing $V(x, y; g)$ for $V(x, y)$ we have

$$V(x, y; g) = \frac{y}{\gamma_2 - \gamma_1}\left\{\gamma_2(\frac{x}{g})^{\gamma_1} - \gamma_1(\frac{x}{g})^{\gamma_2}\right\}. \tag{6.10}$$

Case 2. $r = \frac{1}{2\sigma^2}(\mu - \frac{\sigma^2}{2})^2, \mu > \frac{\sigma^2}{2}$. Now the indicial equation has equal roots γ and

$$V(x, y; g) = y\left\{(\frac{x}{g})^{\gamma} - \gamma(\frac{x}{g})^{\gamma}\log\frac{x}{g}\right\}. \tag{6.11}$$

Note that (6.11) can be obtained by applying L'Hôpital's rule to the right-hand side of (6.10).

Case 3. $r > \frac{1}{2\sigma^2}(\mu - \frac{\sigma^2}{2})$. Now we have complex conjugate roots $\gamma, \overline{\gamma}$ with $Im(\gamma) > 0$ and

$$V(x, y; g) = \frac{y}{\gamma - \overline{\gamma}}\left\{\gamma(\frac{x}{g})^{\overline{\gamma}} - \overline{\gamma}(\frac{x}{g})^{\gamma}\right\}. \tag{6.12}$$

Case 4. $r = 0, \mu > \frac{\sigma^2}{2}$. This is the case when $g(y) \equiv \infty$. To see this, observe that $Y_t \downarrow Y_\infty$ (say) as $t \to \infty$ and it is clear that there is no finite stopping time τ such that $E_{x,y}Y_\tau = V^*(x, y)$. In fact, we have

$$V^*(x, y) = E_{x,y}Y_\infty. \tag{6.13}$$

Turning to the indicial equation with the modified conditions (6.8′) and solving the equation $\frac{\sigma^2 x^2}{2}\frac{\partial^2 V}{\partial x^2} + \mu x \frac{\partial V}{\partial x} = 0$, we obtain

$$V(x, y) = B(y) - \frac{A(y)}{\beta - 1} \cdot \frac{1}{x^{\beta-1}},$$

where A and B are constants of integration and $\beta = \frac{2\mu}{\sigma^2} > 1$. Since $V(x, y) = y$ for all x, $y = \lim_{x\to\infty} V(x, y) = B(y)$. Next,

$$\frac{\partial V}{\partial y}|_{x=y} = 1 - \frac{A'(y)}{\beta - 1} \cdot \frac{1}{y^{\beta-1}} = 0,$$

(A' being the derivative of A). Hence $A(y) = (\beta - 1)\frac{y^\beta}{\beta}$. We thus have

$$\begin{aligned} V(x, y) &= y - \frac{y^\beta}{\beta} \cdot \frac{1}{x^{\beta-1}} \\ &= y\left\{1 - \left(\frac{y}{x}\right)^{2\mu/\sigma^2 - 1} \cdot \frac{\sigma^2}{2\mu}\right\}; \quad y \le x < \infty. \end{aligned} \tag{6.14}$$

The proof that $V^*(x, y) = V(x, y)$ will be deferred until after the first three cases have been discussed. Cases 1–3 are essentially the non-trivial cases. We shall show that, in all three cases, the FBP (6.6)–(6.8) has a unique solution with $g(y) = \theta y$ where θ is the value obtained by Shepp and Shiryaev for the three cases. We shall derive the result only for Case 2 since the other two cases can be treated similarly.

Case 2. Differentiating V with respect to y and using the second condition in (6.7) we get

$$g'(y) = \frac{\gamma(\frac{y}{g})^\gamma \log(\frac{y}{g}) - (\frac{y}{g})^\gamma}{\gamma^2(\frac{y}{g})^{\gamma+1}\log(\frac{y}{g})}, \quad 0 < y < g(y). \tag{6.15}$$

As $y \downarrow 0$ it follows that $\frac{g(y)}{y} \to g'(0) = k$, say where k is finite from (6.15),

$$k = \frac{\gamma \cdot \frac{1}{k^\gamma}\log\frac{1}{k} - \frac{1}{k^\gamma}}{\gamma^2(\frac{1}{k})^{\gamma+1}\log\frac{1}{k}} = e^{\frac{1}{\gamma(\gamma-1)}}.$$

Note that $k > 1$ since $\gamma < 0$. Thus from (6.15) we conclude that $g'(y)$ is continuous and bounded in a closed interval $[0, y_0]$ where y_0 is arbitrary. We then have

$$g(y) = \int_0^y g'(u)du \le Ly \tag{6.16}$$

where L is an upper bound for g'. Much of the ensuing argument is similar to that in the put option case. If we define

$$\hat{g}(y) = ky \tag{6.17}$$

(note that $k = \theta$ in Shepp and Shiryaev's notation), then $\hat{g}$ satisfies all the conditions for it to be a free boundary and is indeed the free boundary obtained by them. Our aim is to show that $\hat{g}$ is the unique boundary which will then imply the uniqueness of the optimal stopping time for our option pricing problem. Let us denote by g, any free boundary which will solve the FBP. Our aim is to show that

$$V^*(x,y) = V(x,y;g). \tag{6.18}$$

Write

$$Z_t = e^{rt}V(X_t, Y_t; g). \tag{6.19}$$

In the continuation region $0 < Y_t \leq X_t \leq g(Y_t)$, (arguing as in the put option case), Y_t grows only when $X_t = Y_t$, that is, $dY_t = 0$ a.s. on the set $\{Y_t < X_t\}$.

Let $\sigma_n \uparrow \infty$ be a sequence of stopping times such that the integral below is a martingale. We then have

$$Z_{t\wedge\sigma_n} = Z_0 + \int_0^{t\wedge\sigma_n} \sigma e^{rs}\frac{\partial V}{\partial x}(X_s, Y_s)X_s dW_s$$

where (by choice of σ_n), $Z_{t\wedge\sigma_n}$ is a positive martingale. The integral on the right-hand side being a martingale we obtain upon replacing t by any finite stopping time τ and taking expectations

$$E_{x,,y}(Z_{\tau\wedge\sigma_n}) = Z_0 = V(x,y;g). \tag{6.20}$$

Making $n \to \infty$, by Fatou's Lemma,

$$E_{x,y}Z_\tau \leq V(x,y;g). \tag{6.21}$$

Now define the stopping time

$$\hat{\tau} := \inf\{t \geq 0 : X_t = g(Y_t)\}. \tag{6.22}$$

Since $g(Y_t) \leq LY_t$,

$$\begin{aligned} P(\hat{\tau} > T) &\leq P\Big\{\sigma(W_t - W_u) + (\mu - \frac{\sigma^2}{2})(t-u) \\ &\leq \log\frac{g(y)}{x} \leq \log\frac{Ly}{x}, \quad 0 \leq u \leq t \leq T\Big\} \\ &\to 0 \quad \text{as} \quad T \to \infty \end{aligned}$$

we have that $P(\hat{\tau} < \infty) = 1$. Hence

$$\begin{aligned} Z_{\hat{\tau}} &= e^{r\hat{\tau}} V(X_{\hat{\tau}}, Y_{\hat{\tau}}) = e^{r\hat{\tau}} V(g(Y_{\hat{\tau}}), Y_{\hat{\tau}}; g) \\ &= e^{r\hat{\tau}} Y_{\hat{\tau}}, \end{aligned}$$

the last equality following from (6.8)(a). From (6.22) we then have

$$E_{x,y} e^{r\hat{\tau}} Y_{\hat{\tau}} \le V(x, y; g)$$

from which it follows that

$$V^*(x, y) \le V(x, y; g). \tag{6.23}$$

To obtain the reverse inequality we use the first condition in (6.7) to get

$$\begin{aligned} E_{x,y} e^{r(t\wedge\tau)} Y_{t\wedge\tau} &\ge E_{x,y} e^{r(t\wedge\tau)} V\left(X_{t\wedge\tau}, Y_{t\wedge\tau}; g\right) \\ &= E_{x,y} Z_{t\wedge\tau}. \end{aligned} \tag{6.24}$$

Since Z is a continuous local martingale, $Z_{t\wedge\tau\wedge\sigma_n}$ is a martingale for some stopping time sequence $\sigma_n \uparrow \infty$ and we have

$$E_{x,y} Z_{t\wedge\tau\wedge\sigma_n} = Z_0 = V(x, y; g).$$

Now (6.24) is true for any finite stopping time τ. So replacing τ by $\tau \wedge \sigma_n$ we have

$$E_{x,y}\left\{e^{r(t\wedge\tau\wedge\sigma_n)} Y_{t\wedge\tau\wedge\sigma_n}\right\} \ge E_{x,y} Z_{t\wedge\tau\wedge\sigma_n} = V(x, y; g). \tag{6.25}$$

By the definition of Y_t, $Y_{t\wedge\tau\wedge\sigma_n} \le y \quad \forall n$ and $t \wedge \tau \wedge \sigma_n$ so that for all n,

$$e^{r(t\wedge\tau\wedge\sigma_n)} Y_{t\wedge\tau\wedge\sigma_n} \le e^{rt} y.$$

The left-hand side of the above inequality $\to e^{r(t\wedge\tau)} Y_{t\wedge\tau}$ a.s. As $n \to \infty$, by the dominated convergence theorem we have, using (6.25)

$$E_{x,y}\left\{e^{r(t\wedge\tau)} Y_{t\wedge\tau}\right\} \ge V(x, y; g), \quad \forall t \ge 0.$$

Again using the dominated convergence theorem as $t \to \infty$ we obtain from the above inequality

$$E_{x,y}\left\{e^{r\tau} Y_\tau\right\} \ge V(x, y; g).$$

Hence

$$V^*(x, y) \ge V(x, y; g). \tag{6.26}$$

From (6.23) and (6.26), the desired equality follows:

$$V^*(x, y) = V(x, y; g), \quad y < x \le g(y). \tag{6.27}$$

Since the free boundary $\hat{g}$ is also a solution of our FBP,

$$V^*(x,y) = V(x,y;\hat{g}), \quad y < x \le \hat{g}(y). \tag{6.28}$$

From (6.27) and (6.28),

$$V(x,y;\hat{g}) = V(x,y;g), \quad y < x \le \min\{g(y), \hat{g}(y)\}. \tag{6.29}$$

Substituting from (6.11) in the above we have

$$y\left\{(\frac{x}{ky})^\gamma - \gamma(\frac{x}{ky})^\gamma \log\frac{x}{ky}\right\} = y\left\{(\frac{x}{g})^\gamma - \gamma(\frac{x}{g})^\gamma \log\frac{x}{g}\right\}.$$

Making $x \to y$,

$$\frac{1}{k^\gamma} - \gamma\frac{1}{k^\gamma}\log\frac{1}{k} = \left\{\frac{y}{g(y)}\right\}^\gamma \left\{1 + \gamma\log\frac{g(y)}{y}\right\}.$$

Writing $h(y) = \frac{g(y)}{y}$ for convenience,

$$\frac{1}{h^\gamma}(1+\gamma\log h) = \frac{1}{k^\gamma} + \frac{\gamma}{k^\gamma}\log k = \text{ constant.}$$

Differentiating with respect to y at $y > 0$, we have

$$\frac{\gamma^2}{h^{\gamma+1}}\log h \cdot \frac{dh}{dy} = 0.$$

which gives $\frac{dh}{dy} = 0$ since $h(y) > 1$ for $y > 0$ and so $\log h(y) \neq 0$. Hence $\frac{g(y)}{y} = c$, a constant for all $y > 0$. From the steps immediately following (6.15) we immediately see that $c = k$. Thus we have shown that if $g(y)$ is any other free boundary solving the FBP (6.6)–(6.8), then $g(y) = ky$. In other words, the solution obtained by Shepp and Shiryaev is unique. The stopping time $\hat{\tau}$ defined in (6.21) then becomes the stopping time obtained by these authors. We have thus proved the following result.

Theorem 3. The F.B.P. (6.6)–(6.8) has a unique solution (V, g) for the three cases considered. The expression for V has already been given above.

For Case 1,

$$g(y) = \theta y \quad \text{where} \quad \theta = \left\{\frac{1-\frac{1}{\gamma_2}}{1-\frac{1}{\gamma_1}}\right\}^{1/(\gamma_2-\gamma_1)};$$

For Case 2,

$$g(y) = \theta y \quad \text{where} \quad \theta = e^{\frac{1}{\gamma(\gamma-1)}};$$

For Case 3, $g(y) = \theta y$ where $\theta = \exp(\frac{\phi}{\beta})$; ϕ is given by $1 - \frac{1}{\gamma} = re^{i\phi}$ and $\beta = Im(\gamma) > 0$.

In all three cases, $\theta > 1$ and the unique optimal stopping rule is given by

$$\hat{\tau} := \inf\{t \geq 0 : X_t = \theta Y_t\}.$$

The value of the option is given by the formula

$$\begin{aligned} V^*(x) &= V(x,y) \quad y < x \leq \theta y \\ &= y \quad \text{for} \quad x > \theta y. \end{aligned} \tag{6.30}$$

To obtain the value of the option claimed in (6.30) when $x > \theta y$ we proceed as follows.

$$V^*(x) = \inf_\tau E_{x,y} e^{r\tau} Y_\tau \leq y \inf_\tau E_{x,y} e^{r\tau} \leq y$$

by arguing as in the put option case and taking the stopping time $\epsilon\tau$ and then making $\epsilon \to 0$. When $X_t > \theta Y_t$, we have $dY_t = 0$. Hence, if $Z_t = e^{rt} Y_t$,

$$dZ_t = re^{rt} Y_t dt, \quad Z_t = Z_0 + \int_0^t re^{rs} Y_s ds \geq Z_0.$$

We then have, replacing t by $t \wedge \tau$,

$$E_{x,y} Z_{t\wedge\tau} \geq E_{x,y} Z_0 = y.$$

Note once again that

$$E_{x,y} Z_{t\wedge\tau} \leq E_{x,y} e^{r\tau} Y_{t\wedge\tau}$$

and that $e^{r\tau} Y_{t\wedge\tau} \downarrow e^{r\tau} Y_\tau$ as $t \to \infty$. We get

$$E_{x,y} e^{r\tau} Y_\tau = \lim_{t\to\infty} E_{x,y} e^{r\tau} Y_{t\wedge\tau} \geq y.$$

Taking the inf over τ on the left-hand side we obtain the reverse inequality

$$V^*(x,y) \geq y$$

thus showing that

$$V^*(x,y) = y \quad \text{if} \quad x > \theta y. \tag{6.31}$$

It remains to complete the proof for Case 4. Since $r = 0$, $Z_t = V(X_t, Y_t)$. From the argument preceding (6.24) we get

$$Z_{t\wedge\sigma_n} = Z_0 + \sigma \int_0^{t\wedge\sigma_n} \frac{\partial V}{\partial x}(X_s, Y_s) X_s dW_s$$

where $\sigma_n \uparrow \infty$ are stopping times such that the integral on the right is a martingale. Then

$$E_{x,y} Z_{t\wedge\sigma_n} = V(x,y).$$

From the first condition in (6.7),

$$E_{x,y}Y_{t\wedge\sigma_n} \geq E_{x,y}Z_{t\wedge\sigma_n} = V(x,y).$$

By the monotone convergence theorem applied twice to the first quantity in this inequality, we have

$$V^*(x,y) = E_{x,y}Y_\infty \geq V(x,y). \tag{6.32}$$

Next,

$$E_{x,y}V(X_{t\wedge\sigma_n}, Y_{t\wedge\sigma_n}) = E_{x,y}Z_{t\wedge\sigma_n} = V(x,y).$$

Hence making $\sigma_n \to \infty$, noting that V is continuous and positive, we have

$$E_{x,y}V(X_t, Y_t) \leq V(x,y). \tag{6.33}$$

Now,

$$X_t = xe^{t\{(\mu-\frac{1}{2}\sigma^2)+\frac{\sigma W_t}{t}\}} \to \infty$$

as $t \to \infty$ since $\mu > \frac{1}{2}\sigma^2$ by assumption. From the formula (6.14) for $V(x,y)$, $V(X_t, Y_t) \to Y_\infty$ as $t \to \infty$ on account of the fact that

$$\frac{Y_t}{X_t} \leq \frac{y}{X_t} \to 0.$$

Hence again by Fatou's Lemma and (6.33),

$$E_{x,y}Y_\infty \leq V(x,y). \tag{6.34}$$

Hence

$$V^*(x,y) \leq V(x,y) \tag{6.35}$$

from (6.13). Thus we have proved that $V^*(x,y) = V(x,y)$ for Case 4.

References

[1] L.A. Shepp and A.N. Shiryaev, The Russian option: Reduced regret, *The Annals of Applied Probability,* **3** (1993), 631–640.

[2] L.A. Shepp and A.N. Shiryaev, A dual Russian option for selling short (1993), DIMACS Technical Report 93-26.

Department of Statistics
University of North Carolina at Chapel Hill
CB #3260 Phillips Hall
Chapel Hill, NC 27599-3260
gk@stat.unc.edu

Cycle Representations of Markov Processes: An Application to Rotational Partitions

S. Kalpazidou

Abstract

The cycle formula asserts that any finite-order recurrent stochastic matrix P is a linear combination of matrices $J_c = (J_c(i,j))$ associated with the cycles c of the graph of P, and defined as follows: $J_c(i,j) = 1$ or 0, according to whether i, j are consecutive points of c or not.

In the present paper we investigate, by using the cycle formula, the asymptotic behavior of the sequence $(t, {}^m\mathcal{S})$, $t \geq 0$, of rotational representations associated to the powers P^m, $m = 0, 1, 2, \ldots$, of an irreducible stochastic matrix on a finite set $S = \{1, \ldots, n\}$, $n \geq 2$. In particular, we give a criterion on the rotational partitions ${}^m\mathcal{S}$, $m = 0, 1, \ldots$, for the sequence $\{P^m\}_m$ to be convergent. A pair $(t, \mathcal{S})$ is a rotational representation of $P = (p_{ij}, i, j = 1, \ldots, n)$ if $\mathcal{S}$ is a partition of $[0, 1)$ into n sets $S_1, \ldots, S_n$, each of positive Lebesgue measure and consisting of a finite union of arcs, such that $p_{ij} = \lambda(S_i \cap f_t^{-1}(S_j))/\lambda(S_i)$. Here f_t is the λ-preserving transformation of $[0, 1)$ onto itself defined by $f_t(x) = (x + t)(\text{mod } 1)$, and λ denotes Lebesgue measure.

1. Preliminaries

Recent research has been devoted to representing the finite-dimensional distributions of stochastic processes as linear or convex decompositions in terms of cycle-passage-matrices $(J_c(i,j))$. The corresponding representation is called the *cycle-decomposition-formula*, and the usual class of stochastic processes, to which it is applied, is that containing Markov processes ξ which admit an invariant measure. If $c = (i_i, i_2, \ldots, i_s)$ is a directed sequence of states of ξ, called also a cycle, then the cycle-passage-matrix $(J_c(i,j))$ is defined as follows: $J_c(i,j) = 1$ or 0, according to whether i, j are consecutive states of c or not.

Investigations of the cycle-decomposition-formula in various contexts are still an active area of research. They are mostly advanced in the framework of stochastic processes (see Qian Minping and Qian Min [12], and author [6]–[11]) and directed networks, with expected developments to harmonic analysis and potential theory on electrical networks (see Soardi [13], Woess [15]), electrical networks (see Zemanian [16]), or random walks in random environment (see Derriennic [4]), etc.

A particularly interesting application of the cycle-decomposition-formula arises when one formulates in terms of Markov chains the coding problem occurring in dynamical systems. This formulation is as follows: find a one-to-one correspondence from the space of $n \times n$ irreducible stochastic matrices into n-partitions. The solution to this problem has its beginning in Cohen's [3] conjecture, according to which any finite-order irreducible stochastic matrix may be represented by rotational partitions as follows: For fixed $n \geq 2$ and $S = \{1, \ldots, n\}$, a stochastic matrix $P = (p_{ij}, i, j \in S)$ is irreducible if and only if there exist a partition $\mathcal{S}$ of $[0, 1)$ into n sets $S_1, \ldots, S_n$, each of positive Lebesgue measure and consisting of a finite union of arcs, and a λ-preserving transformation f_t, $t \geq 0$, of $[0, 1)$ onto itself defined by

$$f_t(x) = (x + t)(\text{mod } 1), \tag{1}$$

such that

$$p_{ij} = \lambda(S_i \cap f_t^{-1}(S_j))/\lambda(S_i), \quad i, j \in S. \tag{2}$$

Here λ denotes Lebesgue measure on Borel subsets of $[0, 1)$. Furthermore, if $\pi = (\pi_1, \ldots, \pi_n)$ denotes the invariant probability row-vector of P, then $\pi_i = \lambda(S_i)$, $i = 1, \ldots, n$. Throughout this paper we shall assume $t = 1/n!$ and $P = (p_{ij}, i, j = 1, \ldots, n)$ will be an irreducible stochastic matrix. When (2) holds, the stochastic matrix P is said to have a rotational representation denoted by $(t, \mathcal{S})$. Solutions to this problem are given for $n = 2$ by Joel E. Cohen [3] and for $n \geq 2$ by S. Alpern [2] (in combinatorial context) and by the author [6], [8] (using either a probabilistic or a homologic argument). Furthermore, the approach of [8] provides a theoretical basis for the definition of finite recurrent matrices by a rotation of the circle and a partition whose elements consist of finite unions of circle-arcs (see also Haigh [5], Rodriquez and Valsero [14]).

From the point of view of homology theory we find similarities between the rotational representations of stochastic matrices and the so-called graph-problem, formulated often as a graph-sphere problem. What we know from the topology of polyhedra is that any finite graph has a geometric realization in Hilbert spaces as $\mathbb{R}^3$, while the theorems of Kuratowski, Whitney, MacLane and Lefschetz provide a collection of necessary and sufficient conditions for a graph to be sketched upon a sphere.

In the present paper we investigate the asymptotic behavior of the sequence of rotational representations $(t, {}^0\mathcal{S})$, $(t, {}^1\mathcal{S}, \dots, (t, {}^m\mathcal{S}), \dots$, associated with the sequence $I, P, \dots, P^m, \dots$, of powers, as $m \to \infty$, where I is the $n \times n$ stochastic matrix with all the entries of the diagonal equal to 1. One preliminary step of this study is the definition of a rule of assignment which associates each sequence $\{P^m\}_{m \geq 0}$ to a sequence $\{{}^m\mathcal{S}\}_m$ of rotational partitions by using a suitable linear decomposition of each matrix $P^m = (p_{ij}^{(m)}, i, j = 1, \dots, n)$, with $P^0 \equiv I$ and $p_{ij}^{(1)} \equiv p_{ij}$, in the form:

$$\pi_i p_{ij}^{(m)} = \sum_{c \in C} {}^m w_c J_c^{(1)}(i, j), \tag{$\mathfrak{c}$}$$

where C is a collection of directed circuits $c = (i_1, \dots, i_s, i_1)$, $s \geq 1$, in S, $w_c > 0$ and $J_c^{(1)}(i, j) = 1$ or 0, according to whether (i, j) is an edge of c or not. We call equation ($\mathfrak{c}$) a cycle decomposition of P^m, and $J_c^{(1)}$ the one-step passage-matrix associated to c.

As known, the basic link between the asymptotics of P^m, $m = 0, 1, \dots$, and the theory of Markov chains is given by the Chapman–Kolmogorov equations, according to which the m-step transition probabilities $\text{Prob}\{\xi_{k+m} = j \mid \xi_k = i)$, $i, j \in S$, $k, m \geq 1$, of the homogeneous Markov chain $\xi = (\xi_k)_k$ on P, are identical to the entries of P^m. That is,

$$p_{ij}^{(k+m)} = \sum_{u \in S} p_{iu}^{(k)} p_{uj}^{(m)}, \quad i, j \in S, \ k, m \geq 1.$$

In Section 2 we give the homologic expression of the Chapman–Kolmogorov equations in terms of cycles by using the m-step passage-matrices. The m-step passage matrix $J_{c_1 \dots c_m}^{(m)}$ generalizes the one step passage-matrix $J_c^{(1)}$ occurring in ($\mathfrak{c}$), when the passages hold on m overlapping directed circuits $c_1, \dots, c_m$. Then, using ($\mathfrak{c}$), the sequence $\{P^m\}_m$ may be assigned by (2) to a rotational system $\{(t, {}^m\mathcal{S}), m = 0, 1, \dots\}$, with $t > 0$ and ${}^m\mathcal{S} \equiv \mathcal{S}(P^m) = \{{}^mS_1, \dots, {}^mS_n\}$, which satisfies the following Chapman–Kolmogorov-type conditions:

$$\begin{aligned} &\lambda({}^{k+m}S_i \cap f_t^{-1}({}^{k+m}S_j)) \\ &\quad = \sum_{u \in S} \lambda({}^mS_u) \lambda({}^kS_i \cap f_t^{-1}({}^kS_u)) \lambda({}^mS_u \cap f_t^{-1}({}^mS_j)), \end{aligned} \tag{$\mathcal{CK}$}$$

for all $i, j = 1, \ldots, n$, and for all $k, m \geq 1$ (see Section 3). The previous assignment $P^m \to (t, {}^m\mathcal{S})$ is basically dependent on the collection $(\mathcal{C}, {}^m w_c)$, which may be provided by many algorithms. Here we shall choose the following definition for $(\mathcal{C}, {}^m w_c)$: the set $\mathcal{C}$ contains all the directed circuits appearing on almost all the trajectories of the Markov chain on P^m, and each ${}^m w_c$ is the mean number of occurrences of c on almost all the trajectories. With this definition, $(\mathcal{C}, {}^m w_c)$ is called a *canonical cycle representation* of P^m, and $(t, {}^m\mathcal{S})$ a *canonical rotational representation* of P^m.

In Theorem 4 it is shown that, given $t > 0$, $S = \{1, \ldots, n\}$ and a countable collection $\{{}^m\mathcal{S},\ m = 1, 2, \ldots\}$ of partitions of $[0, 1)$ with ${}^m\mathcal{S} = \{{}^m S_1, \ldots, {}^m S_n\}$ and satisfying the Chapman–Kolmogorov-type conditions $(\mathcal{CK})$ with respect to a rotation f_t, there exists a sequence $P^{(1)}$, $P^{(2)}, \ldots, P^{(m)}, \ldots$ of recurrent stochastic $n \times n$ matrices which are the powers of the stochastic matrix $P = \lambda({}^1S_i \cap f_t^{-1}({}^1S_j))/\lambda({}^1S_i)$, $i, j = 1, \ldots, n)$, such that $(t, {}^m\mathcal{S}) = (t, \mathcal{S}(P^m))$, $m = 1, 2, \ldots$.

Finally, in Theorem 5 it is proved that, if the sequence $\{P^m\}$ converges (element-wise, or in Cesaro mean) to a stable stochastic matrix Π (according to whether P is aperiodic or not), then the corresponding canonical rotational representations converge to a rotational representation of Π. Furthermore, the same theorem asserts that if a sequence $\{{}^m\mathcal{S}\}_m$ of rotational partitions satisfies the Chapman–Kolmogorov-type equations $(\mathcal{CK})$ and suitable conditions, then it defines a convergent sequence $\{P^m\}_m$ of powers of a stochastic matrix P such that $(t, {}^m\mathcal{S})$ is a rotational representation of P^m, $m = 1, 2, \ldots$.

2. The n-step passage functions

For a finite set S and for $s > 1$, let $i_1, i_2, \ldots, i_s \in S$. Then the collection

$$c = \{(i_1, i_2, \ldots, i_s, i_1), (i_2, i_3, \ldots, i_s, i_1, i_2), \ldots, (i_s, i_1, i_2, \ldots, i_{s-1}, i_s)\}$$

is called a *directed circuit in* S (see [6] and [8]). The directed circuit c may be regarded as an equivalence-class with respect to the following equivalence relation: $c_1 \sim c_2$ if and only if c_2 is obtained from c_1 by a cyclic permutation. Consequently, the directed sequence $(i_1, i_2, \ldots, i_s, i_1)$ completely determines a directed circuit, modulo the cyclic permutations.

Assume that c is determined by $(i_1, i_2, \ldots, i_s, i_1)$ with $s > 1$ and $i_1, i_2, \ldots, i_s \in S$. Then c may be viewed as a periodic function from the set Z of integers into S, that is, $c = (c(n), c(n+1), \ldots, c(n+s-1), c(n+s))$, $n \in Z$, with $c(n) = i_1$, $c(n+1) = i_2, \ldots, c(n+s-1) = i_s$, $c(n+s) = i_1$. The values $c(n)$, $n \in Z$, are called *points* of c, while the directed pairs $(c(n), c(n+1))$, $n \in Z$, are called *directed edges* of c. The ordered sequence (modulo the cyclic permutations) $\hat{c} = (c(n), \ldots, c(n+s-1))$ is called the *directed cycle* associated to the circuit c. The smallest integer $p = p(c) \geq 1$ that satisfies the equation $c(n+p) = c(n)$, for all $n \in Z$, is called the *period of* c. A directed circuit c with $p(c) = 1$ is called a *loop*.

Throughout this paper, we shall assume directed circuits $c = (i_1, i_2, \ldots, i_s, i_1)$ with distinct points $i_1, i_2, \ldots, i_s$.

Definition 1 (The n-step passage-function associated with a circuit). Assuming c is a directed circuit of period $p(c) > 1$, we define the functions $J_c(i)$, $J_c^{(1)}(i,j)$, $J_c^{(2)}(i,j), \ldots, J_c^{(n)}(i,j), \ldots$, as follows:

$$J_c(i) = \begin{cases} 1, & \text{if } i = c(m), \text{ for some } m \in Z, \\ 0, & \text{otherwise;} \end{cases}$$

$$J_c^{(1)}(i,j) = \begin{cases} 1, & \text{if } i = c(m),\ j = c(m+1), \text{ for some } m \in Z, \\ 0, & \text{otherwise;} \end{cases}$$

$$J_c^{(2)}(i,j) = \begin{cases} 1, & \text{if } i = c(m),\ j = c(m+2), \text{ for some } m \in Z, \\ 0, & \text{otherwise;} \end{cases}$$

and, in general,

$$J_c^{(n)}(i,j) = \begin{cases} 1, & \text{if } i = c(m),\ j = c(m+n), \text{ for some } m \in Z, \\ 0, & \text{otherwise;} \end{cases}$$

for any $n > 2$.

For $n \geq 1$, we say that c passes from state i to state j in n-steps, if and only if $J_c^{(n)}(i,j) = 1$, and that c passes through i if and only if $J_c(i) = 1$.

The function $J_c^{(n)}(i,j) : S \times S \to \{0,1\}$, $n \geq 1$, is called the *n-step passage-function associated with the directed circuit c*. We have

Proposition 1. *The n-step passage-functions $J_c^{(n)}$, $n \geq 1$, satisfy the following equations:*

$$J_c^{(n+1)}(i,j) = \sum_{k \in S} J_c^{(n)}(i,k) J_c^{(1)}(k,j), \tag{i}$$

$$J_c^{(n+1)}(i,j) = \sum_{k \in S} J_c^{(1)}(i,k) J_c^{(n)}(k,j), \tag{ii}$$

for any $i, j \in S$.

Proof. Consider $i, j \in S$ such that $J_c^{(n+1)}(i,j) = 1$. Then there exists some integer n such that $i = c(m)$ and $j = c(m+n+1)$. Accordingly, the only n points immediately succeeding i along c are given by $c(m+1), \dots, c(m+n) \equiv k_0$, so that we may write

$$\sum_{k \in S} J_c^{(n)}(i,k) J_c^{(1)}(k,j) = J_c^{(n)}(i,k_0) J_c^{(1)}(k_0,j) = 1.$$

Finally, equation (ii) may be proved by similar arguments. The proof is complete. ■

From Proposition 1, it follows that

$$J_c^{(n)}(i,j) = \sum_{i_1,\dots,i_{n-1} \in S} J_c^{(1)}(i,i_1) J_c^{(1)}(i_1,i_2) \cdot \dots \cdot J_c^{(1)}(i_{n-1},j), \quad n \geq 2, \tag{3}$$

for any $i, j \in S$.

We define the *one-step passage-matrix of* c to be the matrix $J_c = (J_c^{(1)}(i,i_1))_{i,j \in S}$. Analogously, we define the *n-step passage-matrix of* c to be the matrix $J_c^{(n)} = (J_c^{(n)}(i,j))_{i,j \in S}$. Then, from equation (3), it follows that

$$J_c^{(2)} = (J_c)^2.$$

In general, we have

$$J_c^{(n)} = (J_c)^n, \quad n \geq 2,$$

and

$$(J_c)^{n+s} = (J_c)^n (J_c)^s, \quad n, s \geq 1.$$

Then

$$J_c^{(n+s)}(i,j) = \sum_{k \in S} J_c^{(n)}(i,k) J_c^{(s)}(k,j) \tag{4}$$

for any $n, s \geq 1$ and for any $i, j \in S$. We shall call equation (4) the *Chapman–Kolmogorov-type equation* for the n-step passage-functions on c.

The n-step passage-function enjoys a few simple but basic properties as stated by the following lemma.

Lemma 2. *The n-step passage-function $J_c^{(n)}$, $n \geq 1$, satisfies the following balance properties:*

$$J_c(i) = \sum_{j \in S} J_c^{(n)}(i,j) = \sum_{k \in S} J_c^{(n)}(k,i) \tag{5}$$

for any $i \in S$.

Proof. We shall prove equation (5) by induction with respect to n. For $n = 1$, the balance equation (5) becomes

$$J_c(i) = \sum_{j \in S} J_c^{(1)}(i,j) = \sum_{k \in S} J_c^{(1)}(k,i), \quad i \in S.$$

These relations are already proved in [8]. Now suppose that equation (5) is valid for some integer $n > 1$. Then, by applying Proposition 1, we have

$$\begin{aligned}\sum_{j \in S} J_c^{(n)}(i,j) &= \sum_{j \in S} \sum_{k \in S} J_c^{(1)}(i,k) J_c^{(n)}(k,j) = \sum_{k \in S} J_c^{(1)}(i,k) \sum_{j \in S} J_c^{(n)}(k,j) \\ &= \sum_{k \in S} J_c^{(1)}(i,k) J_c(k) = J_c(i).\end{aligned}$$

Analogously,

$$\sum_{k \in S} J_c^{(n+1)}(k,i) = \sum_{k \in S} \sum_{u \in S} J_c^{(n)}(k,u) J_c^{(1)}(u,i) = \sum_{u \in S} J_c(u) J_c^{(1)}(u,i) = J_c(i).$$

The proof is complete. ■

We shall say that two directed circuits c_1 and c_2 have a joint point if and only if there exists some integer m such that $c_1(m) = c_2(m)$. We now introduce the

Definition 2. Given two directed circuits c_1 and c_2, which have at least one joint point, the two-step passage-function $J_{c_1 c_2}^{(2)} : S \times S \to \{0,1\}$ is defined as follows:

$$J_{c_1 c_2}^{(2)}(i,j) = \begin{cases} 1, & \text{if } i = c_1(m),\ c_1(m+1) = c_2(m+1),\ j = c_2(m+2) \\ & \text{for some integer } m, \\ 0, & \text{otherwise} \end{cases}$$

for any $i, j \in S$. In case that c_1 and c_2 have no common point, then $J_{c_1 c_2}^{(2)} \equiv 0$. When $c_1 = c_2 \equiv c$, then $J_{c_1 c_2}^{(2)} \equiv J_c^{(2)}$.

From the very definition of the two-step passage-function $J^{(2)}_{c_1 c_2}$ we have that

$$J^{(2)}_{c_1 c_2}(i,j) = \sum_{k \in S} J^{(1)}_{c_1}(i,k) J^{(1)}_{c_2}(k,j).$$

Let us now consider an ordered sequence (c_1, c_2, c_3) of directed circuits in S and assume that any two consecutive circuits have at least one point in common and that there are consecutive joint points one-edge distant from each other (that is, $c_1(m) = c_2(m)$ and $c_2(m+1) = c_3(m+1)$ for some integer m). In particular, some circuits may be identical to each other. Then, in analogy to the definition of the two-step passage-function, we may define the *three-step passage-function* $J^{(3)}_{c_1 c_2 c_3} : S \times S \to \{0,1\}$ as follows:

$$J^{(3)}_{c_1 c_2 c_3}(i,j) = \begin{cases} 1, & \text{if } i = c_1(m-1),\ c_1(m) = c_2(m),\ c_2(m+1) = \\ & c_3(m+1),\ j = c_3(m+2) \text{ for some integer } m, \\ 0, & \text{otherwise.} \end{cases}$$

In case that the circuits c_1, c_2, c_3 do not satisfy the above assumption on the intersection point, then we shall define $J^{(3)}_{c_1 c_2 c_3} \equiv 0$.

In general, we have

Definition 3 (The n-step passage-function on a sequence of circuits). Assume an ordered sequence $c_1, \ldots, c_n$, $n > 1$, of directed circuits such that $c_1(m) = c_2(m)$, $c_2(m+1) = c_3(m+1), \ldots, c_{n-1}(m+n-2) = c_n(m+n-2)$ for some integer m. Then the n-step passage-function $J^{(n)}_{c_1 \ldots c_n} : S \times S \to \{0,1\}$ is defined as follows:

$$J^{(n)}_{c_1 \ldots c_n}(i,j) = \begin{cases} 1, & \text{if } i = c_1(m-1),\ c_1(m) = c_2(m), \ldots, \\ & c_{n-1}(m+n-2) = c_n(m+n-2),\ j = c_n(m+n-1) \\ & \text{for some integer } m, \\ 0, & \text{otherwise} \end{cases}$$

for any $i, j \in S$. In particular, if $c_1 = \ldots = c_n \equiv c$, then $J^{(n)}_{c_1 \ldots c_n} \equiv J^{(n)}_c$. In case that the circuits $c_1, \ldots, c_n$ do not satisfy the assumption on the joint points, then $J^{(n)}_{c_1 \ldots c_n} \equiv 0$.

Proposition 3. *The n-step passage-functions $J^{(n)}_{c_1 \ldots c_n}(i,j)$, $n \geq 1$, satisfy the following equations:*

$$J^{(n+1)}_{c_1\dots c_n c_{n+1}}(i,j) = \sum_{k\in S} J^{(n)}_{c_1\dots c_n}(i,k) J^{(1)}_{c_{n+1}}(k,j) \tag{6}$$

$$J^{(n+1)}_{c_1 c_2\dots c_{n+1}}(i,j) = \sum_{k\in S} J^{(1)}_{c_1}(i,k) J^{(n)}_{c_2\dots c_{n+1}}(k,j) \tag{7}$$

for any $i, j \in S$.

Proof. Let $i, j \in S$, such that $J^{(n+1)}_{c_1\dots c_{n+1}}(i,j) = 1$. Then, by the definition of $J^{(n+1)}_{c_1\dots c_{n+1}}$, there is some integer m such that $i = c_1(m)$, $c_1(m+1) = c_2(m+1)$, $c_2(m+2) = c_3(m+2), \dots, c_n(m+n) = c_{n+1}(m+n)$ and $c_{n+1}(m+n+1) = j$. Plainly, the n points immediately succeeding i along $c_1, \dots, c_{n+1}$ are uniquely determined and

$$J^{(n)}_{c_1\dots c_n}(i, c_n(m+n)) \cdot J^{(1)}_{c_{n+1}}(c_{n+1}(m+n), j) = 1$$

is the unique nonzero term of the right member of (6). Since, the converse direction uses an analogous argument, equation (6) is now proved. Finally, equation (7) may be proved by similar reasonings. ∎

Equations (6) and (7) imply that

$$J^{(n)}_{c_1\dots c_n}(i,j) = \sum_{i_1,\dots,i_{n-1}\in S} J^{(1)}_{c_1}(i,i_1) J^{(1)}_{c_2}(i_1,i_2)\dots J^{(1)}_{c_n}(i_{n-1},j), \quad n \geq 2$$

for any $i, j \in S$.

We define the *n-step passage-matrix associated with the circuits* $c_1, \dots, c_n$ to be the matrix $J_{c_1\dots c_n} = (J^{(n)}_{c_1\dots c_n}(i,j))_{i,j\in S}$. Then the matrix $J_{c_1 c_2}$ is the product of the matrices J_{c_1} and J_{c_2}, and in general we have

$$J_{c_1\dots c_n} = J_{c_1} \cdot J_{c_2} \cdot \dots \cdot J_{c_n}, \quad n \geq 2.$$

Accordingly, we may write

$$J_{c_1} \dots J_{c_{n+s}} = J_{c_1\dots c_n} \cdot J_{c_{n+1}\dots c_{n+s}}, \quad n, s \geq 1.$$

Then

$$J^{(n+s)}_{c_1\dots c_n\, c_{n+1}\cdots c_{n+s}}(i,j) = \sum_{k\in S} J^{(n)}_{c_1\dots c_n}(i,k) J^{(s)}_{c_{n+1}\dots c_{n+s}}(k,j), \quad n, s \geq 1, \tag{8}$$

for any $i, j \in S$. We shall call equation (8) the *Chapman–Kolmogorov-type relations* for the n-step passage-matrices on sequences of circuits.

3. Rotational representations of the n-step transition matrices

In this section we study the rotational representations $(t, \mathcal{S}(P^m))$ of the powers P^m, $m \geq 1$, of a transition matrix $P = (p_{ij}, i, j = 1, \dots, n)$ which defines an irreducible Markov chain with the state space $S = \{1, \dots, n\}$, $n \geq 2$. Namely, we shall show the existence of a map from the space of sequences $\{P^m\}_{m\geq 0}$ of powers of irreducible $n \times n$ stochastic matrices into a set of sequences $\{{}^m\mathcal{S}\}_{m\geq 0}$ of n-partitions of $[0, 1)$, and the existence of an inverse map from the set of sequences $\{{}^m\mathcal{S}\}_{m\geq 0}$ of n-partitions of $[0, 1)$ which satisfy the Chapman–Kolmogorov-type conditions, into the space of sequences $\{P^m\}_{m\geq 0}$ such that ${}^m\mathcal{S} = \mathcal{S}(P^m)$, $m = 0, 1, \dots$.

To this end we give in the following subsection a rigorous presentation of a general transform which associates with a sequence $\{P^m\}_{m\geq 0}$ a sequence $\{{}^m\mathcal{S}\}_{m\geq 0}$ of n-partitions of $[0, 1)$. An important consequence of this approach will be the characterization of the Chapman–Kolmogorov equations in terms of rotations.

3.1. A stochastic matrix $P = (p_{ij}, i, j = 1, \dots, n)$ is called irreducible if for any row i and any column $j \neq i$, there exists a positive integer k, which may depend on i and j, such that the (i, j)-entry of $P^k = (p_{ij}^{(k)}, i, j = 1, \dots, n)$ is not zero. A matrix P is called recurrent if it is stochastic and has a strictly positive invariant probability distribution. Let $\pi = (\pi_1, \dots, \pi_n)$ be an invariant probability distribution of the recurrent matrix P, that is, $\pi_i > 0$, $i = 1, \dots, n$, $\sum_i \pi_i = 1$, and $\pi P = \pi$. Then π is an invariant probability distribution of the m-step transition matrix $P^m = (p_{ij}^{(m)}, i, j \in S)$, as well.

A general rule which assigns a recurrent matrix P to a rotational partition $\mathcal{S} = \{S_1, \dots, S_n\}$ of $[0, 1)$, consists of two consecutive transforms; the first associates with P a collection $\{\mathcal{C}, w_c\}$ of directed circuits and circuit-weights satisfying a relation of the form ($\mathfrak{c}$), and the second assocites with $\{\mathcal{C}, w_c\}$ a rotational partition $\mathcal{S}$ (see [8], pp. 137; 150–157). The first transform is defined by using one of the many algorithms, which may be either combinatorial or probabilistic, or homologic (see [8], pp. 47–60). Next, we shall use the probabilistic algorithm of [6], which associates with an $n \times n$ irreducible matrix P a canonical cycle representation $\{\mathcal{C}, w_c\}$ as defined in Section 1. This gives us a one-to-one transform $P \to \{\mathcal{C}, w_c\}$. Here, we shall apply this algorithm to any matrix P^m, $m = 1, 2, \dots$. Consequently, we may write the following cycle decomposition:

$$\pi_i p_{ij}^{(m)} = \sum_{k=1}^{N} {}^m w_{c_k} C_{c_k}^{(1)}(i, j), \quad i, j \in S, \ m \geq 1, \tag{9}$$

where $C = \{c_1, \ldots, c_N\}$ is the ordered sequence of all directed circuits in S (with distinct points except for the terminals), $C_{c_k}^{(1)}(i,j) \equiv (1/p(c_k))J_{c_k}^{(1)}(i,j)$ and

$$\begin{aligned} {}^m w_c = p(c)\pi_{i_1} p_{i_1 i_2}^{(m)} \cdot \ldots \cdot p_{i_{s-1} i_s}^{(m)} p_{i_s i_1}^{(m)} \\ \cdot {}^m N(i_2, i_2 \mid i_1)\, {}^m N(i_3, i_3 \mid i_1, i_2) \ldots {}^m N(i_s, i_s \mid i_1, \ldots, i_{s-1}). \end{aligned} \tag{10}$$

Here $c = (i_1, \ldots, i_{s-1}, i_1)$, $s > 1$, $p(c_k)$ denotes as always the period of c_k, and

$${}^m N(i_k, i_k \mid i_1, \ldots, i_{k-1}) = \sum_{n=0}^{\infty} \sum_{j_1, \ldots, j_{n-1} \neq i_1, \ldots, i_{k-1}} p_{i_k j_1}^{(m)} p_{j_1 j_2}^{(m)} \cdots p_{j_{n-1} i_k}^{(m)}$$

is the taboo Green function on P^m. Since P is irreducible, for m sufficiently large, all the entries $p_{ij}^{(m)}$ are strictly positive. Then

$${}^m w_{c_k} > 0, \quad k = 1, \ldots, N.$$

In addition, we have

$$\sum_{k=1}^{N} {}^m w_{c_k} = 1.$$

Furthermore,

$$\pi_i = \sum_{k=1}^{N} (1/p(c_k))\, {}^m w_{c_k} J_{c_k}(i), \quad i \in S, \ m \geq 1. \tag{11}$$

According to the Chapman–Kolmogorov equations we may further write

$$\pi_i p_{ij}^{(2)} = \sum_{k \in S} \frac{1}{\pi_i} (\pi_i p_{ik})(\pi_k p_{kj}), \quad i, j \in S.$$

In general, we have

$$\pi_i p_{ij}^{(m)} = \sum_{j_1,\dots,j_{m-1}} (\pi_{j_1}\cdot\ldots\cdot\pi_{j_{m-1}})^{-1}(\pi_i p_{ij_1})(\pi_{j_1}p_{j_1j_2})\cdot\ldots\cdot(\pi_{j_{m-1}}p_{j_{m-1}j})$$
$$= \sum_{c_1,\dots,c_m\in\mathcal{C}}\Bigg(\sum_{j_1,\dots,j_{m-1}} (\pi_{j_1}\cdot\ldots\cdot\pi_{j_{m-1}})^{-1}C_{c_1}^{(1)}(i,j_1)C_{c_2}^{(1)}(j_1,j_2)$$
$$\cdot\ldots\cdot C_{c_m}^{(1)}(j_{m-1},j)\Bigg)\cdot w_{c_1}\cdot\ldots\cdot w_{c_m}. \tag{12}$$

Introduce

$$C_c^{(1)}(i) \equiv \sum_j C_c^{(1)}(i,j) = \sum_j C_c^{(1)}(j,i),$$

$$C_{c_1c_2}^{(2)}(i,j) = \begin{cases} \sum_k \frac{1}{w_{c_1c_2}\pi_k} C_{c_1}^{(1)}(i,k)C_{c_2}^{(1)}(k,j), & \text{if } c_1 \text{ and } c_2 \\ & \text{have common points,} \\ 0, & \text{otherwise,}\end{cases}$$

where

$$w_{c_1c_2} = \sum_{k\in S}\frac{1}{\pi_k}C_{c_1}^{(1)}(k)C_{c_2}^{(1)}(k), \quad c_1,c_2\in\mathcal{C}.$$

Define

$$C_{c_1\ldots c_m}^{(m)}(i,j) = \sum_{j_1,\dots,j_{m-1}} (w_{c_1\ldots c_m}\pi_{j_1}\cdot\ldots\cdot\pi_{j_{m-1}})^{-1}$$
$$\cdot C_{c_1}^{(1)}(i,j_1)C_{c_2}^{(1)}(j_1,j_2)\cdot\ldots\cdot C_{c_m}^{(1)}(j_{m-1},j),$$
$$\text{if } C_{c_1}^{(1)}(i,j_1)C_{c_2}^{(1)}(j_1,j_2)\cdot\ldots\cdot C_{c_m}^{(1)}(j_{m-1},j)\neq 0,$$
$$\text{for some } j_1,\dots,j_{m-1},$$
$$= 0, \quad \text{otherwise,}$$

where

$$w_{c_1\ldots c_m} = \sum_{j_1,\dots,j_{m-1}} (\pi_{j_1}\cdot\ldots\cdot\pi_{j_{m-1}})^{-1}C_{c_1}^{(1)}(j_1)C_{c_2}^{(1)}(j_1,j_2)$$
$$\cdot\ldots\cdot C_{c_{m-1}}^{(1)}(j_{m-2},j_{m-1})\cdot C_{c_m}^{(1)}(j_{m-1}),$$

for any $c_1,\dots,c_m\in\mathcal{C}$, $m\geq 2$. We call the matrix $C_c^{(1)} = (C_c^{(1)}(i,j),\ i,j = 1,\dots,n)$ the *one-step cycle matrix* associated to c. Similarly, $C_{c_1\ldots c_m}^{(m)} = (C_{c_1\ldots c_m}^{(m)}(i,j),\ i,j = 1,\dots,n)$ is called the *m-step cycle matrix* associated to $c_1, \dots, c_m$, $m \geq 2$. These matrices are invariant to the changes of the circuit-weights. The cycle matrices enjoy the following property:

$$\sum_{i,j \in S} C^{(m)}_{c_1 \dots c_m}(i,j) = 1, \quad c_1, \dots, c_m \in \mathcal{C}.$$

According to relations (12), we may write

$$\pi_i p_{ij}^{(m)} = \sum_{c_1, \dots, c_m \in \mathcal{C}} w_{c_1 \dots c_m} {}^1w_{c_1} \cdot \ldots \cdot {}^1w_{c_m} C^{(m)}_{c_1 \dots c_m}(i,j), \qquad (13)$$

for all $i, j \in S$, where

$$\sum_{c_1, \dots, c_m \in \mathcal{C}} w_{c_1 \dots c_m} {}^1w_{c_1} \cdot \ldots \cdot {}^1w_{c_m} = 1.$$

We now conclude that for each ordering providing all the possible cycles in $S = \{1, \dots, n\}$ and for any choice of the representatives of these cycles, we may assign the sequence $\{P, P^2, \dots, P^m, \dots\}$ of powers of an irreducible matrix P to a unique sequence $\{({}^1\mathcal{C}, {}^1w_c), \dots, ({}^m\mathcal{C}, {}^mw_c), \dots\}$ of cycle representations which satisfy equations (9) and (10). As in Section 1, we call $\{{}^m\mathcal{C}, {}^mw_c\}$ the *canonical cycle representation* of P^m, and equation (9) the *canonical decomposition.*

With these preparations we shall further define a rotational representation for any P^m, $m \geq 1$, by extending the procedure of the author [6] and [8].

Let $M = n!$ and put $t = 1/M$. Partition the circumference of a circle c into N equal consecutive directed arcs $V_1, \dots, V_N$ such that each V_k is associated with the rth circuit c_k of $\mathcal{C}$. Next partition each arc V_k, $k = 1, \dots, N$, into M equal circle-arcs denoted by $\alpha_{k1}, \alpha_{k2}, \dots, \alpha_{kM}$. Further, consider the following rule of assignment:

the index of each α_{k1} along the circle c is given by the pair (k, l) of the $\hat{c}_k(l)$ occurring in the sequence $\hat{c}_k(1), \dots, \hat{c}_k(p(c_k))$, $\hat{c}_k(1 + p(c_k)), \dots, \hat{c}_k(M)$,

where $\hat{c}_k$ is the cycle associated to c. Then we shall call c the *index-circle* of our procedure.

Now, consider another circle and let r_τ be the rotation of length $\tau = 2\pi/M$ ($\pi = 3.14$) of this circle. Divide this circle into M equal arcs of length $2\pi/M$. Let $\tilde{A}$ be one of these arcs. Partition $\tilde{A}$ into N consecutive equal arcs $\tilde{A}_1, \dots, \tilde{A}_N$. Define $\tilde{A}_{k1} = r_\tau^{l-1}(\tilde{A}_k)$, $k = 1, \dots, N$; $l = 1, \dots, M$. Let A and, for each m, let ${}^mA_1, \dots, {}^mA_N$ be the homeomorphs of $\tilde{A}$ and $\tilde{A}_1, \dots, \tilde{A}_N$ in $[0, 1)$ defined as follows: $A = [0, 1/M)$, ${}^mA_1$ starts at 0 and ${}^mA_1, \ldots, {}^mA_N$ are consecutive disjoint subintervals of the form $[\alpha, \beta)$ having relative lengths given by the coordinates of the probability distribution $({}^mw_{c_1}, {}^mw_{c_2}, \ldots, {}^mw_{c_N})$ occurring in the canonical decomposition (9), that is, $\lambda({}^mA_k)/\lambda(A) = {}^mw_{c_k}$, $k = 1, \ldots, N$. (Here λ denotes Lebesgue measure.) Define, for each $m \geq 1$, ${}^mA_{kl} = f_t^{l-1}({}^mA_k)$, $k = 1, \ldots, N$; $l = 1, \ldots, M$, where f_t is the shift (1) of the length $t = 1/M$ on the interval $[0, 1)$. Then $\lambda({}^mA_{kl}) = (1/M)^m w_{c_k}$, $k = 1, \ldots, N$; $l = 1, \ldots, M$. Define also ${}^mU_k = \bigcup_{l=1}^M {}^mA_{kl}$ for $k = 1, \ldots, N$ and $l = 1, \ldots, M$. Then for each $m \geq 1$, ${}^mU_1, \ldots, {}^mU_N$ are the homeomorphs of $V_1, \ldots, V_N$ with $\lambda({}^mU_k) = {}^mw_{c_k}$. Define the sets

$$ {}^mS_i = \bigcup_{(k,l)} {}^mA_{kl}, \quad i = 1, \ldots, n; \; m \geq 1, \tag{14} $$

where the label of each A_{kl} in the union is given by any pair $(k, l) = (k_i, l_i)$ defined as follows:

(i) k_i *is the index of a chosen representative of a class-circuit* c_k, $k = \{1, \ldots, N\}$, *which passes through the pre-given point* i *and which occurs in the decomposition* (9);

(ii) l_i *denotes those ranks* $n \in \{1, \ldots, M\}$ *of all the points* $\hat{c}_k(n)$ *which are identical to* i *in the* $M/p(c_k)$ *repetitions of the cycle* $\hat{c}_k = (\hat{c}_k(1), \hat{c}_k(2), \ldots, \hat{c}_k(p(c_k))$ *associated to the representative of the circuit* c_k *chosen at* (i) *above, i.e., if for some* $s \in \{1, \ldots, p(c_k)\}$ *we have* $\hat{c}_k(s) = \hat{c}_k(s + p(c_k)) = \cdots = \hat{c}_k(s + (M/p(c_k) - 1)p(c_k)) = i$, *then* $l_i \in \{s, s + p(c_k), \ldots, s + (M/p(c_k) - 1)p(c_k)\}$. (15)

The rth repetition of $\hat{c}_k$, with $r \in \{1, \ldots, M/p(c_k)\}$, is meant to be the sequence $(\hat{c}_k(1+(r-1)p(c_k)), (\hat{c}_k(2+(r-1)p(c_k)), \ldots, (\hat{c}_k(p(c_k)+(r-1)p(c_k))$. If (i, j) is an edge of c_k, then $\lambda({}^mS_i \cap f_t^{-1}({}^mS_j) \cap {}^mU_k) = (1/p(c_k))w_{c_k}$. Accordingly, the entries of the one-step cycle matrices $C_{c_1}^{(1)}, \ldots, C_{c_N}^{(1)}$ are the following conditioned probabilities:

$$ C_{c_k}^{(1)}(i, j) = \lambda({}^mS_i \cap f_t^{-1}({}^mS_j) \mid {}^mU_k), \quad i, j \in S; \; m \geq 1. \tag{16} $$

Representations (16) may be interpreted in terms of rotations as follows:

if one rotation along the index-circle holds, starting at i *and ending at* j *in* V_k, *then one rotation* r_τ, $\tau = 2\pi/M$, *of some arcs of* ${}^mS_j$, *occurs with probability* $C_c^{(1)}(i, j)$.

Then we conclude that $\{C_{c_k}^{(1)}(i, j), \; i, j = 1, \ldots, n\}$ is the probability distribution ("uniform", on the positive coordinates) of the circulation of the edges (i, j) through the directed circuits c_k. Also, relations (16) show that the conditioned rotational representations $\lambda({}^mS_i \cap f_t^{-1}({}^mS_j \mid {}^mU_k)$, $k = 1, \ldots, N$; $m \geq 1$, are independent of the representative circuit-weights in the decomposition (9).

The conditioned rotational representation (16) is the basic step to the rotational representation of P^m, $m \geq 1$, since

$$\pi_i p_{ij}^{(m)} = \sum_{k=1}^{N} {}^m w_{c_k} C_{c_k}^{(1)}(i,j) = \sum_{k=1}^{N} \lambda({}^mU_k)\lambda({}^mS_i \cap f_t^{-1}({}^mS_j) \mid {}^mU_k).$$

Then

$$\pi_i p_{ij}^{(m)} = \lambda({}^mS_i \cap f_t^{-1}({}^mS_j)) \tag{17}$$

and

$$\pi_i = \lambda({}^mS_i)$$

for all $i, j \in \{1, \ldots, n\}$ and for any $m \geq 1$. Therefore, for each $m \geq 1$ $(1/M, {}^m\mathcal{S})$ with ${}^m\mathcal{S} = \{{}^mS_1, \ldots, {}^mS_n\}$ is the unique rotational representation of P^m such that the rotational representation process has the same distribution of cycles as the probabilistic cycle distribution (10) on P^m. We call $(1/M, {}^m\mathcal{S})$ the *canonical rotational representation* of P^m.

Now, let $(1/M, {}^1\mathcal{S} = \{{}^1S_1, \ldots, {}^1S_n\})$ be the canonical rotational representation of P. Then

$$C_{c_1c_2}^{(2)}(i,j) = \sum_k \frac{1}{w_{c_1c_2}\pi_k}\lambda({}^1S_i \cap f_t^{-1}({}^1S_k) \mid {}^1U_{c_1})\lambda({}^1S_k \cap f_t^{-1}({}^1S_j) \mid {}^1U_{c_2})$$

if the circuits c_1 and c_2 overlap, where ${}^1U_{c_1}$ and ${}^1U_{c_2}$ are given by the above rotational procedure on P. Then the Chapman–Kolmogorov equations on P^2 have the following expression:

$$\lambda({}^2S_i \cap f_t^{-1}({}^2S_j)) = \sum_{c_1,c_2 \in C} \Bigg(\sum_k \frac{1}{w(k)}\lambda({}^1S_i \cap f_t^{-1}({}^1S_k) \mid {}^1U_{c_1}) \cdot \lambda({}^1S_k \cap f_t^{-1}({}^1S_j) \mid {}^1U_{c_2})\Bigg)_{{}^1w_{c_1}{}^1w_{c_2}} {}^1w_{c_1}{}^1w_{c_2}$$

according to which, they enjoy the following characterization in terms of rotations:

to two independent collections of rotations along the index-circle, one starting at i and the other ending at j, there corresponds one *rotation r_τ, with $\tau = 2\pi/M$, of some arcs of ${}^2S_i$, which rotating, overlap some other arcs of ${}^2S_j$ with probability $\lambda({}^2S_i \cap f_t^{-1}({}^2S_j))$.*

3.2. Let us now introduce the following definition.

Definition 4. A sequence $\{{}^m\mathcal{S}\}_{m\geq 1}$, of n-partitions ${}^m\mathcal{S} = \{{}^mS_1, \dots, {}^mS_n\}$ of $[0,1)$, is called a stochastic sequence of n-partitions if and only if it satisfies the following Chapman–Kolmogorov conditions with respect to a rotation f_t:

$$\lambda({}^{k+m}S_i \cap f_t^{-1}({}^{k+m}S_j)) = \sum_{u\in S} \lambda({}^mS_u)\lambda({}^kS_i \cap f_t^{-1}({}^kS_u))\lambda({}^mS_u \cap f_t^{-1}({}^mS_j)), \quad (18)$$

for all $i, j = 1, \dots, n$, and for all $k, m \geq 1$.

We are now prepared to prove the main result of this section.

Theorem 4. *Assume $n \geq 2$, an ordering of all possible cycles in $S = \{1, \dots, n\}$ and a choice of the representatives of these cycles. Then the following statements hold:*

(i) *There exists a map from the space of sequences $\{P^m\}_{m\geq 1}$ of powers of $n \times n$ irreducible stochastic matrices P into stochastic sequences $({}^m\mathcal{S} = \{{}^mS_i,\ i = 1, \dots, n\})_{m\geq 1}$ of n-partitions of $[0,1)$ such that the distributions of cycles of the rotational representation processes $(f_t, \{{}^mS_i\})$ with $t = 1/n!$ match the canonical cycle distributions on P^m, $m \geq 1$.*

(ii) *There exists a map from the space of stochastic sequences $\{{}^m\mathcal{S}\}_{m\geq 1}$ of n-partitions of $[0,1)$ into sequences $\{P^m\}_{m\geq 1}$ of powers of $n \times n$ recurrent stochastic matrices P.*

Proof. The proof of statement (i) follows from the rotational procedure of Subsection 3.1 applied to the canonical cycle decomposition (9). Accordingly, under the assumptions of the theorem, for any sequence $\{P, P^2, \dots, P^m, \dots\}$ one may define a unique stochastic sequence $\{{}^1\mathcal{S}, {}^2\mathcal{S}, \dots, {}^m\mathcal{S}, \dots\}$ of n-partitions of $[0,1)$ such that, for each m, the distribution of cycles of the rotational representation process $\{f_t, \{{}^mS_i\}\}$, with $t = 1/n!$, matches the canonical cycle distribution on P^m (see also Theorem 3.5.1 of [8]).

To prove statement (ii), let us consider a sequence $\{{}^m\mathcal{S}\}_{m\geq 1}$ of n-partitions of $[0,1)$, which is stochastic with respect to a rotation f_t, $t > 0$. Define the matrices $P^{(m)} = (p_{ij}^{(m)},\ i, j = 1, \dots, n)$, $m \geq 1$, by

$$p_{ij}^{(m)} = \lambda({}^mS_i \cap f_t^{-1}({}^mS_j))/\lambda({}^mS_i), \quad i, j = 1, \dots, n,$$

where ${}^m\mathcal{S} = \{{}^mS_1, \ldots, {}^mS_n\}$. Denote $P = P^{(1)}$. Then each matrix $P^{(m)}$ is recurrent and is represented by the rotational system $(t, {}^m\mathcal{S})$.

Finally, since the sequence $\{{}^m\mathcal{S}\}_{m\geq 1}$ is stochastic, it follows that $P^{(m)} = P^m$, $m \geq 1$. The proof is complete. ■

4. Asymptotic behavior of rotational representations

In the present section we study the convergence of the sequence $\{t, \mathcal{S}(P^m)\}_{m\geq 1}$ of canonical rotational representations associated, by Theorem 4, to the stochastic matrices $P, P^2, \ldots, P^m, \ldots,$ as $m \to \infty$. Before proceeding, we recall the following definitions. Let $P = (p_{ij}, i, j = 1, \ldots, n)$ be an irreducible stochastic matrix. If, for a state $i \in \{1, \ldots, n\}$ the entry $p_{ii}^{(k)}$ of P^k is strictly positive, designate by d_i the greatest common divisor of all $k \geq 1$ that satisfy the inequality $p_{ii}^{(k)} > 0$. An irreducible stochastic matrix $P = (p_{ij}, i, j = 1, \ldots, n)$ is called *periodic* (or *aperiodic*) if $d_i > 1$ (or $d_i = 1$) for one state i (and then for all states). A stochastic matrix is called *stable* if all its row-vectors are identical.

We shall say that a collection C of directed circuits in $S = \{1, \ldots, n\}$ satisfies condition $(\mathcal{T})$ when

$(\mathcal{T})$ *any two distinct integers i and j of S are circuit-edge-connected, that is, there exists a sequence of directed circuits $c_1, \ldots, c_m, c_{m+1}, \ldots, c_s$ of C such that i lies on c_1 and j on c_n, and any pair of consecutive circuits c_m and c_{m+1} have at least one common point.*

We are now prepared to prove

Theorem 5. *Under the assumptions of Theorem 4, the following statements hold:*

(i) (1) *If $P = (p_{ij}, i, j = 1, \ldots, n)$ is an irreducible aperiodic stochastic matrix, then the canonical rotational partitions $\mathcal{S}(P^m)$ associated to P^m, $m = 1, 2, \ldots,$ converge, as $m \to \infty$, to a rotational partition $\mathcal{S}$ such that $(t, \mathcal{S})$ with $t = 1/n!$ is a rotational representation of the stable matrix $\Pi = \lim_{m\to\infty} P^m$. The convergence of the n-partitions of $[0, 1)$ is understood with respect to the metric d defined as*

$$d(\mathcal{S}(P^m), \mathcal{S}) = \sum_i \lambda({}^mS_i + S_i), \tag{19}$$

where λ denotes Lebesgue measure on Borel subsets of $[0, 1)$ and $+$ denotes symmetric difference.

(i) (2) *If P is an irreducible periodic stochastic matrix with period d, then the canonical rotational partitions associated with $(I + P + \cdots +$*

$P^{m-1})/m$, $m = 1, 2, \dots$, converge, as $m \to \infty$, to a rotational partition $\tilde{\mathcal{S}}$ such that $(t, \tilde{\mathcal{S}})$ with $t = 1/n!$ is a rotational representation of the stable matrix $\tilde{\Pi} = \lim_{m\to\infty}(I + P + \cdots + P^{m-1})/m$.

(ii) *Let $\{{}^m\mathcal{S}\}_m$ be a stochastic sequence of n-partitions of $[0, 1)$ with respect to a rotation f_t such that the collection of cycles of any cycle distribution of the rotational representation process $(t, {}^m\mathcal{S})$ for some $m \geq 1$ contains a loop and satisfies condition $(\mathcal{T})$. Then there exists a convergent sequence $\{P^m\}_m$ of powers of a stochastic matrix P such that $(t, {}^m\mathcal{S})$ is a rotational representation of P^m, $m = 1, 2, \dots$.*

Proof. (i)(1) Let $P = (p_{ij}, i, j = 1, \dots, n)$ be an irreducible and aperiodic stochastic matrix. Then, according to Theorem 4, we may associate the sequence $\{P^m\}_{m\geq 1}$ with the sequence $(1/n!, {}^m\mathcal{S})$, $m = 1, 2, \dots$ of canonical rotational representations of P^m, $m = 1, 2, \dots$. Let $\mathcal{C}$ be the ordered collection of directed circuits occurring in the canonical cycle decompositions of P^m, $m = 1, 2, \dots$. According to the well-known theorem of Markov, the sequence $\{P^m\}_{m\geq 1}$ converges, as $m \to \infty$, to a strictly positive stable matrix $\Pi = (\pi_{ij}, i, j \in S)$. Furthermore, all the row-vectors of Π are identical to the invariant probability distribution of P. Then, taking the limit in the corresponding equations of type (9), we get a cycle decomposition for Π, which, by the procedure of Subsection 3.1, will determine a rotational representation $(1/n!, \mathcal{S})$ for Π. Since the above sequence $\{{}^m\mathcal{S}\}_m$ of n-partitions is stochastic, the metric d defined by (19) has the expression

$$d({}^m\mathcal{S}, \mathcal{S}) = 2\sum_i \lambda({}^m S_i \setminus S_i) = 2\sum_i \lambda(S_i \setminus {}^m S_i).$$

Finally, convergence in metric d of the canonical rotational partitions ${}^m\mathcal{S}$ to the rotational partition $\mathcal{S}$ of Π, follows from the convergences of $\lambda({}^m A_k) = (1/n!)^m w_{c_k}$, $c_k \in \mathcal{C}$, as $m \to \infty$, provided by the procedure of Subsection 3.1 on the canonical cycle decompositions of P^m, $m = 1, 2, \dots$.

(i)(2) If P is a periodic irreducible stochastic matrix of period d, then $\{P^m\}_m$ converges in Cesaro mean, as $m \to \infty$, to a strictly positive stable matrix $\tilde{\Pi}$. Specifically, if $c_0, c_1, \dots, c_{d-1}$ denote the cycle subclasses of $S = \{1, \dots, n\}$ then the matrix P^d has the form

$$P^d = \begin{array}{c} \\ c_0 \\ c_1 \\ \vdots \\ c_{d-1} \end{array} \!\!\begin{array}{c} \begin{array}{ccc} c_0 & c_1 & \cdots \; c_{d-1} \end{array} \\ \begin{pmatrix} X & & \\ & X & \mathbf{0} \\ & \mathbf{0} & \ddots & \\ & & & X \end{pmatrix} \end{array},$$

where $\mathbf{0}$ contains only entries equal to zero. Then, applying Markov's theorem to each positive submatrix on c_r, $r = 0, \ldots, d-1$, it follows that $P^{nd} \to \Pi^*$, as $n \to \infty$, where

$$P^d = \begin{array}{c} \\ c_0 \\ c_1 \\ \vdots \\ c_{d-1} \end{array} \begin{array}{c} \begin{array}{cccc} c_0 & c_1 & \cdots & c_{d-1} \end{array} \\ \begin{pmatrix} \Pi_0 & & & \mathbf{0} \\ & \Pi_1 & & \\ & \mathbf{0} & \ddots & \\ & & & \Pi_{d-1} \end{pmatrix} \end{array}$$

Here each Π_r, $r = 0, \ldots, d-1$, is a stochastic matrix whose row-vectors are all identical to a strictly positive probability row-distribution π_r. Accordingly, the sequence $\{P^m\}_m$ has d convergent subsequences, which in turn imply the convergence of the arithmetic mean $(I + P + \cdots + P^{m-1})/m$ to the stochastic matrix

$$\tilde{\Pi} = d^{-1} \sum_{r=0}^{d-1} P^r \Pi^* = d^{-1} \sum_{r=0}^{d-1} \Pi^* P^r.$$

Furthermore, the invariant probability distribution of $\tilde{\Pi}$ is the row-vector $\pi = d^{-1}(\pi_0, \pi_1, \ldots, \pi_{d-1})$, where each π_r occurs in Π_r, $0 \leq r \leq d-1$. Then, since the canonical cycle decompositions still are defined for periodic irreducible matrices, to prove (i)(2) we may apply to the sequence $\{(I + P + \cdots + P^{m-1})/m\}$ the same reasonings of (i)(1) above.

(ii) Let $\{{}^m\mathcal{S}\}_m$ be a stochastic sequence on n-partitions of $[0, 1)$ with respect to a rotation f_t, which satisfies the hypothesis of statement (ii). Then, according to Theorem 4 (ii) there exists a sequence $\{P^m\}$ of powers of a stochastic matrix $P = (p_{ij}, i, j = 1, \ldots, n)$ whose entries are defined as

$$p_{ij} = \lambda({}^1S_i \cap f_t^{-1}({}^1S_j))/\lambda({}^1S_i)$$

and $(t, {}^m\mathcal{S})$ is the rotational representation of P^m, $m = 1, 2, \ldots$. Furthermore, condition $(\mathcal{T})$ implies that, for some m (and then for any $s \geq m$) the stochastic matrix P^m is irreducible. Also, by hypothesis, the matrix P^m is aperiodic.

Let us prove that the sequence $\{P^m\}_m$ is convergent, as $m \to \infty$. To this end we shall consider the canonical rotational representations $(1/n!, {}^m\tilde{\mathcal{S}})$ for the matrices P^m, $m = 1, 2, \ldots$. Then the common invariant probability distribution of P^m, $m = 1, 2, \ldots$, is the row-vector $\pi = (\pi_1, \ldots, \pi_n)$ given by

$$\pi_i = \lambda({}^m\tilde{S}_i), \quad i = 1, \ldots, n; \; m = 1, 2, \ldots,$$

where ${}^m\tilde{\mathcal{S}} = \{{}^m\tilde{S}_1, \dots, {}^m\tilde{S}_n\}$. Since for m sufficiently large each set ${}^m\tilde{S}_i$ is indexed by the same set I_i of indices (k,l) according to labeling (15), it follows that

$$\pi_i = \sum_{(k,l)\in I_i} \lambda({}^m A_{kl}), \quad i = 1, \dots, n,$$

or,

$$\pi_i = \sum_{(k,l)\in I_i} (1/n!)^m w_{c_k}, \quad i = 1, \dots, n.$$

Therefore, $\lim_{m\to\infty} {}^m w_c$ exists and is finite for all circuits c in S. Then, taking the limit in the corresponding equations of type (10), it follows that the limits $\lim_{m\to\infty} p_{ij}^{(m)}$, $i, j \in S$, exist and are finite, as well. The proof is complete. ■

An immediate consequence of Theorem 5 is the proof in terms of rotations of the Markov theorem concerning convergence of $\{P^m\}_m$. Accordingly, we may state

Corollary 6 (Markov's theorem in terms of rotations). *If P is an aperiodic irreducible $n \times n$ stochastic matrix, then the sequence $\{P^m\}_m$ and sequence $\{\mathcal{S}(P^m)\}_m$ of canonical rotational partitions converge.*

Acknowledgments. I would like to express my gratitude to Professor Ioannis Karatzas, who has promoted the organization of the Conference in the memory of Stamatis Cambanis (Athens, December 18–19, 1995) and the edition of the present commemorative volume. I also thank Professor Ioannis Karatzas for his valuable comments which improved the presentation of the paper, and Professor Persi Diaconis for interesting discussions on cycle representations.

References

[1] D. Aldous, P. Diaconis, and M. J. Steele, *Discrete Probability and Algorithms*, Springer-Verlag, New York, 1995.

[2] S. Alpern, Rotational representations of stochastic matrices, *Ann. Probability* **11** (3), (1983), 789–794.

[3] J. E. Cohen, A geometric representation of stochastic matrices; theorem and conjecture, *Ann. Probability* **9** (1981), 899–901.

[4] Y. Derriennic, Ergodic problems on random walks in random environment, in *Selected Talks Delivered at the Department of Mathematics of*

the Aristotle University, S. Kalpazidou ed., Aristotle University Press, Thessaloniki, 1993.

[5] J. Haigh, Rotational representation of stochastic matrices, *Ann. Probability* **13** (1985), 1024–1027.

[6] S. Kalpazidou, Rotational representations of transition matrix functions, *Ann. Probability* **22** (2), (1994), 703–712.

[7] S. Kalpazidou, On the rotational dimension of stochastic matrices, *Ann. Probability* **23** (2), (1995), 966–975.

[8] S. Kalpazidou, *Cycle Representations of Markov Processes*, Springer-Verlag, New York, 1995.

[9] S. Kalpazidou and J.E. Cohen, Orthogonal cycle transforms of stochastic matrices, *Circ. Sys. Sig. Proc.*, **16**(2), (1997), 363–374.

[10] S. Kalpazidou and N. Kassimatis, Markov chains in Banach spaces on cycles, *Circ. Sys. Sig. Proc.*, to appear.

[11] S. Kalpazidou, From network problem to cycle processes, Proceedings of the 2nd World Congress of Nonlinear Analysts, Athens, 1996, Elsevier, part 4, 2041–2049.

[12] Qian Minping and Qian Min, Circulation for recurrent Markov chain, *Z. Wahrsch. Verw. Gebiete* **59** (1982), 203–210.

[13] P. M. Soardi, *Potential Theory on Infinite Networks*, Lecture Notes in Mathematics, Springer-Verlag, New York, 1994.

[14] P. Del Tio Rodriguez and M. C. Valsero Blanco, A characterization of reversible Markov chains by a rotational representation, *Ann. Probability* **19** (2), (1991), 605–608.

[15] W. Woess, Random walks on infinite graphs and groups – A survey on selected topics, *Bull. London Math. Soc.* **26** (1994), 1–60.

[16] A. H. Zemanian, *Infinite Electrical Networks*, Cambridge University Press, Cambridge, 1991.

Department of Mathematics
Aristotle University of Thessaloniki
54006 - Thessaloniki
Greece

On Extreme Values in Stationary Random Fields

M.R. Leadbetter and Holger Rootzén

Summary

This paper develops distributional extremal theory for maxima $M_{\mathbf{T}} = \max(X_{\mathbf{t}} : 0 \leq \mathbf{t} \leq \mathbf{T})$ of a stationary random field $X_{\mathbf{t}}$. A general form of "extremal types theorem" is proven and shown to apply to $M_{\mathbf{T}}$ under very weak dependence restrictions. That is, any non-degenerate distributional limit for the normalized family $a_{\mathbf{T}}(M_{\mathbf{T}} - b_{\mathbf{T}})$ $(a_{\mathbf{T}} > 0)$ must be one of the three classical types. Domain of attraction criteria are discussed.

The dependence structure used here for fields involves a potentially very weak type of strong-mixing, "Coordinatewise (Cw) mixing") using mild individual "past-future" conditions in each coordinate direction. Together with careful control of numbers and sizes of sets involved, this avoids the over-restrictive nature of common generalizations of mixing conditions to apply to random fields. Futher, the conditions may be readily adpated to deal with other quite general problems of Centeral Limit type (cf. [6]).

1. Introduction

Classical extreme value theory primarily concerns the distribution of the maximum $M_n = \max(X_1, X_2 \ldots X_n)$ of i.i.d. random variables X_i, and its centerpiece, the "extremal types theorem" asserts that if $P\{a_n(M_n - b_n) \leq x\} \to G(x)$, non-degenerate, (some $a_n > 0, b_n$) then G must be one of the three types

Type I	$G(x) = \exp(-e^{-x})$	$-\infty < x < \infty$
Type II	$G(x) = \exp(-x^{-\alpha})$	$x \geq 0 \ (\alpha > 0)$
Type III	$G(x) = \exp(-(-x)^{\alpha})$	$x \leq 0 \ (\alpha > 0)$

in which x may be replaced by $ax + b$ for real $a > 0, b$.

It is known [5] that this result holds for dependent (stationary) sequences X_n satisfying a very weak mixing condition, and also for limits $P\{a_T(M_T - b_T) \leq x\} = G(x)$ for the maximum $M_T = \{\sup X(t) : 0 \leq t \leq T\}$, of a stationary process $\{X_t,\ t \geq 0\}$, again under a weak mixing condition. A basic reason for this diverse validity is that in each case

*Research supported by The Office of Naval Research, grant N00014-93-1-0043

the relevant maximum may be regarded as the maximum of an *approximately independent sequence* of submaxima. For a stationary sequence these are the maxima of successive groups of (locally dependent) consecutive terms, and for continuous parameter processes, the maxima over intervals. For example $M_T = \sup \{X_t : 0 < t \le T\} = \max \{Z_1, Z_2 \dots Z_n\}$ where $Z_i = \sup \{X_t : (i-1)/n < t \le i/n\}$ for integer values $T = n$, and approximately with $n = [T]$ otherwise. Long range dependence restrictions may then be used to guarantee sufficient independence of the Z_i to employ the classical theory.

For stationary Gaussian random fields it is known (cf. [1, 7]) that the maximum has a Type I limit under very weak covariance conditions paralleling those of Berman, Pickands (see e.g. [5]) for normal processes. In view of the above remarks it is thus not surprising that an extremal types theorem should hold for stationary random fields under appropriate long–range dependence restrictions.

As will be described in Section 2, the heart of the method can be stated as a simple general result which can be applied to the cases referred to above, and also to the random field context. This will be illustrated specifically in Sections 3 and 4 for random fields in two dimensions — the generalizations to higher dimensions and more abstract parameter situations being evident. A dependence restriction for random fields ("Cw–mixing") is introduced in Section 3. This is a new form of codnition of (array) strong mixing type but replacing a single condition for pairs of separated but "interlaced" sets by one in each coordinate direction involving simple "past and future" separation. Futher, the condition involves far fewer events — indeed only those which are directly germane to the extremal problem rather than σ–fields generated by the random field, and "control sizes" of certain sets in the conditions, in an analogous way to that sometimes used in discrete parameter contexts (cf. [2, 4, 3]). While verification of mixing conditions in particular cases is beyond the scope of this study, it is clear that these economies also give the potential for significant weakening of traditional mixing conditions for random fields.

Section 4 concerns the extremal types theorem for random fields based on Cw-mixing condition, and Section 5 discusses domains of attraction to the classical extreme value limits.

2. A general result on extremal types

The maximum $M_n = \max (X_1, X_2 \dots X_n)$ for i.i.d. X_i has the obvious and simple property that for any $k = 1, 2 \dots$ the distribution of M_{kn} is the kth power of that of M_n, i.e.

$$P\{M_{kn} \le x\} = P^k\{M_n \le x\}. \tag{2.1}$$

This is clear since M_{kn} is the maximum of k independent submaxima of groups of n r.v.'s X_j, e.g. $M_{kn} = \max(Z_1, Z_2, \ldots Z_k)$ with $Z_i = \max\{X_j : (i-1)n < j \leq in\},\ 1 \leq i \leq k$.

The thread common to other dependent and continuous parameter cases is that (2.1) holds approximately as n increases, with $x = x_n$ appropriately chosen for each n. This may be roughly regarded as saying that there may be high local dependence (the group of X_j's defining a given Z_i may be quite dependent) but small long–range dependence which gives approximate independence of Z_i for the k different groups.

More specifically, the method may be summarized in the following general "Extremal types theorem," couched here in a continuous parameter context.

Proposition 2.1 *Let M_T, $T > 0$, be r.v.'s such that*

$$P\{a_T(M_T - b_T) \leq x\} \to G(x), \tag{2.2}$$

non–degenerate, some $a_T > 0, b_T$, as $T \to \infty$. Suppose that for each real x, $u_T = x/a_T + b_T$ the following "mix–max" requirement holds:

$$P\{M_{kT} \leq u_{kT}\} - P^k\{M_{\phi_k(T)} \leq u_{kT}\} \to 0, \text{ as } T \to \infty \tag{2.3}$$

for each $k = 1, 2 \ldots$ and some continuous strictly increasing functions $\phi_k(T)$ ($\to \infty$ as $T \to \infty$). Then G is of extreme value type.

Proof. It follows from (2.2) and (2.3) that $P\{M_{\phi_k(T)} \leq u_{kT}\} \to G^{1/k}(x)$ so that $P\{M_T \leq u_{k\phi_k^{-1}(T)}\} \to G^{1/k}(x)$, giving

$$P\{\alpha_{k,T}(M_T - \beta_{k,T}) \leq x\} \to G^{1/k}(x)$$

where $\alpha_{k,T} = a_{k\phi_k^{-1}(T)}, \beta_{k,T} = b_{k\phi_k^{-1}(T)}$

The fact that M_T converges in distribution under the potentially different normalizations implies (by "Khintchine's Lemma" – cf. [5, Theorem 1.2.3]) that the limits are of the same type in the sense that $G^{1/k}(x) = G(\alpha_k x + \beta_k)$ some $\alpha_k > 0,\ \beta_k,\ k = 1, 2 \ldots$, i.e. G is *max stable* and hence of extreme value type (cf [5, Theorem 1.4.1]). ■

As noted above (2.3) is trivially true in the case (with $T = n,\ \phi_k(T) = T$) where M_n arises as partial maximum of an i.i.d. sequence. We refer to it as a "mix–max" condition since equality of the terms would most naturally hold when M_{kT} is the maximum of k independent terms with the same distribution as M_T, whereas the approximation suggests decay of dependence and a form of mixing. Specific mixing requirements will be discussed next for random fields.

3. Asymptotic independence in stationary random fields

This section primarily concerns the approximate independence of maxima of a random field in disjoint sets, using, as indicated earlier, a new and weak form of mixing which is readily adapted to a variety of other (central limit) contexts. We focus for simplicity on a random field (r.f.) $\{X_{\mathbf{t}} : \mathbf{t} = (t_1, t_2), t_1, t_2 \geq 0\}$ on the two–dimensional index space $(0, \infty) \times (0, \infty)$, assuming a.s. continuity of $X_{\mathbf{t}}$ in $\mathbf{t}$ and define for subsets B of this space,

$$M(B) = M_{\mathbf{T}}(B) = \sup \{X_{\mathbf{t}} : \mathbf{t} \in B\}$$

and for a vector $\mathbf{T} = (T_1, T_2),\ T_1, T_2 > 0$

$$M_{\mathbf{T}} = M(E_{\mathbf{T}})$$

where $E_{\mathbf{T}}$ (here and throughout) denotes the rectangle $(0, T_1] \times (0, T_2]$. Dependence conditions will be given in this section under which the extremal types theorem holds for $M_{\mathbf{T}}$, i.e. any non–degenerate limit G for $M_{\mathbf{T}}$ normalized as in (2.2), must be of extreme value type. Essentially the same conditions will suffice for Poisson limits for exceedances and for central limit problems of e.g. [6].

Typical "strong mixing" conditions restrict dependence by limiting $|P(E \cap F) - P(E)P(F)|$ when E, F are events determined by the field in suitably separated sets A, B. Substantial weakening may be achieved by focusing not on all such events, but just on those pertinent to the problem at hand, e.g. $E = \{M(A) \leq u\}$ (for each constant u) in the case of maxima. This works well in one dimension where the sets A, B are typically in the "past" and "future," respectively (e.g. $A \subset (0, t),\ \ B \subset (t + l, T)$ for an "observation period" T. However, for higher dimensions, without natural "past" and "future," it can lead to restrictions on many more sets than are relevant and hence to overly strict requirements. Here $E_{\mathbf{T}}$ is regarded as the "observation set," i.e. the values of $X_{\mathbf{t}}$ are available for $\mathbf{t} \in E_{\mathbf{T}}$. It will be assumed without comment that other sets (usually rectangles and their unions) are subsets of $E_{\mathbf{T}}$. Thus dependence conditions will have an "array form" within the observation set rather than involving entire "future" or "past" values.

One useful means of reduction of conditions is to control the size of the sets considered in the spirit of conditions for fields defined on lattices (cf. [2, 3, 4]) and in particular to consider only appropriately small sets A. Another is to require the sets A, B to lie on opposite sides of some half plane. Here we show that it is sufficient to use just simple separation by past and future separately in each coordinate direction. The conditions differ somewhat in each direction since they are applied sequentially. But this is a minimal complication compared with the substantial economy of conditions which results.

More specifically, for the plane, the conditions involve constants $r_1 = r_{1,\mathbf{T}} = o(T_1)$, $r_2 = r_{2,\mathbf{T}} = o(T_2)$ to be used as the lengths of sides of "blocks" (rectangles) $B_{ij} = ((i-1)r_1,\ ir_1] \times ((j-1)r_2,\ jr_2]$ which will be used for subdivision of $E_{\mathbf{T}}$. Writing $k_i = k_{i,\mathbf{T}} = [T_i/r_i]$, $i = 1,2$ ($[\cdot]$ denoting integer part), it is evident that $E_{\mathbf{T}}$ contains $k_{\mathbf{T}} = k_1 k_2$ such complete blocks, and no more than $(k_1 + k_2 - 1)$ incomplete ones. The sets B_{ij} will be referred to throughout as "standard blocks."

With this notation the random field $X_{\mathbf{t}}$ will be said to satisfy be Cw-mixing for a given family of constants (levels) $u_{\mathbf{T}}$ if for each $\mathbf{T}$, constants r_1, r_2 as above and some "separation constants" $l_i = l_{i,\mathbf{T}}$ $l_i = o(r_i)$, $i = 1,2$ exist such that

(i) (x–direction condition):

For each $h, 0 < h \leq T_2$, rectangles $B_1 = (0,a] \times (0mh]$, $0 < a < T_1$, and $B_2 = (b,c] \times (0,h]$ with $c - b \leq r_1$, $c < T_1$ lying at least a distance ℓ_1 to the right of B_1 (i.e. $b \geq a + \ell_1$)

$$\begin{aligned}|P\{M(B_1) \leq u_{\mathbf{T}},\ M(B_2) \leq u_{\mathbf{T}}\} - P\{M(B_1) \leq u_{\mathbf{T}}\}\\ P\{M(B_2) \leq u_{\mathbf{T}}\}| \leq \alpha_1(r_1, l_1)\end{aligned} \tag{3.1}$$

where the "x–mixing function" α_1 satisfies $k_1\alpha_1(r_1, l_1) \to 0$.

(ii) (y–direction condition):

For rectangles $B_1 = (0, r_1] \times (0,a]$, $0 < a \leq T_2$ and $B_2 = (0, r_1] \times (b,c]$ with $c - b \leq r_2$, $c \leq T_2$ lying at least a distance ℓ_2 above B_1 (i.e. $b \geq a + \ell_2$)

$$\begin{aligned}|P\{M(B_1) \leq u_{\mathbf{T}}, M(B_2) \leq u_{\mathbf{T}}\} - P\{M(B_1) u_{\mathbf{T}}\}\\ P\{M(B_2) \leq u_{\mathbf{T}}\}| \leq \alpha_2(r_1, r_2, l_2)\end{aligned} \tag{3.2}$$

where the "y–mixing function" α_2 satisfies $k_1 k_2 \alpha_2(r_1, r_2, l_2) \to 0$

These conditions may seem involved but become self–evident by drawing a picture for each.

The following result essentially shows that the maxima of $X_{\mathbf{t}}$ in the $k_{\mathbf{T}}$ blocks B_{ij} have a degree of asymptotic independence.

Lemma 3.1 *Let the stationary field $X_{\mathbf{t}}$ be Cw-mixing for a family of levels $u_{\mathbf{T}}$, $\mathbf{T} = (T_1, T_2)$. Then with the above notation, writing $B_{\mathbf{T}} = (0, k_1 r_1] \times (0, k_2 r_2]$ $(= \cup_{i=1}^{k_1} \cup_{j=1}^{k_2} B_{ij})$,*

$$P\{M(B_{\mathbf{T}}) \leq u_{\mathbf{T}}\} = P^{k_{\mathbf{T}}}\{M(J) \leq u_{\mathbf{T}}\} + o(1) \tag{3.3}$$

as $\mathbf{T} \to \infty$, *where* $J = B_{11} = (0, r_1] \times (0, r_2]$.

Proof. With the above notation write

$$J_i = ((i-1)r_1, ir_1] \times (0, k_2 r_2], \quad J_i' = ((i-1)r_1 + l_1, ir_1 - l_1] \times (0, k_2 r_2], \quad J_i^* = J_i - J_i'$$

Then clearly $B_{\mathrm{T}} = \cup_{i=1}^{k_1} J_i$ and for $2 \le k \le k_1$

$$\begin{aligned} 0 \le P\{M(\cup_1^{k-1} J_i) &\le u_{\mathrm{T}}, \ M(J_k') \le u_{\mathrm{T}}\} - P\{M(\cup_1^k J_i) \le u_{\mathrm{T}}\} \\ &\le P\{M(J*_1) > u_{\mathrm{T}}\} \end{aligned} \tag{3.4}$$

(using stationarity). Further

$$\begin{aligned} |P\{M(\cup_1^{k-1} J_i) &\le u_{\mathrm{T}}, \ M(J_k') \le u_{\mathrm{T}}\} - P\{M(\cup_1^{k-1} J_i) \le u_{\mathrm{T}}\} \\ &P\{M(J_k') > u_{\mathrm{T}}\}| \le \alpha_1(r_1, l_1) \end{aligned} \tag{3.5}$$

and since also

$$0 \le P\{M(J_k') \le u_{\mathrm{T}}\} - P\{M(J_k) \le u_{\mathrm{T}}\} \le P\{M(J_k^*) > u_{\mathrm{T}}\} \tag{3.6}$$

it follows using stationarity that

$$\begin{aligned} |P\{M(\cup_1^k J_i) \le u_{\mathrm{T}}\} &- P\{M(\cup_1^{k-1} J_i) \le u_{\mathrm{T}}\} P\{M(J_1) \le u_{\mathrm{T}}\}| \\ &\le \alpha_1(r_1, l_1) + 2P\{M(J_1^*) > u_{\mathrm{T}}\}. \end{aligned}$$

Applying (3.6) repeatedly, Cw-mixing gives

$$|P\{M(B_T) \le u_{\mathrm{T}}\} - P^{k_1}\{M(J_1) \le u_{\mathrm{T}}\}| \le 2k_1 P\{M(J_1^*) > u_{\mathrm{T}}\} + o(1). \tag{3.7}$$

It will now be shown that

$$\gamma_{\mathrm{T}} = P\{M(B_{\mathrm{T}}) \le u_{\mathrm{T}}\} - P^{k_1}\{M(J_1) \le u_{\mathrm{T}}\} \to 0\ . \tag{3.8}$$

It is sufficient for this to show (3.8) as $\mathbf{T} \to \infty$ in any manner such that $P^{k_1}\{M(J_1^*) \le u_{\mathrm{T}}\}$ converges to some ρ, $0 \le \rho \le 1$. If $\rho = 1$ then $P\{M(J_1^*) > u_{\mathrm{T}}\} \to 0$ and since $k_1 \log P\{M(J_1^*) \le u_{\mathrm{T}}\} \to 0$ it follows that $k_1 P\{M(J_1^*) > u_{\mathrm{T}}\} \to 0$ and $\gamma_{\mathrm{T}} \to 0$ by (3.7).

On the other hand, if $\rho < 1$, since $k_1 \alpha_1(r_1, l_1) \to 0$ and $l_1 = o(r_1)$ there exists $\theta_{\mathrm{T}} \to \infty$ such that $k_1 \theta_{\mathrm{T}} \alpha_1(r_1, l_1) \to 0$ and $\theta_{\mathrm{T}} l_1 = o(r_1)$. Hence for sufficiently large $\mathbf{T}$, θ_{T} rectangles congruent to J_1^* may be chosen in J_1', all mutually separated by at least l_1 in the x–direction. Arguments parallel to (3.4)–(3.7), using only the left hand inequalities in the counterparts of (3.4), (2.6) then show that

$$P\{M(J_1) \le u_{\mathrm{T}}\} \le P^{\theta_{\mathrm{T}}}\{M(J_1^*) \le u_{\mathrm{T}}\} + \theta_{\mathrm{T}} \alpha_1(l_1, r_1) \tag{3.9}$$

so that

$$P^{k_1}\{M(J_1) \le u_T\} \le P^{k_1\theta_T}\{M(J_1^*) \le u_T\} + \theta_T k_1 \alpha_1(l_1, r_1) \quad (3.10)$$
$$= (\rho + o(1))^{\theta_T} + o(1) \to 0$$

since $\rho < 1$. Hence the second term of the difference in (3.8) tends to zero.

Finally, it follows similarly that

$$P\{M(B_T) \le u_T\} \le P^{k_1}\{M(j_1') \le u_T\} + k_1 \alpha_1, (l_1, r_1)$$

which tends to zero since (3.9) and hence (3.10) also apply with $M(J_1')$ in place of $M(J_1)$. Hence both terms of (3.8) tend to zero if $\rho < 1$ and (3.8) again holds.

Now to prove (3.3) it is sufficient to establish that

$$P^{k_1}\{M(J_1) \le u_T\} - P^{k_T}\{M(J) \le u_T\} \to 0.$$

However, this follows by entirely similar reasoning, splitting the rectangle J_1 into rectangles $(0, r_1] \times ((j-1)r_2, jr_2]$, $1 \le j \le k_2$. ■

Proposition 3.2 *Let the stationary field X_t satisfy $\mathcal{H}(u_T)$ for some family $\{u_T\}$. Let $I = (0, a_1T_1] \times (0, a_2T_2]$ $0 \le a_1, a_2 \le 1$, be a subrectangle of E_T, where a_1, a_2 may change with $\mathbf{T}$ but $a_1a_2 \to a > 0$. Then with the established notation*

(i) $P\{M(I) \le u_T\} - P^{ak_T}\{M(J) \le u_T\} \to 0$

(ii) $P\{M(I) \le u_T\} - P^{a}\{M_T \le u_T\} \to 0$

For simplicity of notation (i) will be shown with $a_1 = a_2 = 1$ and hence $I = E_T$. It then follows in general by putting a_iT_i in place of T_i, with the same block sizes and separations, replacing only k_i by $k_i^* = [a_iT_i/r_i] \sim a_ik_i$ $i = 1, 2$. With the previous notation, it is thus sufficient by the lemma to show that

$$\gamma_T = P\{M(B_T) \le u_T\} - P\{M(E_T) \le u_T\} \to 0\,. \qquad (3.11)$$

Note first that

$$0 \le \gamma_T \le (k_1 + k_2)P\{M(J) > u_T\} \qquad (3.12)$$

since there are less than $k_1 + k_2$ incomplete rectangles B_{ij} in E_T. Now it is sufficient to show (3.11) as $T \to \infty$ in such a way that $P^{(k_1+k_2)}\{M(J) \le u_T\}$ converges to some limit ρ, $0 \le \rho \le 1$. If $\rho = 1$, clearly $P\{M(J) > u_T\} \to 0$ and it then follows again by taking logs that $(k_1 + k_2)P\{M(J) > u_T\} \to 0$ so that $\gamma_T \to 0$ by (3.12).

On the other hand, if $\rho < 1$ then by Lemma 3.1,

$$\begin{aligned} P\{M(B_T) \le u_T\} &= P^{k_T}\{M(J) \le u_T\} + o(1) \\ &= [P^{(k_1+k_2)}\{(M(J) \le u_T\}]^{k_1k_2/(k_1+k_2)} + o(1) \\ &= (\rho + o(1))^{\theta_T} + o(1) \end{aligned}$$

where $\theta_T = k_1k_2/(k_1+k_2) \to \infty$, so that $P\{M(B_T) \le u_T\} \to 0$. Also $P\{M(E_T) \le u_T\} \le P\{M(B_T) \le u_T\}$ so that both terms of (3.11) tend to zero and (3.11) again follows, giving (i) when $a_1 = a_2 = 1$ and hence in general.

Finally, it follows from (i) that

$$\begin{aligned} P\{M(I) \le u_T\} &= P^{ak_T}\{M(J) \le u_T\} + o(1) \\ &= [P\{M_{\mathbf{T}} \le u_T\} + o(1)]^a + o(1) \end{aligned} \tag{3.13}$$

from which (ii) follows at once, since it holds as $\mathbf{T} \to \infty$ in any way for which $P\{M_{\mathbf{T}} \le u_{\mathbf{T}}\}$ converges to a limit.

It follows at once from (ii) that if $P\{M_{\mathbf{T}} \le u_T\}$ has a limit $G(x)$ (for $u_T = x/a_T + b_T$), then $P\{M(I) \le u_T\}$ has the limit $G^a(x)$ which will be used to show max–stability of G in the proof of the extremal types theorem in the next section.

4. Extremal types theorem for stationary random fields

The extremal types theorem for the stationary field will be couched in terms of the limiting distribution of $M_{\mathbf{T}}$ as $\mathbf{T} \to \infty$ along a "continuous monotone curve," i.e. $\mathbf{T} = (T, \psi(T))$ for some strictly increasing continuous function ψ on $(0, \infty)$ with $\psi(T) \to \infty$ as $T \to \infty$.

Theorem 4.1 *Suppose $P\{a_T(M_T - b_T) \le x\} \to G(x)$, non-degenerate as $\mathbf{T} \to \infty$ along the continuous monotone curve $\mathbf{T} = (T, \psi(T))$ defined above. If $X_{\mathbf{t}}$ is Cw-mixing for each $u_T = x/a_T + b_T$, $\infty < x < \infty$, then G is of extreme value type.*

Proof. Write $f(T)$ for the continuous strictly increasing function $T\psi(T)$, and

$$\phi_k(T) = f^{-1}(\frac{1}{k}f(kT))$$

so that for a given T, $T^* = \phi_k(T/k)$ satisfies $kf(T^*) = f(T)$. For $\mathbf{T} = (T, \psi(T))$ let $I = (0, T^*] \times (0, \psi(T^*)]$. Then the conditions of Corollary 3.2 hold for this I with $a_1 = T^*/T$ $a_2 = \psi(T^*)/\psi(T)$, $a = 1/k$ and hence

$$P\{M_{(T,\psi(T))} \le u_{(T,\psi(T))}\} - P^k\{M_{(T^*,\psi(T^*))} \le u_{(T,\psi(T))}\} \to 0$$

Write now $M_{(T,\psi(T))} = M_T$, $u_{(T,\psi(T))} = u_T$ so that

$$P\{M_T \le u_T\} - P^k\{M_{T*} \le u_T\} \to 0$$

from which (2.3) is obtained by writing $T^* = \phi_k(T/k)$ and replacing T by kT. Hence the result now follows from Proposition (2.1) ■

5. Domain of attraction criteria

For i.i.d. sequences (or stationary sequences with non–zero "extremal index") the type of limiting distribution for maxima is determined by the tail behavior of the common marginal d.f. for each term, via the classical domain of attraction criteria. These procedures may be adapted for continuous parameter processes and fields by considering the tail of the distribution of the maximum over a (small) fixed interval e.g. for a process X_t by writing

$$M_n = \max\{X_t : 0 \le t \le n\} = \max(\zeta_1 \cdots \zeta_n)$$

where

$$\zeta_i = \max(X_t \ : \ i-1 < t \le i).$$

If the sequence $\{\zeta_j\}$ has a non–zero extremal index (see [5]) then the asymptotic distribution of M_n (and hence of M_T) is determined by inserting the tail distribution $P\{\zeta > x\}$ in the classical domain of attraction criteria. This approach works especially well for stationary Gaussian processes, where fixed intervals "capture the local dependence at high levels." Somewhat more specifically, an excursion above a high level is very likely to be totally included in the interval $J_i = (i-1, i]$ in which it starts, an interval almost never contains more than one such excursion, and excursion occurrences in different J_i are approximately independent. In particular, the ζ_j have extremal index 1. However, for non–Gaussian cases the relevant local dependence features may extend beyond a fixed interval and it may thus be advantageous to consider intervals of relatively small but increasing length. In the random field context the rectangles J_i of a standard partition of E_T play the same role as the intervals for real parameter processes and the tail of the distribution of $M(J_1)$ determines the extremal type.

Specifically the following simply proved sufficient criteria hold. In this for $\mathbf{T} = (T_1, T_2)$, using familiar notation $r_i = r_{i,\mathbf{T}} = o(T_i)$, $i = 1, 2$ are families of constants, $k_i = k_{i,\mathbf{T}} = [T_i/r_i]$, $k_{\mathbf{T}} = k_1 k_2$, J is the rectangle $(0, r_1] \times (0, r_2]$, and $\gamma_{\mathbf{T}}$ is written for the $(1 - k_{\mathbf{T}}^{-1})-$ percentile of $M(J)$, i.e. $P\{M(J) > \gamma_{\mathbf{T}}\} = k_{\mathbf{T}}^{-1}$.

Proposition 5.1 *Let $X_{\mathbf{t}}$ be a stationary random field and suppose that with the above notation* *(i)* $P\{M(J) > \gamma_T + a_T^{-1}x\}/P\{M(J) > \gamma_T\} \to H(x)$ *for some constants* $a_T > 0$ *and some non-increasing function* $H(x)$ $(H(x) \to \infty$ (0) *as* $x \to -\infty$, (∞) *respectively).*

(ii) X_t *for* $u_T = \gamma_T + a_T^{-1}x$, *each real* x, *(with the given constants* r_1, r_2 *and some "separation constants"* l_1, l_2*).*

Then $P\{a_T(M_T - \gamma_T) \le x\} \to G(x) = e^{-H(x)}$ *as* $\mathbf{T} \to \infty$.

Proof. It follows at once from Corollary 3.2 and (i) that

$$P\{M_T \le \gamma_T + a_T^{-1}x\} = P^{k_T}\{M(J) \le \gamma_T + a_T^{-1}x\} + o(1)$$

$$= [1 - H(x)P\{M(J) > \gamma_T\}(1 + o(1))]^{k_T} + o(1)$$

$$\to e^{-H(x)}$$

as required, since $P\{M(J) > \gamma_T\} = k_T^{-1}$. ■

Condition (i) could be written as a standard tail criterion $(1 - F(u + xg(u)))/(1 - F(x)) \to H(x)$ as $u \to \infty$ for some function $g(u) > 0$ where F is the d.f. of $M(J)$, if J were fixed. For increasing J_1 as envisaged this is still possible but less natural and more complicated in view of the coordination of level and parameter $\mathbf{T}$. As indicated above, the use of a fixed J is possible under mild assumptions (via extremal index considerations). However, there is certainly some utility to be gained by using expanding intervals which independently capture high level events of interest.

References

[1] Adler, R.A., *The geometry of random fields*, John Wiley, New York, 1981.

[2] Bolthausen, E., On the central limit theorem for stationary mixing random fields, *Ann. Prob.* **10**, (1982) 1047–1050.

[3] Doukhan, P., *Mixing: Properties and Examples*, Springer Lecture Notes in Statistics #85, 1995.

[4] Guyon, X., *Random Fields on a Network: Modeling, Statistics and Applications*, Springer-Verlag, 1995.

[5] Leadbetter M.R., Lindgren G., Rootzén H., *Extremes and Related Properties of Random Sequences and Processes*, Springer-Verlag, New York, 1983.

[6] Leadbetter MR, Rootzén H, Choi H, Coordinatewise mixing and Central Limit Theory for additive random set functions on $\mathbf{R}^d$, in preparation.

[7] Piterbarg, V.I., Asymptotic methods in the theory of Gaussian processes and fields, *Trans. of Math. Monographs* **148**, American Mathematical Society, 1996.

M.R. Leadbetter
Department of Statistics
University of North Carolina at Chapel Hill
CB #3260 Phillips Hall
Chapel Hill, NC 27599-3260

Holger Rootzén
Mathematics Department
Chalmers University of Technology
Gothenburg, Sweden

Norming Operators for Operator-Self-Similar Processes

Makoto Maejima

1. Introduction and statements

There are many similarities between the theory of operator-stable distributions and that of operator-self-similar processes as discussed by Mason [6].

Let End $(\mathbf{R}^d)$ be the set of all linear operators on $\mathbf{R}^d$ and let Aut $(\mathbf{R}^d)$ be that of all invertible linear operators in End $(\mathbf{R}^d)$. A probability distribution μ on $\mathbf{R}^d$ is said to be operator-stable if it is infinitely divisible, and for every $t > 0$ there exist $B(t) \in$ Aut $(\mathbf{R}^d)$ and $b(t) \in \mathbf{R}^d$ such that

$$\mu^t = B(t)\mu * \delta(b(t)), \tag{1}$$

where μ^t is the distribution whose characteristic function is the t-th power of that of μ, $B(t)\mu = \mu \circ B(t)^{-1}$, $*$ is the convolution operation, and $\delta(b(t))$ is the unit mass at $b(t)$. We say that μ is full if it is not concentrated on a proper hyperplane in $\mathbf{R}^d$.

An $\mathbf{R}^d$-valued stochastic process $\{X(t)\}_{t\geq 0}$ is said to be operator-self-similar if it is continuous in law, and for every $c > 0$, there exist $B(c) \in$ End $(\mathbf{R}^d)$ and $b(c) \in \mathbf{R}^d$ such that

$$X(ct) \stackrel{d}{=} B(c)X(t) + b(c), \tag{2}$$

where $\stackrel{d}{=}$ means the equality for all finite-dimensional distributions. We say that $\{X(t)\}_{t\geq 0}$ is proper, if for each $t > 0$ the distribution of $X(t)$ is full.

We need more notation. For $A \in$ End $(\mathbf{R}^d)$ and $t > 0$, $t^A \in$ End $(\mathbf{R}^d)$ is defined by $\sum_{k=0}^{\infty} (k!)^{-1}(\log t)^k A^k$.

We first list below some known similar results between operator-stable distributions and operator-self-similar processes for the sake of reference, although they are listed by Mason [6].

I. Existence of exponents.

Theorem IA. (Sharpe [11]) *If μ is full and operator-stable on $\mathbf{R}^d$, then there exists $B \in Aut(\mathbf{R}^d)$ such that $B(t) = t^B$ in (1).*

Theorem IB. (Hudson and Mason [4]) *If $\{X(t)\}$ is proper and operator-self-similar on $\mathbf{R}^d$, then there exists $D \in Aut(\mathbf{R}^d)$ such that $B(c) = c^D$ in (2). (Without the assumption that $\{X(t)\}$ is proper, the existence of $D \in End(\mathbf{R}^d)$ was also proved by Sato [10].)*

We call B and D in the above theorems the exponents of μ and $\{X(t)\}$, respectively, and they are not necessarily unique.

II. The set of all exponents and the uniqueness.

Define

$$\mathcal{E}(\mu) = \{B \in \text{Aut}\,(\mathbf{R}^d) | \mu^t = t^B \mu * \delta(b(t)) \text{ for some } b(t) \in \mathbf{R}^d, t > 0\},$$

$$\mathcal{E}(\{X(t)\}) = \{D \in \text{Aut}\,(\mathbf{R}^d) | X(ct) \stackrel{d}{=} c^D X(t) + b(c) \text{ for some } b(c) \in \mathbf{R}^d, c > 0\},$$

$$\mathcal{S}(\mu) = \{A \in \text{Aut}\,(\mathbf{R}^d) | \mu = A\mu * \delta(b) \text{ for some } b \in \mathbf{R}^d\}$$

and

$$\mathcal{S}(\{X(t)\}) = \{A \in \text{Aut}\,(\mathbf{R}^d) | X(t) \stackrel{d}{=} AX(t) + b \text{ for some } b \in \mathbf{R}^d\}.$$

$\mathcal{S}(\mu)$ and $\mathcal{S}(\{X(t)\})$ are called the symmetry groups of μ and $\{X(t)\}$, respectively, and it is known that they are closed subgroups of Aut $(\mathbf{R}^d)$, (see Billingsley [1], Hudson and Mason [4]). For a closed subgroup H of Aut $(\mathbf{R}^d)$, the tangent space TH is defined by the set of all linear operators A on $\mathbf{R}^d$ such that $A = \lim_{n\to\infty} d_n^{-1}(D_n - I)$ for some sequence $\{D_n\}$ in H and $\{d_n\}$ in $(0, \infty)$, where $d_n \to 0$ and I is the identity operator on $\mathbf{R}^d$.

Theorem IIA. (Holmes et al. [2]) *Let μ be full and operator-stable on $\mathbf{R}^d$.*

(i) *For any $B_0 \in \mathcal{E}(\mu)$, $\mathcal{E}(\mu) = B_0 + T(\mathcal{S}(\mu))$.*

(ii) *μ has exactly one exponent if and only if $\mathcal{S}(\mu)$ is discrete.*

Theorem IIB. (Hudson and Mason [4]) *Let $\{X(t)\}$ be proper and operator-self-similar on $\mathbf{R}^d$.*

(i) *For any $D_0 \in \mathcal{E}(\{X(t)\})$, $\mathcal{E}(\{X(t)\}) = D_0 + T(\mathcal{S}(\{X(t)\}))$.*

(ii) *$\{X(t)\}$ has exactly one exponent if and only if $\mathcal{S}(\{X(t)\})$ is discrete.*

III. Connection to scaling limits.

Theorem IIIA. (Sharpe [11]) *Let μ be a full distribution on $\mathbf{R}^d$. If there exist i.i.d. $\mathbf{R}^d$-valued random vectors $\{X_n\}$ and if there exist $\{A_n\} \subset$ Aut$(\mathbf{R}^d)$ and $\{a_n\} \subset \mathbf{R}^d$ such that*

$$\mathcal{L}\Big(A_n(X_1 + \cdots + X_n) + a_n\Big) \to \mu \quad \text{as } n \to \infty, \tag{3}$$

then μ is operator-stable, where $\mathcal{L}(X)$ stands for the law of X.

Theorem IIIB. (Hudson and Mason [4]) *Let $\{X(t)\}$ be a proper $\mathbf{R}^d$-valued stochastic process which is continuous in law. If there exists a stochastic process $\{Y(t)\}$ and if there exist $\{A(\lambda)\} \subset Aut(\mathbf{R}^d)$ and $\{a(\lambda)\} \subset \mathbf{R}^d$ such that*

$$A(\lambda)Y(t\lambda) + a(\lambda) \stackrel{d}{\Rightarrow} X(t) \quad \text{as } \lambda \to \infty, \tag{4}$$

where $\stackrel{d}{\Rightarrow}$ means the convergence of all finite-dimensional distributions, then $\{X(t)\}$ is operator-self-similar.

IV. Norming operators.

Recently Meerschaert [8] proved the following.

Theorem IVA. (Meerschaert [8]) *Let μ be full and operator-stable on $\mathbf{R}^d$ and let $B \in \mathcal{E}(\mu)$ be fixed. Then the norming operators A_n in (3) can always be chosen so that for each $s > 0$,*

$$A_{[sn]}A_n^{-1} \to s^{-B} \quad \text{as } n \to \infty.$$

Any other sequence of norming operators is of the form $G_n A_n$ where $G_n \in \mathcal{S}(\mu)$.

The purpose of this paper is to prove the following analogue of Theorem IVA for operator-self-similar processes.

Theorem IVB. *Let $\{X(t)\}$ be a proper operator-self-similar process and let $D \in \mathcal{E}(\{X(t)\})$ be fixed. Then the norming operators $A(\lambda)$ in (4) can always be chosen so that, for each $s > 0$,*

$$A(s\lambda)A(\lambda)^{-1} \to s^{-D} \quad \text{as } \lambda \to \infty. \tag{5}$$

Any other sequence of norming operators is of the form $G(\lambda)A(\lambda)$, where $G(\lambda) \in \mathcal{S}(\{X(t)\})$.

Remark. (*A historical remark on Theorem IVB.*) Laha and Rohatgi [5] first extended the notion of "self-similar" to "operator-self-similar" to allow scaling by a class of linear operators on $\mathbf{R}^d$ and proved the results corresponding to Theorems IB and IIIB. In their paper, they also proved that all norming operators satisfy (5). The difference between their problem and ours is the following. As pointed out by Hudson and Mason [4], the only scalings that are allowed by Laha and Rohatgi [5] are invertible positive-definite self-adjoint linear operators on $\mathbf{R}^d$. However, their family of processes is not closed under general affine transformations. Properties of positive-definite self-adjoint linear operators played an important role in their work. Our setting is the same as in Hudson and Mason [4]. They claimed that $A(\lambda)$ in (4) varies regularly under some mild assumptions, (see Hudson and Mason [4] p. 283, lines 1–2 from the bottom.) However, no proof has been published. This is a reason for treating this problem once again in this paper.

2. Proof of the main result

The proof of Theorem IVB can be carried out in parallel to that of Theorem IVA which was proved by Meerschaert [8].

Lemma 1. *Let D be any exponent of the limiting operator-self-similar process $\{X(t)\}$ in Thorem IIIB. Then $\{A(\lambda)\}$ in (4) can be expressed as*

$$A(s\lambda) = s^{-D} I(\lambda, s) G(\lambda, s) A(\lambda),$$

where $I(\lambda, s) \to I$ as $\lambda \to \infty$ and $G(\lambda, s) \in \mathcal{S}(\{X(t)\})$.

Lemma 2. *There exists a $D \in \mathcal{E}(\{X(t)\})$ which commutes every symmetry $A \in \mathcal{S}(\{X(t)\})$.*

Once we establish the above two lemmas, the theorem can be proved by exactly the same way as in Meerschaert [8], but we first prove the theorem in this section for the completeness. The proofs of the lemmas will be given in the next section.

Since $\mathcal{S}(\{X(t)\})$ is a compact subgroup of $\mathrm{Aut}\,(\mathbf{R}^d)$, there exists an inner product on $\mathbf{R}^d$ which makes every element of $\mathcal{S}(\{X(t)\})$ orthogonal. In the following we always take this inner product and the norm based on the inner product. Then if $G \in \mathcal{S}(\{X(t)\})$ and $A \in \mathrm{Aut}\,(\mathbf{R}^d)$, then $\|GA\| = \|AG\| = \|A\|$. For an $A \in \mathrm{Aut}\,(\mathbf{R}^d)$ and a compact subset $\mathfrak{C}$ of $\mathrm{Aut}\,(\mathbf{R}^d)$,

$$\|A - \mathfrak{C}\| = \min\{\|A - C\| : C \in \mathfrak{C}\}$$

denotes the distance of A and $\mathfrak{C}$.

By Lemma 1, for all $s > 0$ and all $D \in \mathcal{E}(\{X(t)\})$

$$\|A(s\lambda)A(\lambda)^{-1} - s^{-D}\mathcal{S}(\{X(t)\})\| \to 0 \tag{6}$$

as $\lambda \to \infty$. This convergence is uniform on compact subsets of $s > 0$. This can be shown by the same way as for Theorem 2.2 of Meerschaert [7].

By Lemma 2, we can choose the exponent D that commutes every $A \in \mathcal{S}(\{X(t)\})$, and thus

$$s^{-D}\mathcal{S}(\{X(t)\}) = \mathcal{S}(\{X(t)\})s^{-D} \quad \text{for all } s > 0.$$

First assume that D is commuting. Take any norming operators $\{A(\lambda)\}$. Let $B(1) = A(1)$. For $\lambda > 1$, define k as $2^k < \lambda \leq 2^{(k+1)}$ and let $s = \frac{\lambda}{2^k} \in (1,2]$. Define $B(\lambda) = G(\lambda)^{-1}A(\lambda)$ such that $G(\lambda) \in \mathcal{S}(\{X(t)\})$ minimizes the distance $\|A(\lambda)B(2^k)^{-1} - s^{-D}\mathcal{S}(\{X(t)\})\|$. Note that $\{B(\lambda)\}$ is another sequence of norming operators for (4), which is assured by the convergence of types.

We now have

$$\begin{aligned}
&\|B(\lambda)B(2^k)^{-1} - s^{-D}\| \\
&\quad = \|G(\lambda)^{-1}A(\lambda)B(2^k)^{-1} - s^{-D}\| \\
&\quad = \|G(\lambda)^{-1}(A(\lambda)B(2^k)^{-1} - s^{-D}G(\lambda))\| \\
&\quad = \|A(\lambda)B(2^k)^{-1} - s^{-D}G(\lambda)\| \\
&\quad = \|A(\lambda)B(2^k)^{-1} - s^{-D}\mathcal{S}(\{X(t)\})\| \\
&\quad = \min\{\|A(\lambda)A(2^k)^{-1}G(2^k) - s^{-D}G\| \; : \; G \in \mathcal{S}(\{X(t)\})\} \\
&\quad = \min\{\|A(\lambda)A(2^k)^{-1}G(2^k) - s^{-D}GG(2^k)\| \; : \; G \in \mathcal{S}(\{X(t)\})\} \\
&\quad = \min\{\|(A(\lambda)A(2^k)^{-1} - s^{-D}G)G(2^k)\| \; : \; G \in \mathcal{S}(\{X(t)\})\} \\
&\quad = \min\{\|A(\lambda)A(2^k)^{-1} - s^{-D}G\| \; : \; G \in \mathcal{S}(\{X(t)\})\} \\
&\quad = \|A(\lambda)A(2^k)^{-1} - s^{-D}\mathcal{S}(\{X(t)\})\| \to 0
\end{aligned}$$

as $\lambda \to \infty$ by the uniform convergence in (6). Namely,

$$B(s2^k)B(2^k)^{-1} \to s^{-D} \quad \text{as } k \to \infty$$

uniformly with respect to $s \in [1,2]$.

We next show that for every $\alpha > 0$,

$$B(\alpha\lambda)B(\lambda)^{-1} \to \alpha^{-D} \tag{7}$$

as $\lambda \to \infty$. Let $\alpha \in [1,2]$. Then

$$\begin{aligned}
&B(\alpha\lambda)B(\lambda)^{-1} \\
&\quad = (B(\alpha s2^k)B(2^{k+1})^{-1})(B(2^{k+1})B(2^k)^{-1})(B(2^k)B(s2^k)^{-1}) \\
&\quad \to \alpha^{-D}
\end{aligned} \tag{8}$$

as $\lambda \to \infty$.

If $\alpha > 2$, then $\alpha = \beta 2^j$ for some integer j and some $\beta \in [1,2]$ and hence

$$\begin{aligned}
B(\alpha\lambda)B(\lambda)^{-1} &= B(\beta 2^j\lambda)B(\lambda)^{-1} \\
&= (B(\beta 2^j\lambda)B(2^j\lambda)^{-1})(B(2^j\lambda)B(2^{j-1}\lambda)^{-1}) \cdots (B(2\lambda)B(\lambda)^{-1}) \\
&\to \beta^{-D} \cdot 2^{-D} \cdots 2^{-D} = \alpha^{-D}
\end{aligned} \tag{9}$$

as $\lambda \to \infty$.

If $\alpha < 1$, let $\gamma = \alpha^{-1}$. Then by (8) and (9), we obtain

$$\begin{aligned} B(\alpha\lambda)B(\lambda)^{-1} &= B(\alpha\lambda)B(\gamma\cdot\alpha\lambda)^{-1} \\ &= (B(\gamma\cdot\alpha\lambda)B(\alpha\lambda)^{-1})^{-1} \\ &\to (\gamma^{-D})^{-1} = \alpha^{-D} \end{aligned} \tag{10}$$

as $\lambda \to \infty$, where we have used the fact that the inverse operation is continuous. (7) is now given from (8)–(10).

We next assume that D is not commuting. By Theorem IIB, D can be expressed as $D = D_0 + Q$, where D_0 is commuting and $Q \in \mathcal{S}(\{X(t)\})$. Construct $\{B(\lambda)\}_{\lambda>0}$ such as in (7) with the replacement D by D_0 and define $C(\lambda) = \lambda^{-Q}B(\lambda)$. Since $\lambda^{-Q} \in \mathcal{S}(\{X(t)\})$, $\{C(\lambda)\}$ is another sequence of norming operators. We also have

$$\begin{aligned} &\|C(s\lambda)C(\lambda)^{-1} - s^{-D}\| \\ &= \|(s\lambda)^{-Q}B(s\lambda)B(\lambda)^{-1}\lambda^{Q} - s^{-Q}s^{-D_0}\| \\ &= \|s^{-Q}\{\lambda^{-Q}B(s\lambda)B(\lambda)^{-1}\lambda^{Q} - s^{-D_0}\}\| \\ &= \|\lambda^{-Q}B(s\lambda)B(\lambda)^{-1}\lambda^{Q} - s^{-D_0}\| \\ &= \|\lambda^{-Q}\{B(s\lambda)B(\lambda)^{-1}\lambda^{Q} - \lambda^{Q}s^{-D_0}\| \\ &= \|\{B(s\lambda)B(\lambda)^{-1} - s^{-D_0}\}\lambda^{Q}\| \\ &= \|B(s\lambda)B(\lambda)^{-1} - s^{-D_0}\| \to 0 \end{aligned}$$

as $\lambda \to \infty$ by the regular variation of the $\{B(\lambda)\}$ with index D_0. The last statement of Theorem IVB is obvious from above considerations. This concludes the proof of Theorem IVB.

3. Proof of lemmas

Proof of Lemma 1. By the assumption, we have

$$A(\lambda)Y(\lambda) + a(\lambda) \xrightarrow{d} X(1)$$

and

$$A(s\lambda)Y(\lambda) + a(s\lambda) = A(s\lambda)Y\Big(\frac{1}{s}\cdot s\lambda\Big) + a(s\lambda) \xrightarrow{d} X\Big(\frac{1}{s}\Big),$$

where $\xrightarrow{d}$ means the convergence of distributions. By Billingsley's convergence of types theorem (Billingsley [1]) or its version by Michaliček (Lemma 2 in Michaliček [9]),

$$A(s\lambda) = s^{-D}I(\lambda, s)G(\lambda, s)A(\lambda),$$

where $I(\lambda, s) \to I$ as $\lambda \to \infty$ and $G(\lambda, s) \in \mathcal{S}(X(1))$, $\mathcal{S}(X(1))$ being the symmetry group of the distribution of $X(1)$. This is almost close to the conclusion of Lemma 1 except that the above assures only that $G(\lambda, s)$ is taken from $\mathcal{S}(X(1))$, not necessarily from $\mathcal{S}(\{X(t)\})$. However, if we check the original proof of Billingsley (Theorem 2 in Billingsley [1]), we see that $G(\lambda, s)$ can be taken from the set $\mathcal{F}_s$ defined by

$$\mathcal{F}_s = \{\text{all limit points of } s^D A(s\lambda)A(\lambda)^{-1} \text{ as a function of } \lambda\}.$$

Therefore, to conclude the lemma, it remains to show that $\mathcal{F}_s \subset \mathcal{S}(\{X(t)\})$. Let $V(s)$ be any limit point of $s^D A(s\lambda)A(\lambda)^{-1}$, for $\lambda \to \infty$, namely

$$s^D A(s\lambda')A(\lambda')^{-1} \to V(s) \quad \text{as } \lambda' \to \infty \tag{11}$$

for some subsequence $\{\lambda'\}$. If we let

$$Z_\lambda(t) := s^D A(s\lambda)A(\lambda)^{-1}(A(\lambda)Y(t\lambda) + a(\lambda)),$$

then by (4) and (11),

$$Z_{\lambda'}(t) \overset{d}{\Rightarrow} V(s)X(t). \tag{12}$$

On the other hand,

$$Z_\lambda(t) = J_s(\lambda, t) + p(\lambda, s), \tag{13}$$

where

$$J_s(\lambda, t) = s^D \left(A(s\lambda)Y\left(\frac{t}{s} \cdot s\lambda\right) + a(s\lambda)\right)$$

and

$$p(\lambda, s) = s^D(A(s\lambda)A(\lambda)^{-1}a(\lambda) - a(s\lambda)).$$

It follows from (4) that

$$J_s(\lambda, t) \overset{d}{\Rightarrow} s^D X\left(\frac{t}{s}\right) \quad \text{as } \lambda \to \infty.$$

This together with (12) implies that $p(\lambda, s)$ in (13) must converge to a limit ($p(s)$, say) along with $\lambda = \lambda'$, and thus

$$Z_{\lambda'}(t) \overset{d}{\Rightarrow} s^D X\left(\frac{t}{s}\right) + p(s) \quad \text{as } \lambda' \to \infty. \tag{14}$$

It follows from (12) and (14) that

$$V(s)X(t) \overset{d}{=} s^D X\left(\frac{t}{s}\right) + p(s) \overset{d}{=} X(t) + s^D b\left(\frac{1}{s}\right) + p(s),$$

where we have used the operator-self-similarity of $\{X(t)\}$ with exponent D. The final relation means that $V(s) \in \mathcal{S}(\{X(t)\})$, implying $\mathcal{F}_s \subset \mathcal{S}(\{X(t)\})$. This concludes the lemma. ∎

Proof of Lemma 2. The proof is similar to that for Theorem 2 of Hudson et al. [3]. We first observe that if $A \in \mathcal{S}(\{X(t)\})$ and $D \in \mathcal{E}(\{X(t)\})$, then

$$ADA^{-1} \in \mathcal{E}(\{X(t)\}). \tag{15}$$

Since

$$\begin{aligned} AX(ct) &\stackrel{d}{=} A(c^D X(t) + b(c)) \\ &= Ac^D X(t) + Ab(c) \\ &= c^{ADA^{-1}} AX(t) + Ab(c) \\ &\stackrel{d}{=} c^{ADA^{-1}}(X(t) - a) + Ab(c) \end{aligned}$$

for some $a \in \mathbf{R}^d$, we have for some $a' \in \mathbf{R}^d$

$$\begin{aligned} X(ct) &\stackrel{d}{=} AX(ct) + a' \\ &\stackrel{d}{=} c^{ADA^{-1}} X(t) + a' - c^{ADA^{-1}} a + Ab(c) \\ &= c^{ADA^{-1}} X(t) + b, \end{aligned}$$

where $b = a' - c^{ADA^{-1}} a + Ab(c)$. Hence we have $ADA^{-1} \in \mathcal{E}(\{X(t)\})$.

Following the proof of Theorem 2 of Hudson et al. [3], let H be a Haar probability measure on the compact group $\mathcal{S}(\{X(t)\})$ and let $D \in \mathcal{E}(\{X(t)\})$. Define

$$M = \int_{\mathcal{S}(\{X(t)\})} sDs^{-1} dH(s).$$

By (15), $M \in \mathcal{E}(\{X(t)\})$. Take any $A \in \mathcal{S}(\{X(t)\})$. Then by the invariance property of Haar measure, we have

$$\begin{aligned} AMA^{-1} &= \int_{\mathcal{S}(\{X(t)\})} AsBs^{-1}A^{-1} dH(s) \\ &= \int_{\mathcal{S}(\{X(t)\})} (As)B(As)^{-1} dH(s) \\ &= \int_{\mathcal{S}(\{X(t)\})} sBs^{-1} dH(s) = M. \end{aligned}$$

Hence M is commuting with every symmetry $A \in \mathcal{E}(\{X(t)\})$. This completes the proof of Lemma 2. ■

Acknowledgment. The author wishes to thank M. Meerschaert for his helpful comments and discussions.

References

[1] P. Billingsley, *Convergence of types in k-spaces*, Z. Wahrsch. verw. Gebiete **5** (1966), 175–179.

[2] J.P. Holmes, W.N. Hudson and J.D. Mason, *Operator-stable laws: Multiple exponents and elliptical symmetry*, Ann. Probab. **10** (1982), 602–612.

[3] W.N. Hudson, Z.J. Jurek and J.A. Veeh, *The symmetry group and exponents of operator stable probability measures*, Ann. Probab. **14** (1986), 1014–1023.

[4] W.N. Hudson and J.D. Mason, *Operator-self-similar processes in a finite-dimensional space*, Trans. Amer. Math. Soc. **273** (1982), 281–297.

[5] R.G. Laha and V.K. Rohatgi, *Operator self-similar stochastic processes in R_d*, Stoch. Proc. Appl. **12** (1982), 73–84.

[6] J.D. Mason, *A comparison of the properties of operator-stable distributions and operator-self-similar processes*, Colloquia Mathematicia Societaris János Bolya **36** (1982), 751–760.

[7] M.M. Meerschaert, *Regular variation in R^k*, Proc. Amer. Math. Soc. **102** (1988), 341–348.

[8] M.M. Meerschaert, *Norming operators for generalized domain of attraction*, J. Theoret. Proab. **7** (1994), 793–798.

[9] J. Michaliček, *Der Anziehungsbereich von operator-stabilen Verteilungen im R_2*, Z. Wahrsch. verw. Gebiete **25** (1972), 57–70.

[10] K. Sato, *Self-similar processes with independent increments*, Probab. Th. Rel. Fields **89** (1991), 285–300.

[11] M. Sharpe, *Operator-stable probability distributions on vector groups*, Trans. Amer. Math. Soc. **136** (1969), 51–65.

Department of Mathematics,
Keio University
Hiyoshi, Yokohama 223, Japan
maejima@math.keio.ac.jp

Multivariate Probability Density and Regression Functions Estimation of Continuous-Time Stationary Processes from Discrete-Time Data

Elias Masry *

To the memory of Stamatis:
A lifelong friend, research collaborator, and colleague

Abstract

Let $(Y, X) = \{Y(t), X(t), -\infty < t < \infty\}$ be a real-valued continuous-time jointly stationary processes and let $\{t_j\}$ be a renewal point process on $[0, \infty)$, with finite mean rate independent of (Y, X). Given the observations $\{Y(t_j), X(t_j), t_j\}_{j=1}^n$ and a measurable function ψ, we estimate the multivariate probability density $f(x_0, x_1, \dots, x_m; \tau_1, \dots \tau_m)$ of $X(0), X(\tau_1), \dots, X(\tau_m)$ and the regression function $r(x_0, x_1, \dots, x_m; \tau_1, \dots, \tau_m)$ of $\psi(Y(\tau_m))$ given $X(0) = x_0, X(\tau_1) = x_1, \dots, X(\tau_m) = x_m$ for arbitrary lags $0 < \tau_1 < \dots < \tau_m$. We present consistency and asymptotic normality results for appropriate estimates of f and r.

1. Introduction

Regression function estimation is an important problem in data analysis with a wide range of applications in filtering and prediction, pattern recognition, and econometrics (Härdle (1990) and Tjostheim (1994)). There is extensive literature on the estimation of regression function for discrete-time processes $\{Y_i, X_i\}$ for which the regression function is defined by

$$r(x_1, \dots, x_m) = E[Y_m | X_1 = x_1, \dots, X_m = x_m].$$

Nonparametric estimation of $r(x_1, \dots, x_m)$ for time series has been addressed in the literature under a variety of settings. We mention, in particular, Robinson (1983), Collomb and Härdle (1986), Roussas (1990), and Tran (1993). In

*Research partially supported by NSF Grant DMS-97-03876.

the context of regression with errors-in-variables, we mention Fan and Masry (1992) and Masry (1993). Recent works using local polynomial fitting include Fan (1992, 1993), Fan and Gijbels (1992), Ruppert and Wand (1994), in an i.i.d. setting, and Masry (1996a, 1996b) in a time series context.

Much less attention has been paid to the regression estimation problem for continuous-time processes. Let $(Y, X) = \{Y(t), X(t), -\infty < t < \infty\}$ be real-valued continuous-time jointly stationary processes on a probability space $(\Omega, \mathcal{F}, P)$. Let $0 < \tau_1 < \tau_2 < \ldots < \tau_m$ be m instants of time and let $\psi(x)$ be a measurable function. Let $f(x_0, x_1, \ldots, x_m; \tau_1, \ldots \tau_m)$ be the joint probability density of the random variables $X(0), X(\tau_1), \ldots, X(\tau_m)$. Also, define the regression function

$$\begin{aligned} & r(x_0, \ldots, x_m; \tau_1, \ldots, \tau_m) \\ & \triangleq E[\psi(Y(\tau_m)) | X(0) = x_0, X(\tau_1) = x_1, \ldots, X(\tau_m) = x_m] \,. \end{aligned} \tag{1}$$

In practice one wishes to estimate the probability density function $f(x_0, x_1, \ldots, x_m; \tau_1, \ldots \tau_m)$ and the regression function $r(x_0, \ldots, x_m; \tau_1, \ldots, \tau_m)$ from actual data. Such estimates could be based on a continuous-time observation $\{Y(t), X(t), 0 \le t \le T\}$. Here we are concerned with the more practical case where the observations are taken at discrete-instants of time $\{t_j\}$. If the data is equally-spaced $\{Y(i\Delta), X(i\Delta)\}$, where $\Delta > 0$, consistent estimates of f and of r can only be obtained for lags $(\tau_1, \ldots, \tau_m)$ which are integer multiples of Δ. We do not wish to let $\Delta \to 0$ since this would be equivalent to having continuous-time observations. We seek appropriate non-equally spaced sampling schemes $\{t_j\}$ which could provide consistent estimates for f and r for *arbitrary* lags $0 < \tau_1 < \tau_2 < \ldots < \tau_m$. Indeed there are practical situations where the data is inherently not equally-spaced (see Moore et al. (1988), and the collection of papers in Parzen (1983)).

The estimation of the multivariate probability density function $f(x_0, x_1, \ldots, x_m; \tau_1, \ldots \tau_m)$ from nonequally-spaced data was considered in Masry (1988). Here we focus our attention on the estimation of the regression function r. We show that for a fairly broad class of (random) sampling schemes $\{t_j\}$ with a *fixed* and *finite* mean sampling rate β, consistent estimates of $r(x_0, \ldots, x_m; \tau_1, \ldots, \tau_m)$ for arbitrary lags $0 < \tau_1 < \ldots < \tau_m$, can be formulated on the basis of the discrete-time data $\{Y(t_j), X(t_j), t_j\}_{j=1}^n$. We establish weak consistency for suitable estimates of $r(x_0, \ldots, x_m; \tau_1, \ldots, \tau_m)$ as well as a central limit theorem for these estimates including explicit expressions for the mean and variance (of the asymptotic distribution). We show that the consistency result holds for *all* values of the mean sampling rate $\beta > 0$.

The introduction of the arbitrary transformation ψ allows us to include some important special cases of the general regression model (1):

(a) $\psi(Y) = I\{Y \leq y\}$ corresponds to estimating the conditional distribution of $Y(\tau_m)$ given $(X(0), X(\tau_1), \ldots, X(\tau_m))$.

(b) $\psi(Y) = Y^d$ corresponds to estimating the conditional moment of $Y(\tau_m)$ given $(X(0), X(\tau_1), \ldots, X(\tau_m))$.

The organization of the paper is as follows: Basic notation and assumptions, and the form of the estimators are given in Section 2. Quadratic-mean analysis is presented in Section 3. Asymptotic normality is established in Section 4. Proofs are collected in Section 5. Section 6 provides a discussion on extensions of this work in various directions. The main results of the paper are Theorem 3.2 (bias), Theorem 3.4 (consistency with rates of convergence), and Theorem 4.2 (asymptotic normality).

2. Preliminaries

Let

$$\underline{x} = (x_0, \ldots, x_m), \quad \underline{\tau} = (\tau_1, \ldots, \tau_m) \ . \tag{2}$$

Write $f(\underline{x}; \underline{\tau})$ for the joint probability density of

$$\underline{X}^\circ = (X(0), X(\tau_1), \ldots, X(\tau_m))$$

and write the regression function r in the form

$$r(\underline{x}; \underline{\tau}) \equiv E[\psi(Y(\tau_m)) | X(0) = x_0, \ldots, X(\tau_m) = x_m] \ . \tag{3}$$

Let $g(\underline{x}; \underline{\tau})$ be defined by

$$g(\underline{x}; \underline{\tau}) = r(\underline{x}; \underline{\tau}) f(\underline{x}; \underline{\tau}) \ . \tag{4}$$

Let $\{t_j\}_{j=0}^{\infty}$ be a renewal process on $[0, \infty)$ which is independent of (Y, X) with $t_0 \equiv 0$,

$$t_j = \sum_{\ell=1}^{j} T_\ell, \quad j = 1, 2, \ldots \tag{5}$$

where the inter-arrival times $\{T_\ell\}_{\ell=1}^{\infty}$ are independent identically distributed random variables with a common bounded probability density function $p(u)$ on $[0, \infty)$. It is assumed that $p(u) > 0$ on $(0, \infty)$ and $E(T_\ell) = \beta^{-1} < \infty$, so that β is the mean sampling rate.

We would like to estimate $r(\underline{x}; \underline{\tau})$ of (3) from the discrete-time data $\{Y(t_j), X(t_j), t_j\}_{j=1}^{n+m}$ for all lags $0 < \tau_1 < \ldots < \tau_m$. Let

$$\begin{aligned}
\underline{X}_j &= (X(t_j), \ldots, X(t_{j+m})) \\
\underline{D}_j &= (t_{j+1} - t_j, \ldots, t_{j+m} - t_j), \\
a(\underline{\tau}) &= p(\tau_1) \prod_{i=1}^{m-1} p(\tau_{i+1} - \tau_i), \qquad (6)
\end{aligned}$$

and weight functions $K(\underline{u})$, $\underline{u} \in \mathbf{R}^{m+1}$, and $W(\underline{\tilde{u}})$, $\underline{\tilde{u}} \in \mathbf{R}^m$, which are bounded and nonnegative satisfying

$$\int_{R^{m+1}} K(\underline{u})d\underline{u} = 1, \quad \int_{R^m} W(\underline{\tilde{u}})d\underline{\tilde{u}} = 1$$

$$\lim_{||\underline{\tilde{u}}|| \to \infty} ||\underline{\tilde{u}}||^m W(\underline{\tilde{u}}) \to 0, \quad \lim_{||\underline{u}|| \to \infty} ||\underline{u}||^{m+1} K(\underline{u}) \to 0 . \tag{7}$$

The bandwidth b_n is assumed to satisfy $b_n \to 0$ as $n \to \infty$. Define the averaging kernels

$$K_n(\underline{u}) = \frac{1}{b_n^{m+1}} K(\underline{u}/b_n),$$

$$W_n(\underline{\tilde{u}}) = \frac{1}{b_n^m} W(\underline{\tilde{u}}/b_n) . \tag{8}$$

We estimate $r(\underline{x}; \underline{\tau})$ by

$$r_n(\underline{x}; \underline{\tau}) = g_n(\underline{x}; \underline{\tau}) / f_n(\underline{x}; \underline{\tau}) \tag{9}$$

with

$$g_n(\underline{x}; \underline{\tau}) = \frac{1}{na(\underline{\tau})} \sum_{j=1}^{n} W_n(\underline{\tau} - \underline{D}_j) \psi(Y(t_{j+m})) K_n(\underline{x} - \underline{X}_j) \tag{10}$$

and

$$f_n(\underline{x}; \underline{\tau}) = \frac{1}{na(\underline{\tau})} \sum_{j=1}^{n} W_n(\underline{\tau} - \underline{D}_j) K_n(\underline{x} - \underline{X}_j) . \tag{11}$$

The estimator (9) is a version of the Nadaraya–Watson estimator and the key idea of regressing $Y(t_{j+m})$ on the vector $\underline{X}_j$ as well as on the vector $\underline{D}_j$ of observation-times differences is due to Masry (1988). See Section 6 for a discussion on local polynomial fitting.

In this paper we assumed that $a(\underline{\tau})$ is known. If $a(\underline{\tau})$ is unknown then it can be estimated by

$$a_n(\underline{\tau}) = \hat{p}(\tau_1) \prod_{i=1}^{m-1} \hat{p}(\tau_{i+1} - \tau_i) \tag{12}$$

with

$$\hat{p}(t) = \frac{1}{n} \sum_{j=1}^{n} W_n(t - T_j) . \tag{13}$$

(In (13) the kernel $W_n(u) = (1/b_n)W(u/b_n)$ is one dimensional.) Under regularity conditions on $p(u)$, $\hat{p}(u)$ converges in quadratic mean to $p(u)$ at all points of continuity of p with $nb_n \to \infty$ as $n \to \infty$.

3. Quadratic mean analysis

The bias of $f_n(\underline{x}, \underline{\tau})$ is given in Theorem 5.1 of Masry (1988):

Proposition 3.1

(a) $E[f_n(\underline{x}; \underline{\tau})] \rightarrow f(\underline{x}, \underline{\tau})$ *as* $n \rightarrow \infty$ *at all points of continuity of* $f(\underline{x}, \underline{\tau})a(\underline{\tau})$.

(b) *Assume that* $f(\underline{x}, \underline{\tau})a(\underline{\tau})$ *is twice differentiable in* $(\underline{x}, \underline{\tau})$ *and its second partial derivatives are bounded and continuous on* $\mathbf{R}^{2m+1}$. *Let* $Q(\underline{v}) = K(\underline{v}')W(\underline{v}'')$, $\underline{v} = (\underline{v}', \underline{v}'')$, *such that*

$$\int_{R^{2m+1}} v_j Q(\underline{v}) d\underline{v} = 0, \quad j = 1, \ldots, 2m+1 \tag{14}$$

$$\int_{R^{2m+1}} ||\underline{v}||^2 Q(\underline{v}) d\underline{v} < \infty \, . \tag{15}$$

Then

$$\frac{1}{b_n^2} \, \text{bias}[f_n(\underline{x}, \underline{\tau})] \rightarrow \frac{1}{2a(\underline{\tau})} \int_{R^{2m+1}} \underline{v} G''(\underline{x}, \underline{\tau}) \underline{v}^T Q(\underline{v}) d\underline{v}$$

where $G''(\underline{x}, \underline{\tau})$ *is the Hessian matrix of* $f(\underline{x}, \underline{\tau})a(\underline{\tau})$.

The bias of the estimate $g_n(\underline{x}, \underline{\tau})$ is given by

Theorem 3.1.

(a) *Assume that* $g(\underline{x}; \underline{\tau})a(\underline{\tau}) \in L_1$. *Then* $E[g_n(\underline{x}; \underline{\tau})] \rightarrow g(\underline{x}; \underline{\tau})$ *as* $n \rightarrow \infty$ *at all points of continuity of the function* $g(\underline{x}, \underline{\tau})a(\underline{\tau})$.

(b) *Assume that* $g(\underline{x}, \underline{\tau})a(\underline{\tau})$ *is twice continuously differentiable in* $(\underline{x}, \underline{\tau})$ *and its second partial derivatives are bounded and continuous on* $\mathbf{R}^{2m+1}$. *Assume that* K *and* W *satisfy* (14), (15). *Then*

$$\frac{1}{b_n^2} \, \text{bias} \, [g_n(\underline{x}; \underline{\tau})] \rightarrow \frac{1}{2a(\underline{\tau})} \int_{R^{2m+1}} \underline{y} L''(\underline{x}, \underline{\tau}) \underline{y}^T Q(\underline{y}) d\underline{y}.$$

where L'' *is the Hessian matrix of* $g(\underline{x}, \underline{\tau})a(\underline{\tau})$.

We next establish the asymptotic properties of the variance/covariance of our estimators. Let $\mathcal{F}_a^b$ be the σ-algebra of events generated by $\{Y(t), X(t) : t \in [a, b]\}$ and $L_2(\mathcal{F}_a^b)$ denote the collection of all second order random variables which are $\mathcal{F}_a^b$ measurable. The stationary processes (Y, X) are said to be strongly mixing (Rosenblatt (1956)) if

$$\sup_{\substack{A \in \mathcal{F}_{-\infty}^0 \\ B \in \mathcal{F}_\tau^\infty}} |P(AB) - P(A)P(B)| = \alpha(\tau) \downarrow 0 \text{ as } \tau \rightarrow \infty \tag{16}$$

and are asymptotically uncorrelated or ρ-mixing (Kolmogorov and Rozanov (1960)) if

$$\sup_{\substack{U \in L_2(\mathcal{F}_{-\infty}^0) \\ V \in L_2(\mathcal{F}_\tau^\infty)}} \frac{|\mathrm{cov}(U,V)|}{\mathrm{var}^{\frac{1}{2}}(U)\mathrm{var}^{\frac{1}{2}}(V)} = \rho(\tau) \downarrow 0 \text{ as } \tau \to \infty\,. \tag{17}$$

$\alpha(\tau)$ is the strongly mixing coefficient and $\rho(\tau)$ is the maximal correlation coefficient. We have $\alpha(\tau) \le \frac{1}{4}\rho(\tau)$ (Hall and Heyde (1980)). Denote by $p_k(x)$ the kth fold convolution of $p(x)$ with itself and by

$$h(x) = \sum_{k=1}^{\infty} p_k(x), \quad x \ge 0 \tag{18}$$

the renewal density function of the process $\{t_j\}$. We first state the asymptotic variance/covariance expression for the density estimate $f_n(\underline{x}, \underline{\tau})$ given in Theorems 5.2 and 5.4 of Masry (1988):

Assumption 3.1

(a) (i) $\int_{R^m} f(\underline{x}, \underline{\tau}) a(\underline{\tau}) d\underline{\tau} < \infty$

(ii) The $2m+2$-order probability density function $f(\underline{v}; s_1, \dots, s_{2m+1})$ of $X(0), X(s_1), \dots, X(s_{2m+1})$, $0 < s_1 < \dots < s_{2m+1}$, exists and satisfies

$$\begin{aligned} a(\underline{s}_1)a(\underline{s}_2) \textstyle\int_0^\infty \Big[& f(\underline{v}_1, \underline{v}_2; \underline{s}_1, s_{1,m} + w + r, s_{1,m} + w + r + \underline{s}_2) \\ & - f(\underline{v}_1; \underline{s}_1) f(\underline{v}_2; \underline{s}_2)\Big] p(r) dr \\ & \le M(\underline{v}_1, \underline{v}_2; \underline{s}_1, \underline{s}_2) \end{aligned}$$

uniformly in w where $M \in L_1(\mathbf{R}^{4m+2})$ and is continuous at the point $(\underline{x}, \underline{x}, \underline{\tau}, \underline{\tau})$.

(b) For ρ-mixing processes, $\int_0^\infty \rho(\tau) h(\tau) d\tau < \infty$.

(c) For strongly mixing processes $\alpha(t)$ satisfies, for some $\delta > 0$,

$$b_n^{-\delta/(2+\delta)} \sum_{k=\gamma_n}^{\infty} \int_0^\infty [\alpha(t)]^{\delta/(2+\delta)} p_{k-m+1}(t) dt \to 0 \text{ as } n \to \infty$$

where γ_n is a positive integer such that $\gamma_n \to \infty$ and $\gamma_n b_n^{2m+1} \to 0$ as $n \to \infty$.

Proposition 3.2 (Masry (1988)). *Under Assumption 3.1 and* $nb_n^{2m+1} \to \infty$ *we have*

$$\lim_{n\to\infty} nb_n^{2m+1}\text{cov}(f_n(\underline{x},\underline{\tau}), f_n(\underline{y},\underline{\tau}))$$

$$= \begin{cases} (f(\underline{x},\underline{\tau})/a(\underline{\tau}))\int_{R^{m+1}} K^2 \int_{R^m} W^2; & \underline{x}=\underline{y} \\ 0 & ; \quad \underline{x}\neq\underline{y} \end{cases}$$

where $(\underline{x},\underline{\tau})$ *is a point of continuity of the functions* $\{f(\underline{v};\underline{s}), a(\underline{s})\}$.

Propositions 3.1 and 3.2 imply the quadratic-mean convergence of $f_n(\underline{x},\underline{\tau})$ to $f(\underline{x},\underline{\tau})$ as $n\to\infty$ at continuity points.

We now consider the quadratic-mean analysis of a centered version of the regression estimate $r_n(\underline{x},\underline{\tau})$. Following Fan and Masry (1992) we define a centralizing parameter $B_n(\underline{x};\underline{\tau})$

$$B_n(\underline{x};\underline{\tau}) = \frac{\{E[g_n(\underline{x};\underline{\tau})] - g(\underline{x};\underline{\tau})\} - r(\underline{x};\underline{\tau})\{E[f_n(\underline{x};\underline{\tau})] - f(\underline{x};\underline{\tau})\}}{E[f_n(\underline{x};\underline{\tau})]} . \quad (19)$$

$B_n(\underline{x},\underline{\tau})$ may be viewed as the bias of the estimator $r_n(\underline{x},\underline{\tau})$. By Proposition 3.1 and Theorem 3.1 we have the following asymptotic rate and structure of $B_n(\underline{x};\underline{\tau})$.

Theorem 3.2.

(a) *Assume that* $g(\underline{x};\underline{\tau})a(\underline{\tau}) \in L_1$. *Then* $B_n(\underline{x};\underline{\tau}) \to 0$ *as* $n\to\infty$ *at all points of continuity of the functions* $\{g(\underline{x},\underline{\tau})a(\underline{\tau}), f(\underline{x},\underline{\tau})a(\underline{\tau})\}$.

(b) *Under the second-order differentiability conditions on* f *and* g *given in Proposition 3.1 and Theorem 3.1 and the kernels' integrability conditions* (14) *and* (15), *we have*

$$\frac{1}{b_n^2}\ B_n(\underline{x};\underline{\tau}) \to A(\underline{x};\underline{\tau})$$

with

$$A(\underline{x};\underline{\tau}) = \frac{1}{2f(\underline{x};\underline{\tau})a(\underline{\tau})}$$
$$\times\left\{\int_{R^{2m+1}} \underline{y}L''(\underline{x},\underline{\tau})\underline{y}^T Q(\underline{y})d\underline{y} - r(\underline{x};\underline{\tau})\int_{R^{2m+1}} \underline{y}G''(\underline{x},\underline{\tau})\underline{y}^T Q(\underline{y})d\underline{y}\right\}$$

where L'' *and* G'' *are the Hessian matrices of* $g(\underline{x},\underline{\tau})a(\underline{\tau})$ *and of* $f(\underline{x},\underline{\tau})a(\underline{\tau})$ *respectively.*

We can write

$$r_n(\underline{x};\underline{\tau}) - r(\underline{x};\underline{\tau}) - B_n(\underline{x};\underline{\tau})$$
$$= \frac{(g_n(\underline{x};\underline{\tau}) - E[g_n(\underline{x};\underline{\tau})]) - (r(\underline{x};\underline{\tau}) + B_n(\underline{x};\underline{\tau}))\,(f_n(\underline{x};\underline{\tau}) - E[f_n(\underline{x};\underline{\tau})])}{f_n(\underline{x};\underline{\tau})} . \quad (20)$$

By Theorem 3.2(a) we have that $B_n(\underline{x};\underline{\tau}) = o(1)$ at continuity points. Hence

$$r_n(\underline{x},\underline{\tau}) - r(\underline{x};\underline{\tau}) - B_n(\underline{x};\underline{\tau}) = \frac{C_n(\underline{x};\underline{\tau})}{f_n(\underline{x};\underline{\tau})}(1 + o_p(1)) \tag{21}$$

where

$$C_n(\underline{x};\underline{\tau}) \equiv g_n(\underline{x};\underline{\tau}) - E[g_n(\underline{x};\underline{\tau})] - r(\underline{x};\underline{\tau})\{f_n(\underline{x};\underline{\tau}) - E[f_n(\underline{x};\underline{\tau})]\}\,. \tag{22}$$

We establish the quadratic mean-convergence and asymptotic normality of $C_n(\underline{x};\underline{\tau})$. Define

$$U_{n,j}(\underline{x};\underline{\tau}) = \frac{1}{a(\underline{\tau})}[\psi(Y(t_{m+j})) - r(\underline{x};\underline{\tau})]K_n(\underline{x} - \underline{X}_j)W_n(\underline{\tau} - \underline{D}_j) \tag{23}$$

and

$$Z_{n,j}(\underline{x};\underline{\tau}) = b_n^{m+\frac{1}{2}}\{U_{n,j}(\underline{x};\underline{\tau}) - E[U_{n,j}(\underline{x};\underline{\tau})]\}\,. \tag{24}$$

Then

$$nb_n^{m+\frac{1}{2}}C_n(\underline{x};\underline{\tau}) = \sum_{j=1}^{n} Z_{n,j}(\underline{x};\underline{\tau})\,. \tag{25}$$

We make the following assumptions to establish the asymptotic expression for the variance/covariance of $C_n(\underline{x};\underline{\tau})$.

Assumption 3.2.

(a) $r(\underline{x};\underline{\tau})f(\underline{x};\underline{\tau})a(\underline{\tau}) \in L_1$.

(b) $E[\psi^2(Y(0))] < \infty$.

(c) For any $0 < s_1 < \ldots < s_m < s_\ell < \ldots < s_{\ell+m}$ with $\ell \geq m+1$, the conditional density of $(X(0), X(s_1), \ldots, X(s_m); X(s_\ell), \ldots X(s_{\ell+m}))$ given $(Y(s_m), Y(s_{\ell+m}))$ exists and is bounded by a constant A_1. Also, the conditional density of $(X(0), X(s_1), \ldots, X(s_m))$ given $(Y(s_m), Y(s_{\ell+m}))$ is bounded by a constant A_2.

(d) $W(\underline{u})$ has a compact support.

Assumption 3.3

(a) For ρ-mixing processes

(i) The maximal correlation coefficient $\rho(\tau)$ satisfies

$$\int_0^\infty \rho(\tau)h(\tau)d\tau < \infty$$

(ii) $H_2(\underline{v}) \equiv \int_{R^m} \sigma^2(\underline{v};\underline{s})f(\underline{v};\underline{s})a(\underline{s})d\underline{s}$ is continuous at $\underline{v} = \underline{x}$ where

$$\sigma^2(\underline{x};\underline{\tau}) = \text{var}[\psi(Y(\tau_m))|X(0) = x_0, \ldots, X(\tau_m) = x_m]\,. \tag{26}$$

(b) For strongly mixing processes

(i) The coefficient $\alpha(\tau)$ satisfies, for some $\delta > 0$

$$\frac{1}{b_n^{\delta/(2+\delta)}} \sum_{\ell=\gamma_n}^{\infty} \int_0^{\infty} [\alpha(\tau)]^{\delta/(2+\delta)} p_{\ell-m+1}(\tau) d\tau \to 0 \text{ as } n \to \infty$$

where γ_n is a positive integer such that $\gamma_n b_n^{2m+1} \to 0$ as $n \to \infty$.

(ii) $E[|\psi(Y(0))|^{2+\delta}] < \infty$

(iii) $H_{2+\delta}(\underline{v}) \equiv \int_{R^m} V_{2+\delta}(\underline{v}; \underline{s}) f(\underline{v}; \underline{s}) a(\underline{s}) d\underline{s}$ is continuous at $\underline{v} = \underline{x}$ where

$$V_{2+\delta}(\underline{x}; \underline{\tau}) = E[|\psi(Y(\tau_m)) - r(\underline{x}; \underline{\tau})|^{2+\delta} \mid X(0) = x_0, \ldots, X(\tau_m) = x_m] \, .$$

Define

$$\theta^2(\underline{x}; \underline{\tau}) = \frac{1}{a(\underline{\tau})} \sigma^2(\underline{x}; \underline{\tau}) f(\underline{x}, \underline{\tau}) \int_{R^{m+1}} K^2 \int_{R^m} W^2 \, . \tag{27}$$

Theorem 3.3. *Under Assumptions 3.2 and 3.3 we have,* $nb_n^{2,+1} \to \infty$ *as* $n \to \infty$,

(i)

$$\lim_{n\to\infty} \text{cov}\{Z_{n,0}(\underline{x}; \underline{\tau}), Z_{n,0}(\underline{y}; \underline{\tau})\} = \begin{cases} \theta^2(\underline{x}; \underline{\tau}) & \text{if } \underline{x} = \underline{y} \\ 0 & \text{if } \underline{x} \neq \underline{y} \end{cases}$$

(ii) $\sum_{\ell=1}^{n-1} |\text{cov}\{Z_{n,0}(\underline{x}; \underline{\tau}), Z_{n,\ell}(\underline{y}; \underline{\tau})\}| = o(1)$

(iii)

$$\lim_{n\to\infty} nb_n^{2m+1} \text{cov}\{C_n(\underline{x}; \underline{\tau}), C_n(\underline{y}; \underline{\tau})\} = \begin{cases} \theta^2(\underline{x}; \underline{\tau}) & \text{if } \underline{x} = \underline{y} \\ 0 & \text{if } \underline{x} \neq \underline{y} \end{cases}$$

where $(\underline{x}, \underline{\tau})$ *is a point of continuity of the functions* $\{\sigma^2(\underline{v}; \underline{s}) f(\underline{v}; \underline{s}) a(\underline{s}), r(\underline{v}; \underline{s}) f(\underline{v}; \underline{s}) a(\underline{s})\}$ *and* $\theta^2(\underline{x}; \underline{\tau})$ *is given in* (27).

The combination of Theorems 3.2(a) and 3.3 and Proposition 3.1 yield the consistency of the regression estimate $r_n(\underline{x}; \underline{\tau})$ in view of the relationship (21). We have one of the main results of the paper:

Theorem 3.4. *Under the assumptions of Theorems 3.2(a), 3.3, and Propositions 3.1(a) and* $(nb_n^{2m+1})/(\log n) \to \infty$, *we have*

$$((nb_n^{2m+1})/(\log n))^{1/2} [r_n(\underline{x}; \underline{\tau}) - r(\underline{x}; \underline{\tau}) - B_n(\underline{x}; \underline{\tau})] \xrightarrow{P} 0 \text{ as } n \to \infty \tag{28}$$

and in particular

$$r_n(\underline{x}; \underline{\tau}) \xrightarrow{P} r(\underline{x}; \underline{\tau}) \text{ as } n \to \infty$$

whenever $f(\underline{x};\underline{\tau}) > 0$. If, in addition, f and g are twice continuously differentiable, as in Theorem 3.2(b), and $(nb_n^{2m+5})/(\log n) \to 0$, then

$$((nb_n^{2m+1})/(\log n))^{1/2}[r_n(\underline{x};\underline{\tau}) - r(\underline{x};\underline{\tau})] \xrightarrow{P} 0 \text{ as } n \to \infty .$$

The consistency hold for **all** *values of the mean sampling rate $\beta > 0$.*

4. Asymptotic normality

In this section we establish the asymptotic normality of the regression function estimate $r_n(\underline{x};\underline{\tau})$. The asymptotic normality of the density estimate $f_n(\underline{x};\underline{\tau})$ is given in Theorems 5.3 and 5.5 of Masry (1988).

Assumption 4.1. Let $\{c_n\}$ be a sequence of positive integers $c_n \to \infty$ with $c_n = o((nb_n^{2m+1})^{1/2})$.

a. For ρ-mixing processes, assume that the mixing coefficient $\rho(\tau)$ satisfies

$$\left(\frac{n}{b_n^{2m+1}}\right)^{1/2} \rho_{c_n} \to 0 \text{ as } n \to \infty \tag{29}$$

where

$$\rho_j = \int_0^\infty \rho(\tau) p_{j-m+1}(\tau) d\tau . \tag{30}$$

b. For strongly mixing processes, assume that the mixing coefficient $\alpha(\tau)$ satisfies

$$\left(\frac{n}{b_n^{2m+1}}\right)^{1/2} \alpha_{c_n} \to 0 \text{ as } n \to \infty \tag{31}$$

where

$$\alpha_j = \int_0^\infty \alpha(\tau) p_{j-m+1}(\tau) d\tau . \tag{32}$$

Assumption 4.2. The conditional distribution $G(y|\underline{u})$ of $\psi[Y(\tau_m)]$ given $\underline{X}^\circ = (X(0), X(\tau_1), \ldots, X(\tau_m))$ is continuous at the point $\underline{u} = \underline{x}$.

Assumption 4.2 is needed in the proof of Theorem 4.1 where a truncation argument is employed and the continuity of a conditional second moment of a truncated $\psi(Y(\tau_m))$ given $\underline{X}^\circ = \underline{u}$ is required at $\underline{u} = \underline{x}$.

Theorem 4.1. *Under Assumptions 3.2, 3.3, 4.1, and 4.2 we have with $nb_n^{2m+1} \to \infty$ as $n \to \infty$,*

$$(nb_n^{2m+1})^{1/2}\, C_n(\underline{x};\underline{\tau}) \xrightarrow{L} N(0, \theta^2(\underline{x};\underline{\tau}))$$

at all points of continuity $(\underline{x}, \underline{\tau})$ of the functions $\{f(\underline{v};\underline{s}), r(\underline{v};\underline{s}), \sigma^2(\underline{v};\underline{s}), a(\underline{s})\}$ where $\theta^2(\underline{x};\underline{\tau})$ is given by (27).

Theorem 4.1 along with the quadratic-mean convergence of $f_n(\underline{x};\underline{\tau})$ (Cf. Propositions 3.1 and 3.2) give the second main result for the regression estimator $r_n(\underline{x};\underline{\tau})$.

Theorem 4.2. *Under Assumptions 3.1, 3.2, 3.3, 4.1, and 4.2 we have with* $nb_n^{2m+1} \to \infty$ *as* $n \to \infty$

$$(nb_n^{2m+1})^{1/2}\{r_n(\underline{x};\underline{\tau}) - r(\underline{x};\underline{\tau}) - B_n(\underline{x};\underline{\tau})\} \xrightarrow{L} N(0, \theta^2(\underline{x};\underline{\tau})/f^2(\underline{x};\underline{\tau}))$$

at all points of continuity $(\underline{x},\underline{\tau})$ *of the functions* $\{f(\underline{v};\underline{s}), r(\underline{v};\underline{s}), \sigma^2(\underline{v},\underline{s}), a(\underline{s})\}$ *with* $f(\underline{x};\underline{\tau}) > 0$ *where* $\theta^2(\underline{x};\underline{\tau})$ *is given in* (27).

We remark that if $f(\underline{x};\underline{\tau})a(\underline{\tau})$ and $g(\underline{x};\underline{\tau})a(\underline{\tau})$ are twice differentiable, then by Theorem 3.2(b), $B_n(\underline{x};\underline{\tau}) = O(b_n^2)$ and we then have

Corollary 4.1. *If in addition to the assumptions of Theorem 4.2,* $g(\underline{x};\underline{\tau})a(\underline{\tau})$ *and* $f(\underline{x};\underline{x})a(\underline{\tau})$ *are twice continuously differentiable and* b_n *satisfies* $nb_n^{2m+5} \to 0$ *as* $n \to \infty$, *then*

$$(nb_n^{2m+1})^{1/2}\{r_n(\underline{x};\underline{\tau}) - r(\underline{x};\underline{\tau})\} \xrightarrow{L} N(0, \theta^2(\underline{x};\underline{\tau})/f^2(\underline{x};\underline{\tau}))\ .$$

Remark 4.1. It is seen from Theorem 4.2 that the variance of the asymptotic distribution is given by

$$\text{var}[r_n(\underline{x};\underline{\tau})] = \frac{\theta^2(\underline{x};\underline{\tau})}{f^2(\underline{x};\underline{\tau}))}\frac{1}{nb_n^{2m+1}}$$

whereas by Theorem 3.2(a), the bias is of the form

$$\text{bias}[r_n(\underline{x};\underline{\tau})] = b_n^2 A(\underline{x};\underline{\tau})$$

and thus the asymptotically optimal bandwidth, which minimizes the sum of the squared bias and variance, is given by

$$b_n = \left\{\frac{(2m+1)\theta^2(\underline{x};\underline{\tau})}{4f^2(\underline{x};\underline{\tau})A^2(\underline{x};\underline{\tau})}\right\}^{1/(2m+5)} \frac{1}{n^{1/(2m+5)}}\ .$$

The issue of data-driven bandwidth selection is beyond the scope of this work and the reader is directed to the vast literature on cross-validation and plug-in methods in the context of bandwidth selection for density/regression estimation.

Remark 4.2. We briefly remark on the choice of the density p of the inter-arrival times $\{T_i\}$. A suitable choice is the Gamma family

$$p(u) = \frac{(\beta k)^k}{(k-1)!}u^k e^{-k\beta u} 1_{[0,\infty)}(u)$$

for some $k \geq 2$. Advantages of this choice are: a) exponential decay which helps to ensure the various integrability conditions imposed in the paper, b) the convolution of p with itself is again Gamma; the convolved densities appear in the integrands in Assumptions 3.3(b) and 4.1. c) the corresponding renewal density $h(u)$ of (18) is then uniformly bounded which simplifies Assumption 3.3(a)(i). d) the fact that the density $p(u)$ vanishes at the origin is needed to ensure integrability near the origin (e.g. in Assumption 3.2 (a)).

Remark 4.3. Condition (29) on the mixing coefficient is more restrictive than the integrability condition on $\rho(\tau)$ in Assumption 3.3. Let $b_n = n^{-\mu}$ for some $0 < \mu < 1/(2m+1)$ and $c_n = n^{(1-(2m+1)\mu)/2}(\log(n))^{-1}$. Then condition (29) is satisfied if $\rho_j = O(j^{-s})$ with $s > \frac{3+(2m+1)\mu}{1-(2m+1)\mu}$.

5. Derivations

Proof of Theorem 3.1. Follows in the manner of the proof of Theorem 5.1 in Masry (1988). ■

Proof of Theorem 3.3. Part (i) follows from (24), using Lemma 2.2 of Masry (1988). For Part (ii), write

$$\begin{aligned}\sum_{\ell=1}^{n-1} |\text{cov}\{Z_{n,0}(\underline{x};\underline{\tau}), Z_{n,\ell}(\underline{y};\underline{\tau})\}| &= \left(\sum_{\ell=1}^{m} + \sum_{\ell=m+1}^{\gamma_n} + \sum_{\ell=\gamma_n+1}^{n-1}\right) I_{n,\ell}(\underline{x},\underline{y};\underline{\tau}) \\ &\equiv J_1 + J_2 + J_3 \qquad (33)\end{aligned}$$

where $\gamma_n \to \infty$ such that $\gamma_n b_n^{2m+1} \to 0$. For J_1 one uses Assumption 3.2(c)(d), the boundedness of $p(u)$, and the compact support of W to show that $I_{n,\ell}(\underline{x},\underline{y};\underline{\tau}) = O(b_n^{\ell})$ from which one obtains $J_1 = O(b_n)$. For J_2 we exploit Assumption 3.2(c) to show that $I_{n,\ell}(\underline{x},\underline{y};\underline{\tau}) \leq M(\underline{x},\underline{y};\underline{\tau}) b_n^{2m+1}$ from which one obtains $J_2 = O(b_n^{2m+1}\gamma_n) = o(1)$. For J_3 for strongly mixing processes, one uses Davydov's lemma (Hall and Heyde (1980)) to show that

$$J_3 \leq \frac{M(\underline{x},\underline{y},\underline{\tau})}{b_n^{\delta/(2+\delta)}} \sum_{\ell=\gamma_n}^{\infty} \int_0^{\infty} [\alpha(t)]^{\delta/(2+\delta)} p_{\ell-m+1}(t)dt$$

and $J_3 = o(1)$ by Assumption 3.3(b)(i). The bounding of J_3 for ρ-mixing processes is simpler. Part (iii) of the theorem follows from Parts (i) and (ii). ■

Proof of Theorem 4.1.

We employ the big block - small block procedure. With $Z_{n,j}$ defined as in (24), let

$$S_n = \sum_{i=1}^{n} Z_{n,i} = nb_n^{m+1/2} C_n(\underline{x}; \underline{\tau}) \tag{34}$$

and by Theorem 3.3,

$$\frac{1}{n} E[S_n^2] = nb_n^{2m+1} \mathrm{var}[C_n(\underline{x}; \underline{\tau})] \to \theta^2(\underline{x}; \underline{\tau}) \ . \tag{35}$$

It suffices to show that

$$\frac{1}{\sqrt{n}} S_n \xrightarrow{L} N(0, \theta^2(\underline{x}; \underline{\tau})) \ . \tag{36}$$

Partition the set $\{1, \ldots, , n\}$ into $2k+1$ subsets with large blocks of size $d = d_n$ and small blocks of size $c = c_n$ where

$$k = k_n = \lfloor \frac{n}{c_n + d_n} \rfloor \ . \tag{37}$$

Define the random variables

$$\eta_j = \sum_{i=j(c+d)+1}^{j(c+d)+d} Z_{n,i}, \quad 0 \leq j \leq k-1 \tag{38}$$

$$\xi_j = \sum_{i=j(c+d)+d+1}^{(j+1)(c+d)} Z_{n,i}, \quad 0 \leq j \leq k-1 \tag{39}$$

and

$$\zeta_k = \sum_{i=k(c+d)+1}^{n} Z_{n,i} \ . \tag{40}$$

Write

$$S_n = \sum_{j=0}^{k-1} \eta_j + \sum_{j=0}^{k-1} \xi_j + \zeta_k \equiv S_n' + S_n'' + S_n''' \ . \tag{41}$$

We show that as $n \to \infty$,

$$\frac{1}{n} E[S_n'']^2 \to 0, \quad \frac{1}{n} E[S_n''']^2 \to 0 \tag{42a}$$

$$|E[\exp(itS_n')] - \prod_{j=0}^{k-1} E[\exp(it\eta_j)]| \to 0 \tag{42b}$$

$$\frac{1}{n} \sum_{j=0}^{k-1} E[\eta_j^2] \to \theta^2(\underline{x}; \underline{\tau}) \tag{42c}$$

$$\frac{1}{n}\sum_{j=0}^{k-1} E[\eta_j^2 I\{|\eta_j| > \epsilon\theta(\underline{x};\underline{\tau})\sqrt{n}\}] \to 0 \tag{42d}$$

for every $\epsilon > 0$.

We first choose the large block size d_n. Assumption 4.1 implies that there exist integers $q_n \to \infty$ such that

$$q_n c_n = o((nb_n^{2m+1})^{1/2}), \quad q_n(n/b_n^{2m+1})^{1/2}\alpha_{c_n} \to 0, \quad q_n(n/b_n^{2m+1})^{1/2}\rho_{c_n} \to 0 \tag{43}$$

as $n \to \infty$. Now define the large-block size d_n by

$$d_n = [(nb_n^{2m+1})^{1/2}/q_n] \, . \tag{44}$$

Property (42a) follows by parts (i) and (ii) of Theorem 3.3 using the argument detailed in Fan and Masry (1992) or in Masry (1996b). In order to establish (42b) we make use of the following extension of a result by Volkonskii and Rozanov (1959); the proof is similar to that of Lemma 4.1 in Masry (1988).

Lemma 4.1. *Let* $V_j = F_j(\eta_j), j = 0, 1, \ldots, N$ *where* η_j *is defined by* (38) *and* $|F_j(x)| \le 1$ *for all* j. *Let* H *be the* σ*-algebra generated by the renewal process* $\{t_j\}$ *and let* (Y, X) *be a strongly mixing processes with mixing coefficient* $\alpha(\tau)$. *Then*

$$|E[\prod_{j=0}^{N} V_j] - \prod_{j=0}^{N} E[V_j]| = |E\{E[\prod_{j=0}^{N} V_j|H] - \prod_{j=0}^{N} E[V_j|H]\}| \tag{45}$$

$$\le 4N\int_0^\infty \alpha(\tau)p_{c-m+1}(\tau)d\tau \, . \tag{46}$$

With $F_j(u) = e^{itu}$ and $N = k-1$, (42b) follows from (46) using (43) and (44).

Next we establish (42c): We have by parts (i) and (ii) of Theorem 3.3

$$var[\eta_j] = var[\eta_0] = d_n\theta^2(\underline{x};\underline{\tau})(1 + o(1))$$

so that

$$\frac{1}{n}\sum_{j=0}^{k-1} E[\eta_j^2] = \frac{k_n d_n}{n}\theta^2(\underline{x};\underline{\tau})(1 + o(1)) \to \theta^2(\underline{x};\underline{\tau}).$$

It remains to establish (42d). We employ a truncation argument since ψ is not necessarily a bounded function. Let

$$a_L(y) = yI\{|y| \le L\} \tag{47}$$

where L is a fixed truncation point. Put

$$r_L(\underline{x};\underline{\tau}) = E[a_L(\psi(Y(\tau_m)))|X(0) = x_0, \ldots, X(\tau_m) = x_m] \, , \tag{48}$$

and

$$\sigma_L^2(\underline{x};\underline{\tau}) = E\left[(a_L(\psi(Y(\tau_m))) - r_L(\underline{x};\underline{\tau}))^2 | X(0) = x_0, \ldots, X(\tau_m) = x_m\right], \tag{49}$$

$$\theta_L^2(\underline{x};\underline{\tau}) = \frac{\sigma_L^2(\underline{x};\underline{\tau}) f(\underline{x};\underline{\tau})}{a(\underline{\tau})} \int_{R^{m+1}} K^2 \int_{R^m} W^2 . \tag{50}$$

Put

$$U_{n,i}^L = \frac{1}{a(\underline{\tau})} W_n(\underline{\tau} - \underline{D}_i)\left(a_L[\psi(Y(t_{i+m}))] - r_L(\underline{x};\underline{\tau})\right) K_n(\underline{x} - \underline{X}_i) \tag{51a}$$

and

$$Z_{n,i}^L = b_n^{m+1/2}\{U_{n,i}^L - E[U_{n,i}^L]\} . \tag{51b}$$

Finally put

$$S_n^L = \sum_{i=1}^n Z_{n,i}^L, \quad \tilde{S}_n^L = \sum_{i=1}^n (Z_{n,i} - Z_{n,i}^L) . \tag{51c}$$

Since W and K and a_L are bounded, we have

$$|Z_{n,i}^L| \le \frac{\text{const.}}{b_n^{m+1/2}} .$$

This implies by (38) (with η_j^L defined by using $Z_{n,i}^L$) that

$$\max_{0\le j\le k-1} |\eta_j^L|/\sqrt{n} \le \text{const.} \frac{d_n}{(nb_n^{2m+1})^{1/2}} \to 0 .$$

Hence when n is large, the set $\{|\eta_i^L| \ge \theta_L(\underline{x};\underline{\tau})\epsilon\sqrt{n}\}$ becomes an empty set and thus (42d) holds. Consequently, (42a)–(42d) hold for S_n^L so that

$$\frac{1}{\sqrt{n}} S_n^L \xrightarrow{L} N(0, \theta_L^2(\underline{x};\underline{\tau})) . \tag{52}$$

In order to complete the proof, namely to establish (42d) in the general case, it suffices to show (see the proof of Theorem 2.1 in Fan and Masry (1992)) that

$$\frac{1}{n} \operatorname{var}[\tilde{S}_n^L] \to 0 \text{ as first } n \to \infty \text{ and then } L \to \infty . \tag{53}$$

By the argument of Theorem 3.3,

$$\begin{aligned} &\lim_{n\to\infty} \frac{1}{n} \operatorname{var}[\tilde{S}_n^L] \\ &= E\Big[\psi^2(Y(\tau_m)) I\{|\psi(Y(\tau_m))| > L\} - [r(\underline{x};\underline{\tau}) - r_L(\underline{x};\underline{\tau})]^2 \mid \\ &\qquad X(0) = x_0, \ldots, X(\tau_m) = x_m\Big] (f(\underline{x},\underline{\tau})/a(\underline{\tau})) \int_{R^m} W^2 \int_{R^{m+1}} K^2. \end{aligned}$$

By dominated convergence the right side converges to 0 as $L \to \infty$. This establishes (42d) for general ψ and completes the proof of Theorem 4.1.

6. Discussion

The primary goal of this paper was to formulate simple estimators for the regression function $r(\underline{x};\underline{\tau})$ of (1) of continuous-time processes on the basis of discrete-time observations $\{Y(t_j), X(t_j), t_j\}_{j=1}^{n}$ and establish their consistency (with rates) and asymptotic normality as $n \to \infty$ including expressions for the bias and variance of the asymptotic distribution. This is a hybrid estimation problem whereby the underlying processes are continuous in time but the observations are discrete in time. A principal constraint is that the mean sampling rate β must remain finite. Indeed, the established results provide for the consistency (and asymptotic normality) of the estimator $r_n(\underline{x};\underline{\tau})$ for *any* value of $\beta > 0$. Such results cannot be achieved by equally-space observations unless the sampling rate is allowed to diverge to infinity. Also, the results hold for a fairly broad class of point processes $\{t_j\}$. These were the principal goals of the paper.

The model of the regression function (1) and its estimator that is developed in this paper is particularly useful in the context of nonlinear filtering. It should be noted that, unlike regression estimation for discrete-time processes, the prediction problem for continuous-time processes cannot be incorporated in (1). However, the necessary modifications are quite obvious: Replace (1) by

$$r(x_0, \ldots, x_{m-1}; \tau_1, \ldots, \tau_m) \triangleq E[\psi(Y(\tau_m))|X(0) = x_0, X(\tau_1) = x_1, \ldots, X(\tau_{m-1}) = x_{m-1}].$$

Let $f(\tilde{\underline{x}};\tilde{\underline{\tau}})$ be the density of $X(0), X(\tau_1), \ldots, X(\tau_{m-1})$ where $\tilde{\underline{x}} = (x_0, \ldots, x_{m-1})$ and $\tilde{\underline{\tau}} = (\tau_1, \ldots, \tau_{m-1})$. We can then write $g(\tilde{\underline{x}};\underline{\tau}) = r(\tilde{\underline{x}};\underline{\tau})f(\tilde{\underline{x}};\tilde{\underline{\tau}})$ and formulate estimators for f and g analogous to (10) and (11) with an obvious adjustment for the dimension of the kernels. Similar results to those given in this paper hold for the prediction model.

Because of space limitations in this volume, we adopted a simple estimation scheme (9)–(11) based on the Nadaraya–Watson approach. One could also adopt the recently popular local polynomial approach, as was done in Masry (1996a, 1996b) for discrete-time processes. In fact, a work in progress (Masry 1997) treats the hybrid regression problem using local polynomial fitting of arbitrary order; the analysis, however, is much more involved and lengthy given the hybrid nature of the problem, and the inherent dependence structure of continuous-time processes.

References

[1] P. Hall and C.C. Heyde (1980), *Martingale limit theory and its applications.* New York: Academic Press.

[2] G. Collomb and W. Härdle (1986), Strong uniform convergence rates in robust nonparametric time series analysis and prediction: Kernel regression estimation from dependent observations, *Stochastic Processes and their Applics.*, vol. 23, pp. 77–89.

[3] J. Fan (1992), Design-adaptive nonparametric regression, *Jour. Amer. Statist. Assoc.*, vol. 87, pp. 998–1004.

[4] J. Fan (1993), Local linear regression smoothers and their minimax efficiency, *Ann. Statist.*, vol. 21, pp. 196–216.

[5] J. Fan and I. Gijbels (1992), Variable bandwidth and local linear regression smoothers, *Ann. Statist.*, vol. 20, pp. 2008–2036.

[6] J. Fan and E. Masry (1992), Multivariate Regression Estimation With Errors-in-Variables: Asymptotic Normality for Mixing Processes, *J. Multivariate Analysis*, vol. 43, pp. 237–271.

[7] W. Härdle (1990), *Applied Nonparametric Regression*, Boston, Cambridge University Press.

[8] A.N. Kolmogorov, and Yu. A. Rozanov (1960), On strong mixing conditions for stationary Gaussian processes, *Theory Prob. Appl.*, vol. 52, pp. 204–207.

[9] E. Masry (1988), Random sampling of continuous-time stationary processes: statistical properties of joint density estimators, *J. Multivariate Analysis*, vol. 26, pp. 133–165.

[10] E. Masry (1993), Multivariate Regression Estimation with Errors-in-Variables for Stationary Processes, *Nonparametric Statistics*, vol. 3, pp. 13–36.

[11] E. Masry (1996a), Multivariate local polynomial regression for time series: uniform strong consistency and rates, *J. Time Series Analysis*, vol. 17, pp. 571–599.

[12] E. Masry (1996b), Multivariate regression estimation: Local polynomial fitting for time series, *J. Stochastic Processes and their Applics.*, vol. 65, pp. 81–101.

[13] E. Masry (1997), Multivariate regression estimation of continuous-time processes from sampled-data: Local polynomial fitting approach. Manuscript.

[14] M.I. Moore, P.J. Thompson, and T.G.L. Shirtcliffe (1988), Spectral analysis of ocean profiles from unequally spaced data, *J. Geophys. Res.* vol. 93, pp. 655–64.

[15] E. Parzen (1983), *Time series analysis of irregularly observed data.* Lecture Notes in Statistics, Vol. 25. New York: Springer-Verlag.

[16] P. M. Robinson (1983), Nonparametric estimators for time series, *J. Time Series Anal.*, vol. 4, pp. 185–297.

[17] M. Rosenblatt (1956), A central limit theorem and strong mixing conditions, *Proc. Nat. Acad. Sci.*, vol. 4, pp. 43–47.

[18] G. G. Roussas (1990), Nonparametric regression estimation under mixing conditions, *Stochastic Processes and their Applics.*, vol. 36, pp. 107–116.

[19] D. Ruppert and M.P. Wand (1994), Multivariate weighted least squares regression, *Ann. Statist.*, vol. 22, pp. 1346–1370.

[20] D. Tjostheim (1994), Non-linear time series: A selective review, *Scandinavian J. of Statistics*, vol. 21, pp. 97–130.

[21] L. T. Tran (1993), Nonparametric function estimation for time series by local average estimators, *Ann. Statist.*, vol. 21, pp. 1040–1057.

[22] V.A. Volkonskii and Yu. A. Rozanov (1959), Some limit theorems for random functions, *Theory Prob. Appl.*, vol. 4, pp. 178–197.

Department of Electrical and Computer Engineering
University of California, San Diego
La Jolla, CA 92093-0407
email: masry@ece.ucsd.edu

Tracing the Path of a Wright–Fisher Process with One-way Mutation in the Case of a Large Deviation

F. Papangelou

Abstract

The large deviations theory developed by Wentzell for discrete time Markov chains is used to show that if the state of a Wright–Fisher process modeling the frequency of an allele in a biological population undergoes a transition from one value to another over a number of generations which is large but much smaller than the size of the population, then there is a preferred path which the process follows closely with near certainty in the intervening time. This path was identified in [8] in the case in which there is only random drift acting on the genes. The case in which one-way mutation is added to the drift was cursorily mentioned in [8] and is fully treated here. The preferred path in this case is shown to be an exponential, a parabola, a hyperbolic cosine or a trigonometric cosine, depending on the mutation parameter and the boundary conditions involved.

AMS 1991 Subject classification: primary 60F10; secondary 60J20.

Keywords: Wright–Fisher process; mutation; large deviations; action functional; calculus of variations.

1. Introduction

The Wright–Fisher process, modeling the way in which gene frequencies in a finite biological population change from generation to generation, occupies a venerable position in the history of mathematical population genetics [2]. Despite the simplicity of its structure, it provides a good insight into the manner in which random genetic drift, as well as mutation and selection forces, act in heredity, and it also admits a straightforward diffusion approximation which sheds much light on its large-population, large-time asymptotics. This approximation is obtained by scaling the (discrete) time parameter in such a way that a unit of time corresponds to a number of generations equal to the size of the population [2],[5].

In [8] the large deviations behavior of this process under a different scaling was investigated. Specifically, it was shown that if the time scale is such that the number of generations in a unit of time, though large, is much smaller than the size of the population, then Wentzell's results on the large deviations of discrete time Markov processes [9],[10] can be brought to bear, opening up the way to the solution of the corresponding variational problems and leading ultimately to the determination of the preferred paths of the process. The main area of application of the celebrated Freidlin–Wentzell theory of large deviations is the theory of diffusions with small diffusion coefficient [3] but here the theory is applied directly to the discrete-time process itself and not to its diffusion approximation, where its relevance would be limited in the present context.

As will be seen below, under appropriate conditions a functional can be identified which very nearly plays the role of an action functional in determining the rough order of magnitude of the probabilities with which the scaled process follows the various paths available to it. As is to be expected, this means that we can in principle determine the preferred paths of the process, i.e. given the state of the scaled process at "time" 0 and its state at "time" T (the boundary conditions) we can determine explicitly the approximate path followed by the process with overwhelming probability during the intervening time interval. This path is simply the one that minimizes the action functional [3]. It was shown in [8] that, if there is no mutation or selection so that change occurs only through random sampling, the path followed is an arc of a cosine. The treatment of this case did not require recourse to the Euler–Lagrange theory of the calculus of variations. If one-way mutation is added as an ingredient, then the picture becomes much more involved and intriguing: the paths minimizing the action functional turn out to be of several types (exponentials, parabolas, hyperbolic or trigonometric cosines, depending on the mutation parameter and the boundary conditions) and can be obtained as "extremals," i.e. as solutions of the Euler equation for our problem. The extremals were listed at the end of [8] without proof of their minimizing properties or reference to any probabilistic statement that can be made about them. It is the purpose of the present paper to present the complete picture for this case. The main theorem, asserting weak convergence of the conditional distribution of the process, given the boundary conditions, to the deterministic extremal path satisfying the same boundary conditions, is stated and established in the last section.

The contents of the paper are as follows: In Section 2, which retraces the steps of [8] in our present more general context, we define our processes and introduce the appropriate scaling and calculate the asymptotics of their cumulants. Unlike the contents of subsequent sections which are devoted to the case of one-way mutation, the framework adopted here allows not only for mutation but also for selection and possibly other effects. The crucial functional for the variational problem, which emerges in this section, involves two singularities and thus fails to satisfy the hypotheses under which Wentzell's results were established and cannot be treated straight away as an action functional. The arguments required in overcoming the presence of these singularities in the case of one-way mutation are set out in Section 3, which also contains some groundwork extending the conventional treatment of Euler's equation from the realm of continuously differentiable functions to that of absolutely continuous functions, needed in the sequel. This enables us to show that the paths minimizing the functional are necessarily extremals, i.e. solutions of Euler's equation. The extremals are then given explicitly in Section 4 and the main theorem about them, referred to above, is proved in Section 5.

2. Scaling the process

We will be concerned below with untypically large changes of gene frequencies in a one-species population modeled by the Wright–Fisher process. To introduce this process, we consider a gene at a particular locus on the genetic blueprint of this species, which can take one of two forms (alleles) say, A and a. For a diploid population an organism's genotype corresponding to this locus can be AA, Aa or aa. The Wright–Fisher process models the changes in genotype frequencies by treating them as the result of binomial sampling [2],[5]. The random sampling mechanism can be described in two different but equivalent ways ("random mating of parents is equivalent to random union of genes") and this enables us to focus on the proportion of, say, the A genes in each generation: if the size of the population is N and there are u, v, w individuals of genotypes AA, Aa, aa respectively ($u + v + w = N$), then the state of the Wright–Fisher process is taken to be $y = \frac{i}{2N}$, where $i = 2u + v$ is the total number of A genes in the population. If there are no mutations or selection effects present, the gene population of the next generation is simply produced by sampling $2N$ times with replacement from the pool of $2N$ genes of the "current" generation.

The resulting process is a Markov chain on the state space $\left\{0, \frac{1}{2N}, \frac{2}{2N}, \dots, 1\right\}$ in which the probability of a one-step transition from state $y = \frac{i}{2N}$ to state $\tilde{y} = \frac{j}{2N}$ is

$$P(y, \tilde{y}) = \binom{2N}{j} y^j (1-y)^{2N-j}. \tag{2.1}$$

Effects of mutation or selection can be incorporated in the above scheme by replacing y in (2.1) by an appropriately modified value p.

Suppose now that the proportion of A genes in an "initial" generation is y_0 and that after a large number of generations n it is found to be y_1. As shown in [8] for the case of no mutation or selection, large deviations theory implies that if the size N of the population is very much larger than n, then we can trace the history of the change from y_0 to y_1 with near certainty. More precisely, it can be asserted that, with high probability, the path traced by the state of the process in the course of the intervening time "follows closely" a specific path which can be determined as the solution of a corresponding variational problem. The exact theorem obtained is of course a limit theorem and to prepare the ground for its extension to the case where mutation is possible we introduce next a sequence of time-scaled Wright–Fisher processes

$$Y_t^{(n)}, t \geq 0; \quad n = 1, 2, 3, \ldots$$

in continuous time as follows. For each $n \geq 1$, the process $Y_t^{(n)}$ jumps at times $\frac{1}{n}, \frac{2}{n}, \frac{3}{n}, \ldots$ and is constant on any interval of the form $\left[\frac{k}{n}, \frac{k+1}{n}\right), k = 0, 1, 2, \ldots$. The skeleton process $Y_0^{(n)}, Y_{\frac{1}{n}}^{(n)}, Y_{\frac{2}{n}}^{(n)}, \ldots$ is a Wright–Fisher process in which the gene population size is $2N$, where $N = N(n)$ depends on n. The conditional probability that $Y_{\frac{k+1}{n}}^{(n)} = \frac{j}{2N} = \tilde{y}$, given that $Y_{\frac{k}{n}}^{(n)} = y \ (0 \leq y \leq 1)$ is

$$P(y, \tilde{y}) = \binom{2N}{j} p_{y,n}^j (1 - p_{y,n})^{2N-j}$$

where

$$p_{y,n} = y + \frac{g(y)}{n} + o\left(\frac{1}{n}\right) \tag{2.2}$$

uniformly in $y \in [0, 1]$ as $n \to \infty$. It is assumed that $0 \leq p_{y,n} \leq 1$, that $g(y)$ is a continuous function on $[0, 1]$ and that $Nn^{-1} \to \infty$ as $n \to \infty$.

Here are two special cases indicating the scope of the model. If in the n-th process, $Y_t^{(n)}$, A mutates to a at rate $\frac{\gamma_1}{n}$ per generation (γ_1 per unit of "time") while a mutates to A at rate $\frac{\gamma_2}{n}$ per generation, we can build this into the model by taking

$$p_{y,n} = y + \frac{(1 - y)\gamma_2 - y\gamma_1}{n}. \tag{2.3}$$

If instead there are selective forces in action, favoring the survival of A, we may take

$$p_{y,n} = \frac{(1 + \sigma n^{-1})y}{(1 + \sigma n^{-1})y + (1 - y)} = y + \frac{\sigma y(1 - y)}{n} + o\left(\frac{1}{n}\right)$$

where $\sigma > 0$.

The large deviations behavior of the above sequence of processes is determined by the asymptotic behavior of their cumulants [9],[10]. The cumulant of $Y^{(n)}$ is

$$\begin{aligned} G^n(y,z): &= n \log E_y \exp\left\{z\left(Y^{(n)}_{\frac{1}{n}} - y\right)\right\} \\ &= -nzy + 2nN \log\left\{1 + \left(\exp\frac{z}{2N} - 1\right) p_{y,n}\right\} \end{aligned}$$

where E_y denotes conditional expectation given $Y_0^{(n)} = y$, and the sequence we need to investigate here is $n(2N)^{-1}G^n(y, 2Nn^{-1}z)$ ([10], p. 61). As in [8],

$$n(2N)^{-1}G^n(y,2Nn^{-1}z) = -nzy + n^2 \log\left\{1 + (e^{z/n} - 1)p_{y,n}\right\}$$

which is independent of N. Define

$$G(y,z) = g(y)z + \frac{1}{2}y(1-y)z^2.$$

As a function of z this is the cumulant generating function of the Gaussian distribution with mean $g(y)$ and variance $y(1-y)$. Using the expansions $\log(1+s) = s - \frac{s^2}{2} + O(s^3)$, $e^s - 1 = s + \frac{s^2}{2} + O(s^3)$ for $s \to 0$, and (2.2) we can prove that

$$\lim_{n\to\infty} n(2N)^{-1}G^n(y,2Nn^{-1}z) = G(y,z) \tag{2.4}$$

$$\lim_{n\to\infty} \frac{\partial}{\partial z}(n(2N)^{-1}G^n(y,2Nn^{-1}z)) = \frac{\partial}{\partial z}G(y,z), \tag{2.5}$$

the convergence being in both cases uniform in $y \in [0,1]$ and $z \in K$, if K is a bounded set. We can also prove that if K is a bounded subset of $\mathbb{R}$, there is a constant $C < \infty$ such that for all sufficiently large n, all $y \in [0,1]$ and all $z \in K$

$$\frac{\partial^2}{\partial z^2}(n(2N)^{-1}G^n(y,2Nn^{-1}z)) \le C. \tag{2.6}$$

In fact it is easy to see that

$$\begin{aligned} n(2N)^{-1}G^n(y,2Nn^{-1}z) &= -nzy + n^2\Big\{\frac{zy}{n} + \frac{z^2y}{2n^2} + \frac{zg(y)}{n^2} - \frac{z^2y^2}{2n^2} + o\Big(\frac{1}{n^2}\Big)\Big\} \\ &= g(y)z + \frac{1}{2}y(1-y)z^2 + o(1). \end{aligned}$$

For (2.5) note that $\frac{\partial}{\partial z}(n(2N)^{-1}G^n(y,2Nn^{-1}z))$ is equal to

$$[n(p_{y,n} - y) + p_{y,n}(1-y)n(e^{z/n} - 1)]/[1 + (e^{z/n} - 1)p_{y,n}],$$

the denominator of which tends to 1 as $n \to \infty$, while the numerator tends to $g(y) + y(1-y)z$. (2.6) follows by inspection once the second derivative is calculated.

The Legendre transform of $G(y,z)$ for fixed $0 < y < 1$ is

$$H(y,u) = \sup_z [zu - G(y,z)] = \frac{1}{2}\frac{(u - g(y))^2}{y(1-y)}.$$

Because of the singularities of this function at $y = 0$ and $y = 1$, the large deviations results of [9],[10] cannot be applied directly to the sequence of processes $Y_t^{(n)}, n = 1, 2, \ldots$. We therefore modify these processes (in a manner similar to [8]) and formulate the large deviations property of the resulting modified processes.

Given $\epsilon \in \left(0, \frac{1}{2}\right)$ we introduce for each $n = 1, 2, \ldots$ a new process $\tilde{Y}_t^{(n)}, t \geq 0$ as follows. The process $\tilde{Y}_t^{(n)}$ has state space $(-\infty, \infty)$. It jumps at times $\frac{1}{n}, \frac{2}{n}, \ldots$ and is constant on any interval of the form $\left[\frac{k}{n}, \frac{k+1}{n}\right), k = 0, 1, 2, \ldots$. The sequence $\tilde{Y}_0^{(n)}, \tilde{Y}_{\frac{1}{n}}^{(n)}, \tilde{Y}_{\frac{2}{n}}^{(n)}, \ldots$ is a Markov chain and if $\epsilon \leq y \leq 1 - \epsilon$ its one-step transition probabilities out of y are those of $Y_0^{(n)}, Y_{\frac{1}{2}}^{(n)}, Y_{\frac{2}{n}}^{(n)}, \ldots$. If $-\infty < y < \epsilon$, then the jump out of y has a Gaussian distribution with mean $\frac{g(\epsilon)}{n}$ and variance $\frac{\epsilon(1-\epsilon)}{2N}$, while for $1 - \epsilon < y < \infty$ this jump has a Gaussian distribution with mean $\frac{g(1-\epsilon)}{n}$ and variance $\frac{\epsilon(1-\epsilon)}{2N}$.

The cumulant $\tilde{G}^n(y,z)$ of $\tilde{Y}_t^{(n)}$ is equal to $G^n(y,z)$ if $\epsilon \leq y \leq 1 - \epsilon$ and to $g(\epsilon)z + \frac{n}{4N}\epsilon(1-\epsilon)z^2$ or $g(1-\epsilon)z + \frac{n}{4N}\epsilon(1-\epsilon)z^2$ if $y < \epsilon$ or $y > 1 - \epsilon$ respectively. Define $\tilde{g}(y)$ to be $g(y)$, $g(\epsilon)$ or $g(1-\epsilon)$ according as $\epsilon \leq y \leq 1 - \epsilon$, $y < \epsilon$ or $1 - \epsilon < y$, and $V(y)$ to be $y(1-y)$ or $\epsilon(1-\epsilon)$ according as $\epsilon \leq y \leq 1 - \epsilon$ or $y \notin [\epsilon, 1-\epsilon]$. If we set $\tilde{G}(y,z) = \tilde{g}(y)z + \frac{1}{2}V(y)z^2$ then (2.4)–(2.6) remain valid if we replace G^n and G by $\tilde{G}^n$ and $\tilde{G}$ respectively. The Legendre transform of $\tilde{G}$ is

$$\tilde{H}(y,u) = \frac{1}{2}\frac{(u - \tilde{g}(y))^2}{V(y)}.$$

The functions $\tilde{G}$ and $\tilde{H}$ satisfy Wentzell's conditions (A)–(E) on pp. 56 and 61 of [10] (see p. 79 in [10]) and Theorem 3.2.3′ on p. 68 of [10] then implies that, for fixed $T > 0$, the sequence of processes $\tilde{Y}_t^{(n)}, 0 \leq t \leq T$ has a large deviations property, uniformly with respect to the starting point, with action functional $\frac{2N}{n}\tilde{S}_{0,T}(\phi)$, where $\tilde{S}_{0,T}(\phi)$ is equal to $\int_0^T \tilde{H}(\phi(t), \phi'(t))dt$ if ϕ is an absolutely continuous function on $[0,T]$, and equal to ∞ if ϕ is not absolutely continuous. Here $\phi'(t)$ denotes the derivative of $\phi(t)$.

What this means for our considerations below is the following. Let y_0 be a fixed number in $(0,1)$ and denote by Φ the space of functions ϕ on $[0,T]$ with $\phi(0) = y_0$ which are right-continuous with left-hand limits ("cad-lag"), furnished with the uniform topology, i.e. the topology with metric $\rho(\phi,\psi) = \sup_{0 \leq t \leq T} |\phi(t) - \psi(t)|$. If F is any closed subset and G any open subset of Φ, then

$$\limsup_{n\to\infty} n(2N)^{-1}\log P\left\{\tilde{Y}^{(n)}\in F\right\} \leq -\inf\left\{\tilde{S}_{0,T}(\phi):\phi\in F\right\} \tag{2.7}$$

$$\liminf_{n\to\infty} n(2N)^{-1}\log P\left\{\tilde{Y}^{(n)}\in G\right\} \geq -\inf\left\{\tilde{S}_{0,T}(\phi):\phi\in G\right\} \tag{2.8}$$

where $\tilde{Y}^{(n)}$ stands for the restriction of the process $\tilde{Y}_t^{(n)}$ to $0\leq t\leq T$ and $P(\tilde{Y}^{(n)}\epsilon\,\cdot)$ is the distribution of $\tilde{Y}^{(n)}$ under the assumption $\tilde{Y}_0^{(n)}=y_0$.

As far as the original sequence $Y_t^{(n)}$, $n=0,1,2,\ldots$ is concerned, these statements imply the following (in which it is again assumed that under P we have $Y_0^{(n)}=y_0$). See [8] for the pure sampling case.

Proposition 2.1. *Assume that $N=N(n)$ depends on n in such a way that $Nn^{-1}\to\infty$ as $n\to\infty$. If $\phi(t), 0\leq t\leq T$ is an absolutely continuous function such that $\phi(0)=y_0$ and $0<\phi(t)<1$ for all $t\in[0,T]$, then*

$$\begin{aligned}\frac{1}{2}\int_0^T \frac{(\phi'(t)-g(\phi(t)))^2}{\phi(t)(1-\phi(t))}dt & \\ = -\lim_{\delta\to 0}\lim_{n\to\infty} n(2N)^{-1}\log P&\left(\sup_{0\leq t\leq T}|Y_t^{(n)}-\phi(t)|<\delta\right).\end{aligned} \tag{2.9}$$

To see this choose $\epsilon>0$ such that $\epsilon<\phi(t)<1-\epsilon$ for all $t\in[0,T]$ and construct the corresponding modified process $\tilde{Y}_t^{(n)}$. Since (2.9) holds for $\tilde{Y}_t^{(n)}$ ([1], p. 61), it follows easily that it holds for $Y_t^{(n)}$ as well. We may express (2.9) informally by saying that the logarithm of the probability that $Y_t^{(n)}$, $0\leq t\leq T$ follows closely the path $\phi(t), 0\leq t\leq T$ is of order

$$-Nn^{-1}\int_0^T \frac{(\phi'(t)-g(\phi(t)))^2}{\phi(t)(1-\phi(t))}dt.$$

This however does not in itself amount to a large deviations principle for the original processes $Y_t^{(n)}$. In the following sections we will investigate the implications for $Y_t^{(n)}$ of the large deviations principle for $\tilde{Y}_t^{(n)}$, embodied in (2.7) and (2.8), in the case where there is one-way mutation.

3. The variational problem

From this point on we consider only the case of one-way mutation, in which A mutates to a at rate $\frac{\gamma}{n}$ per generation ($\gamma>0$) in the n-th process $Y_t^{(n)}$. In this case $g(y)=-\gamma y$ (see (2.3)) and

$$H(y,u)=\frac{1}{2}\frac{(u+\gamma y)^2}{y(1-y)}.$$

Consider the strip $\Delta = \{(t,y) : t \geq 0, 0 < y < 1\}$ in $\mathbb{R}^2$. For $T > 0$ and any absolutely continuous function $\phi(t), 0 \leq t \leq T$ such that $0 < \phi(t) < 1$ for all $t \in [0,T]$ define

$$S_{0,T}(\phi) = \int_0^T H(\phi(t), \phi'(t))dt = \frac{1}{2}\int_0^T \frac{(\phi'(t) + \gamma\phi(t))^2}{\phi(t)(1-\phi(t))}dt.$$

For convenience, if $\phi(0) = y_0$ and $\phi(T) = y_1$, we will call $S_{0,T}(\phi)$ the "integral of H along the graph of ϕ from $(0, y_0)$ to (T, y_1)." A similar terminology will be used for integrals of the form $\int_{T_1}^{T_2} H(\phi(t), \phi'(t))dt$. The graph of any function of the form $Ce^{-\gamma t}, t \geq 0$, where C is positive but not necessarily less than 1, will be called a γ-exponential. Note that the integral of H along any arc of a γ-exponential which lies in Δ is 0.

To see how $S_{0,T}(\phi)$ behaves for functions ϕ whose graph gets extremely close to $y = 1$ we need the following lemma.

Lemma 3.1. *Let ϕ be an absolutely continuous function such that $\phi(u) = 1-\epsilon$ and $\phi(v) = 1 - \frac{\epsilon}{2}$, where $0 \leq u < v$ and $0 < \epsilon < 1$. Then*

$$\int_u^v H(\phi(t), \phi'(t))dt \geq 2\gamma\frac{1-\epsilon}{1-\frac{\epsilon}{2}}.$$

Proof. Define $s_2 = \min\{t > u : \phi(t) = 1 - \frac{\epsilon}{2}\}$, $s_1 = \max\{t < s_2 : \phi(t) = 1-\epsilon\}$. Then $s_1 < s_2$ and $1-\epsilon \leq \phi(t) \leq 1 - \frac{\epsilon}{2}$ for all $t \in [s_1, s_2]$. Now

$$\begin{aligned}\int_u^v H(\phi(t), \phi'(t))dt &\geq \int_{s_1}^{s_2} \frac{(\phi'(t)+\gamma\phi(t))^2}{\phi(t)(1-\phi(t))}dt \\ &\geq \frac{1}{\left(1-\frac{\epsilon}{2}\right)\epsilon}\int_{s_1}^{s_2}(\phi'(t)+\gamma\phi(t))^2 dt.\end{aligned}$$

By the Cauchy–Schwarz inequality, the last integral is greater than or equal to

$$\begin{aligned}&\left(\int_{s_1}^{s_2} 1^2 dt\right)^{-1}\left(\int_{s_1}^{s_2}(\phi'(t)+\gamma\phi(t))dt\right)^2 \\ &\quad = (s_2-s_1)^{-1}\left[\phi(s_2)-\phi(s_1)+\gamma\int_{s_1}^{s_2}\phi(t)dt\right]^2 \\ &\quad \geq (s_2-s_1)^{-1}\left[\frac{\epsilon}{2}+\gamma(1-\epsilon)(s_2-s_1)\right]^2 \\ &\quad \geq (s_2-s_1)^{-1}\cdot 2\gamma\epsilon(1-\epsilon)(s_2-s_1) \\ &\quad = 2\gamma\epsilon(1-\epsilon)\end{aligned}$$

by the inequality $(a+b)^2 \geq 4ab$. This proves the lemma.

If in this lemma we have $\phi(v) = 1 - \frac{\epsilon}{2^k}$ instead of $\phi(v) = 1 - \frac{\epsilon}{2}$ (where k is a positive integer) then, by considering the intermediate levels $1 - \frac{\epsilon}{2}, 1 - \frac{\epsilon}{2^2}, \ldots, 1 - \frac{\epsilon}{2^{k-1}}$ between $1-\epsilon$ and $1 - \frac{\epsilon}{2^k}$ we deduce from the lemma that $\int_u^v H(\phi(t), \phi'(t))dt \geq 2k\gamma(1-\epsilon)$. Hence we have proved the following.

Corollary 3.2. *Let $0 < y_0 < 1$. Given $K > 0$, there exists an $\eta \in (0,1)$ such that the integral of H along any graph joining the point $(0, y_0)$ with any point (t, y) having $y > 1 - \eta$ is greater than K.*

We now turn to the variational problem of minimizing $S_{0,T}(\phi)$ subject to $\phi(0) = y_0$, $\phi(T) = y_1$. Choose first an arbitrary ϕ^* joining $(0, y_0)$ with (T, y_1) and then a $K > S_{0,T}(\phi^*)$. Let η be as in Corollary 3.2. Next choose an $\epsilon > 0$ which is less than all three of η, y_1 and $y_0 e^{-\gamma T}$ and consider the processes $\tilde{Y}_t^{(n)}, n = 1, 2, \ldots$. From large deviations theory we know that sets of the form $\{\phi : \tilde{S}_{0,T}(\phi) \leq c\}$ are compact in Φ and the functional $\tilde{S}_{0,T}$ is lower semi-continuous. It follows that $\tilde{S}_{0,T}$ attains its minimum, in the sense that there is at least one absolutely continuous $\phi_0(t), 0 \leq t \leq T$ such that $\phi_0(0) = y_0$, $\phi_0(T) = y_1$ and $\tilde{S}_{0,T}(\phi_0) \leq \tilde{S}_{0,T}(\phi)$ for every other ϕ joining $(0, y_0)$ with (T, y_1). The graph of such a ϕ_0 lies between the γ-exponential passing through $(0, y_0)$ and the one passing through (T, y_1); otherwise we would be able to make $\tilde{S}_{0,T}(\phi_0)$ smaller by replacing a section of the graph of ϕ_0 by an arc of a γ-exponential. Trivially then $\min_{0 \leq t \leq T} \phi_0(t) > \epsilon$. Corollary 3.2 shows that we also have $\max_{0 \leq t \leq T} \phi_0(t) < 1 - \epsilon$. These facts imply that $\tilde{S}_{0,T}(\phi_0) = S_{0,T}(\phi_0)$ and that ϕ_0 minimizes $S_{0,T}$ as well, subject to the same boundary conditions.

The simple argument used above to show that $\min_{0 \leq t \leq T} \phi_0(t) > \epsilon$ implies the following stronger statement which we will need in the last section.

Remark 3.3. Suppose $0 < y_0 < 1$, $T > 0$ and $0 < y_1 < y_0 e^{-\gamma T}$ and let ϕ_0 be a function on $[0, T]$ minimizing $S_{0,T}(\cdot)$. If a point (t, y) with $0 < t < T$, $y > 0$ is below the γ-exponential passing through (T, y_1), then the integral of H along any graph joining $(0, y_0)$ with (t, y) is greater than $S_{0,T}(\phi_0)$.

If we knew that a minimizing ϕ_0 is continuously differentiable, we would conclude immediately from the classical Euler–Lagrange theory of the calculus of variations that ϕ_0 satisfies Euler's equation for the functional H ([4], p. 15). Showing this, however, requires a little more work. To write down Euler's equation we follow standard practice and write $H(y, y')$ for $H(y, u)$. Then, in notation whose meaning should be obvious, we can write Euler's equation in the form

$$H_y - \frac{d}{dt} H_{y'} = 0 \tag{3.1}$$

i.e.

$$y''H_{y'y'} + y'H_{y'y} - H_y = 0$$

where H_y, $H_{y'}$ etc. denote partial derivatives. In our present case this equation is

$$y'' = \frac{(2y-1)y'^2 + \gamma^2 y^2}{y(1-y)}. \tag{3.2}$$

We know from the outset that a minimizing ϕ_0 is absolutely continuous but in order to show that such a ϕ_0 is in fact continuously differentiable we need the following extension of a classical lemma.

Lemma 3.4. *If $\alpha(t), \beta(t)$ are integrable functions on $a \le t \le b$ such that*

$$\int_a^b [\alpha(t)h(t) + \beta(t)h'(t)]dt = 0$$

for every absolutely continuous function $h(t), a \le t \le b$ with $h(a) = h(b) = 0$, then there exists a constant c such that

$$\beta(t) = \int_a^t \alpha(s)ds + c$$

for almost every $t \in [a, b]$.

The proof is an easy adaptation of that of Lemma 4 on p. 11 of [4]. The integration by parts required in it is justified by, for instance, Satz 5 on p. 301 of [7]. To apply the lemma, one next follows the classical argument leading to Euler's equation ([4], pp. 13-15) which shows that if $S_{0,T}(\cdot)$ has a minimum at ϕ_0, then

$$\int_0^T [H_y(\phi_0(t), \phi_0'(t))h(t) + H_{y'}(\phi_0(t), \phi_0'(t))h'(t)]dt = 0$$

for every absolutely continuous function $h(t)$ on $[0, T]$ with $h(0) = h(T) = 0$. The lemma immediately implies that there exists a constant c such that

$$H_{y'}(\phi_0(t), \phi_0'(t)) = \int_0^t H_y(\phi_0(s), \phi_0'(s))ds + c \tag{3.3}$$

for almost all $t \in [0, T]$. The left-hand side of this is $\frac{\phi_0'(t)+\gamma\phi_0(t)}{\phi_0(t)(1-\phi_0(t))}$. The right-hand side is an absolutely continuous function of t which we may denote by $f(t)$. Thus

$$\frac{\phi_0'(t) + \gamma\phi(t)}{\phi_0(t)(1-\phi_0(t))} = f(t)$$

almost everywhere and since ϕ_0 is absolutely continuous we deduce that

$$\phi_0(t) = y_0 + \int_0^t \phi_0'(s)ds = y_0 + \int_0^t (-\gamma\phi_0(s) + \phi_0(s)(1-\phi_0(s))f(s))ds$$

which shows that ϕ_0 is continuously differentiable. It follows that (3.3) is an identity and that ϕ_0 is a solution of Euler's equation (3.1). Solutions of Euler's equation are called *extremals* of the functional H.

4. The extremals of the functional H

Even in as simple a case as that of one-way mutation, the extremals of the functional H present a fascinating picture. They were listed in summary form in [8] and we will discuss them in greater detail here.

First note that the case of no mutation ($\gamma = 0$) was fully settled in [8], where it was shown that the corresponding extremals are monotone arcs of curves of the form

$$\phi(t) = \frac{1}{2} - \frac{1}{2}\cos(ct + k), \quad 0 \leq t \leq T. \tag{4.1}$$

This is close to our intuition, since it shows that change is more rapid where the fluctuations are higher, i.e. where the "variance" $y(1-y)$ is higher. As seen from [8], the formal link between (4.1) and this variance arises from the fact that the indefinite integral of $[y(1-y)]^{-1/2}$ is $\cos^{-1}(1-2y)$. Fisher's angular transformation $y = \frac{1}{2} - \frac{1}{2}\cos\theta$ ([6], p. 23) makes the variance independent of location and although it introduces an unwelcome drift, the latter is asymptotically eliminated.

To obtain the extremals in the case $\gamma > 0$ we make use of the well-known fact ([4], pp. 18-19) that along each extremal the "energy" $y'H_{y'} - H$ is constant. In our present case this means that for each extremal ϕ there is a constant c such that

$$\frac{\phi'(t)^2 - \gamma^2\phi(t)^2}{\phi(t)(1-\phi(t))} = c \tag{4.2}$$

for all t in the appropriate domain. The converse is also true: every non-singular solution of an equation of the form (4.2) satisfies Euler's equation (3.2). Solutions of (4.2) fall into six categories according to the value of the constant c. Suppose we fix $y_0 \in (0,1)$ and require $\phi(0) = y_0$. As pointed out in [8], c is determined by $\phi(0)$ and $\phi'(0)$, its smallest possible value is $-\frac{\gamma^2 y_0}{1-y_0}$ and to each admissible value of c there correspond two values of $\phi'(0)$ unless $\phi'(0) = 0$.

Keeping y_0 fixed, we will describe the six types of extremals in terms of the values of $\phi'(0)$. In each case we will give a more or less explicit formula, with the understanding that each extremal is terminated at the point where it exits from the strip $\Delta = \{(t,y) : t \geq 0, 0 < y < 1\}$.

Case 1. If $\phi'(0) = 0$ $\left(c = -\frac{\gamma^2 y_0}{1-y_0}\right)$ then the extremal is the hyperbolic cosine

$$\phi(t) = \frac{1}{2}y_0\left[\cosh\frac{\gamma t}{\sqrt{1-y_0}} + 1\right] \tag{4.3}$$

which exits from Δ at the point t_0 where $\phi(t_0) = 1$ and is increasing on $[0, t_0]$.
Case 2. If $0 < \phi'(0) < \gamma y_0$ or $-\gamma y_0 < \phi'(0) < 0$ $\left(-\frac{\gamma^2 y_0}{1-y_0} < c < 0\right)$, then

$$\phi(t) = \frac{-c}{2(\gamma^2 - c)}\left[\cosh\left(\left(\sqrt{\gamma^2 - c}\right)(t-k)\right) + 1\right] \tag{4.4}$$

where the constant k depends on c (since $\phi(0) = y_0$ is given) and is negative in the first subcase and positive in the second. In either subcase the extremal exits from Δ at level $y = 1$; in the former the extremal is increasing while in the latter it is decreasing on $0 < t < k$, increasing for $t > k$ and re-crosses the horizontal line $y = y_0$ at $t = 2k$.

Case 3. If $\phi'(0) = \gamma y_0$ then the extremal is the exponential $\phi(t) = y_0 e^{\gamma t}$, while if $\phi'(0) = -\gamma y_0$ then $\phi(t) = y_0 e^{-\gamma t}$. (In both cases $c = 0$). The latter curve stays within Δ for all $t > 0$.

Case 4. If $\gamma y_0 < \phi'(0) < \gamma\sqrt{y_0}$ or $-\gamma\sqrt{y_0} < \phi'(0) < -\gamma y_0$ $(0 < c < \gamma^2)$, then

$$\phi(t) = \frac{c}{2(\gamma^2 - c)}\left[\cosh\left(\left(\sqrt{\gamma^2 - c}\right)(t-k)\right) - 1\right]$$

where the constant $k = k(c)$ is negative in the first subcase and positive in the second. In the former the (increasing) extremal exits from Δ at level $y = 1$, while in the latter the (decreasing) extremal terminates at $t = k$ where it is tangent to the t-axis.

Case 5. If $\phi'(0) = \gamma\sqrt{y_0}$ then the extremal is the parabola $\phi(t) = \frac{1}{4}(\gamma t + 2\sqrt{y_0})^2$ while if $\phi'(0) = -\gamma\sqrt{y_0}$ then $\phi(t) = \frac{1}{4}(\gamma t - 2\sqrt{y_0})^2$. (Here $c = \gamma^2$). In the second subcase the parabola terminates at $t = \frac{2\sqrt{y_0}}{\gamma}$ where it is tangent to the t-axis.

Case 6. If $\phi'(0) > \gamma\sqrt{y_0}$ or $\phi'(0) < -\gamma\sqrt{y_0}$ $(c > \gamma^2)$ then the extremal is a monotone arc of the trigonometric cosine

$$\phi(t) = \frac{c}{2(c - \gamma^2)}\left[1 - \cos\left(\left(\sqrt{c - \gamma^2}\right)(t-k)\right)\right],$$

with $k = k(c) < 0$ in the first subcase and > 0 in the second. In the latter, ϕ terminates at $t = k$ where it is tangent to the t-axis.

We must still establish the existence and uniqueness of an extremal joining any two given points. First note that no two of the above extremals meet inside the strip Δ except at the point $(0, y_0)$. Suppose $\phi_1(t)$, $\phi_2(t)$ are extremals corresponding to values c_1, c_2 of c in (4.2). If $\phi_1'(0) < \phi_2'(0)$, then the graph of ϕ_1 in Δ must be below the graph of ϕ_2. Otherwise, at the smallest positive t^* at which $\phi_1(t^*) = \phi_2(t^*)$ we would have $\phi_1'(t^*) \geq \phi_2'(t^*)$, contradicting $\phi_i'(t^*)^2 = \gamma^2\phi_i(t^*)^2 + c_i(\phi_i(t^*) - \phi_i(t^*)^2)$ $(i = 1, 2)$, as can easily be seen by looking at the different cases involved.

For any point (T, y_1) with $T > 0$, $0 < y_1 < 1$, one (exactly) of the extremals given above joins $(0, y_0)$ with (T, y_1). To see this note that if the point (T, y_1) is above the graph of (4.3), then such an extremal should have $\phi'(0) > 0$ and be increasing. By (4.2), we should have

$$T = \int_{y_0}^{y_1} \frac{dy}{\sqrt{(\gamma^2 - c)y^2 + cy}}. \tag{4.5}$$

The integrand is real and positive for all $y \in (y_0, y_1)$ and the integral on the right decreases from ∞ to 0 as c increases from $-\frac{\gamma^2 y_0}{1-y_0}$ to ∞. Hence there exists a c for which (4.5) holds, proving the existence of an extremal passing through (T, y_1). The case where (T, y_1) is below the γ-exponential $y_0 e^{-\gamma t}$ can be treated similarly. If (T, y_1) is above this γ-exponential but not above (4.3) then an extremal through it should be of the form given in subcase $-\gamma y_0 < \phi'(0) < 0$ of Case 2. In this case the dependence of k on c makes the dependence of ϕ on c rather complicated and it is easier to work by determining the point(s) at which (4.4) crosses the horizontal line $y = y_1$. A continuity argument as above shows that there is a c such that the corresponding extremal passes through (T, y_1).

It was shown in Section 3 that a function $\phi_0(t), 0 \leq t \leq T$ which minimizes $S_{0,T}(\cdot)$ subject to $\phi(0) = y_0$, $\phi(T) = y_1$ is necessarily an extremal. We now know that there is exactly one extremal joining $(0, y_0)$ with (T, y_1), hence this unique extremal must be the minimizing function.

5. The past of the Wright–Fisher process

All the ingredients are now in place for the proof of our main result. Suppose that the process $Y_t^{(n)}, 0 \leq t \leq T$ is in state y_0 when $t = 0$ and close to state y_1 when $t = T$. Given this, large deviations theory suggests that for large n the path followed by the process over the interval $0 \leq t \leq T$ is, with high probability, uniformly close to the function $\phi_0(t)$, $0 \leq t \leq T$ which minimizes the action functional $S_{0,T}(\cdot)$ subject to the conditions $\phi(0) = y_0$, $\phi(T) = y_1$. More precisely, we can prove a limit theorem which establishes weak convergence (as $n \to \infty$) of the conditional distribution of $Y_t^{(n)}, 0 \leq t \leq T$ to the unit mass δ_{ϕ_0} at ϕ_0. If the only boundary condition imposed is $Y_0^{(n)} = y_0$, then the distribution of $Y_t^{(n)}, 0 \leq t \leq T$ converges weakly to the unit mass at the function $y_0 e^{-\gamma t}, 0 \leq t \leq T$. This exponential represents "mean behavior" for our process and is the curve obtained in the corresponding deterministic model of gene frequency change by mutation.

The following theorem was stated and proved in [8] for the case $\gamma = 0$, i.e. the case in which the Wright–Fisher process is only subject to random drift.

Theorem 5.1. *Suppose $0 < y_0 < 1$, $T > 0$ and $y_0 e^{-\gamma T} \leq y_1 < 1$ and let $\phi_0(t), 0 \leq t \leq T$ be the unique extremal such that $\phi_0(0) = y_0$ and $\phi_0(T) = y_1$. Assume that $Y_0^{(n)} = y_0$ for all n and denote by P the probability measure arising from these initial conditions and the transition structure of $Y_t^{(n)}, t \geq 0$. Then, for every $\delta > 0$*

$$\lim_{n\to\infty} P\left(\sup_{0\leq t\leq T} |Y_t^{(n)} - \phi_0(t)| < \delta \mid Y_T^{(n)} \geq y_1\right) = 1.$$

Proof. We assume that $\delta < \min_{0\leq t\leq T} \phi_0(t)$ and $\delta < 1 - \max_{0\leq t\leq T} \phi_0(t)$ and choose $\epsilon > 0$ so small that (i) $\epsilon < \phi_0(t) - \delta < \phi_0(t) + \delta < 1 - \epsilon$ for all $t \in [0, T]$ and (ii) the integral of H along any graph which exits from $[\epsilon, 1-\epsilon]$ is greater than $S_{0,T}(\phi_0)$. This last condition is possible by Corollary 3.2 and Remark 3.3. With this ϵ we may then construct the modified processes $\tilde{Y}_t^{(n)}, n = 1, 2, \ldots$ of Section 2 in such a way that $\tilde{Y}_t^{(n)}$ agrees with $Y_t^{(n)}$ up to the moment of exit from $[\epsilon, 1-\epsilon]$. Note that if $y_1 < y_1^* < 1 - \epsilon$ and ϕ is any function such that $\phi(0) = y_0$ and $\phi(T) = y_1^*$ then $S_{0,T}(\phi) > S_{0,T}(\phi_0)$ as can easily be seen by considering integrals of H along sections of the γ-exponential passing through (T, y_1). Thus ϕ_0 minimizes $S_{0,T}(\phi)$ subject to $\phi(0) = y_0$, $\phi(T) \geq y_1$.

Denote now by V the set of functions ϕ on $[0, T]$ that are right-continuous with left-hand limits ("cadlag") and satisfy $\phi(0) = y_0$ and $\sup_{0\leq t\leq T} |\phi(t) - \phi_0(t)| < \delta$. Likewise let U be the set of cadlag functions with $\phi(0) = y_0$ and $\phi(T) \geq y_1$. Using $Y^{(n)}$ to denote the restriction $Y_t^{(n)}, 0 \leq t \leq T$ we then have

$$\begin{aligned} &P\Big(\sup_{0\leq t\leq T} |Y_t^{(n)} - \phi_o(t)| < \delta | Y_T^{(n)} \geq y_1\Big) \\ &\quad = \frac{P(Y^{(n)} \in V \cap U)}{P(Y^{(n)} \in V \cap U) + P(Y^{(n)} \in V^c \cap U)} \\ &\quad \geq \frac{P(Y^{(n)} \in V \cap U^0)}{P(Y^{(n)} \in V \cap U^0) + P(Y^{(n)} \in \overline{V^c} \cap U)} \end{aligned}$$

where V^c is the complement of V and ^{0}, $^-$ denote interior and closure respectively. By (2.8)

$$\begin{aligned} &\liminf_{n\to\infty} n(2N)^{-1} \log P(Y^{(n)} \in V \cap U^0) \\ &\quad = \liminf_{n\to\infty} n(2N)^{-1} \log P(\tilde{Y}^{(n)} \in V \cap U^0) \\ &\quad \geq -\inf\left\{\tilde{S}_{0,T}(\phi) : \phi \in V \cap U^0\right\} \\ &\quad = -\inf\left\{S_{0,T}(\phi) : \phi \in V \cap U^0\right\} \end{aligned}$$

and the last term is equal to $-S_{0,T}(\phi_0)$ since one can easily construct a sequence $\phi_n \in V \cap U^0$ such that $\phi_n \to \phi_0$ uniformly and $S_{0,T}(\phi_n) \to S_{0,T}(\phi_0)$. On the other hand

$$\begin{aligned} P(Y^{(n)} \in \overline{V^c} \cap U) \\ &= P(Y^{(n)} \in \overline{V^c} \cap U \text{ and } Y^{(n)} \text{ does not exit from } [\epsilon, 1-\epsilon]) \\ &\quad + P(Y^{(n)} \in \overline{V^c} \cap U \text{ and } Y^{(n)} \text{ exits from } [\epsilon, 1-\epsilon]) \\ &= a_n + b_n, \end{aligned}$$

say.

By (2.7), $\limsup_{n\to\infty} n(2N)^{-1} \log a_n$ is less than or equal to $-\inf\{S_{0,T}(\phi) : \phi \in \overline{V^c} \cap U$ and ϕ does not exit from $[\epsilon, 1-\epsilon]\}$, since we can replace $Y^{(n)}$ by $\tilde{Y}^{(n)}$. This upper bound is in turn strictly less than $-S_{0,T}(\phi_0)$ for otherwise, by a compactness argument, there would exist an element ϕ^* of $\overline{V^c} \cap U$ such that $S_{0,T}(\phi^*) = S_{0,T}(\phi_0)$, violating the uniqueness of ϕ_0.

Again, by (2.7)

$$\begin{aligned} &\limsup_{n\to\infty} n(2N)^{-1} \log b_n \\ &\quad \le -\inf\{\tilde{S}_{0,T}(\phi) : \phi(0) = y_0, \quad \phi \text{ exits from } (\epsilon, 1-\epsilon)\} \end{aligned}$$

which is strictly less than $-S_{0,T}(\phi_0)$ by the choice of ϵ. Taken together, all these inequalities establish the theorem.

An analogous theorem can be established for the case $y_1 \le y_0 e^{-\gamma T}$.

POSTSCRIPT. The case of genetic selection is treated in a companion paper "*Elliptic and other functions in the large deviations behavior of the Wright–Fisher process*," Annals of Applied Probability.

References

[1] Deuschel, J.-D. and Stroock D.W.: Large Deviations. Academic Press, Boston, 1989.

[2] Ewens, W.J.: Mathematical Population Genetics. Biomathematics Vol. 9. Springer-Verlag, New York, 1979.

[3] Freidlin, M.I. and Wentzell, A.D.: Random Perturbations of Dynamical Systems. Springer-Verlag, New York, 1984.

[4] Gelfand, I.M. and Fomin, S.V.: Calculus of Variations. Prentice-Hall, Englewood Cliffs, 1963.

[5] Karlin, S. and Taylor, H.M.: A Second Course in Stochastic Processes. Academic Press, New York, 1981.

[6] Kimura, M.: Diffusion Models in Population Genetics. Methuen's Review Series in Applied Probability, Methuen, London, 1964.

[7] Natanson, I.P.: Theorie der Funktionen einer Reellen Veränderlichen. Akademie-Verlag, Berlin, 1961.

[8] Papangelou, F.: Large deviations of the Wright–Fisher process. Proceedings of the Athens Conference on Applied Probability and Time Series Analysis, Vol.1: Applied Probability (eds. C.C. Heyde, Yu. V. Prohorov, R. Pyke, S.T. Rachev). Lecture Notes in Statistics **114**, 245–252, Springer-Verlag, New York, 1996.

[9] Wentzell, A.D.: Rough limit theorems on large deviations for Markov processes, Th. Probab. Appl. **21** (1976), 227–242 and 499–512; **24** (1979), 675–692; **27** (1982), 215–234.

[10] Wentzell, A.D.: Limit Theorems on Large Deviations for Markov Stochastic Processes. Kluwer Academic Publishers, Dordrecht, 1990.

Department of Mathematics
University of Manchester
Manchester M13 9PL
U.K.

A Distribution Inequality For Martingales with Bounded Symmetric Differences

Loren D. Pitt

Abstract

We discuss (with some generalizations) the question of optimal strategies for the game of red-and-black with a fixed goal, with fixed subfair odds, and with both the size of legal bets allowed and the playing time limited.

1. Introduction

In their classic monograph [DS65] Dubins and Savage gave substantial discussion to those circumstances when a bold strategy is an optimal gambling strategy. In particular they discussed the game of red-and-black with fixed subfair odds and no legal limits on the size of wagers. In Chapter 5 they showed that when the goal is to win a fixed amount, then the bold strategy of staking all you possess, or enough to reach the target, whichever is least, is an optimal strategy when the playing time is unlimited. On page 92 of the same monograph the authors presented a separate argument due to Dvoretzky that establishes the same result when the playing time is limited. In the case when bet sizes are limited Dubins and Savage mention only a flawed treatment by Coolidge [C08-9] and they concluded with the remark on page 4, "Whether ... optimality is correct when there is a legal limit, we do not know."

Since [DS65] was published, two articles by Klawe [K79] and Yang [Y90] treated the case of optimal strategies for red-and-black with fixed subfair odds when the funds available are unlimited but the size of the bets is limited by 1. In this setting, Klawe showed in the fair case and Yang in the subfair case that for limited time the bold strategy is the optimal strategy for reaching a *fixed integer valued objective.* Our main result in this note is Theorem 1, a supermartingale inequality for a special class (p-symmetric) of supermartingales, which implies the main results in [K79] and [Y90] for this broader class of processes, and which shows that while the bold strategy may not be optimal for noninteger valued objectives it is never far from optimal.

Related works of other authors are Heath and Kertz [HK88] and McBeth and Weerasinghe [MW96] where optimal strategies for leaving a symmetric interval in a fixed amount of time are discussed for variants of red-and-black, including continuous time variants. These results generalize the works of Klawe and Yang in directions different from ours and they do not imply and are not implied by our results. Other references may be found in [MS96]. Finally, I mention that while the present note is closely related to the literature on optimal strategies, in concept and approach it is perhaps more closely related to works on best constants and sharp inequalities for martingale transforms and martingale subordination. Of this literature, I mention only Burkholder's Wald lectures [B84].

2. Statement of Results

We present our results in the form of inequalities for a class of processes $\mathcal{M}_{p,1}$, $0 \le p \le \frac{1}{2}$, that we term *p-symmetric supermartingales with uniformly bounded jumps*. For $0 \le p \le 1$, an adapted process $\{Z\} = \{Z_n, \mathcal{F}_n : n \ge 0\}$ with $Z_0 = 0$ will be called p-symmetric provided the difference sequence $\{Y_k = Z_k - Z_{k-1}\}$ satisfies the p-symmetry condition: *For each n and each $\lambda \ge 0$ the conditional probability distribution of Y_{n+1} given $\mathcal{F}_n$ satisfies*

$$(1-p)P\{Y_{n+1} \ge \lambda \| \mathcal{F}_n\} = pP\{Y_{n+1} \le -\lambda \| \mathcal{F}_n\}.$$

The class $\mathcal{M}_{p,1}$ consists of those $\{Z\}$ that are p-symmetric and which satisfy

$$\sup_n ||Y_n||_\infty \le 1.$$

For $p \le \frac{1}{2}$ each $\{Z\}$ in $\mathcal{M}_{p,1}$ is a supermartingale. The game of red-and black with legal limits of size 1 falls into the class of processes $\mathcal{M}_{p,1}$. For $Z \in \mathcal{M}_{p,1}$ and λ real we set

$$V_{Z,N}(k) = P\{\max_{0\le n\le N} Z_n \ge \lambda\}.$$

Our first results are sharp upper bounds for the function

$$V_N(\lambda) = \sup_{Z\in\mathcal{M}_{p,1}} V_{Z,N}(\lambda).$$

We state these results in terms of the p-symmetric random walk $S_n = \sum_{k=0}^n X_k$ where the X_k are i.i.d. Bernoulli variables with $p = P\{X_k = 1\} = 1 - P\{X_k = -1\}$. We set $S_n = 0$. For nonnegative integers k and N, let $g_N(k) = P\{\max_{n\le N} S_n \ge k\}$, and let $U_N(\lambda)$ be the continuous function on $[0,\infty)$ that is linear on each interval $[k, k+1]$ and satisfies $U_N(k) = g_N(k)$

for each nonnegative integer k. Both functions $U_N(\lambda)$ and $V_N(\lambda)$ are clearly decreasing in λ and increasing in N. Letting $[\lambda]$ be the greatest integer *less* than λ we show

Theorem 1. *For $0 < p \leq 1/2$ and for all $\lambda \geq 0$ and all integers $N \geq 0$,*

$$U_N([\lambda]+1) \leq V_N(\lambda) \leq U_N(\lambda). \tag{1}$$

Remark. For integers n, $[n]+1 = n$ so Theorem 1 implies that $U_N(n) = V_N(n)$. Equality will hold for other values of λ, N and p only in exceptional cases. In Section 4, after the proof is completed, we present an example with noninteger λ, ($N = 3$ *and* $\lambda = 5/4$), for which $U_N(\lambda) < V_N(\lambda)$.

3. Proof of Theorem 1

3.1.(Derivative from [DS65], Chapter 5). First note that the left-hand side inequality in (1) is elementary, for if $k < \lambda \leq k+1$ then $U_N([\lambda]+1) = U_N(k+1) = g_N(k+1) \leq V_N(k+1) \leq V_N(\lambda)$. For the right-hand side we begin with two lemmas. Here, for convenience, we let $\lambda^* = \{1 \wedge \lambda\}$.

Lemma 2. *For each $N \geq 0$ the function $U_N(\lambda)$ is convex and decreasing in $\lambda \geq 0$, and satisfies*

$$U_{N+1}(\lambda) \geq pU_N(\lambda - s) + (1-p)U_N(\lambda + s), \text{ for all } s \text{ with } 0 \leq s \leq \lambda^*. \tag{2}$$

Lemma 3. *For each N and $\lambda \geq 0$,*

$$V_{N+1}(\lambda) = \sup_{0 \leq s \leq \lambda^*} \{pV_N(\lambda - s) + (1-p)V_N(\lambda + s)\}. \tag{3}$$

Theorem 1 follows directly by induction from Lemmas 2 and 3. Note that (1) holds for $N = 0$ and for $\lambda = 0$. Then using (3) and assuming (1) holds for N, and finally applying (2), we have

$$\begin{aligned} V_{N+1}(\lambda) &= \sup_{0 \leq s \leq \lambda^*} \{pV_N(\lambda - s) + (1-p)V_N(\lambda + s)\} \\ &\leq \sup_{0 \leq s \leq \lambda^*} \{pU_N(\lambda - s) + (1-p)U_N(\lambda + s)\} \\ &\leq U_{N+1}(\lambda). \end{aligned} \tag{4}$$

3.2. Proof of Lemma 2. To establish that $U_N(\lambda)$ is decreasing and convex first note that $g_N(n)$ is decreasing in n. Thus $U_N(\lambda)$ is decreasing in λ

and will be convex iff the sequence $g_N(n)$ is convex in $n \geq 0$. But this is equivalent to showing that for all $n \geq 0$, the second difference of $g_N(n)$ is nonnegative for all n. That is, we must show

$$\Delta_2\, g_N(n) = g_N(n+2) + g_N(n) - 2g_N(n+1) \geq 0. \tag{5}$$

We proceed by induction on N. Assume for a given N that (5) holds for all n. Then for all $n \geq 1$

$$g_{N+1}(n) = pg_N(n-1) + (1-p)g_N(n+1).$$

We consider the two cases in (5) when $n = 0$ and $n > 0$. When $n = 0$ we replace N by $N+1$ to obtain

$$\begin{aligned}\Delta_2 g_{N+1}(0) &= pg_N(1) + (1-p)g_N(3) + g_{N+1}(0) - 2pg_N(0) - 2(1-p)g_N(2)\\ &= (1-p)[g_N(3) + g_N(1) - 2g_N(2)] + (1-2p)[1 - g_N(1)]\\ &\geq 0,\end{aligned}$$

since by hypothesis $[g_N(3)+g_N(1)-2g_N(2)] \geq 0$ while $(1-2p)[1-g_N(2)] \geq 0$. In fact, $1-2p \geq 0$ holds by assumption and $1 - g_N(1) \geq 0$ holds because $g_N(n)$ is decreasing while $g_N(0) = 1$. For $n > 0$ we have

$$\Delta_2 g_N(n) = p\Delta_2 g_N(n-1) + (1-p)\Delta_2 g_N(n+1) \geq 0$$

by the induction hypothesis. This proves that $U_N(\lambda)$ is convex.

To establish (2), fix $N \geq 0$ and λ and consider the function of s,

$$W_N(\lambda, s) = pU_N(\lambda - s) + (1-p)U_N(\lambda + s).$$

Since $U_N(\lambda)$ is convex, $W_N(\lambda, s)$ is a convex function of s, $0 \leq s \leq \lambda^\star$.

$$pU_N(\lambda - s) + (1-p)U_N(\lambda + s) \leq pU_N(\lambda - \lambda^\star) + (1-p)U_N(\lambda + \lambda^\star),\ 0 \leq s \leq \lambda^\star.$$

But for $\lambda \geq 1$, $\lambda^\star = 1$ and

$$W_N(\lambda, 1) = U_{N+1}(\lambda), \tag{6}$$

as is easily checked. In fact, when $N = 0$ this is true and when λ is an integer this states that $pU_N(\lambda - 1) + (1-p)U_N(\lambda + 1) = U_{N+1}(\lambda)$ which is true by definition. Now for any $N \geq 1$ and any noninteger $\lambda > 1$, we set $\alpha = \lambda - [\lambda]$ and $\beta = 1 - \alpha$. Then

$$\begin{aligned}U_{N+1}(\lambda) &= \beta U_{N+1}([\lambda]) + \alpha U_{N+1}([\lambda]+1) \\ &= \beta[pU_N([\lambda]-1) + (1-p)U_N([\lambda]+1)] \\ &\qquad + \alpha[pU_N([\lambda]) + U_N([\lambda]+2)] \\ &= pU_N(\lambda-1) + (1-p)U_N(\lambda+1) \\ &= W_N(\lambda, 1).\end{aligned}$$

But $W_N(\lambda, s)$ is convex and its value at 0 is $W_N(\lambda, 0) = U_N(\lambda)$ which is less than its value $W_N(\lambda, 1) = U_{N+1}(\lambda)$ at 1. Thus, for $\lambda \geq 1$ the inequality (2) holds.

For $\lambda \leq 1$, $\lambda^* = \lambda$ and

$$pU_N(\lambda - \lambda^*) + (1-p)U_N(\lambda + \lambda^*) = (1-p)U_N(2\lambda) + p.$$

But the function $(1-p)U_N(2\lambda) + p$ is convex on $[0, 1]$ and agrees with the linear function $U_{N+1}(\lambda)$ when λ equals 0 or 1. Thus for λ in the interval $[0, 1]$, $(1-p)U_N(2\lambda) + p \leq U_{N+1}(\lambda)$. Thus for all λ we have that $W(\lambda, \lambda^*) \leq U_{N+1}(\lambda)$ and the proof of (2) is complete.

3.3. Proof of Lemma 3. The proof of (3) again proceeds inductively. It is clearly true when $N = 0$. Let $\{Z\}$ be a p-symmetric supermartingale in $\mathcal{M}_{p,1}$, and let $F(\lambda) = F(\lambda; \omega)$ denote the conditional probability $P\{max_{n \leq N} Z_n \geq \lambda | \mathcal{F}_1\}$. Then almost surely,

$$F(\lambda; \omega) \leq V_N(\lambda - Y_1),$$

with equality whenever $Y_1 \geq \lambda$. Hence, if $d\mu$ denotes the distribution of Y_1 and $d\nu$ denotes the distribution of $|Y_1|$ we have

$$\begin{aligned}P\{\max_{n \leq N+1} Z_n \geq \lambda\} &= E\{F(\lambda)\} \\ &\leq \int V_N(\lambda - y)\, d\mu(y) \\ &= \int [pV_N(\lambda - s) + (1-p)V_N(\lambda + s)]\, d\nu(s) \\ &\leq \sup_{0 \leq s \leq \lambda^*} \{pV_N(\lambda - s) + (1-p)V_N(\lambda + s)\}.\end{aligned}$$

Since this is true for all $\{Z\}$ in $\mathcal{M}_{p,1}$ the inequality,

$$\sup_{0 \leq s \leq \lambda^*} \{pV_N(\lambda - s) + (1-p)V_N(\lambda + s)\} \leq V_{N+1}(\lambda)$$

follows. The proof is complete.

4. Optimal strategies

When bet sizes are limited together with the games duration it is easy to construct examples showing that a bold strategy may not be optimal. Using the equation

$$V_{N+1}(\lambda) = \sup_{0\le s\le \lambda^*} \{pV_N(\lambda - s) + (1-p)V_N(\lambda + s)\}$$

from (4), an idea going back to Dvoretzky, one may calculate the function $V_N(\lambda)$ recursively. For example we have

$$V_0(\lambda) = I_{(-\infty,0]}(\lambda),$$

$$V_1(\lambda) = I_{(-\infty,0]}(\lambda) + pI_{(0,1]}(\lambda),$$

$$V_2(\lambda) = I_{(-\infty,0]}(\lambda) + (2p-p^2)I_{(0,1/2]}(\lambda) + pI_{(1/2,1]}(\lambda) + p^2 I_{(0,1]}(\lambda).$$

By the same process we find $V_3(5/4) = 3p^2 - 2p^3$ but the probability of reaching 5/4 in three steps with a bold strategy is seen to be $2p^2 - p < 3p^2 - 2p^3$. Here an optimal strategy is to stake \$0.75 on the first play and, if the first play was won, stake \$0.50 on the second play. Then either the goal has been reached or \$1.00 more is required, in which case you must stake \$1.00 on the third play. If, in the contrary case, the first play is lost you then need to win \$2.00 in two plays. Here the obvious strategy is to stake \$1.00 each on the second and third plays.

This model analysis will enable the computation of an optimal strategy for any fixed N and λ, but in general it does not appear to allow a simple summary description.

Acknowledgments. I thank the Department of Mathematics at the University of British Columbia for their support and hospitality during the academic year 1994–95 when the essential work in this paper was completed. I also thank the referee for drawing my attention to important references and for pointing out several embarrassing errors in our original manuscript.

References

[B84] D. L. Burkholder, Boundary value problems and sharp inequalities for martingale transforms, *Ann. Probab.* **12** 1984, 647–702.

[C08-9] J. L. Coolidge, The Gamblers ruin, *Annals of Math.* **10**, 1908-09, 181–192.

[DS65] L. E. Dubins and L. J. Savage, *How To Gamble If You Must: Inequalities for Stochastic Processes.* McGraw-Hill, New York, 1965.

[HK88] D. Heath and R. Kertz, Leaving an interval in limited playing time. *Adv. Appl. Probab.* **20** 1988, 635–645.

[K79] M. Klawe, Optimal strategies for a fair betting game. *Discrete Appl. Math.* **1** 1979, 105–115.

[MS96] A. Maitra and W. Suddreth, *Discrete Gambling and Stochastic Games.* Springer, New York, 1996.

[MW96] D. McBeth and A. Weerasinghe, Finite-time optimal control a process leaving an interval. *Jour. Appl. Probab.* **33** 1996, 714–728.

[Y96] Z. Yang, Optimal strategies for a betting game. *Discrete Appl. Math.* **28** 19790, 157–169.

Department of Mathematics
Kerchof Hall
University of Virginia
Charlottesville, VA 22903 (USA)
email: ldp@virginia.edu

Moment Comparison of Multilinear Forms in Stable and Semistable Random Variables with Application to Semistable Multiple Integrals

B. S. Rajput, K. Rama-Murthy, and X.R. Retnam

Abstract

Let $1 < \alpha < 2$. We provide a uniform comparison of the tail probabilities of (non-symmetric) strictly α-semistable random variables with the tail probabilities of their symmetrized counterparts as well as of their "associated" strictly & symmetric α-stable random variables. We use this to obtain a uniform comparison between the moments of the multilinear forms in (non-symmetric) strictly α-semistable random variables on the one hand and in their symmetrized counterparts as well as in their "associated" strictly, & symmetric α-stable random variables on the other. In turn, using this and following the approach of Krakowiak and Szulga in the stable case, we construct strictly and symmetric α-semistable multiple stochastic integrals of Banach space-valued integrands.

1991 Mathematics Subject Classification. Primary 30D55, 60G25, 62M20; Secondary 60E07, 60G10.

Keywords and phrases: Tail probabilities and moment comparisons, stable and semistable random measures, stable and semistable multiple integrals.

1. Introduction

We provide a uniform comparison between the tail probabilities of (non-symmetric) strictly α-semistable random variables with the tail probabilities of their symmetrized counterparts as well as of their "associated" strictly and symmetric α-stable random variables (Theorem 3.3) (in this Introduction, unless stated otherwise, α is taken to be in the interval $(1\ ,\ 2)$). The non-trivial part of this comparison depends on another result (Theorem 3.1) where we show that the value at zero of the distribution functions of all non-degenerate strictly α-semistable random variables lie between two universal constants C_0, C_1 satisfying $0 < C_0 < C_1 < 1$. Using the first result and a contraction principle due to Kwapień [7] and an inequality of Krakowiak and Szulga [5], we provide a uniform comparison between the moments of multilinear forms in (non-symmetric) strictly α-semistable random variables on the one hand and in their symmetrized counterparts and also in their "associated" strictly and symmetric α-stable random variables on the other (Theorem 3.4). This result and an estimate due to Krakowiak and Szulga [6] comparing the moment of a symmetric α-stable muiltiple integral of real valued simple functions with a certain L_q-norm/quasinorm of these functions allow us to obtain similar estimates for (non-symmetric) strictly and symmetric α-semistable multiple integrals of such functions (Remark 4.2). This and Theorem 3.4 in turn provide us with the necessary tools to obtain an extension result related to the product of the general (not necessarily product type) (non-symmetric) strictly and symmetric α-semistable random measures and to construct a multiple integral of Banach valued functions relative to these extended product random measures (Theorem 4.3, Definition 4.4).

This approach of multiple integration is akin to the Lebesgue-Dunford type integral relative to a L_p-valued vector measure and is similar to the one initiated and effectively exploited by Krakowiak and Szulga in a series of three interesting papers [3, 6] where they develop a symmetric α-stable, $0 < \alpha < 2$, as well as a strictly α-stable, $0 < \alpha < 2$, $\alpha \neq 1$, multiple stochastic integral. The α-semistable multiple integrals described above extend similar integrals constructed by Retnam [12] under the restrictive condition that the random measures be product type. Theorem 3.1 is important to us in that it is used, via Theorem 3.3, to obtain the above described uniform comparison of the moments of the multilinear forms crucial for developing the semistable multiple integrals. This theorem under the assumption that the family of random variables comes from a product-type α-semistable random measure was obtained by Retnam [12] using a cicuitous approach involving Fourier analytic tools; these do not seem to apply in the present general setting. Here we present a simple, transparent and probabilistic proof which, contrary to the proof in [12], additionally has the advantage of providing an analog of this result for the case, $0 < \alpha < 1$.

The organization of the rest of the paper is as follows: Section 2 contains the preliminaries. Section 3 contains the moment and tail probability comparison results; and Section 4 contains the results related to the development of semistable multiple integrals.

2. Preliminaries and notation

Throughout, r will denote a number in the interval (0,1) and α a number in the interval (0,2), $\alpha \neq 1$; the symbols $\mathbf{R}$ and $\mathbf{A}(\equiv \mathbf{A}(r, \alpha))$ will respectively denote the set of all real numbers and the annulus $\{x \in \mathbf{R} :$

$r^{\frac{1}{\alpha}} < |x| \leq 1\}$. For $x \in \mathbf{R}$, $x \neq 0$, define

$$g_\alpha(x) = \begin{cases} |x|^{-\alpha} \sum r^{-n}\{1 - \ cos(r^{\frac{n}{\alpha}} x) + i(r^{\frac{n}{\alpha}} x - \sin(r^{\frac{n}{\alpha}} x))\} & , \text{if } 1 < \alpha < 2, \\ |x|^{-\alpha} \sum r^{-n}\{1 - \ cos(r^{\frac{n}{\alpha}} x) - i \sin(r^{\frac{n}{\alpha}} x))\} & , \text{if } 0 < \alpha < 1, \end{cases}$$

where $\sum$ stands for $\sum_{n=-\infty}^{\infty}$; and set $\overline{g}_\alpha(x) = \Re(g_\alpha(x))$. When $x = 0$, we shall set $|x|^\alpha g_\alpha(x) = |x|^\alpha \overline{g}_\alpha(x) = 0$. Next, define

$$h_\alpha(x) = 1 + i \tan\left(\frac{\pi\alpha}{2}\right)(sgn(x)), \quad x \in \mathbf{R}.$$

Following the standard convention, we shall use the notation r-SS(α) and $S(\alpha)$ for " r *– semistable index* α" and "*stable index* α", respectively. Recall that a random variable ξ is a strictly r-SS(α) if the characteristic (ch.) function $\hat{L}_\xi$ of ξ is given by

$$\hat{L}_\xi(t) = \exp\{-\int_{\mathbf{A}} |tx|^\alpha g_\alpha(tx)\sigma_\xi(dx)\}, \tag{2.1}$$

where σ_ξ is a finite Borel measure on $\mathcal{B}(\mathbf{A})$, the Borel σ-algebra of $\mathbf{A}$; σ_ξ is referred to as the *spectral measure* of ξ. A random variable ξ is a strictly S(α) if $\hat{L}_\xi$ is given by

$$\hat{L}_\xi(t) = \exp\{-\int_{\mathbf{U}} |tx|^\alpha h_\alpha(tx)\sigma_\xi(dx)\}, \tag{2.2}$$

where σ_ξ is a finite measure on $\mathbf{U} \equiv \{-1, 1\}$, which is referred to as the *spectral measure* of ξ. In either of the two cases if ξ is symmetric, then we can and will assume that σ_ξ is symmetric. In this case g_α in (2.1) (respectively h_α in (2.2)) can be replaced by $\overline{g}_\alpha$ (respectively by 1).

Let $(\mathbf{S}, \sigma(\mathbf{S}))$ denote a measurable space; and let $\Lambda \equiv \Lambda(r, \alpha)$ be *an independently scattered strictly* r-SS(α) *random measure* ($\equiv$ *a strictly* r-SS(α) *random measure*) on $(\mathbf{S}, \sigma(\mathbf{S}))$ [10, 11]. We recall that it follows from [11] that there exists a finite measure Γ on $(\mathbf{S} \times \mathbf{A}, \sigma(\mathbf{S}) \times \mathcal{B}(\mathbf{A}))$ such that the characteristic function $\hat{L}_{\Lambda(A)}$ of $\Lambda(A)$, $A \in \sigma(\mathbf{S})$, is given by (2.1) with σ_ξ replaced by $\Gamma_A(\cdot) = \Gamma(A \times \cdot)$. The symmetrized version of Λ will be denoted by $\bar{\Lambda}$; it follows that the ch. function of $\bar{\Lambda}(A)$, $A \in \sigma(\mathbf{S})$, is given by (2.1) with σ_ξ replaced by $\bar{\Gamma}_A$ where $\bar{\Gamma}_A(\cdot) = \Gamma_A(\cdot) + \Gamma_A(- \ \cdot)$. We recall that the measure $\lambda(\cdot) \equiv \Gamma(\cdot \times \mathbf{A})$ (respectively $\bar{\lambda}(\cdot) \equiv \bar{\Gamma}(\cdot \times \mathbf{A}) = 2\lambda(\cdot)$ where $\bar{\Gamma}$ is the extension of $\bar{\Gamma}.(\cdot)$ to $\sigma(\mathbf{S}) \times \mathcal{B}(\mathbf{A})$) is referred to as the *control measure* of Λ (respectively of $\bar{\Lambda}$).

Given a random measure $\Lambda \equiv \Lambda(r, \alpha)$ as above. Let Γ_s be a finite measure on $\sigma(\mathbf{S}) \times \sigma(\mathbf{U})$ such that $\Gamma_s(A \times \mathbf{U}) = \lambda(A)$, for all $A \in \sigma(\mathbf{S})$, where $\sigma(\mathbf{U})$ is the power set of $\mathbf{U}$; then $\Lambda_s \equiv \Lambda_s(\alpha)$ will always denote that *independently scattered strictly* $\mathrm{S}(\alpha)$ *random measure* ($\equiv$ *strictly* $\mathrm{S}(\alpha)$ *random measure*) for which the characteristic function of $\Lambda_s(A)$, $A \in \sigma(\mathbf{S})$, is given by (2.2) with σ_ξ random replaced by $\Gamma_{s,A}(\cdot) \equiv \Gamma_s(A \times \cdot)$. (Both the measures Λ and Λ_s will be assumed to be defined on the same probability space). Similarly, for a given nondegenerate strictly r-SS(α) random variable ξ with spectral measure σ, we shall associate a $\mathrm{S}(\alpha)$ random variable θ; note that the spectral measure σ_s of θ (on $\sigma(\mathbf{U})$) satisfies the condition $\sigma(\mathbf{A}) = \sigma_s(\mathbf{U})$.

The symmetrized version of Λ_s will be denoted by $\bar{\Lambda}_s$. It follows that the characteristic function of $\bar{\Lambda}_s(A)$, $A \in \sigma(\mathbf{S})$, is given by $\hat{L}_{\bar{\Lambda}_s(A)}(t) = \exp\{-\bar{\lambda}(A)|t|^\alpha\}$. The measure λ (respectively $\bar{\lambda}$), as in the semistable case, is referred to as the *control measure* of Λ_s (respectively of $\bar{\Lambda}_s$). If the measure, Γ (respectively Γ_s) is a product measure, then Λ (respectively Λ_s) is said to be of *product type*. Similar conventions for the symmetrized version will be followed. (Note, however, that $\bar{\Gamma}_s$ is always a product measure even if Γ_s is not a product measure; thus $\bar{\Lambda}_s$ is always of product type).

We now recall a definition from [8, p. 66]: Let $\{\xi_i\}$ and $\{\eta_i\}$ be two families of real random variables defined on the same probability space. The family $\{\xi_i\}$ is said to be (κ, τ) – *strongly dominated* by $\{\eta_i\}$ and written $\{\eta_i\} \overset{(\kappa,\tau)}{\rightharpoonup} \{\xi_i\}$ (equivalently, $\{\xi_i\} \overset{(\kappa,\tau)}{\leftharpoondown} \{\eta_i\}$) (or just $\{\eta_i\} \rightharpoonup \{\xi_i\}$, (equivalently, $\{\xi_i\} \leftharpoondown \{\eta_i\}$), for short) if there exist constants $\kappa > 0$ and $\tau > 0$ such that

$$P(|\xi_i| > t) \le \kappa P(\tau|\eta_i| > t),$$

for all $t > 0$ and for all i. If $\{\eta_i\} \overset{(\kappa,\tau)}{\rightharpoonup} \{\xi_i\}$ and $\{\xi_i\} \overset{(\kappa',\tau')}{\rightharpoonup} \{\eta_i\}$, we say that the families $\{\xi_i\}$ and $\{\eta_i\}$ are *strongly equivalent* and write $\{\eta_i\} \underset{(\kappa',\tau')}{\overset{(\kappa,\tau)}{\rightleftharpoons}} \{\xi_i\}$.

Let M_1 and M_2 be two random measures on the same measurable space $(\mathbf{S}, \sigma(\mathbf{S}))$, defined on the same probability space; then we say M_1 and M_2 are *strongly equivalent* and write $M_1 \rightleftharpoons M_2$ if $\{M_1(A)\}_{A\in\sigma(\mathbf{S})} \rightleftharpoons \{M_2(A)\}_{A\in\sigma(\mathbf{S})}$.

3. The moment and tail probability comparison results

We now state and prove the first result of the paper.

Theorem 3.1. *Let $\mathcal{D} \equiv \mathcal{D}(r, \alpha)$ denote a family of non-degenerate strictly r-SS(α) distributions on* $\mathbf{R}$. *Then the following holds: If $1 < \alpha < 2$, then there exist two universal constants $C_1 \equiv C_1(r, \alpha), C_0 \equiv C_0(r, \alpha)$ such that*

$$0 < C_0 \le \inf\{G(0) : G \in \mathcal{D}\} \le \sup\{G(0) : G \in \mathcal{D}\} \le C_1 < 1. \quad (3.1)$$

If $0 < \alpha < 1$, *then*

$$0 < c_0 \equiv \inf\{G(0) : G \in \mathcal{D}\} \leq \sup\{G(0) : G \in \mathcal{D}\} \equiv c_1 < 1 \qquad (3.2)$$

holds iff the family $\mathcal{U} \equiv \{\nu\}$ *of the spectral measures of distributions in* $\mathcal{D}$ *satisfies the condition*

$$0 < \inf\left\{\frac{\nu(\mathbf{A}_-)}{\nu(\mathbf{A}_+)} : \ \nu \in \mathcal{U}\right\} \leq \sup\left\{\frac{\nu(\mathbf{A}_-)}{\nu(\mathbf{A}_+)} : \ \nu \in \mathcal{U}\right\} < \infty, \qquad (3.3)$$

where $\mathbf{A}_+ = \{x : r^{\frac{1}{\alpha}} < x \leq 1\}$ *and* $\mathbf{A}_- = \{x : -1 \leq x < -r^{\frac{1}{\alpha}}\}$.

Proof. If $1 < \alpha < 2$ then the support of an r-SS(α) probability measure γ on $\mathbf{R}$ is $\mathbf{R}$; if $0 < \alpha < 1$, the same result holds provided the support of the spectral measure of γ intersects both the intervals $\mathbf{A}_+$ and $\mathbf{A}_-$ [9]. Therefore, under the hypothesis that $1 < \alpha < 2$ or that $0 < \alpha < 1$ in conjunction with (3.3), we must have $0 < G(0) < 1$, for every $G \in \mathcal{D}$. Let $\{G_n\}$ be an arbitrary sequence from $\mathcal{D}$ such that $\{G_n(0)\}$ converges to c. Under either one of these two hypotheses we will show that $0 < c < 1$. This will prove (3.2) assuming either one of these two hypotheses.

First consider the case $1 < \alpha < 2$. Denote by γ_n the r-SS(α) probability measure with distribution function G_n and by ν_n the spectral measure of G_n. Choose a sequence $\{m_n\}$ of integers so that $r^{m_n}\nu_n(\mathbf{A}) \in [r, 1]$. Regarding the measures $r^{m_n}\nu_n$ on the compact set $\bar{\mathbf{A}}$, the closure of $\mathbf{A}$, and using the Prokhorov's Theorem, we can find a subsequence of $\{r^{m_n}\nu_n\}$, still denoted by $\{r^{m_n}\nu_n\}$, that converges weakly to a finite measure ν_0 on $\bar{\mathbf{A}}$. It follows that $\nu_0(\bar{\mathbf{A}}) \geq \limsup r^{m_n}\nu_n(\bar{\mathbf{A}}) \geq r$; therefore $\nu_0 \not\equiv 0$. From [11], we have that $\exp\{-\int_{\bar{\mathbf{A}}} |tx|^\alpha g_\alpha(tx)\nu_0(dx)\}$ is the characteristic function of a non-degenerate strictly r-SS(α) probability measure γ_0. Hence from the above paragraph, the support of γ_0 is $\mathbf{R}$; in particular, $0 < \gamma_0((-\infty, 0]) < 1$. Now, observe that $\hat{\gamma}_n^{r^{m_n}}(t) = \exp\{-\int_{\mathbf{A}} |tx|^\alpha g_\alpha(tx) r^{m_n}\nu_n(dx) = \exp\{-\int_{\bar{\mathbf{A}}} |tx|^\alpha g_\alpha(tx)\, r^{m_n}\nu_n(dx)$ (here γ^s denotes the sth root of γ). Therefore, since $|x|^\alpha g_\alpha(x)$ is a bounded continuous function on $\bar{\mathbf{A}}$ and $\{r^{m_n}\nu_n\}$ converges weakly to ν_0, we have that $\{\gamma_n^{r^{m_n}}\}$ converges weakly to γ_0. Therefore, using the property ${\gamma_n}^{r^{m_n}} = r^{\frac{m_n}{\alpha}}.\gamma_n$ (see, e.g., [11, p. 454]), we have that $\lim {\gamma_n}^{r^{m_n}}(-\infty, 0]) = \lim \gamma_n((-\infty, 0]) = c = \gamma_0((-\infty, 0])$, recall that $\gamma_0 \ll Leb$. This shows that (3.2) holds if $1 < \alpha < 2$.

Now suppose $0 < \alpha < 1$ and (3.2) hold. Choose a sequence $\{m_n\}$ of integers so that $r^{m_n}\nu_n(\mathbf{A}_-) \in [r, 1]$. Then, in view of (3.2), there exist constants $0 < d_0, d_1 < \infty$ such that $d_0 \leq r^{m_n}\nu_n(\mathbf{A}_+) \leq d_1$. As above, let ν_0 be the weak limit of a suitable subsequence of $\{r^{m_n}\nu_n\}$ (still denoted by $\{r^{m_n}\nu_n\}$). Now noting that the sets $\bar{\mathbf{A}}_-$ and $\bar{\mathbf{A}}_+$ are both open and closed in the metric space $\bar{\mathbf{A}}$, it follows that $\nu_0(\mathbf{A}_-)$ and $\nu_0(\mathbf{A}_+)$ are both positive. Let γ_0 be the non-degenerate r-SS(α) probability measure with spectral measure ν_0, as defined above. Then, using what we noted in the first paragraph above, the support of γ_0 is $\mathbf{R}$. By repeating the argument in the second paragraph above, one sees that $0 < c < 1$. To prove (3.1) let $\mathcal{D}_0$ be the family of all non-degenerate strictly r-SS(α) distributions and set $C_0 = C_0(r, \alpha) \equiv \inf\{G(0) : G \in \mathcal{D}_0\}$, and $C_1 = C_1(r, \alpha) \equiv \sup\{G(0) : G \in \mathcal{D}_0\}$. Then from what we proved above, we have $0 < C_0 \leq c_0 \leq c_1 \leq C_1 < 1$.

To complete the proof of the theorem we need to show that if $0 < \alpha < 1$ and (3.3) is satisfied, then (3.2) must hold. If (3.2) were not true, we can find a sequence $\{\nu_n\}$ from $\mathcal{U}$ such that $\{\frac{\nu_n(\mathbf{A}_-)}{\nu_n(\mathbf{A}_+)}\}$ converges to 0 or ∞. Suppose this sequence converges to 0. Choose a sequence $\{m_n\}$ of integers such that $r^{m_n}\nu_n(\mathbf{A}_+) \in [r, 1]$ which implies that $\{r^{m_n}\nu_n(\mathbf{A}_-)\}$ converges to 0. As before, $\{r^{m_n}\nu_n\}$ has a subsequence, still denoted by $\{r^{m_n}\nu_n\}$, that converges weakly on $\bar{\mathbf{A}}$ to a finite measure ν_0. It follows trivially that $\nu_0(\bar{\mathbf{A}}_+)$ is positive; and, since $\bar{\mathbf{A}}_-$ is open and closed in $\bar{\mathbf{A}}_-$, one sees that $\nu_0(\bar{\mathbf{A}}_-) = 0$. Hence γ_0, the r-SS(α) probability measure with spectral measure ν_0, has the support $[0, \infty)$ [9]. Since $\{\gamma_n\}$ converges to γ_0 weakly, we see that $\lim \gamma_n((-\infty, 0]) = \gamma_0((-\infty, 0]) = 0$, where the measures γ_n are defined as above. This contradicts (3.2). A similar argument shows that $\{\frac{\nu_n(\mathbf{A}_-)}{\nu_n(\mathbf{A}_+)}\}$ cannot converge to ∞. This completes the proof.

Since a strictly S(α) random variable is an r-SS(α) random variable for any $0 < r < 1$, it follows that (3.1) holds for any family of S(α) distributions. An analog of the above result for the stable case when $0 < \alpha < 1$ can also be proved making use of the methods of Theorem 3.1. However, for a strictly S(α) distribution function G, a precise formula for $G(0)$ is known; using this one can obtain more precise bounds in (3.1) and a "nicer" version of the last part of the theorem for the case $0 < \alpha < 1$: It is known [12, 14] that for a strictly S(α) random variable ξ with the parameters $b > 0$ and $|\beta| \leq 1$, (i.e., $\hat{L}_\xi(t) = \exp\{-b|t|^\alpha\{1 + i\beta \tan\frac{\pi\alpha}{2}\mathrm{sgn}(t)\}\}$),

$$G_\xi(0) \equiv G(0) = \frac{1}{2} - \frac{1}{\pi\alpha}\tan^{-1}(\beta \tan\frac{\pi\alpha}{2}). \tag{3.4}$$

It is easy to see that if $1 < \alpha < 2$, then $\frac{1}{\pi\alpha}\tan^{-1}(\beta\tan\frac{\pi\alpha}{2}) \leq \frac{1}{\pi\alpha}\tan^{-1}(-1.\tan\frac{\pi\alpha}{2}) = \frac{1}{\pi\alpha}(\pi - \frac{\pi\alpha}{2}) = \frac{1}{\alpha} - \frac{1}{2}$; hence by (3.4), $G(0) \geq 1 - \frac{1}{\alpha}$. Also $\frac{1}{\pi\alpha}\tan^{-1}(\beta\tan\frac{\pi\alpha}{2}) \geq \frac{1}{\pi\alpha}\tan^{-1}(1.\tan\frac{\pi\alpha}{2}) = \frac{1}{\pi\alpha}(\frac{\pi\alpha}{2} - \pi) = \frac{1}{2} - \frac{1}{\alpha}$; hence, again using (3.4), $G(0) \leq \frac{1}{\alpha}$. When $0 < \alpha < 1$, $\tan\frac{\pi\alpha}{2} > 0$ and $\tan^{-1}(\beta\tan\frac{\pi\alpha}{2})$ increases as β increases from -1 to 1. These observations and an argument similar to the last paragraph of Theorem 3.1 provide a proof of the following:

Proposition 3.2. *Let $\mathcal{D}$ be any family of $S(\alpha)$ distributions $G_{b,\beta}$ with parameters (b,β); then we have the following: If $1<\alpha<2$, then*

$$0<1-\frac{1}{\alpha}\leq \inf\{G_{b,\beta}(0):G_{b,\beta}\in\mathcal{D}\}\leq \sup\{G_{b,\beta}(0):G_{b,\beta}\in\mathcal{D}\}\leq\frac{1}{\alpha}<1. \tag{3.5}$$

If $0<\alpha<1$, then

$$0<c_0\equiv \inf\{G_{b,\beta}(0):G_{b,\beta}\in\mathcal{D}\}\leq \sup\{G_{b,\beta}(0):G_{b,\beta}\in\mathcal{D}\}\equiv c_1<1 \tag{3.6}$$

if and only if

$$-1<\beta_0\equiv\inf\beta\leq\sup\beta\equiv\beta_1<1;$$

and in this case $c_0=\frac{1}{2}-\frac{1}{\pi\alpha}\tan^{-1}(\beta_1\tan\frac{\pi\alpha}{2})$ and $c_1=\frac{1}{2}-\frac{1}{\pi\alpha}\tan^{-1}(\beta_0\tan\frac{\pi\alpha}{2})$.

Recall that for any two i.i.d. random variables ξ and ξ', we have

$$\min(P(\xi>0),P(\xi<0))P(|\xi|>t)\leq P(|\xi-\xi'|>t)\leq 2P(|\xi|>\frac{t}{2})$$

for all $t>0$. Using this fact, (3.1) and (3.5) we get the following corollary. In the rest of this section, we shall write $C\equiv C(r,\alpha)=(\min\{C_0,1-C_1\})^{-1}$ and $v\equiv v(\alpha)=\frac{\alpha}{\alpha-1}$.

Corollary 3.2. *Let $1<\alpha<2$ and let $\{\xi_i\}$ be a family of non-degenerate r-SS(α) (respectively, $S(\alpha)$) random variables and let $\bar{\xi}_i$ be the corresponding symmetrized family of random variables. Then we have*

$$\{\bar{\xi}_i\}\underset{(2,2)}{\overset{(C,1)}{\rightleftharpoons}}\{\xi_i\}\ (\textit{respectively},\ \{\bar{\xi}_i\}\underset{(2,2)}{\overset{(v,1)}{\rightleftharpoons}}\{\xi_i\}). \tag{3.7}$$

Similar arguments as above show that the counterpart of the first equivalence in (3.7) (with C_0 and C_1 replaced respectively by c_0 and c_1) holds for a family of r-SS(α) random variables, $0<\alpha<1$, provided (3.3) is satisfied. In view of (3.6) a similar remark applies to the $S(\alpha)$ case. Note, however, that in the case $0<\alpha<1$ the constants v and C appearing on the top of $\rightleftharpoons$ depend on the family of random variables contrary to the case $1<\alpha<2$.

If the random measure $\bar{\Lambda}$ is of product type (see Section 2) then it is proved by Rosiński [13] that $\bar{\Lambda}_s\rightleftharpoons\bar{\Lambda}$. Using his methods, it is easy to show that this equivalence in fact holds even when Λ is not of product type. In fact, it follows that if $\bar{\xi}$ is a non degenerate symmetric r-SS(α)

random variable and $\bar{\theta}$ is its associated symmetric S(α) random variable, then $\bar{\theta} \underset{(2b,b)}{\overset{(2a,a)}{\rightleftharpoons}} \bar{\xi}$, where $a \equiv a(\alpha, r) = [\alpha u_\alpha / 2(1-r)] + 1$, $b \equiv b(\alpha, r) = [2(1-r)/\alpha u_\alpha r^2] + 1$ (here, $[x]$ denotes the integral part of x) and $u_\alpha = \int_{-\infty}^{\infty} (1 - \cos x)/(1 + |x|^{\alpha+1}) dx$. Note that the constants a and b depend only r and α. Using this result and the above corollary, we immediately get the following equivalence comparing uniformly the tail probabilities of r-SS(α) random variables with the tail probabilities of their associated S(α) random variables. As noted in the Introduction, and this will be clear below, this result is used in comparing the moments of random multilinear forms in (non-symmetric) strictly and symmetric r-SS(α) random variables with those in the "associated" S(α) random variables.

Theorem 3.3. *Let $1 < \alpha < 2$. Let $\{\xi_i\}$ be a family of strictly r-SS(α) random variables and let $\{\theta_i\}$ be a family of their associated strictly S(α) random variables. Let $\{\bar{\xi}_i\}$ $\{\bar{\theta}_i\}$) be the families of their symmetrized versions. Then we have*

$$\{\theta_i\} \underset{(v,1)}{\overset{(2,2)}{\rightleftharpoons}} \{\bar{\theta}_i\} \underset{(2b,b)}{\overset{(2a,a)}{\rightleftharpoons}} \{\bar{\xi}_i\} \underset{(2,2)}{\overset{(C,1)}{\rightleftharpoons}} \{\xi_i\};$$

in particular,

$$\Lambda_s \underset{(v,1)}{\overset{(2,2)}{\rightleftharpoons}} \bar{\Lambda}_s \underset{(2b,b)}{\overset{(2a,a)}{\rightleftharpoons}} \bar{\Lambda} \underset{(2,2)}{\overset{(C,1)}{\rightleftharpoons}} \Lambda.$$

Let $k \geq 2$ be a fixed positive integer and let X be a real Banach space. Recall that a function $F : N^k \to X$ is called *tetrahedronal* if $F((i_1, \ldots, i_k)) = 0$ whenever $i_j \geq i_l$ for some j and l such that $1 \leq j < l \leq k$. Let $\mathcal{F}_k$ denote the set of all finitely supported tetrahedronal functions $F : N^k \to X$. For a sequence $\underline{\eta} = \{\eta_n\}$ of real random variables, and $F \in \mathcal{F}_k$, we write $\langle F; (\underline{\eta})^k \rangle$ to denote the *random* $k-$ *mutilinearform* $\sum_{\mathbf{i}_k} F(\mathbf{i}_k) \eta_{i_1} \eta_{i_2} \ldots \eta_{i_k}$. Let $(\Omega, \mathcal{F}, P)$ be a probability space and $0 < p < \infty$. Let $f : \Omega \to X$ be a Borel measurable function. We shall write $\|f\|_p$ for $(E\|f\|^p)^{\frac{1}{p}}$ and $L^p(X)$ for $\{f : \Omega \to X, \|f\|_p < \infty\}$. Let $\{f_i\}$ and $\{g_i\}$ be families of elements of $L^p(X)$. If there exists $\kappa > 0$ such that $\|g_i\|_p \leq \kappa \|f_i\|_p$ for all i, we shall write $\|f_i\|_p \overset{\kappa}{\rightharpoonup} \|g_i\|_p$ (equivalently, $\|f_i\|_p \overset{\kappa}{\rightharpoonup} \|g_i\|_p$) or just $\|f_i\|_p \rightharpoonup \|g_i\|_p$(equivalently, $\|g_i\|_p \leftharpoonup \|f_i\|_p$). If $\|f_i\|_p \overset{\kappa}{\rightharpoonup} \|g_i\|_p$ and $\|g_i\|_p \overset{\kappa'}{\rightharpoonup} \|f_i\|_p$, then we shall write

$$\|f_i\|_p \underset{\kappa'}{\overset{\kappa}{\rightleftharpoons}} \|g_i\|_p.$$

Let $1 < p < \infty$ and let the random variables $\eta_i's$ be independent satisfying $E|\eta_i|^p < \infty$ and $E\eta_i = 0$. Let $\underline{\varepsilon} = \{\epsilon_i\}$ be a sequence of independent symmetric Bernoulli random variables such that $\underline{\eta} = \{\eta_i\}$ and $\underline{\varepsilon} = \{\epsilon_i\}$ are independent. Then it is shown by Krakowiak and Szulga [5] that there exists a positive constant $c \equiv c(k,p)$ such that

$$c^{-1}\|\langle F;(\underline{\eta})^k\rangle\|_p \le \|\langle F;(\underline{\varepsilon\eta})^k\rangle\|_p \le c\|\langle F;(\underline{\eta})^k\rangle\|_p \tag{3.8}$$

holds for all $F \in \mathcal{F}_k$ where $\underline{\varepsilon\eta} = \{\epsilon_i\eta_i\}$. (Note that it is assumed in [5] that F be symmetric, but since $\langle F;(\underline{\eta})^k\rangle = \langle \tilde{F};(\underline{\eta})^k\rangle$ where $\tilde{F}$ is the symmetrized version of F [4], (3.8) holds for the function F considered here). This result and a contraction principle due to Kwapień [7] and Theorem 3.3 provide easily a proof of the following result which, as noted earlier, is crucial for developing semistable multiple stochastic integrals in the next section. The above noted constant c and those introduced in Theorem 3.3 are used in the following Theorem.

Theorem 3.4. *Let $1 < \alpha < 2$ and $1 \le p < \alpha$. Let $\underline{\xi} = \{\xi_n\}$, $\bar{\underline{\xi}} = \{\bar{\xi}_n\}$, $\underline{\theta} = \{\theta_n\}$, and let $\bar{\underline{\theta}} = \{\bar{\theta}_n\}$ be sequences of random variables as in Theorem 3.3 with the additional assumption that the random variables with in the sequences $\underline{\xi}$ and $\underline{\theta}$ (and hence also within $\bar{\underline{\xi}}$ and $\bar{\underline{\theta}}$) be independent. Then we have*

$$\|\langle F;(\underline{\theta})^k\rangle\|_p \underset{c\,v^k}{\overset{c\,4^k}{\rightleftharpoons}} \|\langle F;(\bar{\underline{\theta}})^k\rangle\|_p \underset{2^k b^{2k}}{\overset{2^k a^{2k}}{\rightleftharpoons}} \|\langle F;(\bar{\underline{\xi}})^k\rangle\|_p \underset{c\,4^k}{\overset{c\,C^k}{\rightleftharpoons}} \|\langle F;(\underline{\xi})^k\rangle\|_p. \tag{3.9}$$

Proof. In view of the above mentioned three facts, the proof of this result is straightforward. We indicate this only for the extreme right equivalence. Let $\underline{\varepsilon} = \{\epsilon_i\}$ be as above with $\underline{\varepsilon}$ independent of $\underline{\xi}$. Since by Theorem 3.3 $\{\bar{\xi}_i\} \overset{(C,1)}{\rightharpoonup} \{\xi_i\}$ and trivially $\{\xi_i\} \overset{(1,1)}{\rightharpoonup} \{\epsilon_i\xi_i\}$, we have by Kwapień's contraction principle [7] $C^k\|\langle F;(\bar{\underline{\xi}})^k\rangle\|_p \ge \|\langle F;(\underline{\varepsilon\xi})^k\rangle\|_p$. Therefore, since from (3.8) $\|\langle F;(\underline{\varepsilon\xi})^k\rangle\|_p \ge c^{-1}\|\langle F;(\underline{\xi})^k\rangle\|_p$, it follows that $c\,C^k\|\langle F;(\bar{\underline{\xi}})^k\rangle\|_p \ge \|\langle F;(\underline{\xi})^k\rangle\|_p$. On the other hand, again by Theorem 3.3, since $\{\bar{\xi}_i\} \overset{(2,2)}{\leftharpoonup} \{\epsilon_i\xi_i\}$, the contraction principle and (3.8) yield $\|\langle F;(\bar{\underline{\xi}})^k\rangle\|_p \le 4^k\|\langle F;(\underline{\varepsilon\xi})^k\rangle\|_p \le c\,4^k\|\langle F;(\underline{\xi})^k\rangle\|_p$. Thus we have proved $\|\langle F;(\bar{\underline{\xi}})^k\rangle\|_p \underset{c\,4^k}{\overset{c\,C^k}{\rightleftharpoons}} \|\langle F;(\underline{\xi})^k\rangle\|_p$.

Remark 3.5. Let $0 < q < \infty$. Recall that a family $\mathcal{D} \subset L_q(X)$ is said to satisfy the Marcinkiewicz-Paley-Zygmund condition with exponent q, written as, $\mathcal{D} \in MPZ(q)$, if there exists a $p \in (0,q)$ such that $\sup_{\eta\in\mathcal{D}} \frac{\|\eta\|_q}{\|\eta\|_p} < \infty$. It is known that $\mathcal{D} \in MPZ(q)$, if and only if for every

$p \in (0,q)$, $\sup_{\eta\in\mathcal{D}} \frac{\|\eta\|_q}{\|\eta\|_p} < \infty$ holds. It follows easily that if $\mathcal{D} \in MPZ(q)$ then $\bar{\mathcal{D}} \in MPZ(q)$, where throughout $\bar{\mathcal{D}}$ denotes the closure in probability ($\equiv L_0(X) \equiv \|.\|_0$) of $\mathcal{D}$; further, the convergence in all $L_p(X)$ ($\equiv \|.\|_p$) on $\mathcal{D}$ (and hence also on $\bar{\mathcal{D}}$) are equivalent for $p \in [0,q]$. We refer to [3, 4, 6] for these and related results.

Let $0 < \alpha < 2$ and $0 < q < \alpha$. Let $\dot{\underline{\theta}}$ be a sequence of iid nondegenerate standard symmetric S(α) random variables, then it is well known that the family $\{\langle F;\ \dot{\underline{\theta}}\rangle : F \in \mathcal{F}_k\} \in MPZ(q)$ [4]. Let $0 < p < q$ and set $\mathbf{c} \equiv \mathbf{c}(\alpha,p,q,k) = sup\{\|\langle F;\dot{\underline{\theta}}\rangle\|_q/\|\langle F;\dot{\underline{\theta}}\rangle\|_p : F \in \mathcal{F}_k\}$. Let now $1 < q < \alpha$; and let $\underline{\xi}$ be a sequences of independent strictly r-SS(α) random variables; and let $F \in \mathcal{F}_k$. Then using (3.9) we have

$$(c2^{3k}b^{2k})^{-1}\|\langle F;(\bar{\underline{\theta}})^k\rangle\|_1 \le \|\langle F;(\underline{\xi})^k\rangle\|_1$$
$$\le \|\langle F;(\underline{\xi})^k\rangle\|_q \le cC^k2^ka^{2k}\|\langle F;(\bar{\underline{\theta}})^k\rangle\|_q,$$

where $\bar{\underline{\theta}}$ is as in Theorem 3.4. Hence using the fact that, for any $0 < s < \alpha$, $\|\langle F;(\bar{\underline{\theta}})^k\rangle\|_s = \|\langle H;\dot{\underline{\theta}}\rangle\|_s$ for a suitable $H \in \mathcal{F}_k$, we have

$$\|\langle F;\underline{\xi}\rangle\|_q/\|\langle F;\underline{\xi}\rangle\|_1 \le c^2(C2^4a^2b^2)^k\mathbf{c} < \infty, \tag{3.10}$$

where recall $\mathbf{c}(\alpha,q,k) \equiv \mathbf{c} = \sup\{\|\langle H;\dot{\underline{\theta}}\rangle\|_q/\|\langle H;\dot{\underline{\theta}}\rangle\|_1 : H \in \mathcal{F}_k\}$. From the results noted above, it follows that the family $\{\langle F;\underline{\xi}\rangle : F \in \mathcal{F}_k\}$ belongs to $MPZ(q)$, for any $0 < q < \alpha$. In fact, since the constant appearing on the right side of (3.10) is independent of $\underline{\xi}$, we have that $\{\langle F;\underline{\eta}\rangle : F \in \mathcal{F}_k,\ \underline{\eta} \in \mathcal{G}\} \in MPZ(q)$, for any $0 < q < \alpha$, where $\mathcal{G}$ is any given family of sequences of independent strictly r-SS(α) random variables. Applying similar arguments, it follows that all the facts noted here hold when $\underline{\xi}$ is replaced by any of the three associated sequences of random variables appearing in Theorem 3.4.

Let now $0 < p < 1 < q \equiv (1+\alpha)/2 < \alpha$; and let $K \equiv K(r,\alpha,k)$ (note that $q = (\alpha+1)/2$) be the constant appearing on the right side of (3.10). Then using (3.10) it follows from Corollary 1.2 and Proposition 1.3 of [4] that $\|\langle F;\underline{\xi}\rangle\|_q/\|\langle F;\underline{\xi}\rangle\|_p \le (2K)^{(1+1/p)}$, where $\underline{\xi}$ is as above and $F \in \mathcal{F}_k$. Similarly, corresponding to the constant appearing at the right side of the last inequality, we get three constants relative to the three associated sequences of random variables appearing in Theorem 3.4. Let now $\mathbf{K} \equiv \mathbf{K}(r,\alpha,p,k)$ denote the maximum of these four constants. Then $\|\langle F;\underline{\eta}\rangle\|_q/\|\langle F;\underline{\eta}\rangle\|_p \le \mathbf{K}$, where η is any of the four sequences of random variables appearing in (3.9). Using this and (3.9) (with p replaced by q), we have

$$\|\langle F;(\underline{\theta})^k\rangle\|_p \underset{\mathrm{K}c\, v^k}{\overset{\mathrm{K}c\, 4^k}{\rightleftharpoons}} \|\langle F;(\bar{\underline{\theta}})^k\rangle\|_p \underset{\mathrm{K}2^k b^{2k}}{\overset{\mathrm{K}2^k a^{2k}}{\rightleftharpoons}} \|\langle F;(\bar{\underline{\xi}})^k\rangle\|_p \underset{\mathrm{K}c\, 4^k}{\overset{\mathrm{K}c\, C^k}{\rightleftharpoons}} \|\langle F;(\underline{\xi})^k\rangle\|_p. \tag{3.11}$$

4. Stable and semi-stable multiple integrals

Let $T \in [0,\infty]$ and let Δ_k denote the tetrahedron $\{(t_1,\ldots,t_k) \in [0,T)^k : 0 \le t_1 < t_2 < \ldots < t_k < T\}$. Let $\mathcal{A}_k = \{A_1 \times \ldots \times A_k \subset \Delta_k : A_1,\ldots,A_k \in \mathcal{I}\}$, where $\mathcal{I}$ is the class of all finite disjoint unions of all subintervals of $[0,T)$. Let $\mathcal{C}_k$ and $\bar{\mathcal{C}}_k$ be, respectively, the ring and algebra generated by $\mathcal{A}_k$. Let $\sigma(\bar{\mathcal{C}}_k)$ denote the $\sigma-$ algebra generated by $\bar{\mathcal{C}}_k$, and let $\mathcal{B}_k$ denote the Borel $\sigma-$ algebra of Δ_k. Let $A \in \bar{\mathcal{C}}_k$. Then it is easy to show that

$$(i)\ \ \sigma(\bar{\mathcal{C}}_k) = \mathcal{B}_k, \qquad (ii)\ \ A = \cup A_n,\ \ A_n \in \mathcal{C}_k,\ \ A_n \uparrow. \tag{4.0}$$

Further, it is easily seen that if $B_1,...,B_n$ are disjoint sets of $\mathcal{C}_k$, then one can choose finitely many Borel sets (in fact intervals) $A_1,\ldots,A_l$ of $[0,T)$ with $A_1 < \ldots < A_l$ and finite disjoint subsets W_j of N^k satisfying $(i_1,...,i_k) \in W_j \Rightarrow 1 \le i_1 < ... < i_k \le l, j = 1,...,n,$ such that

$$B_j = \bigcup\{A_{i_1} \times ... \times A_{i_k}, (i_1,...,i_k) \in W_j\}. \tag{4.1}$$

(The notation $A < B$ means that if $x \in A, y \in B$ then $x < y$). As above let X denote a real Banach space, and let $\mathcal{S}_X(\mathcal{C}_k)$ denote the set of all X-valued $\mathcal{C}_k$-measurable simple functions on Δ_k. Let $f \in \mathcal{S}_X(\mathcal{C}_k)$; say $f = \sum_{j=1}^n x_j\chi_{B_j}$ where $B_j's$ are as above. Defining $F(\mathbf{i}_k) = \sum_{j=1}^n x_j\chi_{W_j}(\mathbf{i}_k)$, trivially one has $F \in \mathcal{F}_k$. Then using (4.1) it follows that

$$f = \sum_{j=1}^n x_j\chi_{B_j} = \sum_{\mathbf{i}_k} F(\mathbf{i}_k)\chi_{A_{i_1}\times\ldots\times A_{i_k}}. \tag{4.2}$$

Details for this and the facts noted above can be found in [6, 12].

Let $\mathbf{S} = [0,T)$ and $\sigma(\mathbf{S})$ = the Borel $\sigma-$algebra of [0,T). Let Λ be the r-SS(α) random measures introduced in Section 2. Let $f \in \mathcal{S}_X(\mathcal{C}_k)$ be as in (4.2); and set

$$I(f) = \sum_{\mathbf{i}_k} F(\mathbf{i}_k)\Lambda(A_{i_1})\ldots\Lambda(A_{i_k}) = \langle F;(\Lambda(A_1),\ldots,\Lambda(A_l),0,0...)^k\rangle. \tag{4.3}$$

Now let $B \in \mathcal{C}_k$, then as above B admits a representation like (4.1). Set

$$\Lambda^k(B) = \sum_{\mathbf{i}_k} \Lambda(A_{i_1}) \dots \Lambda(A_{i_k}),$$

where the sum is over the indices $\mathbf{i}_k$ varying over the set W of N^k as indicated above. Noting that the random variables $\Lambda(A_j)'s$ are independent, standard arguments show that Λ^k is a (well defined) finitely additive vector measure on $\mathcal{C}_k$ with values in $L_p(\mathbf{R})$, for every $p \in [0, \alpha)$, and that I is a (well defined) linear operator from $\mathcal{S}_X(\mathcal{C}_k)$ in to $L_p(X)$ for every $p \in [0, \alpha)$. Denote by $\bar{I}, I_s$ and $\bar{I}_s$ respectively the counterpart of I, and by $\bar{\Lambda}^k$, Λ_s^k and $\bar{\Lambda}_s^k$ that of Λ^k relative to the random measures $\bar{\Lambda}, \Lambda_s$ and $\bar{\Lambda}_s$. Then (4.3), Theorem 3.4 and Remark 3.5 yield the following corollary.

Corollary 4.1. *Let* $1 < \alpha < 2$, $0 < p < \alpha$ *and let* $f \in \mathcal{S}_X(\mathcal{C}_k)$. *Then we have*

$$\|I_s(f)\|_p \rightleftharpoons \|\bar{I}_s(f)\|_p \rightleftharpoons \|\bar{I}(f)\|_p \rightleftharpoons \|I(f)\|_p;$$

and, in particular,

$$\|\Lambda_s^k(B)\|_p \rightleftharpoons \|\bar{\Lambda}_s^k(B)\|_p \rightleftharpoons \|\bar{\Lambda}^k(B)\|_p \rightleftharpoons \|\Lambda^k(B)\|_p,$$

for any set $B \in \mathcal{C}_k$. *The respective constants above and below the equivalences* $\rightleftharpoons$ *are as in (3.9) when* $1 \leq p < \alpha$, *and as in (3.11) when* $0 < p < 1$.

Remark 4.2. Recall that $\bar{\lambda} = 2\lambda$ denotes the control measure of the symmetric S(α) random measure $\bar{\Lambda}_s$. Denote by $\bar{\lambda}^k$ the restriction of the k-fold product of $\bar{\lambda}$ to Δ_k. Krakowiak and Szulga [6] proved the following Theorem: Let $1 < \alpha < 2$, $0 < p < \alpha < q$ and let $f \in \mathcal{S}_{\mathbf{R}}(\mathcal{C}_k)$. Then there exists a constant C (depending only on α, p, q and $\lambda([0,T])$) such that

$$C^{-1} \parallel f \parallel_{L_\alpha(\bar{\lambda}^k)} \leq \parallel I(f) \parallel_p \leq C \parallel f \parallel_{L_q(\bar{\lambda}^k)}, \tag{4.4}$$

where $\parallel f \parallel_{L_r(\bar{\lambda}^k)} \equiv (\int_{\Delta_k} |f|^r d\bar{\lambda}^k)^{1/r}$; in particular,

$$C^{-1}(\bar{\lambda}^k(B))^{1/\alpha} \leq \|\bar{\Lambda}_s^k(B)\|_p \leq C(\bar{\lambda}^k(B))^{1/q}, \ B \in \mathcal{C}_k. \tag{4.5}$$

In fact a more general result than the above theorem is proved in [6]; but (4.4) and (4.5) are enough to extend $\bar{\Lambda}_s^k$ to $\mathcal{B}_k$ and to define a stochastic integral of a suitable class of X−valued functions relative to this random measure. Clearly, the above theorem and Corollary 4.1 yield the analog of (4.4) and (4.5) for the random measures $\bar{\Lambda}^k$, Λ_s^k and Λ^k. This allows one to use the same methods as in the case $\bar{\Lambda}_s^k$ to extend these random measures to $\mathcal{B}_k$ and to define stochastic integrals of certain X-valued functions relative to these extended random measures. In the following we outline this process for one of these three measures; namely for Λ^k. This is done for completeness; and, except possibly for some comments noted in Remark 4.5 and Proposition 4.6, we do not assert any credit for these results and emphasize that the proofs of these are similar to those as pointed out by Krakowiak and Szulga [3, 6] in the case of $\bar{\Lambda}_s^k$. Before stating these results, we point out that the notation used for the constants in the above theorem and in the following do not represent the same values as in the previous section.

Let λ^k be the counterpart of $\bar{\lambda}^k$ for the measure λ; and let C_1 denote the constant that corresponds to C in the analogs of (4.4) and (4.5) for Λ^k. Note that C_1 depends only on α, p, q, r and $\lambda([0,T))$. Now we are ready to state the extension result:

Theorem 4.3. *Let $1 < \alpha < 2$ and $0 < p < \alpha < q$. Then Λ^k extends uniquely to a countably additive $L_p(\mathbf{R})$-valued λ^k-continuous vector measure $\tilde{\Lambda}^k$ on to $\mathcal{B}_k$. Further, the following holds:*

$$C_1{}^{-1}(\lambda^k(A))^{\frac{1}{\alpha}\cdot\min(1,p)} \leq \| \tilde{\Lambda}^k \| (A) \leq C_1(\lambda^k(A))^{\frac{1}{q}\cdot\min(1,p)}, \tag{4.6}$$

for all $A \in \mathcal{B}_k$, where $\| \tilde{\Lambda}^k \|$ is the semi-variation [2, p. 320] of $\tilde{\Lambda}^k$ on $\mathcal{B}_k$.

The proof of this proceeds as follows: Using the analog of (4.5) for Λ^k and property (ii) noted in (4.0), one shows that Λ^k extends to a finitely additive $L_p(\mathbf{R})$ -valued λ^k-continuous vector measure on to the algebra $\bar{\mathcal{C}}_k$. This and the Caratheodory-Hahn-Kluvanek Theorem [1, p. 27] gives the first part of the theorem. The second part is proved using the first part, the definition of the semi-variation, (4.4), (i) of (4.0) and the Caratheodory Approximation Theorem.

Now one defines the stochastic integral of X-valued functions relative to $\tilde{\Lambda}^k$ in the usual way as in [6]. Before stating this we introduce another notation: The space of all X-valued $\mathcal{B}_k$-measurable functions on Δ_k we shall denote by $\mathcal{M}(\mathcal{B}_k)$ (the σ-algebra on X here and in the following is taken to be the Borel σ-algebra; and α is assumed to belong to belong to the interval (1, 2)).

Definition 4.4. Let $f \in \mathcal{M}(\mathcal{B}_k)$ be a simple function: $f = \sum_j x_j I_{B_j}$, $B_j's$ disjoint; and let $B \in \mathcal{B}_k$. Then one sets $\int_B f d\tilde{\Lambda}^k \equiv \sum x_j \tilde{\Lambda}^k(B \cap B_j)$. A general function $f \in \mathcal{M}(\mathcal{B}_k)$ is said to be $\tilde{\Lambda}^k$-integrable if there exists a sequence of simple functions $\{f_n\} \subset \mathcal{M}(\mathcal{B}_k)$ such that $\{f_n\}$ converges to f in λ^k measure and, for every $B \in \mathcal{M}(\mathcal{B}_k)$, $\lim \int_B f_n d\tilde{\Lambda}^k$ exists in $L_0(X)$

(equivalently in $L_p(X)$ for each $p \in (0, \alpha)$, see below). In this case one defines $\int_B f d\tilde{\Lambda}^k \equiv \lim \int_B f_n d\tilde{\Lambda}^k$.

Let $\mathcal{D} \equiv \{\mathbf{I}(f) : f \in \mathcal{S}_X(\mathcal{C}_k)\}$. Using (4.0) and the monotone class theorem it follows easily that $\int_B f d\tilde{\Lambda}^k$ belongs to the $L_0(X)$-closure of $\mathcal{D}$ for every simple function $f \in \mathcal{M}(\mathcal{B}_k)$ and $B \in \mathcal{M}(\mathcal{B}_k)$. Hence, since from the second paragraph of Remark 3.5 we know that $\mathcal{D}$ belongs to $MPZ(p)$ for all $0 < p < \alpha$, it follows from Remark (3.5) that the convergence of the sequence $\{\int_B f_n d\tilde{\Lambda}^k\}$ in Definition 4.4 is equivalent in $L_p(X)$ for all $0 \le p < \alpha$. The proof of the fact that the integral in Definition 4.4 is well defined is standard: In fact when $X = \mathbf{R}$ it follows from [2, p. 324]; the proof is based on the Vitali-Hahn-Saks Theorem [1, p. 158] and a simple property of the semi-variation of $\tilde{\Lambda}^k$. Using this and the Hahn Banach Theorem, one proves this fact when X is a general Banach space.

Remark 4.5. In Definition 4.4 one can replace the sequence of $\mathcal{B}_k$-measurable simple functions by $\mathcal{C}_k$-measurable simple functions without changing the class of $\tilde{\Lambda}^k$-integrable functions. This is a useful fact and is needed, for example, in proving Proposition 4.6; we therefore outline a proof of this in the following: Let $f, \{f_n\}$ be as in Definition 4.4. Then $f_n = \sum_{i=1}^{m_n} x_{n,i}\chi_{B_{n,i}}$, $x_{n,i} \in X$, $B_{n,i} \in \mathcal{B}_k$, $B_{n,i}'s$ disjoint. Let $0 < p < \alpha < q$. By the Caratheodory Approximation Theorem and (4.0)(ii), choose $A_{n,i} \in \mathcal{C}_k$, $i = 1, ..., m_n$, so that

$$\lambda^k(B_{n,i} \triangle A_{n,i})^{\min(1,p)/q} < [2nm_n \max\{\|x_{n,i}\|^p, \|x_{n,i}\| : i = 1, ..., m_n\}]^{-1}, \tag{4.7}$$

where $\triangle$ denotes the symmetric difference. For every $n = 1, 2, ...$, set $g_n = \sum_{i=1}^{m_n} x_{n,i}\chi_{A_{n,i}}$. Then clearly $\{g_n\} \subset \mathcal{S}_k(X)$; and using (4.6), (4.7), the triangle inequality and a property of the semi-variation one gets easily that

$$(i)\ [\|\int_B f_n d\tilde{\Lambda}^k - \int_B g_n d\tilde{\Lambda}^k\|_p]^{\min(1,p)} \le C_1/n,$$
$$(ii)\ [\|f_n - g_n\|_{L_q(\lambda^k)}]^{\min(1,p)} \le 1/n, \tag{4.8}$$

for every $B \in \mathcal{B}_k$, where the constant C_1 is as in (4.6) and $\|.\|_{L_r(\lambda^k)}$ is defined as in Remark 4.2 with an obvious modification. Now let $\delta > 0$; then observing that $\lambda^k\{\|g_n - f\| > \delta\} \le \lambda^k\{\|g_n - f_n\| > \delta/2\} + \lambda^k\{\|f_n - f\| > \delta/2\}$ it follows by Chebyshev's Inequality and (ii) of (4.8) that $\lambda^k\{\|g_n - f\| > \delta\} \le n^{-\{q/\min(1,p)\}}(2/\delta)^q + \lambda^k\{\|f_n - f\| > \delta/2\}$. Hence, since $\{f_n\}$ converges to f in λ^k measure, so does $\{g_n\}$. Now let $\epsilon > 0$ and $B \in \mathcal{B}_k$. Now observing that $[\|\int_B g_n d\tilde{\Lambda}^k - \int_B g_m d\tilde{\Lambda}^k\|_p]^{\min(1,p)} \le [\|\int_B g_n d\tilde{\Lambda}^k - \int_B f_n d\tilde{\Lambda}^k\|_p]^{\min(1,p)} +$

$[\|\int_B f_n d\tilde{\Lambda}^k - \int_B f_m d\tilde{\Lambda}^k\|_p]^{\min(1,p)} + [\|\int_B f_m d\tilde{\Lambda}^k - \int_B g_m d\tilde{\Lambda}^k\|_p]^{\min(1,p)}$, using (i) of (4.8), one has that $[\|\int_B g_n d\tilde{\Lambda}^k - \int_B g_m d\tilde{\Lambda}^k\|_p]^{\min(1,p)} \leq n^{-1}C_1 + [\|\int_B f_n d\tilde{\Lambda}^k - \int_B f_m d\tilde{\Lambda}^k\|_p]^{\min(1,p)} + m^{-1}C_1$. Thus, since $\{\int_B f_n d\tilde{\Lambda}^k\}$ converges in $\|.\|_p$ (i.e., in $L_p(X)$), $\{\int_B f_n d\tilde{\Lambda}^k\}$ also converges in $\|.\|_p$ (as well as in $L_0(X)$, as noted above). This completes the proof of the assertion.

Denote by $\tilde{\bar{\Lambda}}^k$, $\tilde{\Lambda}^k_s$, and $\tilde{\bar{\Lambda}}^k_s$ respectively the extensions of $\bar{\Lambda}^k$, Λ^k_s and $\bar{\Lambda}^k_s$ to $\mathcal{B}_k$; and by $\mathcal{L}(\tilde{\Lambda}^k), \mathcal{L}(\tilde{\bar{\Lambda}}^k), \mathcal{L}(\tilde{\Lambda}^k_s)$ and $\mathcal{L}(\tilde{\bar{\Lambda}}^k_s)$ the X-valued integrable functions relative to the random measures $\tilde{\Lambda}^k, \tilde{\bar{\Lambda}}^k, \tilde{\Lambda}^k_s$ and $\bar{\Lambda}^k_s$ respectively. With these notations we have:

Proposition 4.6. *The following equalities hold:*

$$\mathcal{L}(\tilde{\Lambda}^k) = \mathcal{L}(\tilde{\bar{\Lambda}}^k) = \mathcal{L}(\tilde{\Lambda}^k_s) = \mathcal{L}(\tilde{\bar{\Lambda}}^k_s).$$

Further for, $0 < p < \alpha$, *and for any* $B \in \mathcal{B}_k$ *and* $f \in \mathcal{L}(\tilde{\Lambda}^k)$, *one has*

$$\|\int_B f d\tilde{\Lambda}^k_s\|_p \rightleftharpoons \|\int_B f d\tilde{\bar{\Lambda}}^k_s\|_p \rightleftharpoons \|\int_B f d\tilde{\bar{\Lambda}}^k\|_p \rightleftharpoons \|\int_B f d\tilde{\Lambda}^k\|_p,$$

where the respective constants above and below the equivalences $\rightleftharpoons$ *are as described in Corollary 4.1 (of which this is an extension).*

Proof. We will show that $\mathcal{L}(\tilde{\Lambda}^k) = \mathcal{L}(\tilde{\Lambda}^k_s)$, the proof of other equalities are similar. This will complete the proof of the first part. Let $f \in \mathcal{L}(\tilde{\Lambda}^k)$; then by Remark 4.5 we have a sequence $\{f_n\} \subset \mathcal{S}_k(X)$ such that it converges in λ^k-measure and that $\{\int_B f_n d\tilde{\Lambda}^k\}$ converges in $L_p(X)$ for every $0 \leq p < \alpha$ and $B \in \mathcal{B}_k$. Fix $0 < p < \alpha$ and $B \in \mathcal{B}_k$. Using (4.0) and the Caratheodory Approximation Theorem, choose a sequence of sets $\{A_j\} \subset \mathcal{C}_k$ such that $\lim_j \lambda^k(A_j \triangle B) = 0$. Then it follows from (4.6) and its analog for $\tilde{\Lambda}^k_s$ that

$$\|\int_B (f_n - f_m) d\tilde{\Lambda}^k\|_p = \lim_j \|\int_{A_j} (f_n - f_m) d\tilde{\Lambda}^k\|_p = \lim_j \|\mathbf{I}(\chi_{A_j}(f_n - f_m))\|_p, \tag{4.9}$$

$$\|\int_B (f_n - f_m) d\tilde{\Lambda}^k_s\|_p = \lim_j \|\int_{A_j} (f_n - f_m) d\tilde{\Lambda}^k_s\|_p = \lim_j \|\mathbf{I}_s(\chi_{A_j}(f_n - f_m))\|_p. \tag{4.10}$$

Hence, since by the hypothesis and (4.9) $\lim_{m,n} \|\int_B (f_n - f_m) d\tilde{\Lambda}^k\|_p = \lim_{m,n} \lim_j \|\mathbf{I}(\chi_{A_j}(f_n - f_m))\|_p = 0$, it follows from (4.10) and Corollary 4.1 that $\lim_{m,n} \|\int_B (f_n - f_m) d\tilde{\Lambda}^k_s\|_p = \lim_{m,n} \lim_j \|\mathbf{I}_s(\chi_{A_j}(f_n - f_m))\|_p = 0$.

Thus $\{\int_B f_n d\tilde{\Lambda}_s^k\}$ converges in $L_p(X)$; i.e., $f \in L(\tilde{\Lambda}_s^k)$. We have thus shown that $\mathcal{L}(\tilde{\Lambda}^k) \subset \mathcal{L}(\tilde{\Lambda}_s^k)$; the proof of the reverse inclusion is identical.

Let $0 < p < \alpha$, $f \in L(\tilde{\Lambda}^k)$ and $B \in \mathcal{B}_K$. Using the notation introduced above, the same arguments as above show that $\lim_j \|\mathbf{I}(\chi_{A_j} f_n)\|_p = \| \int_B f_n d\tilde{\Lambda}^k \|_p\|$ and $\lim_n \| \int_B f_n d\tilde{\Lambda}^k\|_p = \| \int_B f d\tilde{\Lambda}^k\|_p$, and also that $\lim_j \|\mathbf{I}_s(\chi_{A_j} f_n)\|_p = \| \int_B f_n d\tilde{\Lambda}_s^k\|_p$ and $\lim_n \| \int_B f_n d\tilde{\Lambda}_s^k\|_p = \lim_n \| \int_B f d\tilde{\Lambda}_s^k\|_p$. These and analogous equalities relative to the other two random measures and Corollary 4.1 complete the proof of the second part.

Remark 4.7. We recall from [6] the fact that $\mathcal{L}(\tilde{\tilde{\Lambda}}_s^k) \subseteq L_\alpha(\lambda^k)$ and that $\bigcup_{p>\alpha} L_p(\lambda^k) \subseteq \mathcal{L}(\tilde{\tilde{\Lambda}}_s^k)$. In view of Proposition 4.6 the same inclusions hold when $\mathcal{L}(\tilde{\tilde{\Lambda}}_s^k)$ is replaced by any one of the three spaces of integrable functions noted in the proposition. The inclusions in the symmetric stable case are consequences of appropriate inequalities similar to those appearing in (4.4) in the Banach space setting.

The analogs of (3.8) and the contraction principle used in Theorem 3.4 are not available in the case $0 < p < 1$. This is one reason why we are not able to obtain a result analogous to Corollary 4.1 for the case $0 < \alpha < 1$. Because of this we are unable to construct α-semistable stochastic integrals when $0 < \alpha < 1$. For the stable case, Krakowiak and Szulga [6] overcame this difficulty by adopting an alternative approach based on decoupling principle; which we are also unable to obtain for the α-semistable case when $0 < \alpha < 1$.

References

[1] Diestel, J., Uhl, J.J. Jr., *Vector Measures*, Amer. Math. Soc., Providence, RI, 1977.

[2] Dunford, J., Schwartz, J.T., *Linear Operators, Part I*, Wiley, New York, 1988.

[3] Krakowiak, W., Szulga, J., *On p-stable multiple integral I, II*, Preprint, Wroclaw University, Poland (1985).

[4] Krakowiak, W., Szulga, J., *Random multilinear forms*, Preprint no 22, Wroclaw University, Poland (1985).

[5] Krakowiak, W., Szulga, J., *Summability and contractivity*, CWRU Preprint (1986).

[6] Krakowiak, W., Szulga, J., *A multiple stochastic integral with respect to a strictly p-stable random measure*, Ann. Probab. **16** (1988), 764-777.

[7] Kwapień, S., *Decoupling inequality for polynomial chaos*, Ann. Probab. **15** (1987), 1062-1071.

[8] Kwapień, S., Woyczyński, W. A., *Random series and stochastic integrals*, Birkhauser, Boston, MA., 1992, pp. 10–20.
[9] Rajput, B. S., Rama-Murthy, K. and Zak T., *Supports of semistable probability measures on locally convex spaces spaces*, J. Theor. Prob. **7** (1994), 931-942.
[10] Rajput, B. S., Rama-Murthy, K., *Spectral representations of semistable processes, and semistable laws on Banach spaces*, J. Multivariate Anal. **21** (1987), 139-157.
[11] Rajput, B. S., Rosiński, J., *Spectral representations of infinitely divisible processes*, Probab. Th. Rel. Fields **82** (1989), 451-487.
[12] Retnam, Xavier R., *On a multiple stochastic integral with respect to strictly semistable random measure*, Dissertation, University of Tennessee, Knoxville (1988).
[13] Rosiński, J., *Bilinear random integrals*, Dissertationes Mathematicae CCLIX (1987).
[14] Zolotarev, V.M., *One dimensional stable distributions*, Amer. Math. Soc., Translations of Math Monographs, Providence RI, 1986.

Balram S. Rajput
Department of Mathematics
University of Tennessee
Knoxville,TN 37996

Kavi Rama-Murthy
Indian Statistical Institute,
Bangalore, India

Xavier R. Retnam
Department of Mathematics
Blue Mountain College,
Blue Mountain, MS 38610

Global Dependency Measure for Sets of Random Elements: "The Italian Problem" and Some Consequences*

*Jordan Stoyanov***

Dedicated to the Memory of Stamatis Cambanis!

Abstract

We suggest a detailed analysis of the classical independence/dependence properties for finite sets of random events or variables. All possible combinations of random elements are considered as a configuration obeying a hierarchical property. We define a function called a *Dependency Measure* taking values in the interval [0, 1] (0 corresponds to mutually independent sets; 1 corresponds to totally dependent sets) and serving as a global measure of the amount of dependency which is contained in the whole set of random elements. This leads to "The Italian Problem" about the existence of a probability model and a set of random elements with any prescribed independence/dependence structure. Some consequences and nonstandard illustrative examples are given. Related properties such as exchangeability and association are also discussed.

1. Introduction

Our main goal is to introduce a function as a global measure of the amount of dependency which is contained in a collection of random events or random variables. The idea is to consider the configuration of all possible combinations of random elements as a hierarchical system. Then we define an easy function to serve as a *Dependency Measure* and discuss its properties and possible applications. In a natural way we arrive at a problem which got the name "The Italian Problem." Its positive solution is of an independent interest but also it justifies our definition of a dependency measure. Some interesting consequences can be derived. We can establish an existence or a lack of a relationship between properties like independence, exchangeability and association for sets of random elements.

*This work was partially supported by Grant # MM-432.

** Author is currently on leave from the Bulgarian Academy of Sciences.

Now let $(\Omega, \mathcal{F}, P)$ be a probability space and let $\mathcal{A}_n = \{A_1, \ldots, A_n\}$ be a collection of $n \geq 2$ random events. We assume here and in what follows that all the events are nontrivial, i.e. $0 < P(A_j) < 1$. The independence property is defined by the following relation called a *product rule*:

$$P(A_{j_1} \cdots A_{j_k}) = P(A_{j_1}) \cdots P(A_{j_k}). \tag{1}$$

Here $k \geq 2$, $(j_1, \ldots, j_k)$ is an arbitrary k-subset of the set $\{1, \ldots, n\}$ with $1 \leq j_1 < j_2 < \ldots < j_k \leq n$.

The relations (1), their number is $2^n - n - 1$, can be naturally incorporated into $n-1$ groups called *levels*, from level 2 to level n. Clearly, level k contains exactly $\binom{n}{k}$ relations. Equivalently, this is the number of all k-subsets of the set $\mathcal{A}_n$ for $k = 2, \ldots, n$.

When studying the dependence property of $\mathcal{A}_n$ we have to count exactly those cases when the product rule (1) is not satisfied. Let us denote the number of the failing relations (1) for level k by d_k. Thus we obtain the $(n-1)$-tuple $(d_2,\ d_3, \ldots, d_n)$ which is called an *independence/dependence structure* of $\mathcal{A}_n$ (i/d-*structure*). Clearly the following restrictions hold:

$$0 \leq d_k \leq \binom{n}{k},\ k = 2,\ 3, \ldots, n. \tag{2}$$

The set $\mathcal{A}_n$ is *mutually independent* if all $2^n - n - 1$ relations (1) are satisfied and obviously in this case $d_k = 0$ for all $k = 2, 3, \ldots, n$. The set $\mathcal{A}_n$ is *totally dependent* if the product rule (1) does not hold for all possible combinations of events and in this case $d_k = \binom{n}{k}, k = 2, 3, \ldots, n$. Finally, $\mathcal{A}_n$ is *partially independent/partially dependent* if (1) holds for a part of the combinations of events and does not hold for the remaining so that $0 < d_k < \binom{n}{k}$, $k = 2, 3, \ldots, n$.

2. Some nonstandard examples

Almost any textbook in probability theory contains illustrative examples showing, for example, that the pairwise independence does not in general imply the mutual independence. The first examples of this kind were proposed by S. Bernstein (1916) and G. Bohlmann (1908). (Thanks to Prof. U. Krengel (Göttingen) who not only mentioned to me the name of G. Bohlmann but also kindly provided me with a copy of his paper.) Their description as originally proposed can be seen, for example, in Stoyanov (1987, 1997). The meaning of these examples is that the independence at a given level does not in general imply the independence at a higher level. Examples of this kind are well-known and frequently used. An interesting phenomenon is observed when going in the opposite direction, from a higher level to a lower level. It is still surprising for some people that we can have $P(ABC) = P(A)P(B)P(C)$ but, for example, $P(AB) \neq P(A)P(B), P(AC) \neq P(A)PC)$, etc. Such an example is given in Section 3 (see Example 6). It seems the first example of this kind was proposed by Crow (1967); see also Stoyanov (1987 and 1997).

In this section we describe a few nonstandard examples with unusual i/d-structure. It turns out examples of this kind are not so well-known. These examples and many others motivate the necessity to look more carefully at the structure of sets of random elements and try to answer important questions such as: How much is the dependency in this or that set? Which one of two sets of random elements is more dependent?

Example 1. Suppose $\Omega = \{1,2,3,4,5,6,7,8\}$ is the space of all possible outcomes whose probabilities are

$$p_1 = \alpha - \frac{1}{8},\ p_2 = p_3 = p_4 = \frac{(7-16\alpha)}{24},\ p_5 = p_6 = p_7 = (1+8\alpha)/24,\ p_8 = \frac{1}{8}.$$

Here $\alpha \in (\frac{1}{8}, \frac{7}{16})$ is arbitrary. We are interested in the events:

$$A_1 = \{2,5,6,8\},\ A_2 = \{3,5,7,8\},\ A_3 = \{4,6,7,8\}.$$

We easily find that for all possible choices of i and j we have:

$$P(A_i) = \frac{1}{2};\ P(A_iA_j) = \frac{1+2\alpha}{6},\ i<j;\ P(A_1A_2A_3) = \frac{1}{8}.$$

It is easy to see that the i/d-structure of $\mathcal{A}_3 = \{A_1, A_2, A_3\}$ is $(d_2, d_3) = (0,0)$ if $\alpha = \frac{1}{4}$, while $(d_2, d_3) = (3,0)$ if $\alpha \neq \frac{1}{4}$. Hence the set $\mathcal{A}_3$ is independent at level 3 for any $\alpha \in (\frac{1}{8}, \frac{7}{16})$. It is independent at level 2 if $\alpha = \frac{1}{4}$ and only in this case these three events are mutually independent.

Example 2. Consider the following probability model:

$$\Omega = \{\omega_1, \omega_2, \omega_3, \omega_4, \omega_5, \omega_6\},\ p_1 = \frac{1}{16},\ p_2 = p_3 = p_4 = p_5 = \frac{7}{48},\ p_6 = \frac{17}{48}.$$

We are interested in the events:

$$A_1 = \{\omega_1, \omega_2, \omega_3, \omega_4\},\quad A_2 = \{\omega_1, \omega_2, \omega_3, \omega_5\},$$

$$A_3 = \{\omega_1, \omega_2, \omega_4, \omega_5\},\quad A_4 = \{\omega_1, \omega_3, \omega_4, \omega_5\}.$$

We find that $P(A_i) = \frac{1}{2}, i = 1,2,3,4$ and for all possible indexes i, j, l :

$$P(A_iA_j) = \frac{17}{48},\ i<j;\ P(A_iA_jA_l) = \frac{5}{24},\ i<j<l;\ P(A_1A_2A_3A_4) = \frac{1}{16}.$$

The i/d-structure of the set $\mathcal{A}_4 = \{A_1, A_2, A_3, A_4\}$ is $(d_2, d_3, d_4) = (6, 4, 0)$. Hence these four events are: (a) independent at level 4; (b) dependent at level 3; (c) dependent at level 2.

Example 3. The sample space Ω contains $|\Omega| = 12$ outcomes denoted by 1, 2, ..., 12. Suppose that "1" has probability $\frac{3}{48}$, each of "2","3","4", "5" has probability $\frac{1}{24}$, each of "6","7", "8", "9", "10", "11" has probability $\frac{5}{48}$, and finally, "12" has probability $\frac{7}{48}$. Consider the following events:

$$A_1 = \{1, 2, 3, 4, 6, 7, 8\}, \ A_2 = \{1, 2, 3, 5, 6, 9, 10\},$$

$$A_3 = \{1, 2, 4, 5, 7, 9, 11\}, \ A_4 = \{1, 3, 4, 5, 8, 10, 11\}.$$

Simple calculations show that each of A_i has probability $\frac{1}{2}$, each pair A_iA_j, $i < j$, has probability $\frac{1}{4}$, each triplet $A_iA_jA_l$, $i < j < l$, has probability $\frac{5}{48}$ and, finally, $A_1A_2A_3A_4$ has probability $\frac{1}{16}$.

Hence $\mathcal{A}_4 = \{A_1, A_2, A_3, A_4\}$ is a set of four random events with i/d-structure $(d_2, d_3, d_4) = (0, 4, 0)$. Thus $\mathcal{A}_4$ is: (a) independent at level 2; (b) dependent at level 3; (c) independent at level 4.

Example 4. Suppose we have at our disposal 192 cards with numbers written on some of them as follows: 110 cards are marked by one "triplet" chosen from $\{123, 124, 125, 134, 135, 145, 234, 235, 245, 345\}$ with each of these "triplets" occurring on 11 cards; 30 cards are marked by a "quartet" from $\{1234, 1245, 1245, 1345, 2345\}$, each "quartet" appearing on six cards; six cards are marked by the "quintet" 12345; the remaining 46 cards are left blank. All 192 cards are put into a box and well mixed. We are interested in the following five events:

$$A_j=\{\text{randomly chosen card contains the number } j\}, \quad j = 1, 2, 3, 4, 5.$$

It is easy to check that for all possible indexes i, j, l, s we have:

$$P(A_i) = \frac{1}{2}; \ P(A_iA_j) = \frac{17}{64}, \ i < j; \ P(A_iA_jA_l) = \frac{23}{192}, \ i < j < l;$$

$$P(A_iA_jA_lA_s) = \frac{1}{16}, \ i < j < l < s; \ P(A_1A_2A_3A_4A_5) = \frac{1}{32}.$$

The i/d-structure of $\mathcal{A}_5 = \{A_1, A_2, A_3, A_4, A_5\}$ is $(d_2, d_3, d_4, d_5) = (10, 10, 0, 0)$. Thus we conclude that $\mathcal{A}_5$ is a set of five random events which are: (a) dependent at level 2; (b) dependent at level 3; (c) independent at level 4; (d) independent at level 5.

3. The combinations of events as a hierarchical system

In mathematical literature the term *hierarchical system* is used on several different occasions. Let us specify the situation we are going to analyze and try to explain in which sense we talk about such a property.

We consider a set $\mathcal{A}_n$ of n random events (or of random variables). We are interested in developing a simple number to measure the amount of independency or dependency of the whole collection $\mathcal{A}_n$. Of course, if the product rule (1) fails to hold, this should make a contribution to the amount of dependency of $\mathcal{A}_n$, so the (future) dependency measure will be a function of $d_2, \ldots, d_n$, and moreover, it will be an *increasing function* of $d_2, \ldots, d_n$. One of the main points of our approach is that we want to make a quantitative distinction between the levels, in the sense that the contribution of d_k to the total amount of dependency of $\mathcal{A}_n$ be a function also of the level k. We assume (some arguments can be seen in Stoyanov (1995)) that if the product rule (1) does not hold at a lower level, this will contribute "more" to the total dependency of $\mathcal{A}_n$ and hence this contribution will be "less" if (1) fails to hold at a higher level. The *hierarchical property* introduced and discussed below is an attempt to formalize the properties of functions from R^{n-1} to R^1 that might be suitable for our purpose.

The arguments given above by words about the desired properties of our dependency measure motivated us to introduce a general class of functions called *hierarchical functions.*

Suppose $H(x_1, \ldots, x_n)$ is a function of n arguments each being a non-negative real number and such that $x_k \leq m_k$, $k = 1, \ldots, n$ for some given $m_1, \ldots, m_n$. Let us choose two levels, say i and j, $1 \leq i < j \leq n$ and let $x_i = a$, $x_j = b$, where $a \leq m_i$, $b \leq m_j$ and $a < b$. We say that H is a *level decreasing function* if the following inequality is satisfied:

$$\begin{aligned} &H(x_1, \ldots, x_{i-1}, a, x_{i+1}, \ldots, x_{j-1}, b, x_{j+1}, \ldots, x_n) \\ &\qquad > \ H(x_1, \ldots, x_{i-1}, \tilde{b}, x_{i+1}, \ldots, x_{j-1}, \tilde{a}, x_{j+1}, \ldots, x_n) \end{aligned}$$

where $\tilde{b} = \min[b, m_i]$, $\tilde{a} = \min[a, m_j]$ and all other arguments are in the domain of H. (Similarly, if the reverse inequality is satisfied, H is a *level increasing function.*)

There is a large class of functions, say $\mathcal{H}_n$, satisfying the above property as well other similar ones. The term *hierarchical functions* is most appropriate for the class $\mathcal{H}_n$. However for our goals here we can restrict ourselves by considering easy functions, for example, linear functions and see how to answer some of the questions we are interested in.

Take the function f of n arguments:

$$f(x_1, x_2, \ldots, x_n) = \lambda_1 x_1 + \lambda_2 x_2 + \cdots + \lambda_n x_n$$

and assume that the arguments $x_1, \ldots, x_n$ are real nonnegative numbers while the coefficients $\lambda_1, \ldots, \lambda_n$ are (strictly) positive numbers. Then obviously f is an increasing function in each of its arguments (for any choice of λ_i). Moreover, if the coefficients $\{\lambda_i\}$ strictly decrease, i.e. $\lambda_1 > \lambda_2 > \cdots > \lambda_n$, then f is a level decreasing function while if $\{\lambda_i\}$ strictly increase, i.e. $\lambda_1 < \lambda_2 < \ldots \lambda_n$, then f is a level increasing function.

The level increasing (decreasing) property of f is easy to get. Take a fixed number $c > 0$ and define $\lambda_k = c^k$, $k = 1, 2, \ldots, n$. Then the function

$$f(x_1, x_2, \ldots, x_n) = cx_1 + c^2 x_2 + \cdots + c^n x_n$$

is level decreasing if $0 < c < 1$ and level increasing if $c > 1$. If $c = 1$ we have $f = x_1 + x_2 + \cdots + x_n$, but this function is not useful for our goals.

Recall that each of the x_k take values in a finite and closed interval: $0 \leq x_k \leq m_k$, $k = 1, 2, \ldots, n$, so let us define the quantity Λ_n by

$$\Lambda_n = 1/\left(cm_1 + c^2 m_2 + \cdots + c^n m_n\right).$$

It is easy to see that the function H defined by

$$H(x_1, x_2, \ldots, x_n) = \Lambda_n \left(cx_1 + c^2 x_2 + \cdots + c^n x_n\right)$$

obeys the following three properties:

(1) $H(\cdot)$ is an increasing function in each of its arguments;

(2) $0 \leq H(\cdot) \leq 1$ with

$$H(x_1, x_2, \ldots, x_n) = 0 \iff x_k = 0 \quad \text{for all} \quad k = 1, \ldots, n,$$

$$H(x_1, x_2, \ldots, x_n) = 1 \iff x_k = m_k \text{ for all } k = 1, \ldots, n.$$

(3) If $0 < c < 1$, the function $H(\cdot)$ is level decreasing and if $c > 1$ it is level increasing.

Finally we need a special property of the function H. We say that H obeys the *separating property,* or that H is a *separating function* if

$$H(x_1, \ldots, x_n) = H(\tilde{x}_1, \ldots, \tilde{x}_n) \iff x_1 = \tilde{x}_1, \ldots, x_n = \tilde{x}_n.$$

Notice that the three properties (1), (2), (3) listed above hold for arbitrary real arguments $x_1, x_2, \ldots, x_n$. It turns out the separating property of

H will be guaranteed if we make the following assumption: The arguments $x_1, x_2, \ldots, x_n$ are nonnegative *integer* numbers where $0 \leq x_k \leq m_k$, $k = 1, 2, \ldots, n$ for some integers $m_1, m_2, \ldots, m_n$. Let us denote

$$m^* = \max[m_1, m_2, \ldots, m_n].$$

Now we can formulate the next property:

(4) If we take $c = 1/(m^* + 1)$ or $c = m^* + 1$, then the function H obeys the separating property. (Recall that H_n is level decreasing if $0 < c < 1$ and level increasing if $c > 1$.)

The properties (1), (2) and (3) are easy to check and the proof of property (4) can be performed in a similar way as in Stoyanov (1995).

4. Dependency Measure

Motivated by the above arguments we want to suggest a function serving as a dependency measure of the whole collection of random events $\mathcal{A}_n$. Let us use the notations D or $D(\mathcal{A}_n)$ for that measure.

As we have explained, D will be a function of the vector $(d_2, d_3, \ldots, d_n)$ representing the i/d-structure of the set $\mathcal{A}_n$ and hence will depend on n, the number of the events in the set $\mathcal{A}_n$. Let us choose D to take its values in the interval [0, 1] such that

$$D(\mathcal{A}_n) = 0 \iff \mathcal{A}_n \text{ is mutually independent;}$$

$$D(\mathcal{A}_n) = 1 \iff \mathcal{A}_n \text{ is totally dependent.}$$

A natural requirement is $D(d_2, d_3, \ldots, d_n)$ to be an increasing function in each of the arguments $d_2, d_3, \ldots, d_n$ and be a level decreasing function. Finally, it is strongly desirable D to obey the *separating property.* This requirement means that two sets, each consisting of n events, will have the same "amount of dependency" (the value of D) if and only if their i/d-structures coincide, i.e.:

$$D(d_2, \ldots, d_n) = D(\tilde{d}_2, \ldots, \tilde{d}_n) \iff d_2 = \tilde{d}_2, \ldots, d_n = \tilde{d}_n.$$

Note that the separating property is essential when comparing different sets of events and answering the question of which one is "more" dependent or equivalently which one is "less" independent.

The conditions described above do not imply that the measure D can be constructed in a unique way. Hence it is natural to suggest that D be of the simplest possible form, for example, D to be a linear function. Let us define ($[x]$ is the integer part of x) the number c_n by

$$c_n = \binom{n}{[(n+1)/2]} + 1, \quad n \geq 2. \tag{3}$$

We define the *Dependency Measure* of $\mathcal{A}_n$, the global measure of the amount of dependency of this collection of events, by

$$\begin{aligned} D(\mathcal{A}_n) &= D(d_2, d_3, \ldots, d_n) \\ &= \frac{1}{(1+1/c_n)^n - (1+n/c_n)} \left\{ \frac{d_2}{c_n^2} + \frac{d_3}{c_n^3} + \cdots + \frac{d_n}{c_n^n} \right\}. \end{aligned} \tag{4}$$

Here $n \geq 2$, c_n is given by (3) and the arguments $d_2, d_3, \ldots, d_n$ are integer numbers satisfying the restrictions (2): $0 \leq d_k \leq \binom{n}{k}, k = 2, 3, \ldots, n$. It is easy to see that D can be written in the following equivalent form:

$$D(d_2, d_3, \ldots, d_n) = \frac{d_2 c_n^{n-2} + d_3 c_n^{n-3} + \cdots + d_{n-1} c_n + d_n}{\binom{n}{2} c_n^{n-2} + \binom{n}{3} c_n^{n-3} + \cdots + \binom{n}{n-1} c_n + 1}.$$

Proposition 1. *The function D_n defined by (4) under the conditions (2) and (3) obeys the properties described above. Hence we can adopt it as a global measure of the amount of dependency of the whole collection of random events $\mathcal{A}_n$.*

A detailed proof and some comments can be found in Stoyanov (1995).

Let us write down the dependency measure D in a few particular cases.

Case $n = 2$. Here the i/d-structure is (d_2) and clearly $d_2 = 0$ or 1. In this case $c_2 = 3$ and we easily find that

$$D(d_2) = d_2,$$

i.e. for two events D is either 0 or 1 which exactly corresponds to the fact that two events are either independent or dependent.

Case $n = 3$. The i/d-structure is (d_2, d_3), $c_3 = 4$ and

$$D(d_2, d_3) = \frac{1}{13}(4 \cdot d_2 + d_3).$$

Case $n = 4$. Here the i/d-structure is (d_2, d_3, d_4), $c_4 = 7$ and

$$D(d_2, d_3, d_4) = \frac{1}{323}\left(7^2 \cdot d_2 + 7 \cdot d_3 + d_4\right).$$

Case $n = 5$. Here the i/d-structure is (d_2, d_3, d_4, d_5), $c_5 = 11$ and hence

$$D(d_2, d_3, d_4, d_5) = \frac{1}{14576}\left(11^3 \cdot d_2 + 11^2 \cdot d_3 + 11 \cdot d_4 + d_5\right).$$

For any of the examples described above we can find the exact value of the measure D. Specifically: In Example 1: $D(\mathcal{A}_3) = D(0,0) = 0$ if $\alpha = \frac{1}{4}$, and $D(\mathcal{A}_3) = D(3,0) = \frac{12}{13}$ if $\alpha \neq \frac{1}{4}$. In Example 2: $D(\mathcal{A}_4) = D(6,4,0) = \frac{322}{323}$. In Example 3: $D(\mathcal{A}_4) = D(0,4,0) = \frac{28}{323}$. In Example 4: $D(\mathcal{A}_5) = D(10,10,0,0) = \frac{14520}{14576}$.

We can establish several interesting properties of the function D. In particular, if V_n is the set of all possible values of $D(d_2, \ldots, d_n)$, then V_n is an ε-net (of $2^n - n - 1$ numbers) in the interval [0, 1] where ε can be written exactly. Clearly, for large n, V_n becomes a dense set in [0, 1] as is the union $V = \cup_{n=2}^{\infty} V_n$. Among the interesting questions here, let us mention the following one. Suppose $\rho \in (0,1)$ is a fixed number. Does there exist a set of random events $\mathcal{A}_n$ such that $D(\mathcal{A}_n) = \rho$ or at least $D(\mathcal{A}_n) \approx \rho$. In other words, we are looking for n and $d_2, d_3, \ldots, d_n$ such that $D(d_2, d_3, \ldots, d_n) = \rho$, exactly or approximately. It follows from (4) that for some values of ρ we can find the exact answer. However, since V is dense in $[0,1]$, we can always find at least a good approximate solution. Another question is to find the "minimal" set $\mathcal{A}_n$ (containing the smallest possible number n of events) such that $D(\mathcal{A}_n) = \rho$ or $D(\mathcal{A}_n) \approx \rho$.

Here is another property which could be called an *invariance property*. If $\mathcal{A}_n = \{A_1, \ldots, A_n\}$, let A_j^* denote either A_j or its complement A_j^c. Then the "new" collection $\mathcal{A}_n^* = \{A_1^*, \ldots, A_n^*\}$ is exactly as much dependent as is the original one $\mathcal{A}_n$, i.e. $D(\mathcal{A}_n^*) = D(\mathcal{A}_n)$. This follows from the easily checked fact that the product rule (1) holds or does not hold simultaneously for both $A_{j_1} A_{j_2} \cdots A_{j_k}$ and $A_{j_1}^* A_{j_2}^* \cdots A_{j_k}^*$ for any choice of the indexes $j_1, j_2, \ldots, j_k$, $k \geq 2$.

For any set of random events $\mathcal{A}_n$, if we are able to determine explicitly its i/d-structure $(d_2, \ldots, d_n)$, we can then use (4) to calculate the value of $D(\mathcal{A}_n)$, i.e. the amount of dependency which is contained in $\mathcal{A}_n$. If $D(\mathcal{A}_n)$ is "close" to 0, we can say that $\mathcal{A}_n$ is "almost" mutually independent; if $D(\mathcal{A}_n)$ is "close" to 1, we can say that $\mathcal{A}_n$ is "almost" totally dependent.

The separating property of D_n allows us to compare easily any two (or more) sets each containing the same number of random events. Thus, if $\mathcal{A}_n$ and $\mathcal{A}_n'$ are such sets and $(d_2, \ldots, d_n)$ and $(d_2', \ldots, d_n')$ are their i/d-structures, we calculate $D(d_2, \ldots, d_n)$ and $D(d'_2, \ldots, d'_n)$ and say that $\mathcal{A}_n$ is "less" dependent than $\mathcal{A}'_n$ if and only if $D(d_2, \ldots, d_n) < D(d'_2, \ldots, d'_n)$. This is equivalent to saying that $\mathcal{A}'_n$ is "more" independent than $\mathcal{A}_n$. Hence, based on D, we can introduce an *order* relation, denoted by $\prec$, among the sets containing the same number of events:

$$\mathcal{A}_n \prec \mathcal{A}'_n \iff D(\mathcal{A}_n) < D(\mathcal{A}'_n).$$

As an illustration let us suppose that $\mathcal{A}_n$ and $\mathcal{A}'_n$ are sets each containing $n = 4$ random events and such that their i/d-structures are $(d_2, d_3, d_4) = (2, 1, 1)$ and $(d'_2, d'_3, d'_4) = (1, 3, 0)$. Obviously $\mathcal{A}_4$ and $\mathcal{A}'_4$ have the same total number of failing product relations (1). However we find that

$$D(2, 1, 1) = \tfrac{106}{323} > \tfrac{70}{323} = D(1, 3, 0) \Rightarrow \mathcal{A}_4 \succ \mathcal{A}'_4.$$

The existence of sets $\mathcal{A}_4$ and $\mathcal{A}'_4$ with such i/d-structures follows from the positive solution of "The Italian Problem" (see the next section). This example also shows that it is not appropriate to ignore the levels and define the measure $D(\mathcal{A}_n) = T_n/(2^n - n - 1)$ where T_n is the *total* number of failing product relations (1). In fact, $T_n = d_2 + \cdots + d_n$ and the "new" normalizing constant is $2^n - n - 1$. If we adopt such a definition, we would not be able to distinguish obviously different sets of events.

The choice (3) for the constant c_n is not so strange. Simply c_n is equal to the largest binomial coefficient (of order n) plus 1. It turns out this is the minimal value of c ensuring the separating property. Look at the property (4) of the hierarchical function H discussed in Section 3. If we take any $\tilde{c} \le c_n - 1$, we can easily describe sets of events with different i/d-structure but the same dependency measure. However we can use any $\bar{c} \ge c_n$, for example, $\bar{c} = n^{[n/2]}$ or $\bar{c} = n^n$. The properties of the dependency measure are preserved but the values are appropriately shifted except the boundary values 0 and 1. The order relation $\prec$ is unchanged. Thus comparing sets of random events we just have to use some fixed $c \ge c_n$.

Let us mention another possibility. The dependency measure D_n can be used to define a *metric* in the set of all sets of n random events. For the two sets above given by their i/d-structures we can define the *distance* between $\mathcal{A}_n$ and $\mathcal{A}'_n$ as follows:

$$\delta(\mathcal{A}_n, \mathcal{A}'_n) = D(|d_2 - d'_2|, |d_3 - d'_3|, \ldots, |d_n - d'_n|).$$

Clearly $\delta(\mathcal{A}_n, \mathcal{A}'_n) = \delta(\mathcal{A}'_n, \mathcal{A}_n)$ which is the symmetric property. Then $\delta(\mathcal{A}_n, \mathcal{A}'_n) = 0$ only if the two sets $\mathcal{A}_n$ and $\mathcal{A}'_n$ have the same i/d-structure. Finally it is not difficult to see that δ satisfies the triangle inequality. Hence δ is a metric. In particular, if $\mathcal{A}'_n$ is totally dependent (its dependency measure is 1), the quantity δ tells us how far $\mathcal{A}_n$ is from being totally dependent. It follows from the above definition that the values of δ are in the interval [0, 1]. We can say immediately that the distance between the mutually independent set and the totally dependent set is 1. But this is only one such a case. It is interesting to note that among all the sets of n events, there are exactly 2^{n-1} pairs of sets such that for each pair the δ-distance between the two sets is 1. It is easy to describe all these pairs.

5. "The Italian Problem" and its solution

The definition of the dependency measure D by (4) under the conditions (2) and (3) naturally leads to the following problem which got the name "The Italian Problem": Does there exist a probability model and a set of random events with any prescribed i/d-structure? This problem is of an independent interest but let us also note that D would be a reasonable measure only if the answer to this question is positive. Our belief that "The Italian Problem" has a positive solution was expressed years ago. This appeared in Stoyanov (1995) as a conjecture supported by several examples. Later an affirmative proof of this conjecture was given by Mori and Stoyanov (1995/1996). A different proof was also given by Mizera and Balek (1995).

The solution of "The Italian Problem" is based on the following statement.

Proposition 2. *Let $\mathcal{M}$ be an arbitrary system of subsets $M \subset I$, where $I = \{1, \ldots, n\}$ and $|M| \geq 2$. Then there is a probability space $(\Omega, \mathcal{F}, P)$ and a set of n random events, $\mathcal{A}_n = \{A_1, \ldots, A_n\}$, such that for $k = 2, \ldots, n$ and $1 \leq i_1 < \ldots < i_k \leq n$, the product rule*

$$P(A_{i_1} \cdots A_{i_k}) = P(A_{i_1}) \cdots P(A_{i_k}),$$

holds if and only if $\{i_1, \ldots, i_k\} \in \mathcal{M}$.

The proof relies on some algebraic arguments concerning the solvability of a system of linear equations. (Details can be found in Mori and Stoyanov (1995/1996).) In general the solution is not unique. The latter fact, however, can be used when constructing sets of events not only with a prescribed i/d-structure but also obeying another specific property.

As a consequence of Proposition 2, we derive the following. If we have given numbers n and $(d_2, \ldots, d_n)$, where $0 \leq d_k \leq \binom{n}{k}$, $k = 2, \ldots, n$ then we can easily construct a probability space and define in it n random events whose i/d-structure is exactly $(d_2, \ldots, d_n)$. Indeed, if $d_2' = \binom{n}{2} - d_2, \ldots, d_n' = \binom{n}{n} - d_n$, we choose $\mathcal{M}$ containing d_k' subsets $(i_1, \ldots, i_k)$ from $\{1, \ldots, n\}$ each having a size k, where k ranges from 2 to n. Symbolically, we can take

$$\mathcal{M} = \mathcal{M}_2 \cup \mathcal{M}_3 \cup \cdots \cup \mathcal{M}_n$$

where $\mathcal{M}_k$ is an arbitrary set of k-subsets of indexes such that $|\mathcal{M}_k| = d'_k$ for $k = 2, \ldots, n$. We have a freedom for this choice, except the case $d'_k = \binom{n}{k}$, when $\mathcal{M}_k$ is the set of all k-subsets of indexes. Note that we can think of all possible combinations of indexes ordered at levels, from level 2 to level n, and follow the so-called lexicographical order for each level.

Remark. Let us note that if we start with the probability space $(\Omega, \mathcal{F}, P)$ then we can consider different sets of random elements, find their i/d-structure and calculate the dependency measure. If the space is "rich enough," we can expect to see sets of random elements with a diverse i/d-structure. It turns out there are finite probability spaces such that any (at least two) nontrivial events are dependent (see, for example, Stoyanov (1997) or Wang (1997)). Moreover, there are even infinite (but countable) spaces in which any set of at least 2 events is totally dependent (see, for example, Stoyanov (1995)). It is not immediately obvious but it is true that if the probability measure P on the measurable space $(\Omega, \mathcal{F})$ is absolutely continuous, then any solution of "The Italian Problem" for random events can be realized on such a probability space. In general this is not true if the measure P is purely discrete, or if P has an absolutely continuous part and a discrete part. (Let us leave aside the singular measures.)

6. Some consequences

The examples in Section 2 and many others can be obtained as possible solutions of "The Italian Problem." We are going to present more examples looking not only for the independence (dependence) property but also for other such as exchangeability and association.

Recall that $\mathcal{A}_n = \{A_1, \ldots, A_n\}$ is *exchangeable* if for any k, $k = 1, 2, \ldots, n-1$, the probability $P(A_{i_1} \cdots A_{i_k})$ is the same for all k-subsets $A_{i_1}, \ldots, A_{i_k}$ of the set $\mathcal{A}_n$. In an obvious way the exchangeability property is defined for an infinite sequence of random events.

In Section 1 we introduced and then used the independence (or dependence) property at a given level k for $k = 2, \ldots, n$. The same idea can be extended by introducing the following new notion.

For $k < n$ we say that the set $\mathcal{A}_n$ is *exchangeable at level* k, if any k of the events in $\mathcal{A}_n$ have the same probability. Hence $\mathcal{A}_n$ is *totally exchangeable,* or simply *exchangeable,* if $\mathcal{A}_n$ is exchangeable at each level $k = 1, 2, \ldots, n-1$. (If $k = n$ we have only one subset of n events.)

Note that if $P(A_1) = \cdots = P(A_n)$ and if $\mathcal{A}_n$ is mutually independent, then clearly $\mathcal{A}_n$ is totally exchangeable. The converse, however, is not true, as seen from the examples below.

Let us recall another notion called *association.* For this purpose take k events $A_{i_1}, \ldots, A_{i_k}, 2 \leq k \leq n$ and denote

$$\Delta(i_1, \ldots, i_k) = P(A_{i_1} \cdots A_{i_k}) - P(A_{i_1}) \cdots P(A_{i_k}).$$

These k events are *positively associated* if $\Delta(\cdot) \geq 0$ and *negatively associated* if $\Delta(\cdot) \leq 0$. The set of events $\mathcal{A}_n = \{A_1, \ldots, A_n\}$ is said to be *positively associated* if this property holds at all levels k, $k = 2, \ldots, n$. Similarly we define the *negative association* of the set $\mathcal{A}_n$.

The relationships between the notions mentioned above will be illustrated by several examples.

Remark. Let us note that regardless of the independence/dependence properties, the random set $\mathcal{A}_3 = \{A_1, A_2, A_3\}$ considered in Example 1 is totally exchangeable for any $\alpha \in (\frac{1}{8}, \frac{7}{16})$. Further, the set $\mathcal{A}_4$ in Example 2, the set $\mathcal{A}_4$ in Example 3 and the set $\mathcal{A}_5$ in Example 4 have quite diverse i/d-structures but each set is totally exchangeable.

Example 5. Let us suppose that $\Omega = \{1, 2, 3, 4, 5\}$ is the sample space of some experiment whose outcomes have probabilities respectively equal to $p - p^2$, $p - p^2$, $p - p^2$, p^2, $1 + 2p^2 - 3p$, where p is an arbitrary number in the interval $(0, \frac{1}{2})$. The events

$$A_1 = \{1 \text{ or } 4\},\ A_2 = \{2 \text{ or } 4\},\ A_3 = \{3 \text{ or } 4\}$$

are such that for all possible choices of i and j we have:

$$P(A_i) = p,\ P(A_iA_j) = p^2,\ i < j;\ P(A_1A_2A_3) = p^2.$$

Obviously the set $\mathcal{A}_3 = \{A_1, A_2, A_3\}$ is independent at level 2 and dependent at level 3. Clearly $\mathcal{A}_3$ is totally exchangeable. This example is of the same kind as the classical Bernstein–Bohlmann examples.

Example 6. Suppose the sample space Ω contains $|\Omega| = 16$ outcomes denoted by $1, 2, \ldots, 16$ each having probability $\frac{1}{16}$. Define the events:

$$A_1 = \{2, 3, 4, 5, 6, 9, 13, 16\},\ A_2 = \{4, 7, 8, 10, 11, 13, 14, 16\},$$

$$A_3 = \{4, 6, 7, 8, 10, 11, 13, 14\},\ A_4 = \{3, 4, 5, 6, 9, 10, 15, 16\}.$$

Then $P(A_i) = \frac{1}{2}$, $i = 1, 2, 3, 4$ and since $A_1A_2A_3A_4 = \{4\}$ we find

$$\frac{1}{16} = P(A_1A_2A_3A_4) = P(A_1)P(A_2)P(A_3)P(A_4)$$

and hence the product rule is satisfied at level 4. Further on, $A_1A_2A_3 = \{4, 13\}$ implying that

$$\frac{1}{8} = P(A_1A_2A_3) = P(A_1)P(A_2)P(A_3)$$

and similarly the product rule holds for the remaining five possible triplets of events. It turns out, however, that the product rule fails to hold for any of the six possible pairs of events. In particular, $A_3A_4 = \{4, 6, 10\}$ and

$$\frac{3}{16} = P(A_3A_4) \neq P(A_3)P(A_4) = \frac{1}{4}.$$

Therefore the set $\mathcal{A}_4 = \{A_1, A_2, A_3, A_4\}$ is independent at level 4, independent at level 3 and (completely) dependent at level 2. Note finally that these events are exchangeable at level 1, exchangeable at level 3 and not exchangeable at level 2. The set $\mathcal{A}_4$ is negatively associated.

Example 7. Let the sample space Ω consist of the outcomes 1, 2, 3, 4, 5, 6 and 7 with probabilities respectively equal to $\frac{4}{16}, \frac{1}{16}, \frac{1}{16}, \frac{3}{16}, \frac{2}{16}, \frac{3}{16}, \frac{2}{16}$. Define the events:

$$A_1 = \{2, 3, 4, 6\},\ A_2 = \{4, 6, 7\},\ A_3 = \{3, 5, 6, 7\}.$$

Then we easily find that $P(A_1) = P(A_2) = P(A_3) = \frac{1}{2}$ and

$$P(A_1A_2) = \frac{6}{16},\ P(A_1A_3) = \frac{4}{16},\ P(A_2A_3) = \frac{5}{16},\ P(A_1A_2A_3) = \frac{3}{16}.$$

Therefore these three events are exchangeable at level 1 and not exchangeable at level 2. They are partially independent/partially dependent at level 2 and are dependent at level 3. Also they are neither positively nor negatively associated.

Example 8. Consider the sample space Ω containing the integer numbers 1, 2,...,8, where each of the outcomes 1 and 8 has probability $\frac{1}{32}$ while each of the remaining outcomes, 2,...,7, is assigned by probability $\frac{5}{32}$.

We are interested in the following events:

$$A_1 = \{2, 5, 6, 8\},\ A_2 = \{3, 5, 7, 8\},\ A_3 = \{4, 6, 7, 8\}.$$

Then we easily find that for all possible choices of i and j we have

$$P(A_i) = \frac{1}{2};\ P(A_iA_j) = \frac{6}{32},\ i < j;\ P(A_1A_2A_3) = \frac{1}{32}.$$

These equalities imply that the set $\mathcal{A}_3 = \{A_1, A_2, A_3\}$ is: (a) totally dependent; (b) totally exchangeable; (c) strictly negatively associated.

7. Further developments

We have concentrated our attention almost entirely on sets of random events. Obviously the same properties, independence, exchangeability and association, should be considered for sets of random variables (r.v.). Also we should like to define a dependency measure in this case as we did for random events. There are some similarities but also there is a little difference. Let us exhibit some details. Suppose

$$\mathcal{R}_n = \{X_1, \ldots, X_n\}$$

is a collection of n r.v.'s defined on a given probability space. We can introduce an analog of the product rule (1) to define the independency property of the whole collection $\mathcal{R}_n$. Let $F(x_1, \ldots, x_n)$, $(x_1, \ldots, x_n) \in R^n$ be the distribution function (d.f.) of the random vector $(X_1, \ldots, X_n)$ and $F_j(x_j)$ be the d.f. of the component X_j. The set $\mathcal{R}_n$ is *independent* if the following relation is satisfied for all $(x_1, \ldots, x_n) \in R^n$:

$$F(x_1, \ldots, x_n) = F_1(x_1) \cdots F_n(x_n).$$

Obviously if $\mathcal{R}_n$ is independent, then any of its subsets (containing at least 2 variables) is also independent. By using our terminology this is equivalent to saying that the independence of r.v.'s at level n implies their independence at any lower level! We have seen above that for random events this is not always true. Hence some almost obvious changes of the "Dependency Measure" has to be involved and "The Italian Problem" has to be reformulated for sets of random variables.

Corollary. *"The Italian Problem" has a positive solution for random variables each taking exactly two different values. More precisely, for given integer numbers $n \geq 2$ and $(d_2, \ldots, d_n)$, with $0 \leq d_k \leq \binom{n}{k}$, we can construct a probability space and define n (simple) random variables, say $\mathcal{R}_n = \{\xi_1, \ldots, \xi_n\}$, each taking only two different values and such that the i/d-structure of $\mathcal{R}_n$ is exactly $(d_2, \ldots, d_n)$.*

Let us emphasize once again that this positive solution of "The Italian Problem" is correct only for two-valued r.v.'s (Bernoulli r.v.'s). In the general case we have to examine the possible values of $d_2, \ldots, d_n$ and only then to look for the solution. Some results from Rüschendorf (1985) and Wang (1991) can be used in this case.

Example 9. Let us denote the n-dimensional cube $[0, 2\pi]^n$ in R^n by Q_n and consider the following function

$$f(x_1,\dots,x_n) = (2\pi)^{-n}[1 - \cos x_1 \cdots \cos x_n],\ (x_1,\dots,x_n) \in Q_n$$

and 0, otherwise. It is easy to check that f is nonnegative and the integral of f over R^n equals 1. Hence f is a probability density function of a random vector in R^n, say $(X_1,\dots,X_n)$. Denoting by $f_j(x_j)$ the marginal density of the component X_j we find that

$$f_j(x_j) = 1/(2\pi),\ 0 \le x_j \le 2\pi$$

and 0, otherwise. Hence X_j is uniformly distributed on the interval $[0, 2\pi]$ and this holds for any (single) r.v. $X_1,\dots,X_n$. The joint density f shows that these n variables are not independent. If, however, we take k of them, we conclude that they are independent for any $k = 2, 3, \dots, n-1$; their joint density is equal to $1/(2\pi)^k$ on the cube $Q_k = [0, 2\pi]^k$ in R^k (and zero otherwise). Therefore $\mathcal{R}_n = \{X_1,\dots,X_n\}$ is a collection of n dependent r.v.'s which are $(n-1)$-wise independent.

Example 10. It is of a more general interest to exhibit examples of sets of n dependent r.v.'s which are m-wise independent, for any $m < n$, not just $m = n-1$ as above. This will mean that any m (and hence fewer but at least 2) variables from such a set will be independent while any $m+1$ (and hence any more but up to n) will be dependent.

The idea can be illustrated well even in the case $n = 5$ and $m = 3$. Let us start with five arbitrary 1-dimensional d.f.'s F_1, F_2, F_3, F_4, F_5 and denote $G_j = 1 - F_j$, $j = 1,\dots,5$. Define the function

$$H_{12345}(x_1, x_2, x_3, x_4, x_5),\ (x_1, x_2, x_3, x_4, x_5) \in R^5$$

as follows:

$$H_{12345} = F_1F_2F_3F_4F_5(1 + \varepsilon_1 G_2G_3G_4G_5 + \varepsilon_2 G_1G_3G_4G_5 + \varepsilon_3 G_1G_2G_4G_5 + \varepsilon_4 G_1G_2G_3G_5 + \varepsilon_5 G_1G_2G_3G_4).$$

We assume that each $\varepsilon_i, i = 1,\dots,5$ is a non-zero number in the interval $(-1, 1)$ and that the sum of their absolute values is less than 1. Under these conditions H_{12345} is a 5-dimensional d.f. of some random vector in R^5, say $(\eta_1, \eta_2, \eta_3, \eta_4, \eta_5)$. We are interested in what kind of independency/dependency there exists among the components of this vector. Clearly we have first to find all k-dimensional marginal distributions for $k = 4, 3, 2, 1$. In particular, if $H_{1234}, H_{123}, H_{12}$ and H_1 are the d.f.'s of $(\eta_1, \eta_2, \eta_3, \eta_4)$, (η_1, η_2, η_3), (η_1, η_2) and η_1, respectively, we easily find that

$$H_{1234} = F_1F_2F_3F_4(1 + \varepsilon_5 G_1G_2G_3G_4),\ H_{123} = F_1F_2F_3,\ H_{12} = F_1F_2,\ H_1 = F_1.$$

Similarly we can write down the d.f.'s in all the remaining cases and arrive at the following conclusions:

(1) η_j has a d.f. equal to $F_j, j = 1, 2, 3, 4, 5$;

(2) any two of the r.v.'s $(\eta_1, \eta_2, \eta_3, \eta_4, \eta_5)$ are independent;

(3) any three of them are also independent;

(4) any four and hence all the five variables are dependent.

Hence $\{\eta_1, \eta_2, \eta_3, \eta_4, \eta_5\}$ is a collection of five dependent r.v.'s which are 3-wise independent. The i/d-structure of these variables is $(d_2, d_3, d_4, d_5) = (0, 0, 5, 1)$. Note that a similar idea works when describing n dependent r.v.'s which are m-wise independent for arbitrary $m < n$. The construction can also be given in terms of probability densities.

Another interesting case is to consider a collection consisting of two groups of r.v.'s such that each group is independent, but the two groups are not. Take, for example, a Gaussian random vector of dimension $m + n$ with zero-mean components and a covariance matrix of a special block-structure representing the independence between the "first" m components, the independence between the "last" n components and the dependence between these two groups. A similar idea works also for more than two groups of r.v.'s.

The dependency measure $D(\mathcal{A}_n)$ is actually a *projection* of the set $\mathcal{A}_n$ on the closed interval $[0,\ 1]$. Clearly it can happen that for different n-sets, say $\mathcal{A}_n$ and $\mathcal{B}_n$ the values $D(\mathcal{A}_n)$ and $D(\mathcal{B}_n)$ coincide. This means that these two n-sets of events have the same i/d-structure and hence from the point of view of the measure D they are indistinguishable. If we want to perform a further analysis of the properties of sets of random elements, beside the i/d-structure, we have to involve some more information. One possibility is to introduce a quantity called the *Generalized Bonferroni Measure.* This is only one of the ideas under current study.

Another possible direction of study is to look more carefully at the exchangeability property of the set $\mathcal{A}_n$. Obviously it is not the same if lower levels are exchangeable and higher are not, and conversely. If we have motivations to accept that the exchangeability of the whole $\mathcal{A}_n$ is more when this property holds at higher levels than at lower levels, then following the same line of reasoning as above we can construct a function, say $E = E(\mathcal{A}_n)$ serving as a *Global Exchangeability Measure* of the set $\mathcal{A}_n$. We can take E to be a function only of the levels $1, 2, \ldots, n-1$ and to be level increasing. Further, we can choose E taking values in the closed interval $[0,\ 1]$ and such that $E(\mathcal{A}_n) = 1$ if and only if $\mathcal{A}_n$ is totally exchangeable and $E(\mathcal{A}_n) = 0$ if and only if $\mathcal{A}_n$ is not exchangeable. Such a function could serve as a global measure of exchangeability of the whole set $\mathcal{A}_n$ and be used to compare any two sets of events and order them saying which one is "more" exchangeable. This topic is also under study.

It is well-known that *independence/dependence* is one of the central concepts in Probability and Statistics with many practical applications. This justifies the strong interest in these topics and the intensive studies in this field during the last decades. Remarkable collections of papers are: Eberlein and Taqqu (1986); Block, Sampson and Savits (1991) and Dall'Aglio, Kotz and Salinetti (1991). In these sources one can find ideas and results which are related to the approach developed in this paper.

8. Concluding remarks

My interest in the topics discussed here arose several years ago when I was working on the book *Counterexamples in Probability.* Some strange and striking examples needed an adequate explanation. Several delicate questions were waiting for answers. The *Dependency Measure* and "The Italian Problem" were announced and discussed for the first time during my public lectures at Università di Bologna, Italy (April '90 and May '91). Lectures on these topics were delivered at several universities in Europe, Canada and the USA. New ideas and results are described in Stoyanov (1997).

The topic of this paper and many others were intensively discussed with Prof. S. Cambanis during our many personal meetings (Sofia '83, Thessaloniki '87, Rome '90, Toronto '92, Amsterdam '93). I feel it is my human and professional obligation to dedicate this paper to the memory of Stamatis Cambanis.

References

Bernstein, S. N. *Theory of Probability.* Gostechizdat, Moscow-Leningrad, 1928. [In Russian; previous lithographic edition 1916.]

Block, H. W., A. R. Sampson and T. H. Savits, eds. *Topics in Statistical Dependence.* IMS Ser., vol. **16**, Inst. Math. Statist., Hayward (CA), 1991.

Bohlmann, G. *Die Grundbegriffe der Wahrscheinlichkeitsrechnung in Ihrer Anwendung auf die Lebensversicherung*, pp. 244-278. In: "Atti dei 4. Congresso Internationale del Matematici (Rome 1908)," ed. G. Castelnouvo, vol. **3**, Rome, 1908.

Crow, E. L. *A counterexample on independent events.* Amer. Math. Monthly **74** (1967), 716–717.

Dall'Aglio, G., S. Kotz and G. Salinetti, eds. *Advances in Probability Distributions with Given Marginals* (Proc. Symp. Rome '90), Kluwer Acad. Publ., Dordrecht, 1991.

Eberlein, E. and M. Taqqu, eds. *Dependence in Probability and Statistics.* Birkhäuser, Boston, 1986.

Mizera, I. and V. Balek. *On the logical independence of the identities defining the stochastic independence of random events*, Statist. & Probab. Letters **31** (1997), 281–284.

Mori, T. F. and J. Stoyanov. *Realizability of a probability model and random events with a prescribed independence/dependence structure*, 1995/1996.

Rüschendorf, L. *Construction of multivariate distributions with given marginals.* Annals Inst. Statist. Math. **37** (1986), 225–233.

Stoyanov, J. *Counterexamples in Probability.* John Wiley & Sons, Chichester-New York, 1987; 2nd ed. 1997.

Stoyanov, J. *Dependency measure for sets of random events or random variables.* Statist. & Probab. Letters **23** (1995), 108–115.

Wang, Y. H. *Dependent random variables with independent subsets - II.* Canad. Math. Bull. **33** (1990), 24–28.

Wang, Y. H. *On the existence of a totally dependent probability space.* J. Appl. Statist. Sci. **5** (1997), to appear.

Instituto de Matematics
Universidade Federal do Rio de Janeiro
Rio de Janeiro 21945-970
Brasil
e-mail: jordan@dme.ufrj.br